Understanding the Properties of Matter

Understanding the Properties of Matter

Michael de Podesta

Birkbeck College, University of London

Taylor & Francis
Publishers since 1798

USA Publishing Office:

Taylor and Francis
1101 Vermont Ave., N. W., Suite 200
Washington, DC 20005
Tel: (202) 289-2174   Fax: (202) 289-3665

Distribution Center:

Taylor and Francis
1900 Frost Road, Suite 101
Bristol, PA 19007-1598
Tel: (215)785-5800   Fax: (215) 785-5515

First published in 1996 by UCL Press

UCL Press Limited
University College London
Gower Street
London WC1E 6BT

The name of University College London (UCL) is a
registered trade mark used by UCL Press with the
consent of the owner.

∞ The paper in this publication meets the requirements of the ANSI Standard Z39.48-1984 (Perma-
nence of Paper)

**Library of Congress Cataloguing-in-Publication
Data** (available upon request from the publisher)

ISBNs: 1-56032-614-X cloth
        1-56032-616-6 paper

Typeset in Times New Roman and Univers.
Printed and bound by
Bookcraft (Bath) Ltd, England.

To Maxwell and his mother

Contents

Appendices

Foreword

Professor Sir Brian Pippard, FRS

One of the greatest transformations of physics began about 70 years ago, when quantum mechanics was formulated. Many of the outstanding puzzles of atomic structure suddenly became clear, and it was not long before a start was made on understanding the chemical bond. Before 1930 the new ideas were applied to solids; one of the first successes was to explain how electrons can move so easily between the closely packed atoms of a metal, and soon after this came the explanation of why some solids conduct electricity well while others are insulators. Since then there has been no slackening of effort; many more phenomena have been revealed and explained, and new technologies have grown up out of the successes, to transform our daily lives. One example will suffice. The transistor was invented in 1947 by solid-state physicists, and without it modern communications and information systems would never have been developed.

There are more people engaged on research into condensed matter, and its applications, than on any other branch of physics. One consequence of this activity is that the assimilation of new facts and theories is an almost overwhelming task, and we are forced to rely on the efforts of those who can put together a readable account of the main themes. It is not surprising that most of the recent textbooks meet the challenge by concentrating on general principles and the theoretical treatment of rather simple examples, to such a degree that a casual reader might conclude that the whole subject is dominated by theory. One cannot deny that without theory we should be left with a miscellaneous ragbag of facts, lacking the means to find any underlying structure. On the other hand, without the challenge of the facts, the deep general theories like quantum mechanics would never have been looked for.

Nowadays most physicists specialize in either experiment or theory, and there are few to rival those of earlier days who, like Maxwell, Rayleigh, Helmholtz and Fermi, were superlatively good at both. Nevertheless, all modern specialists acknowledge their dependence on the different skills of others. A theoretician ignorant of facts is no more to be trusted than an engineer who takes a textbook diagram of a bridge as the basis of a design, never having seen the real thing or tested a model. Unfortunately the prevalence of software packages for almost every conceivable task encourages the mistaken belief that they incorporate all the experience needed for success in their application. If you read Michael de Podesta's book seriously you are unlikely to fall into such a grievous error. Once its lesson has been learnt, it will remain a source of information and, more important still, an incentive to continue with the process of self-education, which is the key to achievement.

Of all the messages that this book conveys there is especially one with which I feel the strongest sympathy. It is not that facts are things that you have to assimilate before you settle down to the interesting theories – rather, it is that the facts themselves, if well presented, are interesting in their own right and raise a host of questions that make you want to find out how everything ties together. All too rarely do most of us stop to think about apparently trivial everyday observations. How often, for example, do we wonder why there are such things as liquids? You will not find a complete answer here because the complete answer is not known (and perhaps least is understood about water, that most extraordinary of liquids) but you will discover the beginning of an answer, and will find that the beautiful simplicities of thermodynamics help you to see what happens when a solid melts, and give hints of the difficulties that one may hope eventually to overcome. And this is only one of the tantalizing problems discussed here in the hope that you will discover (if you have not done so already) both the lovely certainties and the stimulating mysteries that together make physicists enjoy their work, and continue to enjoy it as long as they can carry on with physics.

Acknowledgements

Writing this book has been a highly educational, but mainly solitary, experience; two years or so of late nights and weekends trying to explain things to myself and then attempting to record my explanations coherently. However, several other people have contributed, many in ways they may not have realized.

Each of my colleagues in the Physics Department at Birkbeck College has contributed (at the very least) some small snippet of information or understanding. I should like to thank them all, but especially Stan Zochowski for his kindness and encouragement.

Only a few people have read and critically commented on the book. The most significant of these comments (which caused an entire rewrite and a general improvement) has been from Professor Mike Holmes at the University of Central Lancashire. It is you, the readers, who should thank him more than me. Thanks also to Mark Ellerby and Rod Bateman for their comments.

At UCL Press I must thank Andrew Carrick for the opportunity to write the book, and Kate Williams and Julie Gorman for their attention to detail in its production. Thanks also to Peter Phillips and the Tate Gallery for permission to use *Spectrocoupling* as the cover illustration.

Ten years ago this book would have been impossible for an individual to write. The ability to do this has arisen through the availability of personal computers and superb software. I acknowledge my debt to the authors of the Macintosh system software, Word 5.1, Kaleidagraph 3.0, MacDraw Pro and Claris Draw.

Finally, and most importantly, I am grateful to all those scientists who have faithfully recorded their observations of the physical world. As I have read more, I have learnt that some scientists – without special reward – have taken that extra step in clearly explaining what they have done, or have scrutinized carefully compilations of data for poor results. It is truly only through the attention to detail of these nameless scientists that this book has been written at all.

If you have any comments on this book, good or bad, I would be glad to receive them. A list of suitably edited comments will be placed on my world wide web home page.

Thanks to you all.

Michael de Podesta
March 1996

Physics Department
Birkbeck College
Malet Street
London WC1E 7HX

m.depodesta@physics.bbk.ac.uk

Constants

The fundamental constants are known to much greater precision than in this table, but the precision shown is sufficient for all work in this text, and most work related to the physics of matter. See Kaye & Laby (§1.4.1) for more precise values and uncertainty estimates.

Constant	Symbol	Value	Unit
Speed of light in a vacuum	c	2.9979×10^8	$m\,s^{-1}$
Charge on the proton	e	1.602×10^{-19}	C
Planck constant	h	6.626×10^{-34}	J s
Planck constant ($h/2\pi$)	$\hbar$	1.054×10^{-34}	J s
Mass of proton	m_p	1.673×10^{-27}	kg
Mass of electron	m_e	9.109×10^{-31}	kg
Atomic mass unit	u	1.661×10^{-27}	kg
Electron volt	eV	1.602×10^{-19}	J
Bohr magneton	μ_B	9.274×10^{-24}	$J\,T^{-1}$
Boltzmann constant	k_B	1.381×10^{-23}	$J\,K^{-1}$
Avogadro's number	N_A	6.022×10^{23}	mol^{-1}
Molar gas constant	R	8.314	$J\,K^{-1}mol^{-1}$
Permeability of free space	μ_0	$4\pi \times 10^{-7}$	$H\,m^{-1}$
Permittivity of free space	ε_0	8.854×10^{-12}	$F\,m^{-1}$

CHAPTER ONE

The quest for an understanding of matter

1.1 Welcome

People have wondered about the properties of matter for centuries and have written many excellent books on the subject. Even books that were printed half a century ago are relevant today. The properties of matter, and our theories about matter, have changed relatively little in this time. This is not to say that new phenomena have not been discovered and explained – for example, superfluidity and superconductivity have been exciting areas of advance in understanding. However, the laws of thermodynamics have not changed and the speed of sound in copper is the same as it was 100 years ago. The scientists who have worked during the twentieth century have left you, students of physics, a vast legacy of experimental results and theoretical analysis.

1.2 Personal experience

Half a lifetime ago I was a student of physics. From the manner in which I was I taught the various components that comprise the study of matter, I learned several lessons that have taken a long time to unlearn. First, I learned that it isn't necessary to actually know what the properties of matter *are* in order to debate them earnestly. This is not, of course, the case, but fear of my own ignorance prevented me from realizing this earlier. Secondly, I learned to avoid thinking about awkward subjects. I imagined that cleverer people than myself had sorted these things out. This is also not the case, but fear of my own ignorance still haunts me here.

In hoping that your experience will differ from my own I would urge you to develop your study of the properties of matter along the following lines. First, become familiar with what the properties of matter actually are. These properties have been determined and tabulated for thousands of substances so this is not difficult. I believe that unless you are *familiar* with the properties of matter it is impossible to ever say that you *understand* them. Secondly, and also secondarily, try to think about why things are the way they are. Often the data seem perplexing, but they usually make sense eventually

1.3 A historical perspective

Before embarking on this contemporary study of the properties of matter, it is instructive to spend some time reviewing the historical context in which this study takes place.

1.3.1 Discovery and rediscovery

In all aspects of my own work as a scientist, I am constantly reminded that "others have been here before". Thus although I am discovering particular facts for the first time, these facts are leading me to rediscover the insights of previous generations of scientists. Those scientists lacked the technical resources available today, but in their place they used resourcefulness, dedication and imagination in a truly inspiring combination.

This process of rediscovery is not to be viewed as a failure. Indeed, it was – and is – the essential process through which the health of the scientific body is maintained. Furthermore, I have realized that such rediscovery is not achieved during a degree course, but over a lifetime. If you wish to learn about physics, the only way is to rediscover it for yourself. Unfortunately, rediscovery is as hard as discovery, but without the glory.

1.3.2 Familiarity and overfamiliarity

The initial section of each topic in this book is a discussion of the data available on that topic. Before discussing the *theories of what happens*, I want you to become familiar with *what actually happens*. In this way I hope to allow you to rediscover some of the more striking properties of matter for yourself.

However, in encouraging this familiarization I encounter a severe problem; an improper overfamiliarity with rather sophisticated theories about matter. For example, many of you reading this will have seen images of atoms; you will have heard or even taken part in discussions of their detailed structure. I would go so far as to say that many of you *believe* in atoms. However, I would also venture to suggest that many of you believe in atoms not because of insights you have made yourself, but because you have been told that atoms exist and that they possess certain properties. Being told something is different from convincing yourself it must be so. This is the sense in which I would like you to embark upon the study of matter. Don't believe it until *you* have seen the data! I would like you to be sceptical about the data in this book, about the theories presented here, and about your own credulity.

1.3.3 Consciousness-lowering exercise

In order to help get yourself in the mood to rediscover the properties of matter, you may care to take part in an exercise in "consciousness lowering".

Picture yourself in the late seventeenth century. Look around you. The "stuff" of the world – the matter – is diverse in its properties. You can probably see wood, metal, stone, paper and, most amazingly, animated flesh. Your task is to *categorize* these substances and their experimental properties. What do they have in common? How do they differ? In other words, you must decide on a set of *organizational principles* which will allow you to form an understanding of matter. Note that you cannot include references to concepts such as atoms, or electric charge that we now accept. These concepts must be developed by studying matter – categorizing results, preparing hypotheses, worrying about exceptions – and trying to convince a sceptical group of colleagues that your insights are valid. I think you will find it extraordinarily difficult to know where to begin. The bewildering variety of the properties of matter will provide exceptional cases for just about any scheme of categorization you adopt.

If *you* find this process of categorization difficult, then pity the people from earlier times who reflected on the nature of matter. They suffered from a lack both of reliable data and of validated concepts for understanding the data, and did not have available experimental equipment of the type a 13 year old might use in a modern teaching laboratory. I wonder how long it would have taken *you* to "discover" atoms and determine their properties, to "discover" that there are *two* types of electric charge, to "discover" that heat is not a substantive fluid. All *your* present clarity of vision, such as your wise belief in the existence of atoms, has been constructed on a foundation of centuries of sceptical enquiry.

1.3.4 Organizational principles

What the above exercise is intended to show is that it is not obvious what organizational principles one should use when attempting to categorize matter. However, without the appropriate organizational principles the properties of matter are bewilderingly diverse, and explanations appear arbitrary and unconvincing. So, how should we choose to categorize matter?

Animate/inanimate

The first categorization that suggests itself to me is that between *animate* and *inanimate* matter. Historically, the division of the study of matter in this way allowed the emerging science of physics to become successful at explaining and parameterizing the simpler properties of inanimate matter. By contrast, the study of animate matter is still barely beyond the stage of naming and categorizing.

Solids, liquids and gases

Considering the inanimate world, the next most apparent organizational scheme is the classification of matter into one of three categories: *solid, liquid* or *gas*. This rather natural categorization is now seen to be by no means exhaustive. (Indeed it never was: our own bodies are made from matter that is half liquid and half solid.) *Liquid crystals*, for example, disclose by their very name that the solid/liquid/gas categorization is inadequate. Furthermore, substances normally considered to be in one category (such as "solid" rock) when viewed over geological time-scales show properties associated with the liquid state (such as flow and convection). However, the solid/liquid/gas division is still a useful one, and forms the basis for the structure of this book.

The conceptual division of matter into the distinct categories of solid, liquid, and gas was historically extremely important. Once made, real progress became possible in understanding the properties of the simplest category of matter, i.e. the gaseous state. From studies of gases over 200 years from 1700 to 1900, an enormous number of ideas emerged and were tested. In particular, the nature of four key concepts became distilled into a form somewhat similar to the understanding we hold today. The key concepts were descriptions of the properties of matter in terms of atoms and in terms of the nature of heat, electricity and light. When we examine the properties of gases in Chapter 5 bear in mind that, with the exception of our reference to quantum mechanics, the study would not have been out place a century ago!

1.4 The structure of this book

Chapters 1 to 3 provide, respectively, the context for the book, a mini revision course on some relevant background theory, and a discussion of some aspects of measurement, units and the significance of data thereby obtained. These chapters may be skipped or referred to, as you please.

Chapters 4 to 11 comprise the main text of the book. Gases are considered in Chapters 4 and 5, solids in Chapters 6 and 7, liquids in Chapters 8 and 9 and changes between these phases in Chapters 10 and 11. The first of each pair of chapters outlines relevant background theory, and the second considers the experimental data, and the extent to which they can be understood using the background theory. Where relevant, the theory is extended to understand the data.

The appendices contain detailed points of theory that are important, but which would tend to distract attention from the flow of the text if placed within the chapters. The exception to this is Appendix 4 which contains the text of a computer program. The program simulates the molecular dynamics of two-dimensional solids, liquids and gases with up to 25 molecules. The program can also be downloaded from my world wide web page at:

http://www.bbk.ac.uk/Departments/Physics

1.4.1 Sources of data

The data given in this book have been taken from a variety of sources. All the sources are, however, secondary; that is, they have already been compiled by somebody else from primary data reported in the research literature. These secondary sources are enormously useful in all areas of physics and I would urge anyone contemplating a career in physics to buy and treasure them.

The main sources of data used in this book are:

- G. W. C. Kaye & T. H. Laby, *Tables of physical and chemical constants*, 14th and 15th editions (Harlow, England: Longman; New York: Wiley). The 16th edition of Kaye & Laby is now available (1995) and, on brief inspection, it represents a considerable improvement on earlier editions. However it has not been used in this compilation.
- Weast , *CRC handbook of chemistry and physics*, 65th edition [also known as the "Rubber Bible"] (Chicago, Ill.: Chemical Rubber Publishing Company).
- J. Emsley, *The elements*, (Oxford: Clarendon Press/Oxford University Press).

Data on the elements are also available from world wide web sites such as:

- http://www.shef.sc.uk/uni/academic/a-c/chem/web-elements/chem/periodic-table.html

References to electromagnetic theory are to the excellent:

- B. I. Bleaney & B. Bleaney, *Electricity and magnetism* [2 vols] (Oxford: Oxford University Press).

Other sources are given in the text. Where no reference is given in the text, the data have been compiled from and cross-checked between several sources. I have tried, by means of several techniques, to eliminate erroneous data from my compilations. However, I cannot guarantee this, and the reader is referred to the original compilations and the references given therein.

A collection of corrections to this text, hints on the answers to the more difficult questions and other developments in the project can be found on my world wide web page at:

http://www.bbk.ac.uk/Departments/Physics

1.5 Exercise

Isaac Newton was unaware of developments in electromagnetism, quantum mechanics, genetics and evolutionary theory. However, in my opinion, his view of the world was undoubtedly "modern". Obtain a copy of Newton's work *Opticks* (New York: Dover) and read the final section of Book Three marked "Queries" (Page 339 in the Dover edition) . This consists of a number of questions over which Newton had puzzled, and to which he had arrived at tentative answers. After reading his "Queries", carry out the following exercise.

Imagine that Newton were to return to life in our time, and it fell to you to explain to him the key developments in modern science since his death. Write the script of a half-hour conversation between the two of you. Remember, you haven't got long so (a) don't waste time telling him about what he already knows and (b) be sure to script Newton's part as well as your own. I imagine he would be rather ill-tempered and aggressive, but compulsively curious. So he would be sure to interrupt if he didn't understand the language you were using and would be sure to say things like "Yes, yes, I suspected that all along".

CHAPTER TWO

Background theory

2.1 Introduction

In attempting to understand the properties of matter on a large scale, we need to be clear about what the components of matter are, and what *general principles* we may use to analyze their behaviour. In this context, the principles amount to "tools" for understanding components (the "nuts and bolts") of matter. Historically, the principles were evolved to allow the analysis of particular problems, in rather the same way that mechanical tools have evolved. To the uninitiated, however, it is often difficult to differentiate between the "tools" used to analyze a problem and the "nuts and bolts" of the problem itself – it all looks like "just so much metal".

In the context of this book we consider the following "components" of the world: (a) electrons, neutrons and protons; and (b) the electromagnetic field. The "tools" that we shall use are: (a) classical mechan-

ics and quantum mechanics; and (b) thermodynamics and statistical mechanics.

In what follows I assume that readers will have had some introduction to the topics outlined above, and the presentation here is by way of a summary of the essential results, a knowledge of which will be assumed in later chapters.

2.2 Matter

2.2.1 Electrons, neutrons and protons

In what follows we assume that electrons, protons and neutrons are fundamental particles from which all matter is made. Some of their properties, as deduced from numerous experiments, are listed in Table 2.1. The mass and electric charge of the electron, proton and neutron do not require much further

Table 2.1 The properties of particles which are treated as fundamental in this book. The most important properties of the particles for understanding the properties of matter are the mass and the electric charge. The internal angular momentum (spin) and magnetic moment of particles are discussed in the text.

Property	Units	Electron	Neutron	Proton
Mass	atomic mass units $u = 1.661 \times 10^{-27}$ kg	5.485×10^{-4} $\approx 1/1836$	1.0085	1.0071
Electric charge	proton charge $e = 1.602 \times 10^{-19}$ C	-1	0	$+1$
Magnetic moment	Bohr magneton $\mu_B = 9.274 \times 10^{-24}$ JT^{-1}	1.001	1.0419×10^{-3}	1.521×10^{-3}
Magnetic moment	nuclear magneton $\mu_n = 5.051 \times 10^{-27}$ JT^{-1}	1837.8	1.913	2.793
Intrinsic (spin) angular momentum	Planck constant divided by 2π $\hbar = 1.054 \times 10^{-34}$ Js	½	½	½
Electric dipole moment	Cm	0	0	0
Lifetime		Stable	Stable within nuclei; half-life ≈ 15 min in free space	Stable

(Data from Kaye & Laby; see §1.4.1)

comment, except to note that they are all exceedingly small. Their smallness is significant because we are apt to bring to our study of these particles understandings gained from studies of more familiar particles, which are usually substantially larger. As we will see later in this chapter, because of their size electrons, protons and neutrons must be described using the language of *quantum mechanics*. Although this allows a *technical* description of small particles and their properties, the description can be very difficult to understand.

Spin: the conceptual problems of the microscopic world

The difficulty of understanding the properties of fundamental particles cannot be more clearly illustrated than by considering the property known as the *spin* or *intrinsic angular momentum*. After the identification of the electron as a particle, the idea that electrons possessed the property referred to as the "spin" took another 20 years or so to develop. Experiments showed that electrons possess an *intrinsic magnetic moment*; that is, each electron behaves as if it were a tiny bar magnet, or loop of electric current. The anisotropic nature of this behaviour implies that the (unknown) structure of an electron must be such as to select one direction (or axis) as special. Indeed, we shall see later that the interactions between electrons depend slightly on the relative orientation of their "axes".

It seemed reasonable to infer that the magnetic moment arose from some internal circulation of the electric charge that constitutes the electron. This would be analogous to the way in which magnetic fields are generated by a current flowing in a loop of wire. Experiments showed that, when placed in a magnetic field, electron magnetic moments only orient themselves in one of two different ways. Using the mathematics of angular momentum, this implied that the magnitude of the intrinsic angular momentum would be exactly $\frac{1}{2}\hbar$. It seemed that electrons behaved as if they were "spinning".

However, no model of an electron as a sphere of circulating charge has yet been found which does not require the charge to move faster than the speed of light. Since this is forbidden by the theory of relativity, we can conclude that the intrinsic magnetic moment definitely does *not* arise from circulating electric charge. Thus we have *observations* of electrons which show that:

(a) Electrons possess an intrinsic magnetic moment, the origin of which is not known.

(b) When placed in a magnetic field, the electron orients itself with respect to the field gradient in one of only two ways.

From these observations we infer that:

(c) Electrons have a property which is limited to only two possibilities. This is *as if* it had an intrinsic angular momentum of magnitude $\frac{1}{2}\hbar$.

This discussion of spin is included to show that, although it is relatively straightforward to *describe* the properties of fundamental particles, it is difficult to *understand* these properties in terms of familiar phenomena. In this book we accept the properties listed in Table 2.1 as fundamental, and use them to understand the properties of matter on a larger scale.

But are protons, electrons and neutrons really fundamental?

At the deepest level at which scientists can study them electrons are currently considered to be fundamental particles. In other words, the properties of electrons are not explained in terms of "component" particles. However, the properties of neutrons and protons are explained in terms of a yet deeper layer of structure. They are considered to consist of particles called *quarks*: when certain types of quarks are bound together they create what we call a proton; when other types of quark are bound together they create a neutron. This is imagined to be broadly analogous to the way in which electrons, neutrons and protons are bound into atoms, and cause atoms to have different properties depending on the precise numbers and arrangement of electrons, neutrons and protons within them.

According to the current cosmological view, all the electrons, protons and neutrons in the universe were created shortly after the start of the universe as the temperature of a primordial "gas" fell. Initially, the gas contained free quarks, but as the gas cooled the quarks "condensed", i.e. bound together to form the particles we now know.

2.2.2 Nuclei

The total number of protons and neutrons in the universe (approximately 10^{87}) is not thought to have changed considerably since very shortly after the start of the universe. These protons and neutrons constitute the raw material from which the physical universe is made. The protons and neutrons interact strongly in the regions of high temperature and pressure within the cores of stars, binding together into particles of varying sizes that we refer to as *nuclei*. The number of protons in a *nucleus* is called its *atomic number, Z*. Most of the nuclei formed are unstable and rapidly disintegrate. Thus most nuclei that we find around us now are the relatively stable nuclei which have not yet disintegrated (Fig. 2.1). The stable nuclei have roughly equal numbers of protons and neutrons, usually with a slight excess of neutrons. Nuclei with the same number of protons but different numbers of neutrons are called *isotopes*.

Nuclei are extremely small with typical diameters of the order of 10^{-14} m. They are held together by attractive forces that act between the quarks which make up the protons and neutrons. The attractive forces that hold the nucleus together act roughly equally between *protons and protons, protons and neutrons*, and *neutrons and neutrons*. Thus it is common to use the collective term *nucleon* which does not distinguish between protons and neutrons. The attractive forces are extremely short range ($\approx 10^{-15}$ m), and amount essentially to a contact force between nucleons – as if their surfaces were "sticky". The *mass number* of a nucleus, A, is the total number of nucleons in a nucleus.

No stable nuclei with more than 83 protons have been observed, a fact that is ascribed to the strong electric repulsion between protons. The repulsive

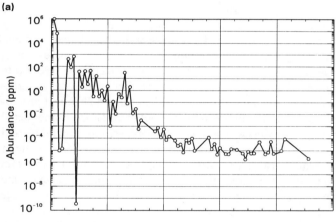

(a)

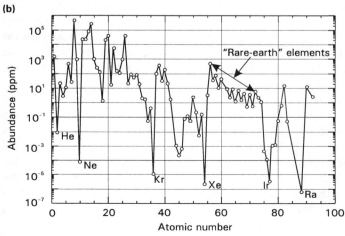

(b)

Figure 2.1 The relative abundances of the elements in the outer part of the Sun (a), and in the solid, outer part of the Earth the "crust" (b). Note that the vertical axis is logarithmic, so that points which are noticeably lower on the graph represent elements that are several orders of magnitude rarer than their neighbours.

The Sun is composed primarily of hydrogen, with all the other elements together accounting for only 6% of the total number of atoms. Furthermore, with one or two exceptions, there is a relatively smooth variation between atomic number and abundance, the heavier nuclei being successively less common than lighter nuclei.

The distribution of elements in the Earth's crust is quite different from that in the Sun. Understanding the distribution of elements on Earth is considerably more difficult than for the Sun. For example, some of the rare elements in the Earth's crust are rare because they are chemically unreactive gases, which are relatively abundant in the Earth's atmosphere. Thus the abundance of an element in the Earth's crust is the product of both the primordial composition of the Earth and other factors, such as whether at the Earth's mean temperature the element is a gas, liquid or solid, and whether it will bind chemically with other elements. The reader's attention is drawn to the fact that "rare-earth" elements are not particularly rare. (Data from Emsley; see §1.4.1)

force on a proton within the nucleus is roughly proportional to $Z(Z-1)$. Thus, for example, the repulsive force on a proton in a nucleus for which $Z=82$ is 74 times bigger than for $Z=10$. Above $Z=82$, the repulsive forces overcome the attractive forces between the nucleons in the nucleus.

At temperatures below about 10 000 K, electrons bind to nuclei until they become electrically neutral, i.e. until the number of electrons equals the number of protons. The combination of a nucleus with sufficient electrons to achieve electrical neutrality is known as an *atom*. It is thought that the universe as a whole is electrically neutral and that there are equal numbers of electrons and protons in existence.

In this book, we restrict our study of matter to the temperature range below a few thousand kelvin. This excludes an explicit study of the state of matter known as a "plasma"; although plasma is by far the most common state of matter in the universe, it is relatively rare on our planet.

2.2.3 Atoms

In later chapters, the arrangement of atoms in the different states of matter will be the main focus of our attention. Only occasionally will I refer separately to the existence of nuclei and their protons or neutrons. The structure of all atoms is broadly similar, consisting of two main features:

(a) A dense nucleus of A nucleons: Z protons and $(A - Z)$ neutrons. The nucleus therefore has an electric charge of $+Ze$, where e is the elementary charge on the proton, and an *approximate* mass of Au, where u is the *atomic mass unit* (see Table 2.1).

(b) A diffuse outer structure of Z electrons of total electric charge $-Ze$ held around the nucleus by electrical attraction. The overall diameter of the electron structure is similar for most atoms and is about 3×10^{-10}m. The diameter of the electron structure around an atom is typically more than 10 000 times larger than the diameter of the nucleus.

A substance composed of only one type of atom is called an *element*. Historically, it was through the realization that some substances were elemental, and the subsequent analysis of their properties, that the basic properties of atoms were deduced. Table 2.2 lists the names and symbols of the elements having atomic numbers up to 105.

Valence electrons

At separations greater than a few atomic diameters, atoms barely interact with one another. The interactions between atoms occur when they are brought close together and are caused only by the electrical interactions between the atoms. This single fact is worth repeating: *the interactions between atoms are caused only by electrical interactions*. The magnitude of the interaction is mainly determined by the *outer part* of the electronic structure around the atom. The electrons in the outer part of the structure are called *valence electrons*. The number and distribution of the valence electrons strongly affects both the *physical* and the *chemical* properties of atoms, and the substances from which they are composed.

Electrons tend to cluster around the nucleus in a sequence of spherically symmetric *shells*. This gives rise to a periodicity in the properties of atoms as function of Z, the number of electrons. For example, atoms with 2, 10, 18, 36, 54 or 86 electrons are able to pack electrons into successive shells with no electrons left over. Their physical and chemical properties are all strikingly similar, being the properties of atoms with no electrons outside a closed shell. Atoms with one electron outside a closed shell, i.e. those with 3, 11, 19, 37, 55 or 87 electrons also behave with striking similarity. For this reason, Mendeleyev and others grouped the elements into a *Periodic Table* (Fig. 2.2) in which elements with similar properties, such as density, melting temperature, or chemical reactivity, were placed in columns. An example of the striking periodicity found in the properties the elements can be seen in the densities of the solid elements, which are considered later (§7.1 & Fig. 7.1, p. 155).

Molecules and compounds

It is unusual on our planet to find matter in elemental form. Normally one encounters matter in which there are many types of atoms. The atoms are held together by electrical interactions between the outer electrons on each atom known as *chemical bonds*. A set of atoms that are bonded together chemically is known

Table 2.2 The elements with atomic numbers up to 105. The names of the elements tell many fascinating stories about their discovery.

Z	Element				Date of discovery and origin of name
1	Hydrogen	H	1766	Greek	*Hydros genes*, meaning "water forming"
2	Helium	He	1895	Greek	*Helios*, meaning "Sun"
3	Lithium	Li	1817	Greek	*Lithos*, meaning "stone"
4	Beryllium	Be	1797	Greek	*Beryllos*, meaning "beryl"
5	Boron	B	1808	Arabic	*Buraq*, meaning "borax"
6	Carbon	C	Old	Latin	*Carbo*, meaning "charcoal"
7	Nitrogen	N	1772	Greek	*Nitron genes*, meaning "nitre forming"
8	Oxygen	O	1774	Greek	*Oxy genes*, meaning "acid forming"
9	Fluorine	F	1886	Latin	*Fluere*, meaning "to flow"
10	Neon	Ne	1898	Greek	*Neos*, meaning "new"
11	Sodium	Na	1807	English	Soda; the symbol comes from the Latin *natrium*
12	Magnesium	Mg	1755	Greek	*Magnesia*, a district in Thessaly
13	Aluminium	Al	1825	Latin	*Alumen*, meaning "alum"
14	Silicon	Si	1824	Latin	*Silicis*, meaning "flint"
15	Phosphorus	P	1669	Greek	*Phosphoros*, meaning "bringer of light"
16	Sulphur	S	Old	Sanskrit	*Sulvere*, meaning "sulphur"
17	Chlorine	Cl	1774	Greek	*Chloros*, meaning "pale green"
18	Argon	Ar	1894	Greek	*Argos*, meaning "inactive"
19	Potassium	K	1807	English	Potash; the symbol comes from the Latin *kalium*
20	Calcium	Ca	1808	Latin	*Calix*, meaning "lime"
21	Scandium	Sc	1879	Latin	*Scandia*, meaning "Scandinavia"
22	Titanium	Ti	1791		Titans, sons of the Earth Goddess
23	Vanadium	V	1801		Vanadis, Scandinavian goddess
24	Chromium	Cr	1780	Greek	*Chroma*, meaning "colour"
25	Manganese	Mn	1774	Latin	*Magnes*, meaning "magnet"
26	Iron	Fe	Old	Saxon	Iron; the symbol comes from the Latin *ferrum*
27	Cobalt	Co	1735	German	*Kobald*, meaning "goblin"
28	Nickel	Ni	1751	German	*Kupfernickel*, meaning either "Devil's copper" or "St Nicholas's copper"
29	Copper	Cu	Old	Latin	*Cuprum*, meaning "Cyprus"
30	Zinc	Zn	1400	German	Zink
31	Gallium	Ga	1875	Latin	*Gallia*, meaning "France"
32	Germanium	Ge	1886	Latin	*Germania*, meaning "German"
33	Arsenic	As	1280	Greek	*Arsenikon*, meaning "yellow orpiment"
34	Selenium	Se	1817	Greek	*Selene*, meaning "moon"
35	Bromine	Br	1826	Greek	*Bromos*, meaning "stench"

Z	Element				Date of discovery and origin of name
36	Krypton	Kr	1898	Greek	*Kryptos*, meaning "hidden"
37	Rubidium	Rb	1861	Latin	*Rubidius*, meaning "deepest red"
38	Strontium	Sr	1790	English	Strontian in Scotland
39	Yttrium	Y	1794		The town of Ytterby in Sweden
40	Zirconium	Zr	1789	Arabic	*Zargun*, meaning "gold colour"
41	Niobium	Nb	1801	Greek	Niobe, a daughter of Tantalus; also called "Colimbium" in the USA
42	Molybdenum	Mo	1781	Greek	*Molybdos*, meaning "lead"
43	Technetium	Tc	1937	Greek	*Technikos*, meaning "artificial"
44	Ruthenium	Ru	1808	Latin	*Ruthenia*, meaning "Russia"
45	Rhodium	Rh	1803	Greek	*Hodon*, meaning "rose"
46	Palladium	Pd	1803		The asteroid "Pallas"
47	Silver	Ag	Old	Saxon	*Siolfur*, meaning "silver"; the symbol comes from the Latin *argentum*
48	Cadmium	Cd	1817	Latin	*Cadmia*, meaning "calomine"
49	Indium	In	1863		Indigo
50	Tin	Sn	Old	Saxon	Tin; the symbol comes from the Latin *stannum*
51	Antimony	Sb	Old	Greek	*Anti+monos*, meaning "not alone"; the symbol is from the Latin *stibium*
52	Tellurium	Te	1783	Latin	*Tellus*, meaning "Earth"
53	Iodine	I	1811	Greek	*Iodes*, meaning "violet"
54	Xenon	Xe	1898	Greek	*Xenos*, meaning "stranger"
55	Caesium	Cs	1860	Latin	*Caesius*, meaning "blue"
56	Barium	Ba	1808	Greek	*Barys*, meaning "heavy"
57	Lanthanum	La	1839	Greek	*Lanthanein*, meaning "to lie hidden"
58	Cerium	Ce	1803		Ceres, an asteroid discovered in 1801
59	Praseodymium	Pr	1885	Greek	*Prasios didymos*, meaning "green twin"
60	Neodymium	Nd	1885	Greek	*Neos didymos*, meaning "new twin"
61	Promethium	Pm	1945	Greek	Prometheus
62	Samarium	Sm	1879		The mineral samarskite
63	Europium	Eu	1901		Europe
64	Gadolinium	Gd	1880		J. Gadolin, a Finnish chemist
65	Terbium	Tb	1843		The town of Ytterby in Sweden
66	Dysprosium	Dy	1886	Greek	*Dysprositos*, meaning "hard to obtain"
67	Holmium	Ho	1878	Latin	*Holmia*, meaning "Stockholm"
68	Erbium	Er	1842		The town of Ytterby in Sweden
69	Thulium	Tm	1879		Thule, meaning "Ancient Scandinavia"; the uttermost north
70	Ytterbium	Yb	1878		The town of Ytterby in Sweden

(continued overleaf)

Z	Element		Date of discovery and origin of name
71	Lutetium	Lu	1907 Latin *Lutetia*, meaning "Paris"
72	Hafnium	Hf	1923 Latin *Hafnia*, meaning "Copenhagen"
73	Tantalum	Ta	1802 Greek Tantalos, the father of Niobe
74	Tungsten	W	1783 Swedish *Tung sten*, meaning "heavy stone"; the symbol comes from the alternative name "wolfram"
75	Rhenium	Re	1925 Latin *Rhenus*, meaning "Rhine"
76	Osmium	Os	1803 Greek *Osme*, meaning "smell"
77	Iridium	Ir	1803 Latin *Iris*, meaning "rainbow"
78	Platinum	Pt	Old Spanish *Platina*, meaning "silver"
79	Gold	Au	Old Saxon Gold (Latin *aurum*)
80	Mercury	Hg	Old Latin The planet Mercury; the symbol comes from the Latin *hydrargyros* meaning "liquid silver"
81	Thallium	Tl	1861 Greek *Thallos*, meaning "green twig"
82	Lead	Pb	Old Saxon Lead; the symbol comes from the Latin *plumbum*
83	Bismuth	Bi	1450 German Wismut
84	Polonium	Po	1898 Poland

Z	Element		Date of discovery and origin of name
85	Astatine	At	1940 Greek *Astatos*, meaning unstable
86	Radon	Rn	1900 Radium
87	Francium	Fr	1939 France
88	Radium	Ra	1898 Latin *Radius*, meaning "ray"
89	Actinium	Ac	1899 Greek *Aktinos*, meaning "ray"
90	Thorium	Th	1815 Thor, the Scandinavian God of War
91	Protactinium	Pa	1917 Greek *Protos*, meaning "first"
92	Uranium	U	1789 The planet Uranus
93	Neptunium	Np	1940 The planet Neptune
94	Plutonium	Pu	1940 The planet Pluto
95	Americium	Am	1944 English America
96	Curium	Cm	1944 Pierre and Marie Curie
97	Berkelium	Bk	1949 English Berkeley
98	Californium	Cf	1950 English California
99	Einsteinium	Es	1952 Albert Einstein
100	Fermium	Fm	1952 Enrico Fermi
101	Mendelevium	Md	1955 Dmitri Mendeleyev
102	Nobelium	No	1958 Alfred Nobel
103	Lawrencium	Lr	1961 Ernest O. Lawrence
104	Unnilquadium	Unq	1964 One-zero-four
105	Unnilpentium	Unp	1967 One-zero-five

* "Old" indicates that the element was known in antiquity.
(Data from Emsley; see §1.4.1)

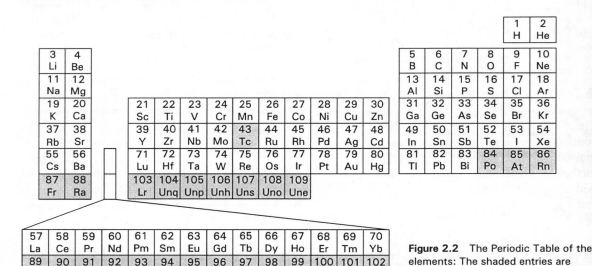

Figure 2.2 The Periodic Table of the elements: The shaded entries are radioactive.

as a *molecule*. Molecules range in size from two atoms (e.g. N_2 and O_2, where the subscript after the chemical symbol indicates the number of atoms of that type in the molecule), through three atoms (e.g. H_2O and CO_2) up to hundreds of thousands of atoms.

The phrase *chemical compound* generally refers to solids consisting of more than one type of atom, but for which there is no small group of atoms that may be considered independently. For example, carbon dioxide, CO_2, in its solid state would not be described as a compound because even in the solid state the molecules are clearly identifiable and interact only weakly with one other. It is thus referred to as a *molecular solid*. On the other hand, sodium chloride, NaCl, in its solid state *is* a chemical compound because although its formula refers to only two atoms, each sodium atom is bonded to six chlorine atoms, and each chlorine atom is bonded to six sodium atoms. The bonding forms a network that extends throughout the solid. These matters are discussed further in Chapter 6.

Complex molecules

I have excluded from this book systematic consideration of materials composed of large complex molecules, i.e. those consisting of more than a small number of atoms. This is not because materials made from such molecules are not interesting; indeed the opposite is true. They are excluded because the study of their physical properties is still in its infancy and the variety of their properties is simply overwhelming. Such materials are referred to in Chapters 8 and 9 on the properties of liquids (and liquid crystals) composed of organic molecules, and in passing at several other points.

2.3 The electromagnetic field

The matter of the world as we encounter it may be considered to be normally composed of atoms and molecules, as described in §2.2.3. The primary method by which these atoms and molecules interact with one another is through the *electromagnetic field*, and in particular the electric aspect of the field.

2.3.1 The concept of a field

Fields are not composed of matter (electrons, protons, neutrons, etc.), but they do exist. Fields are found to affect matter, and matter to affect fields, and it is through these mutual effects that we determine the properties of both matter and fields. For example, consider two particles X and Y. We can imagine the interaction between X and Y as a three-stage process, as indicated below.

1. Particle X affects the field around it.
2. The disturbance of the field around X spreads away from X.
3. A little later, particle Y is affected by the disturbance of the field created by X.

Note that the field itself is not an observable "object". We infer its properties from systematic studies of the way in which one particle affects other particles around it. Currently we can explain all observable phenomena in the universe in terms of three different types of fields: the *gravitational*, *electroweak* and *colour* (or *strong*) field. It is hoped that eventually all these fields will be understood as separate aspects of a single "unified" field.

For our purposes, we need to examine the properties of only one of the above fields, the *electroweak* field. Before leaving the other two fields note that:

- The colour field, while existing everywhere acts only between particles that possess a property called "colour charge" and it is through this field that quarks interact with one another. As mentioned previously, the forces between quarks are extremely short range ($\approx 10^{-15}$ m), and they thus act only within nuclei. In attempting to understand the properties of matter we ignore any processes that take place within nuclei and treat nuclei as if they were point masses with an electric charge $+Ze$.

- The gravitational field acts between all particles that possess *mass*, but its effects are generally negligible unless the masses involved are large, which will not happen over the length scales that we are interested in. We do, however, use the fact that particles possess mass, but only to discuss the way in which their motion is affected under the action of electrical forces, i.e. we use mass in the sense of inertial mass, but not gravitational mass.

Note that we will be unable to understand some quite common phenomena such as *radioactivity*, or *convection*, without recourse to explanations that involve these fields. However, in order to understand most properties of the matter which we find around us, we need concern ourselves only with the electroweak field. In fact, in the regions in which we shall be discussing it, the "weak "part of this field represents only a tiny correction to the "electro" part of the field. The "weak" part of the field causes a

small difference between the interactions of:
- electrons that move with their *internal spin* axis oriented in the same direction as their direction of motion; and
- electrons that move with their *internal spin* axis oriented in the opposite direction to their direction of motion.

We will neglect the small effects caused by the "weak" aspect of the field entirely and concentrate on the "electro" part of the field, conventionally called the *electromagnetic field*.

2.3.2 Electric and magnetic fields and forces

In what follows we consider the electric and magnetic parts of the electromagnetic field separately.

Force on an electric charge q in an electric field

The *electric field* $\mathbf{E}$ in the region of a point charge of magnitude q is given by

$$\mathbf{E} = \frac{q}{4\pi\varepsilon_0 r^2}\hat{\mathbf{r}} \qquad (2.1)$$

where r is the distance from the point charge, $\hat{\mathbf{r}}$ is a unit vector from q to the point under consideration, and ε_0 is a universal constant known as the *permittivity of free space*. The force $\mathbf{F}$ on a charge q_1 in an electric field $\mathbf{E}$ is given by

$$\mathbf{F} = q_1\mathbf{E} \qquad (2.2)$$

and so the force on a point charge q_1 a distance r from q is given by combining Equations 2.1 and 2.2 to give

$$\mathbf{F} = \frac{q_1 q}{4\pi\varepsilon_0 r^2}\hat{\mathbf{r}} \qquad (2.3)$$

The electrical force $\mathbf{F}$ described in Equation 2.3 is sometimes referred to as the *Coulomb force*. The potential energy of the two charges is

$$u = \frac{q_1 q}{4\pi\varepsilon_0 r} \qquad (2.4)$$

The potential energy per unit charge, the *electric potential*, around charge q is u/q_1, i.e.

$$V = \frac{q}{4\pi\varepsilon_0 r} \qquad (2.5)$$

Example 2.1

(a) **Calculate the force, acceleration, and potential energy of an electron a distance 10^{-10} m from a proton. This distance is typical of separations between charges found in atoms. As outlined in Table 2.1, protons possess an electric charge $+e$ and electrons possess an electric charge $-e$, where e has the value 1.6×10^{-19} C.**

The force is given by Eq. 2.3 with $q_1 = -1.6\times10^{-19}$ C, $q = +1.6\times10^{-19}$ C, $\varepsilon_0 = 8.85\times10^{-12}$ and $r = 1\times10^{-10}$ m. We thus have

$$F = \frac{-1.6\times10^{-19}\times1.6\times10^{-19}}{4\pi\times8.85\times10^{-12}\times\left(10^{-10}\right)^2} = -2.31\times10^{-8}\,\text{N}$$

where the negative sign indicates that the force is attractive. The force acts in a straight line between the two particles.

The acceleration of an electron is given by Newton's Third Law $a = F/m$ and so, since $m = 9.1\times10^{-31}$ kg we have

$$a = \frac{F}{m} = \frac{-2.31\times10^{-8}}{9.1\times10^{-31}} = -2.53\times10^{22}\,\text{ms}^{-2}$$

Note the colossal magnitude of this acceleration in comparison with the accelerations we experience daily ($g \approx 10\,\text{ms}^{-2}$).

The potential energy of the electron–proton pair is given by Equation 2.4, and so we have

$$u = \frac{-1.6\times10^{-19}\times1.6\times10^{-19}}{4\pi\times8.85\times10^{-12}\times10^{-10}} = -2.31\times10^{-18}\,\text{J}$$

where the negative sign indicates attraction, i.e. the particles have a lower energy than if they were separated by a large distance. The small energies involved in Coulomb interactions between electrons and protons are often expressed in units of electron volts where $1\,\text{eV} = 1.6\times10^{-19}$ J. In these units the energy of interaction of a proton and electron separated by 10^{-10} m is

$$u = \frac{-2.31\times10^{-18}}{1.6\times10^{-19}} = -14.4\,\text{eV}$$

(b) **What is the electric field a distance 10^{-10} m from a proton?**

The field is given by Equation 2.1 with $q = +1.6\times10^{-19}$ C and $r = 1\times10^{-10}$ m. We thus have

$$E = \frac{1.6\times10^{-19}}{4\pi\times8.85\times10^{-12}\times\left(10^{-10}\right)^2} = 1.44\times10^{11}\,\text{Vm}^{-1}$$

This field is enormous in comparison to the electric fields commonly encountered in laboratories.

Figure 2.3 Conventional illustration of the electric field in the region of a positive point charge. The arrowed lines indicate the direction of the electric field, and their closeness indicates the intensity or strength of the electric field.

Force on a charge q in a magnetic field

The force on a charge q in a magnetic field **B** is given by

$$\mathbf{F} = q\mathbf{v} \times \mathbf{B} \qquad (2.6)$$

where **v** is the velocity of the particle. Thus a magnetic field exerts no force on a stationary particle. A moving particle experiences a force of magnitude

$$F = qvB \sin\theta \qquad (2.7)$$

where θ is the angle between **v** and **B**.

Example 2.2

Calculate the magnitude of the force on an electron with a speed of $10^6\,\mathrm{m\,s^{-1}}$ moving perpendicular to a magnetic field of 1 T.

The force is given by Equation 2.7 with $q = -1.6 \times 10^{-19}\,\mathrm{C}$, $v = 10^6\,\mathrm{m\,s^{-1}}$ and $B = 1\,\mathrm{T}$. We thus have

$$F = qvB = -1.6 \times 10^{-19} \times 10^6 \times 1 = -1.6 \times 10^{-13}\,\mathrm{N}$$

The force acts perpendicular to both the magnetic field and the direction of motion.

Force on a charge q in electric and magnetic fields

The force on a charge q in electric field **E** and magnetic field **B** is given by the sum of Equations 2.2 and 2.6:

$$\boxed{\mathbf{F} = q\mathbf{E} + (q\mathbf{v} \times \mathbf{B})} \qquad (2.8)$$

Equation 2.8 is sometimes referred to as the *Lorentz force*.

2.3.3 Electric dipoles

In the study of matter, one frequently needs to consider electrical charges that are paired together; this pairing means that there is overall electrical neutrality. However, for a variety of reasons, the positive and negative electrical charges may not be distributed such that they have the same centre of symmetry. For example, in an applied electric field, the electrons and nuclei of all atoms move in opposite directions and the atom then acquires a so-called *electric dipole moment*. Also, the distribution of electric charge within some molecules is such that the molecules possess intrinsic electric dipole moments even in the absence of an applied electric field.

The magnitude of an electric dipole moment such as that illustrated in Figure 2.4a is defined as

$$\mathbf{p} = q\mathbf{d} \qquad (2.9)$$

and so has a magnitude

$$p = qd \qquad (2.10)$$

(a)

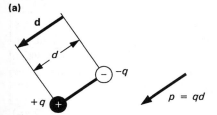

Figure 2.4 (a) An electric dipole consists of two equal and opposite charges and has the magnitude $p = qd$; as a vector the dipole is considered to point from the negative to the positive charge. (b, c) In a uniform applied electric field, an electric dipole experiences no net force, but is subject to a torque $\Gamma = pE\sin\theta$. (d) Electric field lines in the region of an electric dipole.

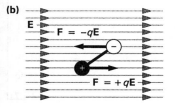

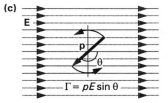

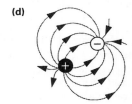

Force on an electric dipole p *in an electric field*

As illustrated in Figure 2.2b the net force on an electric dipole in a uniform electric field is zero

$$\mathbf{F} = 0 \qquad (2.11)$$

because the forces on each component of the dipole cancel.

Torque on an electric dipole p *in an electric field*

As illustrated in Figure 2.2c a uniform electric field exerts a torque Γ perpendicular to both $\mathbf{E}$ and $\mathbf{p}$ given by

$$\Gamma = \mathbf{p} \times \mathbf{E} \qquad (2.12)$$

which has the magnitude

$$\Gamma = pE \sin\theta \qquad (2.13)$$

where θ is the angle between $\mathbf{p}$ and $\mathbf{E}$.

Force on an electric dipole p *in an electric field gradient*

We note for completeness that there is a force on a molecule if its dipole moment is placed in a *non-uniform* electric field. The force is complicated to express, but the x component of it may be written as

$$F_x = p_x\left(\frac{\partial E_x}{\partial x}\right) + p_y\left(\frac{\partial E_y}{\partial x}\right) + p_z\left(\frac{\partial E_z}{\partial x}\right) \quad (2.14)$$

Example 2.3

In a carbon monoxide (CO) molecule, slightly more electric charge clusters around the oxygen atom than the carbon atom due to details of the structure of the electronic orbits around the two atoms. The molecule thus has a "built-in" electric dipole moment of approximately 0.1×10^{-30} Cm. (The unit of 1×10^{-30} Cm is typical of the electric dipole moments found in molecules and is sometimes called a *debye* unit. CO has an electric dipole moment of 0.1 debye.) Estimate the torque on a CO molecule in an electric field of $1000\,\mathrm{V\,m^{-1}}$.

Assuming the moment is oriented perpendicular to the field, then Γ is given by

$$\Gamma \approx pE = 0.1 \times 10^{-30} \times 10^{3} = 10^{-28}\ \mathrm{Nm}$$

Note: $1000\,\mathrm{V\,m^{-1}}$ is not a particularly strong electric field. It corresponds to the electric field between two plates 1 mm apart with a potential difference of 1 V. Further details can be found in Example 5.14, p. 116.

---

with equivalent expressions for the y and z components of the force. Note that, in general, there will also be a torque on the moment as described in Equation 2.12.

Electric field due to an electric dipole

As well as being subject to forces and torques in electric fields, electric dipoles act as sources of electric fields. At distances r greater than d, the separation of the charges in the dipole, the electric potential due to the dipole may be written as

$$V = \frac{\mathbf{p} \cdot \mathbf{r}}{4\pi\varepsilon_0 r^2} \qquad (2.15)$$

where $\mathbf{p}$ is defined in equation 2.9 and $\mathbf{r}$ is a unit vector from the dipole to the position at which the potential is evaluated. The electric field of a dipole (Fig. 2.4d) is related to the gradient of the potential function and varies roughly as $1/r^3$.

2.3.4 Magnetic dipoles

In the study of matter, one often considers electrical charges moving in orbits that effectively amount to a current loop. Such loops are known as *magnetic dipoles*. Most importantly, electric charges moving within atoms form magnetic dipoles, the magnitudes of which are characteristic of their orbits. Magnetic dipoles create a magnetic field around them, and respond to applied magnetic fields in a manner analogous to the response of electric dipole moments to applied electric fields.

The strength of a magnetic dipole is called the *magnetic dipole moment* $\mathbf{m}$. The strength of the magnetic dipole is expressed in terms of the current i enclosing an area A which would create an equivalent magnetic field. The magnetic dipole moment then has the value

$$\mathbf{m} = i\mathbf{A} \qquad (2.16)$$

where the vector $\mathbf{A}$ has magnitude A and points perpendicular to the area in the sense indicated by a right-hand screw rotating in the same sense as the current. Thus the magnetic moment has the magnitude (Fig. 2.5a)

$$m = iA \qquad (2.17)$$

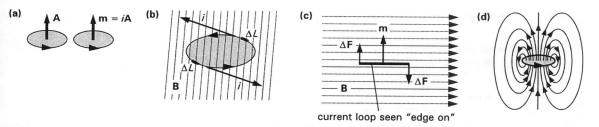

Figure 2.5. (a) A magnetic dipole **m** can be modelled as a loop of current i enclosing area A. (b, c) In a uniform applied magnetic field, a magnetic dipole (current loop) experiences no net force, but is subject to a torque $\Gamma = mB\sin\theta$. (d) The magnetic field in the region of a magnetic dipole.

Force on a magnetic dipole in a magnetic field

As illustrated in Figure 2.5b the net force on a magnetic dipole in a uniform magnetic field is zero, because the force on each small length of the dipole cancels:

$$\mathbf{F} = 0 \qquad (2.18)$$

This result is analogous to Equation 2.11 for an electric dipole moment in a uniform electric field.

Torque on a magnetic dipole in a magnetic field

As illustrated in Figure 2.5c a uniform magnetic field exerts a torque Γ perpendicular to both **B** and **m** given by

$$\Gamma = \mathbf{m} \times \mathbf{B} \qquad (2.19)$$

which has the magnitude

$$\Gamma = mB\sin\theta \qquad (2.20)$$

where θ is the angle between **m** and **B**.

These results are analogous to Equations 2.12 and 2.13 for an electric dipole moment in a uniform electric field.

Force on a magnetic dipole in a magnetic field gradient

We note for completeness that there is a force on a molecule if its magnetic dipole moment is placed in a *non-uniform* magnetic field. The force is complicated to express, but the x component of it may be written as

Example 2.4

In a cerium atom (Ce) , the electronic orbits are such that there is more electric current flowing around the atom in one sense. Thus cerium has a "built-in" magnetic dipole moment of approximately $2.23 \times 10^{-23}\,\text{Am}^2$. (A unit of $9.27 \times 10^{-24}\,\text{Am}^2$ is typical of the magnetic dipole moments found in atoms; it is called a *Bohr magneton* and has the symbol μ_B. Thus cerium has a "built-in" magnetic dipole moment of approximately $2.4\mu_B$.)

Estimate the torque on a cerium atom in a magnetic field of 1 T.

Assuming the moment is oriented perpendicular to the field, then Γ is given by

$$\Gamma \approx mB = 2.23 \times 10^{-23} \times 1 = 2.23 \times 10^{-23}\ \text{Nm}$$

Note: 1 T is a moderately strong magnetic field.

$$F_x = m_x\left(\frac{\partial B_x}{\partial x}\right) + m_y\left(\frac{\partial B_y}{\partial x}\right) + m_z\left(\frac{\partial B_z}{\partial x}\right) \quad (2.21)$$

with equivalent expressions for the y and z components of the force. Note that, in general, there will still be a torque on the dipole moment as well. This result is analogous to Equation 2.14 for an electric dipole moment in a non-uniform electric field.

Magnetic field due to a magnetic dipole

As well as being subject to forces and torques in magnetic fields, magnetic dipoles act as sources of magnetic fields. The magnetic field around a dipole is illustrated in Figure 2.5d.

15

2.3.5 Changes in electromagnetic fields

The speed of light in a vacuum

The speed of electromagnetic disturbances in a region of the electric field in which there is no matter (a vacuum) is a universal constant c defined to be $2.997924580 \times 10^8 \approx 2.998 \times 10^8 \, \text{ms}^{-1}$. The theory of electromagnetic wave propagation predicts that c is related to other constants that characterize the electric field in a vacuum by

$$c = \frac{1}{\sqrt{\varepsilon_0 \mu_0}} \qquad (2.22)$$

where μ_0 is the *permeability* of free space and ε_0 is the *permittivity* of free space. The value of μ_0 is $4\pi \times 10^{-7} \, \text{Hm}^{-1}$ and so ε_0 must have the value $1/\mu_0 c^2 = 8.854 \times 10^{-12} \, \text{F m}^{-1}$.

The speed of light in a medium

The speed v of electromagnetic disturbances through a region of the electric field containing matter varies with the frequency of the disturbance and with the properties of the matter through which the disturbance passes. The theory of electromagnetic wave propagation predicts that v is given by

$$v = \frac{1}{\sqrt{\varepsilon \varepsilon_0 \mu \mu_0}} \qquad (2.23)$$

where μ is the relative magnetic permeability and ε is the relative dielectric permittivity (dielectric constant) of the substance. Commonly, μ is close to unity, in which case Equation 2.23 may be rewritten

$$v = \frac{c}{\sqrt{\varepsilon}} \qquad (2.24)$$

Electromagnetic spectrum

When charged particles are accelerated or decelerated in a vacuum, an electromagnetic disturbance spreads outward from the particle at the speed of light. Commonly the acceleration is periodic, resulting from a simple harmonic motion of the charge. In such cases the electromagnetic disturbance has a well-defined frequency f and wavelength λ. Table 2.3 lists the different names given to disturbances in different frequency ranges.

Photons

The electromagnetic field can gain or lose energy only in small packets called *photons*. The energy E required to create a photon with a frequency f is given by

$$E = hf \qquad (2.25)$$

where h is the *Planck constant* $= 1.054 \times 10^{-34} \, \text{J s}$ ($6.58 \times 10^{-16} \, \text{eV s}$). The magnitude of the energy required to excite photons of different frequencies and wavelengths is summarized in Table 2.3.

Table 2.3 The energies of photons with given frequencies and wavelengths.

Frequency (Hz)	Wavelength (m)	Energy (eV)	Comment
1×10^6	3×10^2	4.136×10^{-9}	Radio broadcasts
1×10^7	3×10^1	4.136×10^{-8}	
1×10^8	3	4.136×10^{-7}	Television broadcasts
1×10^9	3×10^{-1}	4.136×10^{-6}	Microwave
1×10^{10}	3×10^{-2}	4.136×10^{-5}	Infrared
1×10^{11}	3×10^{-3}	4.136×10^{-4}	Infrared; corresponds to processes that occur when atoms vibrate
1×10^{12}	3×10^{-4}	4.136×10^{-3}	Infrared
6.6×10^{12}	4.55×10^{-4}	2.5×10^{-2}	Infrared corresponds to processes that occur typically at around room temperature (290 K)
1×10^{13}	3×10^{-5}	4.136×10^{-2}	Infrared
4×10^{14}	7.5×10^{-7}	1.654	Red light; corresponds to processes that occur to electrons in the outer (valence) shells of atoms.
1×10^{15}	3×10^{-7}	4.136	Blue light; corresponds to processes that occur to electrons in the outer (valence) shells of atoms.
1×10^{16}	3×10^{-8}	4.136×10	Ultraviolet light
1×10^{17}	3×10^{-9}	4.136×10^2	Ultraviolet light
1×10^{18}	3×10^{-10}	4.136×10^3	X-rays
1×10^{19}	3×10^{-11}	4.136×10^4	X-rays
1×10^{20}	3×10^{-12}	4.136×10^5	X-rays; corresponds to processes that occur to electrons in the inner shells of atoms.
1×10^{21}	3×10^{-13}	4.136×10^6	X-rays
1×10^{22}	3×10^{-14}	4.136×10^7	X-rays
1×10^{23}	3×10^{-15}	4.136×10^8	γ-rays; corresponds to processes that occur within nuclei

2.4 Classical and quantum mechanics

2.4.1 Classical mechanics or quantum mechanics?

Classical mechanics concerns itself with the motion of "relatively large" particles of matter. Quantum mechanics concerns itself with the motion of particles of matter of all sizes, but is used mainly where the particles involved are "relatively small". Where to draw the distinction between "relatively small" and "relatively large" is a subject of both physical and philosphical controversy at present, but broadly speaking, particles larger than a few atomic diameters in size ($\approx 10^{-9}$ m to 10^{-8} m) may be treated classically, whereas particles smaller than this must be treated using quantum mechanics. However, even for much smaller particles, classical mechanics can be used to give "a feel" for a problem, and is often useful for obtaining order of magnitude estimates.

2.4.2 Classical mechanics

Newton's laws of motion

Particles are assumed to obey the three laws of motion, which were first enunciated clearly by Newton. The first and second laws may be summarized as

$$\mathbf{F} = m\mathbf{a} \tag{2.26}$$

$\mathbf{F}$ is the force acting on a body of mass m and $\mathbf{a}$ is the resulting acceleration. Equation 2.26 may be restated in terms of momentum $\mathbf{p}$, velocity $\mathbf{v}$, or position $\mathbf{r}$:

$$\mathbf{F} = \frac{d\mathbf{p}}{dt} = m\frac{d\mathbf{v}}{dt} = m\frac{d^2\mathbf{r}}{dt^2} \tag{2.27}$$

If we rewrite the first equality in Equation 2.27 for short times, we obtain an expression that allows us to estimate $\mathbf{F}$:

$$\mathbf{F} \approx \frac{\Delta\mathbf{p}}{\Delta t} \tag{2.28}$$

in terms of the change of momentum $\Delta\mathbf{p}$ in a short time Δt.

Simple harmonic oscillator

The motion of many objects may be described (to some extent at least) as being similar to that of an idealized simple harmonic oscillator. In a simple harmonic oscillator, a particle is attracted to a central position x_0 by a force that increases in magnitude linearly with displacement from x_0 (Fig. 2.6a) i.e.

$$F = -K(x - x_0) \tag{2.29}$$

The proportionality constant K in Equation 2.29 is generally known as the "spring constant" of the oscillator. The larger the value of K, the greater the force attempting to restore the particle to position x_0 (Fig. 2.6b). The potential energy of the particle in a simple harmonic oscillator (Fig. 2.6c) is given by

$$V = \tfrac{1}{2} K(x - x_0)^2 \tag{2.30}$$

Figure 2.6 Simple harmonic oscillator. (a) A particle trapped by springs in a physical situation that would result in simple harmonic motion. (b) The force on a particle as a function of its displacement. The three lines correspond to different values of K, the "spring constant", with the steepest lines representing the stiffest spring. (c) The potential energy of a particle as a function of its displacement. The three curves correspond to different values of K, the narrowest, most strongly curving parabola representing the stiffest spring.

(a)   **(b)**   **(c)**

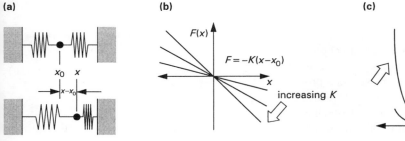

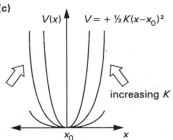

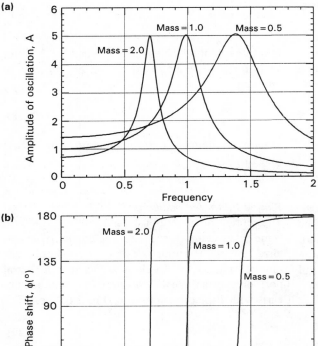

Figure 2.7 The response of a simple harmonic oscillator to an applied oscillating force. The oscillator has a low damping coefficient $\alpha = 0.02$, a spring constant $K = 1$, and the response is depicted for particles of three different masses, $m = 0.5$, 1.0, and 2.0. (a) The amplitude of the oscillations as a function of frequency. Note the increase in the resonant frequency as the mass of the particle decreases, and the increase in the low frequency amplitude as the mass decreases. (b) The phase ϕ of the oscillations as a function of frequency. At frequencies much less than the resonant frequency, the particle oscillates essentially in phase with the applied force. The graphs in (a) and (b) were calculated using a spreadsheet program to evaluate Equations 2.36 and 2.37 at a set of closely spaced frequencies.

Resonance

A particle in an oscillator potential has a *natural frequency* of oscillation f_0 which arises from the combination of the mass m of the oscillating particle and the spring constant K. If the particle is displaced from x_0 to a position x and released, it naturally oscillates with frequency f_0

$$f_0 = \tfrac{1}{2\pi}\sqrt{K/m} \qquad (2.31)$$

and angular frequency ω_0

$$\omega_0 = \sqrt{K/m} \qquad (2.32)$$

If the particle is subject to an externally applied force oscillating with angular frequency ω then the amplitude of the resulting oscillations shows a pronounced maximum when $\omega = \omega_0$ a phenomenon known as *resonance*. In order to calculate the oscillation amplitude at resonance, one must take account of

the loss of energy from the oscillator system, a process that may be fairly represented by a damping force proportional to the particle velocity v:

$$F = -\alpha v \qquad (2.33)$$

If the applied (driving) force has the form

$$F = F_0 \cos(\omega t) \qquad (2.34)$$

then the position of the particle varies as

$$x = A \cos(\omega t + \phi) \qquad (2.35)$$

where the *amplitude* of oscillation A is given by

$$A = \frac{F_0/m}{\left[\left(\omega_0^2 - \omega^2\right)^2 + 4(\alpha/2m)^2\omega^2\right]^{\frac{1}{2}}} \qquad (2.36)$$

and the *phase difference*, ϕ, between the particle

motion and the applied force satisfies

$$\tan \phi = \frac{\alpha\omega}{m\left(\omega_0^2 - \omega^2\right)} = \frac{\alpha f}{2\pi m\left(f_0^2 - f^2\right)} \quad (2.37)$$

Examples of the application of these equations are illustrated in Figure 2.7. There are two key features of these results that are worthy of note. The first feature (Fig. 2.7a) is the large amplitude of the vibrations of the particle when the forcing frequency f is close to the natural frequency of oscillation of the particle $f_0 = (2\pi)^{-1}(K/m)^{\frac{1}{2}}$. The second feature (Fig. 2.7b) concerns the phase difference between the particle motion and the driving force F. If the frequency of the driving force is well below f_0, the particle oscillates essentially in-phase with the driving force. However, if the frequency of the driving force is well above f_0, the particle oscillates essentially 180° out-of-phase with the driving force.

Small simple harmonic oscillators: electrons and atoms

The resonant frequency of a particle of mass m in a simple harmonic oscillator of spring constant K is given by

$$\omega_0 = \sqrt{K/m} \quad (2.38)$$

Because ω_0 is inversely proportional to m, the resonant frequencies of particles of atomic mass may be very large. The "spring constants" that keep electrons and atoms in place in a solid are rather difficult to calculate accurately, but are generally of a similar order of magnitude to those found in common household and industrial springs! (See Example 2.5.)

The forces acting on electrons and ions originate solely from their Coulomb interaction, and since electrons and ions have charges of a similar magnitude ($\pm e$) they have "spring constants" of a similar magnitude. However, atoms or ions are typically 10000 times heavier than electrons, and so their resonant frequencies are approximately $(10000)^{\frac{1}{2}} = 100$ times lower than the resonant frequencies of electrons. Typically, electrons have resonant frequencies of approximately 10^{15} Hz to 10^{16} Hz, in the optical and ultraviolet regions of the spectrum, and ions and atoms have resonant frequencies of approximately

---

Example 2.5

A diatomic molecule of nitrogen consists of two atoms of nitrogen of mass $m = 14 \times 1.66 \times 10^{-27}$ kg. The distribution of electric charge throughout the molecule that has the minimum Coulomb energy is illustrated below.

In equilibrium

Vibrations of such molecules are observed to take place with resonant frequencies of the order of 10^{13} Hz.

During these vibrations, the charge distribution changes and Coulomb forces act to restore the atoms to their minimum-energy position. These vibrations may be modelled as a simple harmonic oscillator with spring constant K. What is the approximate value of K?

Vibrations of the molecule

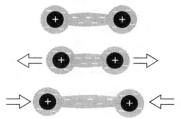

Our model of the molecule is illustrated above, so one can use Equation 2.31 to estimate the resonant frequency f_0.

$$f_0 = \frac{1}{2\pi}\sqrt{\frac{K}{m}}$$

We can rearrange this to yield an expression for K

$$K = 4\pi^2 m f_0^2$$

Substitution into this expression yields

$$K = 4\pi^2 \times 14 \times 1.66 \times 10^{-27} \times \left(10^{13}\right)^2 = 91.7 \,\mathrm{Nm}^{-1}$$

Note: The correct mass to use in the calculation is the so-called *reduced mass* of the particle $m_R = (1/m + 1/m)^{-1} = m/2$. However, this does not affect the order of magnitude of the answer and so we conclude that for resonant frequencies in the region of 10^{13} Hz, K is of the order $100 \,\mathrm{Nm}^{-1}$. It is interesting, and not entirely coincidental, that this value is of a similar order of magnitude to that found in ordinary springs.

10^{13} Hz to 10^{14} Hz, in the infrared region.

However, in order to describe small simple harmonic oscillators *accurately*, one must use the quantum mechanical description outlined in the next section.

2.4.3 Quantum mechanics

Quantum states

Classically, a particle may be described by listing the values of its physical properties, for example its mass, position, momentum, energy and magnetic dipole moment. Taken together, these properties constitute a specification of the *state* of a particle. The laws of classical mechanics describe the way in which the state of a particle changes with time.

In quantum mechanics one uses the concept of a *quantum state* to describe a particle. The laws of quantum mechanics describe which states are physically realistic, and the ways in which a particle changes from one quantum state to another. Each quantum state is characterized by a *unique* set of *quantum numbers* that "index" the quantum state. Sometimes quantum numbers have continuously variable values, but commonly quantum numbers are restricted to a set of discrete values, i.e. values outside the set do not describe physically realistic quantum states. For example, for a particle of mass m trapped inside a cubic box of side L, the physically realistic quantum states have energies given by

$$E\left(n_x, n_y, n_z\right) = \frac{h^2}{8mL^2}\left[n_x^2 + n_y^2 + n_z^2\right] \quad (2.39)$$

where h is the Planck constant and n_x, n_y and n_z are the quantum numbers that uniquely label each particular quantum state.

Importantly, one frequently discusses the properties of a quantum state even if no particle is actually occupying the state (Fig. 2.8a,b.) Although this seems strange at first, it is merely analogous to, say, discussing the potential merits of theatre seats even when no one is sitting them. Thus in addition to specifying the properties of quantum states, one then describes a physical situation by specifying which quantum states are *occupied*.

It is also important to note that a quantum state can describe a *state of motion*. Thus a particle in a quantum state – and remaining in that state – may have a velocity. This is particularly significant when discussing the quantum states of electrons near atoms. Classically, we describe the electrons as "orbiting" the atom – indeed the quantum states are often called *orbitals* – which implies motion around the nucleus. However, quantum mechanically this "state of motion" is described by specifying the quantum state which corresponds to that "state of motion".

Quantum states and energy levels

In any particular physical situation, there are in general many possible quantum states that a particle may occupy (Fig. 2.8c). It is possible to have several quantum states that all have the same energy, although other properties of the quantum states will be different. Quantum states with the same energy are said to belong to the same *energy level*. An energy level with more than one quantum state is, curiously, said to be *degenerate*.

Figure 2.8 A simple representation of quantum states using a line to represent a state, the vertical position of the line to represent its energy, and a filled circle to indicate whether or not it is occupied. (a) A set of quantum states with different energies; (b) The same set of states as in (a) with just one state occupied; (c) A different set of quantum states, again with only one state occupied. Note that at some energy levels there are several distinct quantum states, i.e. some of the energy levels are *degenerate*.

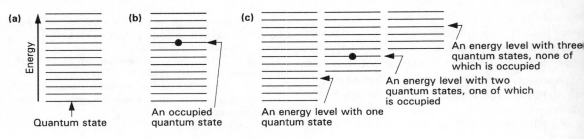

The wave function

Each quantum state has associated with it a *wave function* $\Psi(\mathbf{r},t)$ (Ψ is pronounced *psi*). The wave function is a *complex function*, i.e. it has two components, a real part Ψ_r and an imaginary part Ψ_i. Note that the technical mathematical terms "real" and "imaginary" can be used to describe many two-component quantities; the terms do not refer to the "existence" (ontology) of the components of the wave function. The probability of finding a particle in a region of space of volume $d\mathbf{r}$ around a position $\mathbf{r}$ is given by

$$P(\mathbf{r})d\mathbf{r} = \left|\Psi(\mathbf{r})\right|^2 d\mathbf{r} = \Psi^*(\mathbf{r})\Psi(\mathbf{r})d\mathbf{r} = \left[\Psi_r^2 + \Psi_i^2\right]d\mathbf{r}$$

(2.40)

where $\Psi^*(\mathbf{r})$ is the *complex conjugate* of $\Psi(\mathbf{r})$ Since the particle must be somewhere, the integral of $P(\mathbf{r})$ over all space must equal unity, i.e.

$$\int_{\text{all space}} P(\mathbf{r})d\mathbf{r} = \int_{\text{all space}} \left|\Psi(\mathbf{r})\right|^2 = \int_{\text{all space}} \Psi^*(\mathbf{r})\Psi(\mathbf{r}) = 1$$

(2.41)

This property is a key feature of the quantum mechanical description of matter: the probability of finding a particle near a particular position is related to the value of $|\Psi(\mathbf{r})|^2$ near that position (Fig. 2.9).

The Schrödinger equations

Two equations are relevant to our discussion of wave functions: the *time-independent Schrödinger equation* and the *time-dependent Schrödinger equation*. The first determines what types of wave function correspond to particular allowed quantum states, and the second governs how a wave function evolves over time. In this book we are mainly concerned with the time-independent Schrödinger equation, which is

$$\boxed{\frac{-\hbar^2}{2m}\nabla^2\Psi(\mathbf{r}) + V(\mathbf{r})\Psi(\mathbf{r}) = E\Psi(\mathbf{r})}$$

(2.42)

or in one dimension

$$\frac{-\hbar^2}{2m}\frac{\partial^2}{\partial x^2}\Psi(x) + V(x)\Psi(x) = E\Psi(x)$$

(2.43)

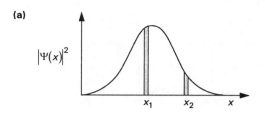

(a)

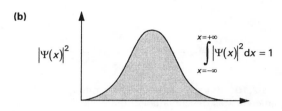

(b)

$$\int_{x=-\infty}^{x=+\infty} \left|\Psi(x)\right|^2 dx = 1$$

Figure 2.9 Illustration of the way in which the modulus of the wave function squared is interpreted. The figures show a possible variation of $|\Psi(x)|^2$. (a) If a particle occupies the quantum state that the wave function represents, it is more likely to found near x_1 than x_2 because $|\Psi(x_1)|^2 > |\Psi(x_2)|^2$. (b) If a particle occupies the quantum state that the wave function represents, then it must be found somewhere between $x = -\infty$ and $x = +\infty$. Therefore the equation shown must hold true.

In Equations 2.42 and 2.43:
- $V(\mathbf{r})$ is the function describing how the potential energy of the particle changes with position, e.g. for a one-dimensional harmonic oscillator $V(x) = \frac{1}{2}Kx^2$;
- $\hbar$ is the Planck constant h divided by 2π;
- m is the mass of the particle under discussion;
- $\partial^2/\partial x^2[\Psi(x)]$ is the second derivative of the wave function, known colloquially as its *curvature*; and
- E is the total energy of the particle under discussion.

The Schrödinger equation determines at each point in space the relative values of the *curvature* of the wave function and its magnitude in terms of the energy E of the wave function. Suppose a given wave function (E fixed) has a certain amplitude Ψ_a at position x_a, in a region of low potential energy. Then if V is larger at position b, the Shrödinger equation (Fig. 2.10) tells us that at b either the wave function must be reduced, or the curvature must be lessened. Finding solutions to the Schrödinger equation is, in general, quite tricky. Below we list a few solutions

21

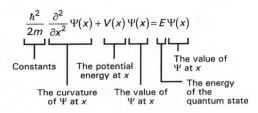

$$\frac{\hbar^2}{2m}\frac{\partial^2}{\partial x^2}\Psi(x) + V(x)\,\Psi(x) = E\,\Psi(x)$$

Constants

The potential energy at x

The curvature of Ψ at x

The value of Ψ at x

The value of Ψ at x

The energy of the quantum state

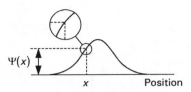

$\Psi(x)$

x   Position

Figure 2.10 At each value of position x, the Schrödinger equation relates the curvature of the wave function at x, the value of the wave function itself at x, the potential energy at x and the value of the energy of the quantum state.

appropriate to some of the situations we shall encounter in later chapters.

Examples of quantum mechanics

A free particle

For a particle in free space the potential energy is a constant V_0 independent of position. If this is the case then the Schrödinger equation 2.43

$$\frac{-\hbar^2}{2m}\frac{\partial^2}{\partial x^2}\Psi(x) + V(x)\Psi(x) = E\Psi(x) \qquad (2.44)$$

becomes

$$\frac{-\hbar^2}{2m}\frac{\partial^2}{\partial x^2}\Psi(x) + V_0(x)\Psi(x) = E\Psi(x) \qquad (2.45)$$

Taking the V_0 over to the right-hand side of the equation, and rearranging the constants results in

$$\frac{\partial^2}{\partial x^2}\Psi(x) = \left[\frac{-2m}{\hbar^2}(E - V_0)\right]\Psi(x) \qquad (2.46)$$

Since all the quantities in square brackets are constants, the equation may be rewritten as

$$\frac{\partial^2}{\partial x^2}\Psi(x) = -A^2\Psi(x) \qquad (2.47)$$

where A is a constant given by

$$A = \sqrt{\frac{2m}{\hbar^2}(E - V_0)} \qquad (2.48)$$

This differential equation has solutions that look like

$$\psi(x) \approx \sin(Ax) \quad \text{and} \quad \psi(x) \approx \cos(Ax) \qquad (2.49)$$

If $E - V_0$ is positive, then $(2m/\hbar^2)(E - V_0)$ will be a positive number, and its square root will be a real positive or negative number. The solutions to the equation are sketched for several different positive values of $E - V_0$ in Figure 2.11.

The wavelength, λ, of the wave function (Equ. 2.49) for the particle is given by

$$A = 2\pi/\lambda \qquad (2.50)$$

The *de Broglie hypothesis* is that the momentum, p, of the particle is given by

$$p = h/\lambda \qquad (2.51)$$

Substituting Equation 2.50 into Equation 2.51, we find

$$p = \frac{hA}{2\pi} = \hbar A \qquad (2.52)$$

$$p = \hbar\sqrt{\frac{2m}{\hbar^2}(E - V_0)} = \sqrt{2m(E - V_0)} \qquad (2.53)$$

Figure 2.11 Portions of wavefunctions $\Psi(x)$ satisfying Equation 2.47 for three different values of $E - V_0$. From (a) to (c) the value of $E - V_0$ is increasing. Notice that for a given V_0, the higher the energy E, the shorter the wavelength of the wavefunction.

(a)

(b)

(c)

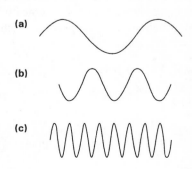

The energy in the case of Figure 2.11 is entirely kinetic energy, and is related to the momentum by the familiar relationship:

$$KE = \tfrac{1}{2}mv^2 = \frac{p^2}{2m} \qquad (2.54)$$

The square-well potential

Figure 2.12 shows the variation of potential energy in a one-dimensional square potential well, the walls of which are extremely high. The concept of this type of potential energy well is of great use in theoretical physics because, although the situation never occurs in nature, it is one of the few problems that it is possible to solve exactly. Furthermore, it often makes a good first approximation to real problems such as the problem of electrons trapped in a metal, or on an atom. It is considered in depth in §6.5 and Chapter 7.

There are several features to note about the wave functions of the two lowest quantum states sketched in Figure 2.12:

- The wave functions satisfy the time-independent Schrödinger equation (Eq. 2.43)
- At the edges of the potential well, the wave function is zero where the potential energy is very large.
- In the central region of the potential well, the potential energy term $V(x)$ is zero, and so this is equivalent to a "small part of free space". The solution in this region will therefore be similar to that described in the Figure 2.11.
- Unlike the case of the particle in free space (Fig.

2.11), the wavelength is restricted to just a few special values that cause the wave function to be zero at the edges of the potential well. The wave functions that satisfy this condition are known as *eigenfunctions*.

- By the de Broglie hypothesis (Eq. 2.51), since the wavelength is restricted to special values, so are the momentum and energy of the particles trapped in the potential well restricted to a few special values – known as *eigenvalues*.

The wavelengths are restricted to the set of values

$$\lambda = \frac{2L}{1}, \frac{2L}{2}, \frac{2L}{3}, \dots, \frac{2L}{n} \qquad (2.55)$$

where n is an integer. The energy of a particle trapped in this well is therefore restricted by Equation 2.54 to be

$$E(n) = \frac{p^2}{2m} = \frac{h^2}{2m[2L/n]^2} \qquad (2.56)$$

or, more simply

$$\boxed{E(n) = \frac{h^2 n^2}{8mL^2}} \qquad (2.57)$$

The two-dimensional box

Figure 2.13 shows the variation of potential energy in a two-dimensional square potential well, the walls of which are extremely high. The two-dimensional

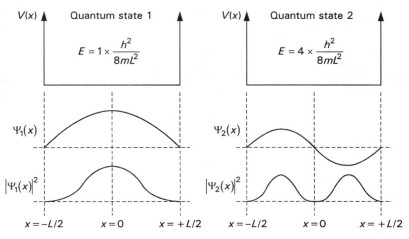

Figure 2.12 The variation of potential energy in a one-dimensional square potential well. The wave functions $\Psi(x)$ and probability densities $|\Psi(x)|^2$ of the two lowest energy quantum states ($n = 1$ and $n = 2$) are also shown.

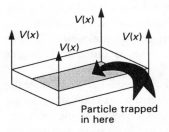

Figure 2.13 A two-dimensional square potential well (see Fig. 2.14).

where n_x and n_y have values 1, 2, 3, . . ., but not zero. In addition to the points noted in the one-dimensional example, the fact that the wave functions and energies of each quantum state are determined by *more than one quantum number* raises the possibility of *degeneracy*, i.e. two distinct quantum states possessing the same energy. For example, referring to the state with $n_x = 1$, $n_y = 2$ as the (1,2) state, Figure 2.14 shows that the (1,2) and (2,1) states have quite distinct wave functions and probability densities, but the same energy. The significance of this degeneracy is discussed more fully in the examples of three-dimensional wave functions that follow.

square well is an interesting intermediary between the one- and three-dimensional square well potentials. The primary purpose of discussing the two-dimensional example is that its wave functions can be visualized relatively easily, whereas the three-dimensional wave functions are rather more difficult to depict. The wave functions in Figure 2.13 have energies given by

$$E(n_x, n_y) = \frac{h^2}{8mL^2}\left[n_x^2 + n_y^2\right] \quad (2.58)$$

The three-dimensional box

A particle confined to a cubic "box" of side L has wave functions described by *three* quantum numbers n_x, n_y and n_z and the wave functions have energies given by an expression similar to Equation 2.58:

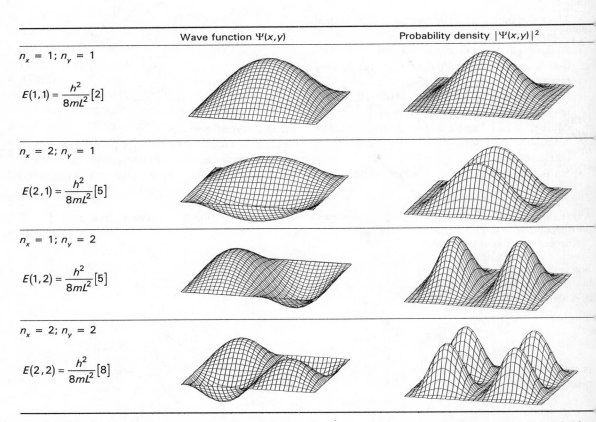

	Wave function $\Psi(x,y)$	Probability density $\lvert\Psi(x,y)\rvert^2$
$n_x = 1; n_y = 1$ $E(1,1) = \dfrac{h^2}{8mL^2}[2]$		
$n_x = 2; n_y = 1$ $E(2,1) = \dfrac{h^2}{8mL^2}[5]$		
$n_x = 1; n_y = 2$ $E(1,2) = \dfrac{h^2}{8mL^2}[5]$		
$n_x = 2; n_y = 2$ $E(2,2) = \dfrac{h^2}{8mL^2}[8]$		

Figure 2.14 The wave functions and probability densities of the four lowest energy quantum states in Figure 2.13.

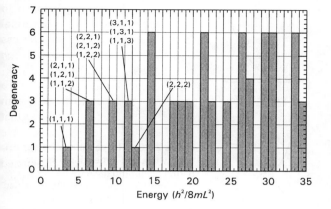

Figure 2.15 Three-dimensional particle in a box problem. The figure shows a histogram of the number of quantum states with a given energy. The energy is indicated in units of $h^2/8mL^2$ and is calculated according to Equation 2.59. For the first few energy levels, the quantum numbers of the states that comprise those energy levels are shown.

$$E\left(n_x, n_y, n_z\right) = \frac{h^2}{8mL^2}\left[n_x^2 + n_y^2 + n_z^2\right] \quad (2.59)$$

where n_x, n_y and n_z may take the integer values 1, 2, 3, . . ., but not zero. Labelling the quantum state with n_x = 3, n_y = 4 and n_z = 1 as (3,4,1), we see that some quantum states, such as the (1,2,2), (2,2,1) and (2,1,2) states all have the same energy

$$E(2,2,1) = \frac{h^2}{8mL^2}\left[2^2 + 2^2 + 1^2\right] = 9 \times \frac{h^2}{8mL^2}$$

Using the terminology mentioned above, one may refer to the $E = 9 \times (h^2/8mL^2)$ energy level as being *three-fold degenerate*. The degeneracy of other energy levels is indicated in Figure 2.15.

The simple harmonic oscillator

The potential energy of a one-dimensional simple harmonic oscillator

$$V(x) = \tfrac{1}{2} K\left(x - x_0\right)^2 \quad (2.60)$$

is shown in Figure 2.16. As in the classical treatment of such an oscillator, the frequency

$$f_0 = \frac{1}{2\pi}\sqrt{\frac{K}{m}} \quad \text{or} \quad \omega_0 = \sqrt{\frac{K}{m}} \quad (2.61)$$

is particularly significant because the energy of a quantum state with quantum number n is given by

$$E(n) = \left(n + \tfrac{1}{2}\right)hf_0 = \left(n + \tfrac{1}{2}\right)\hbar\omega_0 \quad (2.62)$$

The allowed energies of a particle trapped in a three-dimensional harmonic oscillator potential

$$V(x, y, z) = \tfrac{1}{2} K\left[\left(x - x_0\right)^2 + \left(y - y_0\right)^2 + \left(z - z_0\right)^2\right] \quad (2.63)$$

are given by

$$E\left(n_x, n_y, n_z\right) = \left(n_x + n_y + n_z + \tfrac{3}{2}\right)hf_0 \quad (2.64)$$

Note that the energy levels are equally spaced, unlike those of a particle in a three-dimensional box potential where the energy gap between quantum states becomes smaller for higher energy states.

When many particles interact: the exclusion principle

The three fundamental particles that we have considered as our basic set of particles (Table 2.1) belong to a class of fundamental particles called *fermions*. They can be identified as fermions by the fact that their intrinsic spin (in units of $\hbar$) is half an odd integer, in this case ½. Remember that, as we discussed previously (§2.2), it is not known what the property referred to as "spin" actually corresponds to. However, we know that it corresponds to some intrinsic property of a particle, and that it gives the particle an internal "axis" or special direction.

The *Pauli exclusion principle* states that:

> *At any time, no more than one fermion may occupy an individual quantum state.*

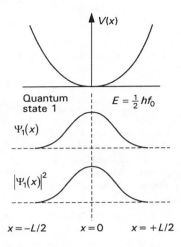
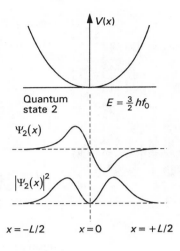

Figure 2.16 The variation of potential energy in a one-dimensional simple harmonic potential well. The wave functions $\Psi(x)$ and probability densities $|\Psi(x)|^2$ of the two lowest energy quantum states ($n = 0$ and $n = 1$) are also shown.

Note that it is possible for many fermions to have the same energy, as long as the energy level has many individual quantum states.

The application of this principle is critical to understanding matter and is discussed further in §2.5. For example, it is a consequence of this principle that only one electron is allowed to occupy the lowest energy quantum state within an atom, and other electrons are forced to occupy higher energy states. Similarly, when atoms collide in a gas, it is the exclusion principle – in combination with the Coulomb repulsion between valence electrons – that causes the atoms to "bounce" off one another.

2.5 Thermodynamics and statistical mechanics

Historically, the approach to understanding the large-scale properties of matter has developed along two rather separate lines.

- In the *thermodynamic* approach, one assumes as little as possible about the microscopic nature of matter. For example, one does not have to posit the existence of atoms. One then derives relationships between measurable large-scale properties of materials.
- In the *statistical mechanical* approach, one first makes assumptions about the properties of the basic microscopic entities, usually atoms or mol-

ecules, involved in a problem. One then tries to deduce the properties of large numbers of these basic entities. This approach takes account of the fact that atoms are very small and the large-scale properties of a substance will be the *average* properties of a very large number of atoms. Statistical mechanics makes calculations of these *average values*.

Temperature
In the statistical mechanical approach, the key property of temperature is related to the *average energy* per basic entity of the substance, usually per molecule or per atom. Thus, if the temperature of a substance is high, its atoms possess relatively large amounts of energy. The relationship between temperature and average energy implies an *absolute zero* of temperature, when the average energy per basic entity is reduced to zero. A more carefully worded definition of temperature is given at the end of this section, after we have described some specialized terminology.

Thermal equilibrium
Thermal equilibrium is the state reached by a material after it has been in contact with a "bath" – essentially a large object – at a constant temperature for a long time. If a substance is in a state of thermal equilibrium, then its temperature is uniform throughout its volume, and does not change with time.

2.5.1 First law of thermodynamics

The first law of thermodynamics is essentially a restatement of the principle of conservation of energy. It defines changes in a property called the *internal energy* of a sample of material in terms of the heat supplied to the sample and the work done on the sample. The first law takes account of the conversion of energy between mechanical forms (work), in which energy is tied up in the coherent motion of molecules, to thermal forms (heat), in which energy is tied up in the random motion of molecules. The first law

Change in internal energy	=	Heat supplied **TO** the object	+	Work done **ON** the object

$$\Delta U = \Delta Q + \Delta W \qquad (2.65)$$

First Law of Thermodynamics

relates changes between one state of thermal equilibrium and another in terms of the *work done on*, and the *heat supplied to*, an object. In its simplest form, the work done on the object includes only mechanical work, but it may be extended to include electrical or magnetic work.

2.5.2 Counting quantum states

Matter as we generally encounter it contains vast numbers of atoms or molecules. A few tens of grams of most substances contain of the order of 10^{23} atoms. The large number of atoms involved in even tiny pieces of matter makes the task of determining the macroscopic (i.e. large scale) properties of matter easier in some ways, and more difficult in others.

Big numbers make life easy
The large numbers of atoms mean that *average properties* may be very well defined. For example, in §4.4 we will see that at around room temperature, gas molecules move with speeds ranging from zero to several thousand metres per second. Despite this variability, the average speed of a gas molecule may be defined with high precision, and shows only tiny fluctuations. If just a few tens of molecules were involved in the gas, the fluctuations would be considerably larger, and the average value of their speed less significant.

Example 2.6

Two joule of heat is dissipated in an object which then expands and does 0.1 J of work. By how much is its internal energy U increased?

We use Equation 2.65 with $\Delta Q = 2\,\text{J}$ and $\Delta W = -0.1\,\text{J}$ (note the negative sign, which is used because the object does work on its environment). Substituting into Equation 2.65, $\Delta U = \Delta Q + \Delta W$ we have

$$\Delta U = 2.0 - 0.1 = 1.9\,\text{J}$$

Big numbers make life difficult
Due to the large number of molecules, in order to calculate the average properties one needs a coherent and consistent framework for counting the particles, their speeds and energies. This framework can be rather intimidating at first, but has an elegance that can be inspiring. In the next section we discuss the systematic approach to counting particles and the quantum states that they occupy

2.5.3 Statistical mechanics

In the context of this book, we consider statistical mechanics to be a scheme or plan for calculating average quantities. The plan has three stages
Stage 1 Make a list of all the quantum states of a system and hence define a *density of quantum states* function.
Stage 2 Consider whether the particles which will occupy these states are fermions or bosons, and hence choose an *occupation function* that describes the way in which the quantum states will be occupied.
Stage 3 Combine the density of quantum states function and the occupation function to yield a *distribution function* that describes the way in which the particles are distributed among quantum states.

Stage 1A: A list of quantum states
Produce a list of the possible quantum states that the individual particles of a substance may occupy. This is usually in the form of a general formula stating how the energy E of a quantum state depends on its other quantum numbers. For example, the possible energies of particles in a three-dimensional box (Eq. 2.59) are

$$E(n_x, n_y, n_z) = \frac{h^2}{8mL^2}\left[n_x^2 + n_y^2 + n_z^2\right]$$

Table 2.4 lists the energies of a few low energy quantum states, and the number of electrons that can be accommodated in these states.

Stage 1B: The density of states function

From the list given in Table 2.4 one can calculate the form of a function that answers the question "How many quantum states are there with energy between E and $E + dE$?". This function is known as the *density of states* function and in this book has the symbol $g(E)$. In §6.5 we show that the density of quantum states for a particle in a box increases in the manner

$$g(E) = A\sqrt{E} \qquad (2.67)$$

However, in order to observe to the dependence one must generally consider a few thousand quantum states. If one considers $g(E)$ on a very fine energy scale, then one will discover "graininess" and fluctuations around this behaviour (Fig. 2.17a). This graininess is illustrated qualitatively in Figure 2.17. The exact detailed behaviour is also illustrated in Figure 2.15, and the transition to the "continuum" is illustrated in Figure 2.17(b).

Stage 2: Fermions or bosons?
The occupation function

In the second stage of the plan to calculate average quantities, we consider the type of particle that will occupy the quantum states and whether they are bosons or fermions. It is a fundamental assumption of statistical mechanics that *at equilibrium, the average number of particles in an individual quantum state*

Table 2.4 The first few energy levels for particles trapped in a box.

Quantum numbers n_x, n_y, n_z	Energy$\times h^2/8mL^2$	Number of quantum states with this energy	Cumulative total of quantum states
(1,1,1)	3	1	1
(1,1,2) (1,2,1) (2,1,1)	5	3	1 + 3 = 4
(1,2,2) (2,1,2) (2,2,1)	9	3	1 + 3 + 3 = 7
(1,1,3) (1,3,1) (3,1,1)	11	3	1 + 3 + 3 + 3 = 10
(2,2,2)	12	1	1 + 3 + 3 + 3 + 1 = 11

Figure 2.17 Three-dimensional particle in a box problem. The figure shows a histogram of the number of quantum states with a given energy. The energy is given in units of $h^2/8mL^2$ and is calculated according to Equation 2.59. The behaviour at the low-energy end of the graph is detailed in Figure 2.15. The histogram shows the number of quantum states within an energy range ($10\times h^2/8mL^2$). If $L\approx1$ m, and m corresponds to, say, a nitrogen molecule ($28\times1.66\times10^{-27}$ kg) then the $10\times h^2/8mL^2$ evaluates to about 10^{-43} J. There are thus about 60 quantum states in an energy range $10h^2/8mL^2$ around an energy of $100h^2/8mL^2$. Thus the average spacing between quantum states in this energy range is roughly $(10h^2/8mL^2)/60\approx10^{-44}$ J.

(a)

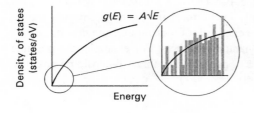

(b)

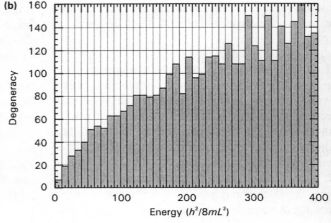

Example 2.7

For a box of volume V the density of quantum states that may be occupied by electrons is $g(E) = A\sqrt{E}$ states per eV. Sketch this function and evaluate how many quantum states there are (a) with energies between 9 eV and 9.1 eV, and (b) with energies between 0 eV and 9 eV.

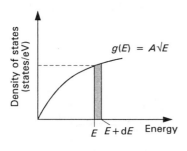

(a) Since the energy range $\Delta E = 0.1\,\text{eV}$ is small in comparison with the total energy, we may use a definition of the density of states function "$g(E)\Delta E$ is the number of quantum states ΔN with energies in the range E to $E + \Delta E$".

In this case, this is given by

$$\Delta N = g(E)\Delta E = A\left(\sqrt{E}\right)\Delta E$$

which evaluates to

$$\Delta N = A\sqrt{9} \times 0.1 = 0.3A \cdot$$

(b) Now, the energy range $\Delta E = 9$ eV is significant in comparison with the total energy, so we must integrate the density of states function using

$$N = \int_{E=0}^{E=9} g(E)\mathrm{d}E = \int_{E=0}^{E=9} AE^{\frac{1}{2}}\mathrm{d}E$$

$$N = A\left[\frac{2}{3}E^{\frac{3}{2}}\right]_{E=0}^{E=9} = A \times \frac{2}{3} \times 9^{\frac{3}{2}} = 18A$$

depends on only three factors:

- the type of particle, which may be either a *fermion* or a *boson*;
- the temperature; and
- the energy of the state.

Thus, for a given type of particle, states of equal energy are equally likely to be occupied, even if the motions associated with these two states are quite different. For example, if a vibrational state of a molecule has the same energy as a rotational state, then both states will be occupied with equal probability.

Fermions have half an odd-integer spin ($\frac{1}{2}\hbar$, $\frac{3}{2}\hbar$, ...) and only a single particle may occupy the same quantum state at the same time. Fermions compose all the particles of matter that we commonly encounter.

Bosons have integer spin and many particles may occupy the same quantum state. The bosons most commonly referred to are *photons* or *phonons*. Photons and phonons are "particles", but they have no rest mass, and are used to describe the state of excitation of the electromagnetic field, or the displacement waves in a solid respectively. Bosons can be also used to describe particles of matter, such as atoms of ^{4}He.

At high temperatures both sets of statistics reduce to a simpler set known as *Maxwell–Boltzmann* or classical statistics (see §2.5.4 and Appendix A1.2).

Based on the type of particle populating the quantum states under discussion, one may use an *occupation function $f(E,T)$*:

$f(E,T)$ yields the average occupancy of an individual quantum state.

For fermions the appropriate occupation function is called the *Fermi–Dirac function*

$$f_{\text{FD}}(E,T) = \frac{1}{\exp\left[(E - \mu)/k_{\text{B}}T\right] + 1} \qquad (2.67)$$

where μ is a characteristic energy for the system of particles. Technically μ is known as the *chemical potential*, but for electrons, to which we shall apply Equation 2.69, the chemical potential is generally referred to as the *Fermi energy E_{F}*, i.e.

$$f_{\text{FD}}(E,T) = \frac{1}{\exp\left[(E - E_{\text{F}})/k_{\text{B}}T\right] + 1} \qquad (2.68)$$

Note (Fig. 2.18) that when $E = E_{\text{F}}$, the probability that a quantum state is occupied is just ½.

For bosons the appropriate occupation function is called the *Bose–Einstein function*

$$f_{\text{BE}}(E,T) = \frac{1}{\exp\left[(E - \mu)/k_{\text{B}}T\right] - 1} \qquad (2.69)$$

(a)

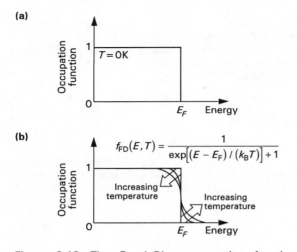

(b)

$$f_{FD}(E,T) = \frac{1}{\exp\left[(E - E_F)/(k_B T)\right] + 1}$$

Figure 2.18 The Fermi–Dirac occupation function shows the average occupancy of an individual quantum state: (a) The function appropriate to absolute zero; (b) How the function changes with increasing temperature. Note that, as the temperature changes, the maximum value of the function may never exceed unity. This is because the function is the "embodiment" of the exclusion principle that forbids multiple occupancy of quantum states. Note also that at E_F, the average occupancy always has the value ½.

Unlike atoms and molecules, *photons* and *phonons* are not conserved – photons may be absorbed or created – and for such particles the chemical potential may be set equal to zero.

$$f_{BE}(E,T) = \frac{1}{\exp\left(E/k_B T\right) - 1} \tag{2.70}$$

This function is sketched in Figure 2.19.

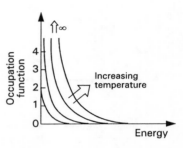

Figure 2.19 The Bose–Einstein occupation function shows the average occupancy of an individual quantum state. Note that the maximum value of the function may exceed unity, and indeed becomes infinite at zero energy.

Density of states (states/eV)

$$g(E) = A\sqrt{E}$$

Occupation function

$$f_{FD}(E,T) = \frac{1}{\exp\left[(E - E_F)/(k_B T)\right] + 1}$$

Distribution function

$$D(E) = g(E)f_{FD}(E)$$

$$= \frac{AE^{\frac{1}{2}}}{\exp\left[(E - E_F)/k_B T\right] + 1}$$

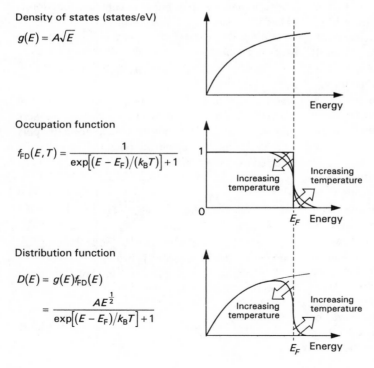

Figure 2.20 An illustration of how the density of states function, the occupation function and the distribution function describe the way in which quantum states are occupied.

Stage 3: The distribution function

In the first stage of our plan to calculate average quantities, we worked out how many quantum states have energies between E and $E + dE$. In the second stage we stated that the occupation factor $f(E,T)$ yields the average occupancy of each individual quantum state of energy E at temperature T. In the final stage of the plan we may calculate the number of particles *occupying* quantum states with energies between E and $E + dE$ by taking the product of the density of states function $g(E)$ and the occupation function $f(E,T)$, to yield the *distribution function* $D(E,T)$:

$$D(E,T)dE = g(E)f(E,T)dE \qquad (2.71)$$

Calculating average quantities

Using the expressions derived above we can define several useful expressions. In particular, the number of particles occupying quantum states with energies in some energy range is given by

$$N = \int_{\substack{\text{minimum} \\ \text{energy}}}^{\substack{\text{maximum} \\ \text{energy}}} D(E,T)dE \qquad (2.72)$$

or, using Equation 2.71 by

$$N = \int_{\text{min. energy}}^{\text{max. energy}} g(E)f(E,T)dE \qquad (2.73)$$

Similarly, the total energy of those particles may be written as

$$E_{\text{Tot}} = \int_{\text{min. energy}}^{\text{max. energy}} g(E)f(E,T)EdE \qquad (2.74)$$

So if we divide Equation 2.74 by Equation 2.72 we obtain the *average energy* of a particle in that energy range

$$\overline{E} = \frac{E_{\text{Tot}}}{N} = \frac{\displaystyle\int_{\text{min. energy}}^{\text{max. energy}} g(E)f(E,T)EdE}{\displaystyle\int_{\text{min. energy}}^{\text{max. energy}} g(E)f(E,T)dE} \qquad (2.75)$$

Some technical terms

Degrees of freedom

The strict definition of the degrees of freedom of a system is:

> *The number of degrees of freedom of a system is the number of independent squared terms that enter into the expression for the energy of the system.*

This definition gives little clue as to the usefulness of the concept. For example, the energy of an isolated atom of mass m is

$$E = \tfrac{1}{2}mv_x^2 + \tfrac{1}{2}mv_y^2 + \tfrac{1}{2}mv_z^2 \qquad (2.76)$$

which has three independent squared terms. One would thus describe such a molecule as having three degrees of freedom. Other degrees of freedom may be associated with vibrational or rotational motions.

Temperature

In thermal equilibrium, the *average* energy associated with each *accessible* degree of freedom of the system is

$$\overline{E} = \tfrac{1}{2}k_B T \qquad (2.77)$$

where k_B is the *Boltzmann constant* ($1.38 \times 10^{-23}\,\text{JK}^{-1}$) and T is a constant of proportionality familiar to us as the *absolute temperature*. The term "accessible" is discussed below.

Accessibility

The *occupation function* $f(E,T)$ determines the average occupancy of a quantum state. In a sense, there is nothing more to be said about the matter of occupying quantum states than is contained in the occupation function. However, there is an additional terminology – that of accessibility – that is commonly used to describe the occupancy of quantum states. When a quantum state has a low average occupancy, it may in certain circumstances be described as *inaccessible*. This occurs when the *separation* ΔE between quantum states is much greater than the $k_B T$. This idea is illustrated in Table 2.5.

Example 2.8

Work out the average energy of a colection of electrons at $T = 0\,K$.

According to Equation 2.73 we can define the number of electrons in our system as

$$N = \int_{0}^{E=\infty} g(E) f_{FD}(E, T) dE$$

At absolute zero, $f_{FD}(E, 0)$ has the value 1 for all energies below E_F, and 0 for all energies above. We may therefore write the integral for N as

$$N = \int_{0}^{E_F} g(E) dE$$

Substituting for g(E) from Example 2.7

$$N = \int_{0}^{E_F} A E^{\frac{1}{2}} dE$$

and integrating

$$N = \left[\frac{2}{3} A E^{\frac{3}{2}} \right]_{0}^{E_F}$$

we arrive at

$$\boxed{N = \frac{2}{3} A E_F^{\frac{3}{2}}}$$

We may follow a similar procedure to arrive at an expression for the total energy. We have

$$E_{Tot} = \int_{0}^{E_F} A E^{\frac{1}{2}} E dE$$

which amounts to

$$E_{Tot} = \int_{0}^{E_F} A E^{\frac{3}{2}} dE$$

and integrates to

$$E_{Tot} = \left[\frac{2}{5} A E^{\frac{5}{2}} \right]_{0}^{E_F}$$

which evaluates to

$$E_{Tot} = \frac{2}{5} A E_F^{\frac{5}{2}}$$

Using Equation 2.75, we can evaluate the average energy of an electron as

$$\bar{E} = \frac{E_{Tot}}{N} = \frac{\frac{2}{5} A E_F^{\frac{5}{2}}}{\frac{2}{3} A E_F^{\frac{3}{2}}} = \frac{3}{5} E_F$$

2.5.4 Classical statistics: Boltzmann factor

If there is a great excess of quantum states over particles, i.e. if the average occupancy of quantum states is low, less than (say) a tenth or so, then both bosons and fermions occupy quantum states in a similar way. One can see how this happens if one considers the situation shown in Figure 2.21. If there is just one particle occupying this system of quantum states, then clearly there are no restrictions placed by its fermion/boson nature on the quantum states that it may occupy. If there are two particles in the same system of quantum states and if they are fermions, there will be a restriction such that a single quantum state may not be doubly occupied. But if the chance of double occupation is low, then the actual distribution of the particles among the available quantum states will not be much affected.

Under the conditions of low occupancy described above and in Figures 2.21 and 2.22, both Fermi–Dirac and Bose–Einstein occupation factors reduce to a rather simpler form known as a *Boltzmann factor*

$$f(E, T) = A \exp(-E / k_B T) \qquad (2.78)$$

where A is a constant determined from the requirement of Equation 2.73 that the total number of particles is given by

$$N = \int_{min.\ energy}^{max.\ energy} g(E) f(E, T) dE \qquad (2.79)$$

The deduction of Equation 2.78 from Equation 2.67 or 2.69 is discussed in Appendix A1.2.

So, under conditions such that the average occupancy of quantum state is small, both quantum occupation functions reduce to a simpler Boltzmann factor. In this regime – the *classical regime* – the probability of occupying a state of particular energy varies exponentially with temperature. The Boltzmann factor manifests itself in many physical properties of substances such as:

- the distribution of the energies of molecules of a gas (Appendix A);
- the excitation of electron carriers in an insulator or semiconductor (§7.5.7 & §7.5.9); and
- the evaporation of molecules from the surface of a liquid (§11.4.2).

Table 2.5 An example of the accessibility of states.

Inaccessible	Marginal accessibility	Fully accessible
$k_{B}T \ll \Delta E$	$k_{B}T \approx \Delta E$	$k_{B}T \gg \Delta E$
e.g. $k_{B}T < 0.1\Delta E$	e.g. $0.1\Delta E < k_{B}T < 1.5\Delta E$	e.g. $\frac{1}{2}k_{B}T > 1.5\Delta E$
In this case only occasionally do molecules make transitions to the higher quantum state and we can consider the degrees of freedom associated with these transitions to be inaccessible	In this case molecules make transitions to the higher quantum state, but detailed calculations are required to assess the extent to which the quantum state can be considered accessible	In this case molecules frequently make transitions to the higher quantum state and we can consider the degrees of freedom associated with these transitions to be fully accessible
In colloquial terms, the process associated with transitions between quantum states occurs so infrequently that it may generally be ignored		In colloquial terms, the process associated with transitions between quantum states occurs so frequently that the quantum nature of the states may generally be ignored

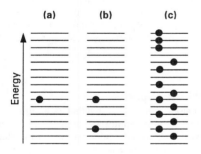

Figure 2.21 Classical and Fermi–Dirac statistics for fermions. In cases (a) and (b) the fermion nature of a particle does not become apparent and classical statistics may be used. In case (c) one must use Fermi–Dirac statistics. (a) There is only a single particle and so the exclusion principle places no restrictions on which state it may occupy. (b) There are two particles, but in most configurations of the particles they do not attempt to occupy the same state. Thus the exclusion principle only slightly affects their distribution amongst the accessible quantum states. (c) The density of particles is so high that the distribution of the particles among the available quantum states is dominated by the fermion nature of the particles.

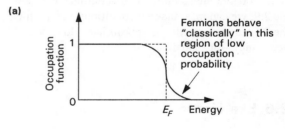

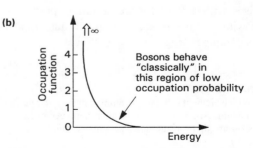

Figure 2.22 Qualitative illustration of the energy ranges in which the Fermi–Dirac (a) and Bose–Einstein (b) occupation factors may be approximated by a Boltzmann factor.

Example 2.9

In the classical regime, what is the relative probability that two particular quantum states with energies E_1 and E_2 will be occupied at temperature T?

The average occupancy of a quantum state with energy E_1 is equal to

$$f(E_1, T) = A \exp(-E_1/k_B T)$$

where A is a constant for the system under consideration. Similarly, the average occupancy of a quantum state with energy E_2 is equal to

$$f(E_2, T) = A \exp(-E_2/k_B T)$$

We can write the ratio of the probability of occupation of states with E_1 and E_2 as

$$\text{Ratio} = \frac{A \exp(-E_2/k_B T)}{A \exp(-E_1/k_B T)}$$
$$= \exp\left[-(E_2 - E_1)/k_B T\right] = \exp(-\Delta E/k_B T)$$

where $\Delta E = E_2 - E_1$.

In fact, wherever we encounter a system the physical properties of which display an exponential dependence on temperature similar to that outlined in Example 2.9, we can infer that the conditions under which the Boltzmann equation is appropriate – low average occupancy – are satisfied.

2.6 Exercises

Exercises with a P prefix are "normal" problems. Those with a C prefix are best solved numerically using a computer program or spreadsheet.

The elements

P1. Which element is the most common in the outer part of the Sun? Approximately how many parts per million of gold are there in the outer part of the Sun? (see Fig. 2.1.)

P2. Why are the noble gases Ne, Kr and Xe comparatively relatively rarer on earth than in the Sun? (see Fig. 2.1.)

P3. Is gold really rare on Earth? What are the relative abundances of copper and gold? (see Fig. 2.1.)

P4. A histogram of the number of elements discovered in a given decade shows that discoveries of elements were clumped in certain decades over the last 300 years. Plot the histogram and suggest why this occurred (see Table 2.2.).

P5. A mnemonic for remembering the first row of the transition elements is, "**K**atie **Can**'t **S**(**c**)kate **Ti**ll **V**ery **C**old **M**orning **F**(**e**)ollows **C**old **N**ight **Cu**pper **Zn**inc". Compose a mnemonic for the second row of the transition metals (see Fig. 2.2.) beginning with rubidium.

Electromagnetism

P6. Example 2.1 shows that the force between an electron and a proton separated by a distance of $0.1\,\text{nm}$ is $2.31 \times 10^{-8}\,\text{N}$. Work out the attractive force between an electron and a proton if they are separated by (a) $0.105\,\text{nm}$ and (b) $0.095\,\text{nm}$. Assuming that the force varies roughly linearly with separation over this small range, show that the "spring constant" K, which describes the variation of electric force with distance, is approximately $K = 460\,\text{Nm}^{-1}$. Estimate the resonant frequency of the electron on the atom based on Equation 2.31, and the wavelength of light with this frequency. Compare this with the result of Example 7.26 and Figure 7.52 based on the analysis of the optical properties of glasses.

P7. The speed of light through a transparent substance is $c/2$. Estimate its refractive index and dielectric constant? (see Eq. 2.24.)

P8. A photon of X-radiation with a wavelength $3 \times 10^{-11}\,\text{m}$ is completely absorbed by an atom within a complex biological molecule. How many photons of light with wavelength $3 \times 10^{-7}\,\text{m}$ would have to be absorbed to impart a similar total energy to the molecule? (see Eq. 2.25.)

Classical mechanics

P9. Two atoms are bonded together in a molecule. They vibrate in a stretching vibration with a frequency $1.13 \times 10^{13}\,\text{Hz}$. If their atomic mass is 14u, what is the spring constant that characterizes the stiffness of their atomic bond? (See Example 2.5.)
(Note. In this case the mass you should use in Equation 2.31 is the *reduced mass*, μ, given by $1/\mu = (1/m + 1/m)$ i.e. $\mu = m/2$. See a classical mechanics text on relative motion or Example 5.14.)

P10. Two particles have the same mass and resonant frequency ω_0, but are subjected to different damping mechanisms. The particles are subjected to the same sinusoidally oscillating force at frequency ω_0, but one particle has an amplitude of motion twice that of the other. Use Equation 2.36 to evaluate the relative magnitudes of the damping constant, α, in Equation 2.33.

P11. Based on the answer to Exercise P10, outline how an apparatus could be constructed and operated to determine the relative viscosities of two liquids.

C12. Program a computer or use a spreadsheet to plot Equations 2.36 and 2.37 and so reproduce Figures 2.7a&b. Change the constant α in these equations to see the effect of damping on the dynamics of the particle.

Quantum mechanics

P13. According to Equation 2.59, what is the minimum kinetic energy of (i) an electron and (ii) a proton confined inside a cubic region of (a) side $L = 0.1\,nm$ and (b) side $L = 1\,nm$?

P14. According to Equation 2.64, what is the minimum energy of (a) an electron and (b) a proton confined by a simple harmonic potential of spring constant $430\,Nm^{-1}$. What is the "resonant frequency", f_o, in each case?

Statistical mechanics

P15. Following Table 2.4 for a particle trapped in a three-dimensional cubic box, how many quantum states have energy $14 \times h^2/8mL^2$? You may check your answer against Figure 2.15.

C16. Following Figure 2.17, write a computer program to calculate the degeneracy of all quantum states of a three-dimensional particle in a box with energies less than $40 \times h^2/8mL^2$.

P17. For a box of volume V, the density of quantum states that may be occupied by electrons is $g(E) = A\sqrt{E}$ states/eV. Derive an expression for the number of quantum states (a) with energies between 4 and 4.01 eV and (b) with energies between 0 and 4.5 eV. (See Example 2.7.)

C18. Use a calculator or a spreadsheet program to make an accurate sketch of the Fermi–Dirac function for μ (i.e. E_F) = 3 eV and temperatures of $T = 10$, 100 and 1000 K.

C19. Use a calculator or a spreadsheet program to make an accurate sketch of the Bose–Einstein function for $\mu = 0$ and temperatures of $T = 10$, 100 and 1000 K.

CHAPTER THREE

Measurement

3.1 Introduction – the importance of measurements

No matter what the context, the basic idea of measurement is to enable *quantitative comparison* of one property or thing with another. The ability to measure quantities systematically may be considered as one of the defining characteristics of a science – without measurement, there is only speculation.

This book contains a chapter on measurement because, in the teaching of physics, the insights achieved as a result of studying measurements are often taught as secondary to insights derived from theoretical approaches to a subject. Indeed, I have attended entire lecture courses where no reference was made to a single measurement in the outside world. However, in practice, it is exceedingly rare that physicists proceed in such a fashion. The beliefs that physicists hold about the world are, in a sense, forced upon them by measurements. The results of measurements act as guides, and constraints for developing explanatory ideas about the world. Thus a familiarity with measurements and the way in which they are made is as important an aspect of physics as is a familiarity with, say, mathematics. Although a student might be forgiven for thinking otherwise, it is measurement, not mathematics, that makes physics science rather than philosophy.

Given the importance of measurement, the techniques used to carry out this process are of some interest, and this chapter provides an overview of some of the more common techniques of measurement.

3.2 Units

Recalling the fundamental idea of a measurement as a comparison, *units* act as *standard comparison quantities*. This notion of a unit is apparent in the names given to historical units of length such as the *foot*. These units referred to the length of a standard object against which all else in the "kingdom" would be compared. Unfortunately, neighbouring kingdoms frequently had different definitions for quantities, which made quantitative comparison – the essence of measurement – exceedingly difficult. For example, in 1686, Newton was attempting to show *quantitatively* that his theory of gravity could explain both phenomena on the earth as well as the motion of the moon and other astronomical bodies. In searching for observations in support of his theory (Book III of *Principia mathematica*) he describes the diameter of the earth and the motion of the moon in terms of *Paris* feet, as distinct from *English* feet used elsewhere in his work. Clearly the units of measurement are getting in the way of comparison rather than making it easier.

Historically, there have been two trends that have taken the science of measurement away from the use of units such as the foot. These trends are:

- towards the use of a system of units that are agreed internationally; and
- towards defining standard quantities in terms of *phenomena* that can be *realized* by anyone. This is a move away from the use of defining objects or *artefacts*.

3.2.1 National standards and international agreements

Each country employs a system of *legal metrology*, the main aim of which is to facilitate trade rather than science, but which serves for both purposes. The UK is currently well on the way to adopting the SI system of measurement units detailed below. The aim of having a coherent system of measurement units is a major scientific and management challenge.

The scientific part of the challenge is tackled by national standards laboratories: in the UK, the National Physical Laboratory (NPL); and in the USA, the National Institute of Science and Technology (NIST). These laboratories manufacture and maintain apparatus that allows the realization of accurate representations of the units (see Tables 3.1 to 3.3). For example, they might have an apparatus that produces a known voltage against which a voltmeter may be calibrated, or an apparatus that produces air of a known humidity with which a hygrometer may be calibrated. These laboratories also work with other national standards laboratories to ensure that their realizations of measurement units are internationally consistent.

The management part of the challenge is tackled by a quality assurance organization. In the UK the organization is called UKAS (the UK Accreditation Service) and the scheme it administers is called NAMAS (the National Accreditation of Measurement and Sampling). This organization ensures that laboratories bearing its seal of approval carry out valid calibrations and do not claim measurement uncertainties lower than they can actually achieve. Part of this validation is to ensure that relevant measuring instruments used in these laboratories are regularly re-calibrated at the NPL. This system maintains a hierarchy of measurement accuracy, and ensures that the lower orders of the hierarchy maintain as much as is practicable of the accuracy achieved at the apex of the hierarchy, the NPL. At each tier of the hierarchy, extra measurement uncertainties are inevitably introduced. This makes it particularly important that the realizations of the base quantities at the National Standards Laboratories have the smallest measurement uncertainties possible.

3.2.2 Artefacts and realizations

There are two ways in which the world can agree on a measurement standard. Countries can agree on either:

- a standard *object* of which there is only one, and against which all others are compared. This object is referred to in measurement circles as an *artefact*; or
- a standard *physical phenomenon* or situation which it is practical for all countries to create. This situation is referred to in measurement circles as a *realization* of a measurement unit.

The reason for the trend away from defining units by artefacts to definitions in terms of realizations may be seen by considering the one measurement unit that is still defined by an artefact: the kilogram.

The kilogram

There is only one kilogram, a unique piece of platinum–iridium alloy kept at the International Bureau of Weights and Measures (BIPM), Paris. There are 41 copies of the kilogram in existence; the UK holds copy number 18. After it was last cleaned, the UK copy of the kilogram weighed $59 \pm 3\,\mu g$ more than the prototype kilogram, and then gained mass at a rate of $1\,\mu g$ per month for a year and now gains mass at a rate of approximately $1\,\mu g$ per year. However, the UK kilogram is so valuable that it spends its days inside three bell jars. Every time it is removed from its protective surrounding its mass changes a little as it reacts slowly with the air and adsorbs a little moisture. Periodically the UK's copy of the kilogram is returned to Paris to make sure that its mass has not changed too much. But what if the international prototype kilogram itself changed its mass? In fact *the kilogram definitely has changed mass*, and a small coterie of scientists keep track of its changes by intercomparing the copies of it.

Thus defining a standard quantity in terms of an object makes life very difficult for everyone concerned, and allows for the possibility of long-term drift in the magnitude of measurement units. Definitions of standard quantities in terms of *artefacts* of any kind have died out leaving "standard metres" as museum pieces.

Other standards are defined in terms of physical phenomena that may, in principle at least, be *realized* by anyone. The phenomena chosen are believed to be well understood, and so the definitions are unlikely to require revision, and hence the system of units will

Table 3.1 The SI base units. Note that, with the exception of the kilogram, the definitions are in terms of physical phenomena and not defining artefacts. Although the definitions seem obtuse, the language is carefully chosen in order to make accurate realizations of the standards feasible.

Quantity and unit (abbreviation)	Definition
Time second (s)	The second is the duration of 9,192,631,770 periods of the radiation corresponding to the transition between two hyperfine levels of the ground state of the caesium-133 atom
Length metre (m)	The metre is the length of the path travelled by light in a vacuum during a time interval 1/299,792,458 of a second (Note: this statement implicitly defines 299,792,458 $m s^{-1}$ as the exact speed of light in a vacuum)
Mass kilogram (kg)	The kilogram is the unit of mass; it is equal to the mass of the international protype of the kilogram
Electric current ampere (A)	The ampere is that constant current which, if maintained in two straight parallel conductors of infinite length, of negligible circular cross-section, and placed 1 m apart in vacuum, would produce between these conductors a force equal to $2 \times 10^{-7} N m^{-1}$ of length
Thermodynamic temperature: kelvin (K)	The kelvin, unit of thermodynamic temperature, is the fraction 1/273.16 of the thermodynamic temperature of the triple point of pure water
Amount of substance mole (mol)	The mole is the amount of substance of a system which contains as many elementary entities as there are atoms in 0.012 kg of carbon-12
Luminous intensity: candela (cd)	The candela is the luminous intensity, in a given direction, of a source that emits monochromatic radiation 540×10^{12} Hz and that has a radiant intensity of (1/683) W per steradian

Reproduced with permission from *SI: the international system of units*, sixth edition (London: HMSO, 1993).

not change with time.

For example, the phenomenon of the coexistence in equilibrium of water vapour, liquid water and ice defines the *triple point* of water (see §10.7.2 for more details). The temperature at which the triple point of pure water occurs is *defined* to be 273.16 K. All one needs to do in order to realize this temperature is to obtain some pure water, manufacture a simple vessel, and stick it into melting ice. The temperature inside is defined to be 273.16 K. This is a physical phenomenon that may be realized with relative ease and which will not change from one year to the next.

3.2.3 The International System of Units

All rational physicists use the system of units known as the *International System of Units*, or *Le Systeme International d'Unités*, or more normally *SI units*. At least they use SI units when discussing subjects other than their own. For discussing their own field of expertise, they frequently use sets of *colloquial units* which, for historical reasons, and reasons of genuine convenience, have not died out. In this book I have chosen to always use the SI units, adding colloquial units where appropriate.

The SI units consist of seven base or fundamental

quantities in terms of which all other quantities are derived. The SI base units and definitions are given in Table 3.1.

There are two other units, called supplementary units, that are used to distinguish between quantities which are of a different nature, but which would otherwise have the same SI units. For example, *angular velocity* is specified in units of radians per second even though the radian is dimensionless. This distinguishes the units of *angular velocity* from those of *frequency*.

Other units are derived from these base units, and have names given in terms of the base units involved. For example, *mass density* is expressed in a unit called the *kilogram per cubic metre* and has the symbol $kg m^{-3}$. Other units derived from these base units are specially named in honour of various scientists. These are given in Table 3.3 below.

Table 3.2 SI supplementary units.

Quantity	Name	Symbol	Expression in terms SI base units
Plane angle	raidan	rad	$m m^{-1} = 1$
Solid angle	steradian	sr	$m^2 m^{-2} = 1$

Reproduced with permission from *SI: the international system of units*, sixth edition (London: HMSO, 1993).

Table 3.3 SI derived units with special names. Note that the names of the units are written with lower case letters (with the exception of degree Celsius), but that the symbols for the units have upper case letters: be careful to distinguish between seimens (S) and seconds (s). The symbol for the ohm, Ω, is the Greek letter "W" and is called omega.

Quantity	Name	Symbol	Expression in terms of other units	Expression in terms of SI base units
Frequency	hertz	Hz		s^{-1}
Force	newton	N		$m\,kg\,s^{-2}$
Pressure Stress	pascal	Pa	$N\,m^{-2}$	$m^{-1}\,kg\,s^{-2}$
Energy Work Quantity of heat	joule	J	$N\,m$	$m^2\,kg\,s^{-2}$
Power Radiant flux	watt	W	$J\,s^{-1}$	$m^2\,kg\,s^{-3}$
Electric charge Quantity of electricity	coulomb	C		$s\,A$
Electrical potential Potential difference Electromotive force	volt	V	$W\,A^{-1}$	$m^2\,kg\,s^{-3}\,A^{-1}$
Capacitance	farad	F	$C\,V^{-1}$	$m^2\,kg^{-1}\,s4\,A^{-1}$
Electric resistance	ohm	Ω	$V\,A^{-1}$	$m^2\,kg\,s^{-3}\,A2$
Electric conductance	siemens	S	$A\,V^{-1}$	$m^2\,kg^{-1}\,s^3\,A^{-1}$
Magnetic flux	weber	Wb	$V\,s$	$m^2\,kg\,s^{-2}\,A^{-1}$
Magnetic flux density	tesla	T	$Wb\,m^{-2}$	$kg\,s^{-2}\,A^{-1}$
Inductance	henry	H	$Wb\,A^{-1}$	$m^2\,kg\,s^{-2}\,A^{-2}$
Celsius temperature	degree Celsius	°C		K
Luminous flux	lumen	lm		$cd\,sr$
Illuminance	lux	lx	$lm\,m^{-2}$	$m^{-2}\,cd\,sr$

Reproduced with permission from *SI: the international system of units*, sixth edition (London: HMSO, 1993).

3.2.4 An example of the realization of units: the mole

As an example of the way that the SI defines the magnitude of a unit, let us consider what SI has to say about the mole: the unit of the *amount of substance*. First, recall the definition from Table 3.1:

The mole is the amount of substance of a system which contains as many elementary entities as there are atoms in 0.012 kilogram of carbon-12.

A supplementary note to the agreement of this definition mentions that it is important to specify which elementary entities are being referred to. They may be ions, electrons, atoms, molecules, other particles or groups of particles.

Note that this definition leaves it up to the scientific community to invent practical realizations of the definition. Let's look at a couple of ways in which, starting at first principles, one might determine the amount of substance contained in a sample of material.

Realization 1

If the substance is a gas then we can use the fact that, as we show in Table 5.3, the volume of 1 mole of any gaseous substance held at standard temperature and pressure (STP), is close to $22.413 \times 10^{-3}\,m^3$. In Section 5.2.2 we see that we can understand *why* this is so based on an analysis of a microscopic theory of the behaviour of a gas. This analysis produces the perfect gas equation $PV = zRT$ (Eq. 4.1) while more sophisticated analyses (§4.5.2) arrive at slightly different equations that enable us to understand the small deviations of the volume at STP from the ideal value. We thus have a piece of physics that contains no "fudge factors" and so we can use this to realize definitions of any of the quantities in the perfect gas equation in terms of other quantities.

Thus by measuring the pressure of the gas at 0°C in a container of known volume we can infer the number of moles z of gaseous substance in the container. (For completeness we note at this point that $PV = zRT$ is more commonly used to realize the definition of temperature T in terms of the pressure P of z

moles of gaseous material in a container of fixed volume V.)

Realization 2

If the material is a solid or liquid and its chemical composition is known, then the number of moles can be determined by weighing. If we consider a material A of chemical composition X_pY_q then we note that the molar mass of A is given by

$$M(A) = \frac{m(A)}{m\left(^{12}C\right)} \times 0.012 \, \text{kg mol}^{-1} \qquad (3.1)$$

where $m(A)$ indicates the mass of a single entity of the substance A, in this case a collection of atoms specified by p atoms of type X and q atoms of type Y, and $m(^{12}C)$ indicates the mass of a single atom of ^{12}C. Note that, although neither of these masses is known with great accuracy, the *ratio* expressed in Equation 3.1 can be measured with high accuracy by using, for example, a *mass spectrograph*. Similarly, the ratios

$$\frac{m(X)}{m\left(^{12}C\right)} \quad \text{and} \quad \frac{m(Y)}{m\left(^{12}C\right)} \qquad (3.2)$$

can also be determined accurately. We can thus state that the mass of 1 mole of A is

$$M(A) = \left[p \times \frac{m(X)}{m\left(^{12}C\right)} + q \times \frac{m(Y)}{m\left(^{12}C\right)} \right] \times 0.012 \, \text{kg mol}^{-1}$$

$$(3.3)$$

For example, consider calcium fluoride (CaF_2), for which X = Ca, Y = F, $p = 1$ and $q = 2$. The ratios of Equation 3.3 are noted by Kaye and Laby (see §1.4.1) to be

$$\frac{m(Ca)}{m\left(^{12}C\right)} = \frac{40.08}{12} = 3.34$$

$$\text{and} \quad \frac{m(F)}{m\left(^{12}C\right)} = \frac{18.9984}{12} = 1.5832 \qquad (3.4)$$

and so the mass of a mole specified according to Equation 3.3 is

$$M(CaF_2) = \left[1 \times 3.34 + 2 \times 1.5832\right] \times 0.012 \, \text{kg mol}^{-1}$$

$$M(CaF_2) = 0.07808 \, \text{kg mol}^{-1} \qquad (3.5)$$

If we have 4.3209 kg of CaF_2 then we can determine the amount of CaF_2 by dividing the sample mass by the molar mass, i.e.

$$\text{Amount of } CaF_2 = \frac{4.3209}{0.07808} = 55.342 \, \text{mol} \qquad (3.6)$$

Note that the definition of a mole does not specify which realization should be used: this is for scientists and engineers to choose. As improvements in our understanding develop, or new techniques become available, the definition need not change, but the realizations of the definition may change, and hopefully become more accurate or more easy to use.

3.3 Key measurement techniques

Certain techniques of measurement occur commonly in a wide variety of situations. The commonness, and hence the importance, of these techniques make them worth mentioning separately.

3.3.1 Time

Time, or its inverse, *frequency*, can be measured with very great accuracy. The UK National Physical Laboratory can realize a second with an accuracy of 1 part in 10^{13} (see §3.6: Exercise P5). In part, this is the result of hard work, but it is also due to the fact that time lends itself to being measured accurately. For example, through what is known as the *piezo-electric effect*, a quartz (SiO_2) crystal mechanically oscillates at a frequency that is dependent on its size, shape, temperature and the applied electric field. If all these quantities are stabilized, it is possible for such a crystal (costing maybe 10 pence) to oscillate with a frequency of about 10^8 Hz stable to a few parts in 10^9 over a few hours. These oscillations can be detected and converted into a stream of electrical pulses. Because we have the technology to count pulses extremely reliably, we can determine how many pulses occur in a given interval of time with an uncertainty of, say, two pulses at most – one at the beginning and one at the end of the timing interval. So, by counting these pulses one can determine elapsed times of the order of one second with a resolution of 2 parts in 10^8. All that remains is to determine the actual frequency of the crystal by comparison with an instrument calibrated at the National Physical Laboratory.

Voltage-to-frequency conversion

The ease with which high resolution is achieved for time or frequency measurement has caused people to seek ways to convert the quantity they wish to measure into a time measurement. The most important example of this is the development in the last 30 years of techniques for measuring voltage. A simplified example of the conversion of a voltage measurement to a time measurement is the *integrating analogue-to-digital converter* (ADC). An integrating ADC (Fig. 3.1) consists of three parts: an *integrator*, a *comparator* and a source of clock pulses such as that described above. There are three stages to the measurement.

1. A circuit generates a current that is accurately proportional to an input voltage. This current is arranged to charge a capacitor for a fixed period of time, i.e. for a fixed number of clock pulses.

2. With the input disconnected, the capacitor is discharged by a constant current circuit until the voltage reaches zero, as determined by a comparator circuit. The time taken for this discharge is measured by counting the number of clock pulses that occur during the discharge.

3. The count of clock pulses required for the capacitor to discharge has been designed to be accurately proportional to the magnitude of the input voltage – the larger the input voltage, the longer the discharge time and hence the more pulses that are counted. Digital circuitry can then convert this count to the correct units. Thus a measurement of time performed by counting clock pulses has been converted into a measurement of voltage.

If the clock pulses occur at a frequency of 10 MHz, and the counting is accurate to ±1 pulse, then the discharge period may be timed with a resolution of approximately 0.1 μs. If the discharge period is, say, approximately 0.1 s, this corresponds to a measurement resolution of the order of 1 part in 10^6 and would allow the detection of changes of only a few microvolts in a voltage of the order of a volt.

Many variations of this technique exist under the general heading of *charge balancing* or *voltage-to-frequency conversion* techniques.

3.3.2 Voltage measurement and transducer technology

The accuracy and ease with which voltage can be measured has caused a trend towards the creation of devices, known as *transducers* or *sensors*, that convert changes in physical quantities into changes in a voltage. Once a voltage has been generated that is related to the physical quantity to be measured, electrical circuits can act on the voltage to actuate a display, or an alarm. Alternatively, once a voltage has been created which corresponds to the quantity to be measured, ADCs (Fig. 3.1) can convert the voltage, and hence the quantity being measured, into a *digital code*. The data can then be conveniently analyzed and stored using ever more powerful computers.

Temperature sensors

In general, temperature sensors operate by passing an

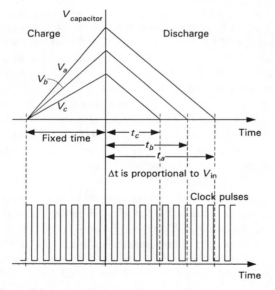

Figure 3.1 An integrating analogue to digital converter.

electric current through a piece of material, and measuring changes in the flow of electric current as the temperature changes.

Platinum resistance thermometer

A platinum resistance thermometer (PRT) consists of a length of thin platinum wire. The electrical resistivity of platinum has been investigated extensively over many years and is now well documented (see Figs 3.2 and 7.34). Ideally, the wire should be strain-free and held delicately, but more commonly the requirement that the PRT be robust and easy to use is valued more highly than the absolute accuracy of the results. Practical PRTs are commonly formed by using a *thin film* of platinum on a thermally conducting, but electrically insulating, substrate. Alternatively they are

wound on a frame and then held in place by a thermally conducting, but electrically insulating, cement.

Platinum is chosen in preference to other metals with similar electrical properties because of its exceptional resistance to chemical corrosion. This resistance means that the diameter of wires made of platinum does not decrease as the wire slowly corrodes. The insensitivity to corrosion allows PRTs to be used up to temperatures as high as 1000°C.

The length and diameter of the platinum wire are commonly chosen to give the PRT a resistance of $100\,\Omega$ at 0 °C.

Thermistor

Thermistors are pieces of semiconducting materials operated in a similar fashion to PRTs, but with a dramatically different dependence on temperature (Fig. 3.2). Thermistors are particularly useful because their sensitivity to temperature changes is much higher than that of a PRT, but only over a limited range of temperatures.

Semiconducting diodes

Semiconducting diodes are used as temperature sensors in almost exactly the same way as resistance thermometers. A constant current, typically $10\,\mu A$, is passed through the diode, and the voltage across the diode is measured (Fig. 3.3). Diodes are considered separately from thermistors and PRTs because at constant temperature the voltage–current characteristic is highly non-linear, or non-ohmic, i.e. V is not pro-

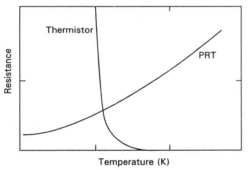

Figure 3.2 Schematic variation of the electrical resistance of a platinum resistance thermometer (PRT) and a thermistor. Figure 7.34 shows a more accurate representation.

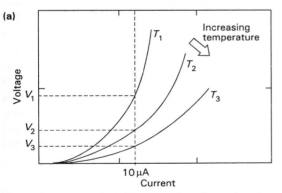

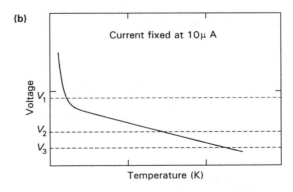

Figure 3.3 Illustration of the operation of a semiconductor diode thermometer. (a) The variation of the voltage across the diode is a non-linear, i.e. non-ohmic, function of the current through the diode. However, if the current through the diode is fixed, the variation of the voltage with absolute temperature (b) is surprisingly linear over a wide temperature range.

portional to I. However, at constant current, the voltage across the diode is highly linear over a wide temperature range.

Thermocouples

A thermocouple is a junction of two dissimilar metals. Electrons are bound within a metal by an energy of typically $4 \pm 1\,\text{eV}$ per electron; however, the precise binding energy differs from metal to metal. Thus at a junction between dissimilar metals, a potential difference of the order of a fraction of a volt arises as electrons redistribute themselves between the metals. Furthermore, this *contact potential* varies with temperature by a few microvolts per degree Celsuis. This potential is normally measured with respect to a second thermocouple held at a reference temperature.

The requirement for this second junction is now routinely eliminated by devices which determine the reference temperature using a diode or resistance sensor, and then simulate the expected thermoelectric voltage. Although thermocouples are still common and relatively cheap, especially for use at moderate temperatures they have been increasingly superseded by the use of resistance sensors. Aside from low cost, their key advantage is that they can be manufactured to have a low mass, and hence to respond rapidly (within a few tens of microseconds) to temperature changes.

Light intensity transducers

The importance of fibre optics for communication systems has led to many developments in the field of transducers that convert variations in light intensity into electrical signals.

Photoresistors

The earliest of these detectors consisted of a resistor made from the element selenium. When illuminated the electrical resistance of selenium falls dramatically, and so if a constant current is passed through a selenium resistor the voltage across the resistor will fall at high light levels and rise at low light levels. The microscopic processes that cause the change in resistance are mentioned in Chapter 7 along with the optical behaviour of the better known semiconducting elements silicon and germanium.

Photodiodes

Currently, photodiodes are the most commonly used elements in light-detecting electrical circuits. The operation of the devices is complex, but may be summarized as follows. A photodiode is manufactured out of a semiconductor in which there is a junction between regions of different impurities. At the interface between the different regions there is a thin region of intense electric field known as the *depletion layer*. In a photodiode this region is arranged to occupy as large a volume as possible.

When light interacts with valence electrons in a semiconductor, the electrons may become detached from a particular atom and thus become available to travel through the semiconductor. Normally, the electron is drawn back to the atom it has come from by the residual positive electric charge which it leaves behind it. However if the valence electron that absorbs the energy is located in a strong electric field, such as that which exists within the depletion layer, then it can be drawn away from its parent atom. This electron then contributes to an electric current that has been induced by the interaction of light with the semiconductor.

If photodiodes could operate perfectly they would generate one electron for each *photon* that was absorbed. By measuring the current generated between the two terminals of the diode one can determine the number of photons arriving at the *depletion layer* between the two impurity regions of the semiconductor.

Photomultipliers

The most sensitive of all light-detecting devices are photomultipliers. These are evacuated glass tubes with a series of metal plates (*cathodes*) held at different voltages. Light illuminates the first plate, called the *photocathode*. By means of the photoelectric effect, a photon incident upon the photocathode can cause the emission of an electron. The electric field around the photocathode is such as to accelerate the emitted electron towards the second cathode. As the electron strikes the second cathode it causes the emission of typically 6 or 7 *secondary electrons*, which are then drawn towards the third cathode, where each of these electrons causes a further 6 or 7 electrons to be emitted, making between $6^2 = 36$ and $7^2 = 49$ electrons, each of which is drawn towards a fourth cathode, and so on. After perhaps ten cathodes, the number of electrons has grown to between $6^{10} \approx 6.0 \times 10^7$ and $7^{10} \approx 2.8 \times 10^8$.

Thus each photon produces a current pulse of, typically, 10^8 electrons. Large though this amplification is, it is still tricky to measure. In terms of voltages it represents a voltage pulse of perhaps 1 mV lasting for a fraction of a microsecond on top of the total voltage across the photomultiplier tube, which is around 1000 V.

3.3.3 Optical interferometry

Distances or motion can be measured by using the techniques of optical interferometry with a resolution of less than a wavelength of light ($\approx 10^{-7}$ m) and almost independent of the distance in question. The use of these techniques is made enormously more straightforward by the existence of cheap and robust sources of coherent light, i.e. lasers.

The basic principle of interferometry is illustrated in Figure 3.4. Light from a source S travels along two (or more) paths, say A and B, to a detector D. Recall that light is the name we give to oscillations of the electric field occurring with frequencies between 400×10^{12} Hz (red) and 750×10^{12} Hz (blue). So, if light from a source S follows a route to point D via path A, then the electric field at D oscillates rapidly with amplitude E_A volts per metre. If light from S also travels to D by a second route, say via B then the electric field at D is the sum of two oscillating electric fields, one with amplitude E_A and the other with amplitude E_B.

If the light from S is derived from several random sources of light within S – such as the atoms within a light bulb – then the electric field at D due to the waves arriving via A and B will sometimes add up, and sometimes subtract, producing on average a value of electric field amplitude greater than either E_A or E_B independently, but not equal to the sum $E_A + E_B$. Importantly, the average resultant amplitude of the electric field oscillation (given by $\sqrt{(E_A^2 + E_A^2)}$ does not depend on the precise lengths of paths A and B, or on the *details* of the conditions experienced along those paths.

However, if the light from S is derived from a *coherent* source of light, i.e. one in which all the atoms in S emit light in-phase with each other, then the waves arriving at D will add up on every oscillation, subtract on every oscillation, or something in between, on every oscillation. Exactly which situation occurs depends on the details of the conditions experienced along paths A and B (for example, the presence or absence of a gas) and on the precise length of the paths. For a given set of conditions along the paths, the amplitude of the electric field oscillations is between 0 and ($E_A + E_B$). Let us say, for example, that under a given set of conditions along paths A and B the electric field oscillations at D have zero amplitude: then either:

- a shift in the position of a mirror by approximately one-half a wavelength of light ($\approx 10^{-7}$m), or
- a delay of the light wave by $\approx 10^{-15}$ s caused by a change in some property of the medium through which the light travels,

is sufficient to change the amplitude of the electric field oscillations at D from zero amplitude to maximum amplitude ($E_A + E_B$).

There are many examples of the use of laser interferometry, but space permits the inclusion of only two. The first example is the detection of changes in

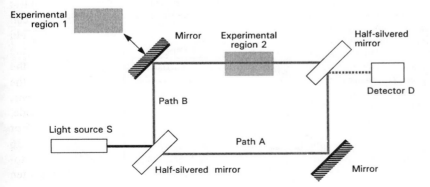

Figure 3.4 The basic principle of interferometry (see text for further details).

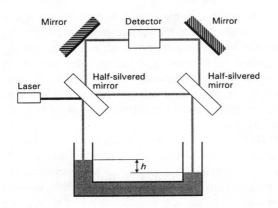

Figure 3.5 An optical interferometer used for the determination of a change in height of a column of mercury.

the position of height of a liquid surface, and the second is the direct measurement of the optical properties of a gas.

- In measurements of *pressure* it is common to use a U-tube filled with liquid as a simple manometer (Fig. 3.5). This translates pressure differences into changes in the heights of liquid in the arms of the manometer. Using an optical interferometer the change in the liquid height can be detected with a resolution of $\approx 10^{-7}$m.

- The speed of light travelling through a gas is very slightly different from the speed of light in a vacuum. As discussed more fully in Section 5.7.1, the speed difference is related to the polarizability of the molecules of the gas. The device shown in Figure 3.6 is designed to meas-

ure small changes in the speed of light as the pressure of the gas in an experimental cell is changed (§3.6: Exercise P8).

3.3.4 Bridges/balances

Many substances show only relatively small changes in properties in response to changes in their environment. Consider the examples given below.

Example A The volume of solids and liquids changes by typically 1 part in 10^5 when the temperature of the sample changes by 1°C. The thermal expansion of most solids is of a similar order of magnitude, and so in order to compare the thermal expansion of different materials we need to detect the length change of 1 part in 10^5 with a resolution of $\approx 1\%$ at least, i.e. we need an apparatus that will detect a length change of 1 part in 10^7. In an experimental sample that might be approximately 1 cm×1 cm×1 cm, this corresponds to detecting a length change of $\approx 10^{-9}$m, or about three atomic diameters.

Example B The magnetic moment that develops in a material when a magnetic field is applied is often tiny. It is not uncommon for the "excess magnetic field" due to the sample to be one-millionth of the magnetic field that caused it.

Example C The determination of heat capacity of materials requires the temperature to be changed by a small amount ΔT at an absolute temperature of T. Ideally ΔT will be a small fraction ($\approx 1\% \, T$). But ΔT must be measured with a resolution of about 1 part in 10^3 in order to determine the heat

Figure 3.6 An optical interferometer used for the determination of the electrical polarizability of a gas. The apparatus is first set to a reference state by evacuating both the reference cell and the experimental cell. The reference cell is then kept under vacuum while the gas under investigation is introduced into the experimental cell. The light travelling through the experimental cell is slowed down slightly by the interaction of its oscillating electric field with the electric charge on each atom. This results in a change in the intensity of light at the detector because the interference condition now depends on the transit time through the experimental cell.

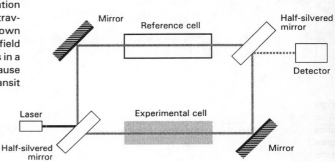

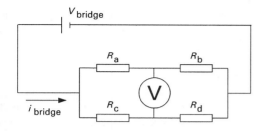

Figure 3.7 The *Wheatstone bridge* circuit. A voltage source drives a current i_bridge through the resistance network illustrated. The detector V need not have, but most commonly will have, an impedance much greater than any of the other resistances in the circuit. The bridge is balanced when $R_a/R_b = R_c/R_d$.

capacity with an accuracy of 0.1%. Thus it seems that the temperature must be measured to 1 part in 10^5.

Situations such as these, and a variety of others, call for the application of *comparison* or *bridge techniques*. What these techniques have in common is that they are sensitive not to the static background quantity (length, magnetic field and temperature in the above examples), but to small changes in it. The simplest example of this is the case of the *Wheatsone bridge* electrical circuit shown in Figure 3.7.

The circuit, consisting of four resistors, a voltmeter and a voltage source, has two key features of its response:

(a) The circuit is said to be *balanced* when $R_a/R_b = R_c/R_d$. At this point there will be no potential difference across the detector **V**. Thus the voltmeter is not sensitive to the magnitude of any of the resistances in the circuit, but it is sensitive to whether the resistances are *balanced*.

(b) Suppose that the circuit is initially balanced, and then R_d changes slightly from its balance condition by an amount ΔR (Fig. 3.8). The voltmeter may be used on its most sensitive range to detect the out of balance voltage which is given by

$$\Delta V = i_\text{bridge}\left[\frac{\Delta R(R_a + R_b)}{R_a + R_b + R_c + R_d}\right] \qquad (3.7)$$

where we have assumed $\Delta R \ll R_a + R_b + R_c + R_d$.

Suppose the voltmeter can resolve a change of $1\,\mu\text{V}$ in $100\,\text{mV}$, a common specification. If R_a, R_b, R_c and R_d were all $1000\,\Omega$, $\Delta R_d = 1\,\Omega$ and $i_\text{bridge} = 10\,\text{mA}$ then ΔV would be $5\,\text{mV}$. The voltmeter would then be able to detect a change of $1\,\Omega$ in a total resistance of $1000\,\Omega$ with a resolution of 1 part in 5000. The bridge circuit would thus be able to just detect resistance changes of the order of one-thousandth of an ohm, i.e. 1 part in 10^6 of the total resistance.

Note that the voltage at the detector in a bridge circuit ($5\,\text{mV}$ in the above example) is less than the voltage that would occur if we passed the entire current, i_bridge, through the resistor R_d.

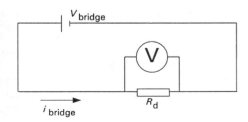

Figure 3.8 The alternative to a Wheatstone bridge circuit. A voltage source drives a current i_bridge through the resistance R_d. The detector V must have an impedance much greater than R_d. If R_d changes by ΔR, then the voltage across R_d changes by $\Delta V = i_\text{bridge}\Delta R$.

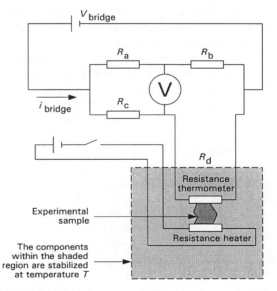

Figure 3.9 The Wheatstone bridge may be used to detect small changes in a resistance thermometer due to a temperature change ΔT.

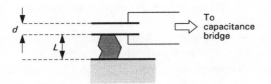

Figure 3.10 A schematic view of a capacitance cell. Small changes in the length of a sample cause changes in the separation between the two plates of a capacitor.

If we did this we would get a change in voltage $\Delta V = i_{\text{bridge}}\Delta R = 10\,\text{mV}$ in the above example. So why bother with the complication of a bridge circuit? When using the bridge, the signal is $5\,\text{mV}$ increased from nothing, i.e. $0\,\text{mV} \to 5\,\text{mV}$, whereas in the direct measurement configuration the $10\,\text{mV}$ is on top of a $1\,\text{V}$ background, i.e. $1000\,\text{mV} \to 1010\,\text{V}$. Although the bridge has halved the voltage due to the resistance change, the $5\,\text{mV}$ signal is easier to measure with high resolution because the background signal has been subtracted (see Exercise 3P4).

This is exactly the technique used for example C above. The element R_d of the circuit is arranged to be a resistance thermometer (Fig. 3.9). The temperature step ΔT is applied by a resistive heater and results in small change ΔR in the value of the resistance thermometer. The magnitude of the change ΔR is inferred from measurement of ΔV, and then converted to an equivalent temperature change ΔT.

Similarly for example A above, changes in the sample length are converted into changes in the separation of two metal plates forming a capacitor (Fig. 3.9). A circuit analogous to the Wheatstone bridge but operating with alternating current called a *capacitance bridge* can detect the changes in capacitance with high resolution.

Finally example, B above (Fig. 3.11) uses inductances that are arranged such that variations in the strong applied magnetic field produce no voltage across the two inductance coils A and B. This is achieved by winding the coils in the opposite sense to each other in order to eliminate the effects of any fluctuations in the applied field.

The examples A, B and C all represent ways of detecting *small changes* in quantities in a way that is insensitive to the magnitude of the quantities themselves.

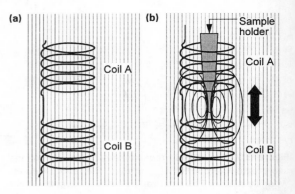

Figure 3.11 The principle of operation of a vibrating sample magnetometer (VSM). Two oppositely wound, but otherwise identical, coils are placed in a large static magnetic field indicated by the lined background in (a). Any small fluctuations in the field induce equal and opposite voltages in the two coils, and so no net signal is induced in the absence of a sample. When a sample is placed in the large magnetic field between the coils (b), it acquires a small magnetic moment. If the sample is now vibrated up and down it induces a net voltage in the coil system. In this way the tiny magnetic field due to the sample is separated from the applied field, which might be one million times larger.

3.4 Environments

As well as being able to measure different physical quantities, we also need to be able to create and maintain environments in which we can perform experiments on samples of matter.

3.4.1 Temperature

High temperatures
Temperatures above room temperature are usually created by ovens or furnaces that are heated electrically. It is not difficult or particularly expensive to create an environment of a volume of about $10\,\text{cm}^3$ or so with temperatures up to $900°\text{C}$. For temperatures above this, special materials must be used to resist the high rates of oxidation that would otherwise reduce the material of the heating filament to a powder. At the highest temperatures, platinum or silicon carbide must be used for the heating elements.

Low temperatures

Low temperature environments are usually created by placing experimental samples in contact with a cold fluid. For temperatures only slightly below room temperature one can use liquid refrigerants that are similar to those found in a domestic refrigerator. For lower temperatures it is common to use liquid nitrogen, which boils at atmospheric pressure at a temperature of around 77 K (–197°C). For still lower temperatures the refrigerant used is liquid helium which boils at atmospheric pressure at a temperature of around 4.2 K (–269°C). By lowering the vapour pressure above the liquid, one can reduce the boiling temperature to ≈1.2 K, which suffices for most purposes.

3.4.2 Pressure

The pressure under which an experiment is performed may be altered from below atmospheric pressure ($\approx 10^5$ Pa) to values as large as 10^{10} Pa (10 GPa). The devices used to change the pressure are all based on, or adapted from, the *piston* (Fig. 3.12).

For gases and liquids it is a straightforward extension of Figure 3.12 to see how pistons can be used to apply pressure. The experimental sample is placed in a suitable container, and a force applied to the piston. The force is often amplified by means of a lever or a hydraulic arrangement. At high pressures it is often a matter of considerable ingenuity to extract measurements from within the high pressure environment, but that need not concern us here.

For solids, it is important to ensure that pressure is applied uniformly in all directions, otherwise sam-

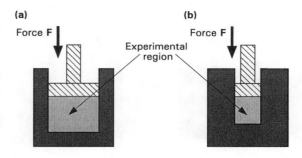

Figure 3.12 Example of the use of piston to create pressure environments: (a) a force F acts over area A to produce a pressure $P_1 = F/A$; (b) the force acts over a smaller area, $A/4$, to produce a pressure, $4P_1$, even though the force used is the same as in (a).

ples are easily damaged. To ensure that pressure is *hydrostatic*, a pressure-transmitting fluid that does not itself solidify under pressure is used (Fig. 3.13).

Achieving low pressures is no less a matter for ingenuity. Figure 3.14 shows schematically how the pressure may be lowered within a chamber by a simple mechanical *vacuum pump*, generally called a *rotary* or *rough pump*. Devices operating on this principle can fairly easily reduce the pressure in a chamber to one-ten-thousandth of atmospheric pressure (i.e. around 10 Pa), i.e. they can remove 99.99% of all the molecules in the chamber. However, there would still be typically 10^{21} molecules left per cubic metre (see Eq. 4.50). A number of devices are available – using different principles to effect the compression of the gas described in Figure 3.14b – which can achieve pressures of 10^{-4} Pa and, with great care

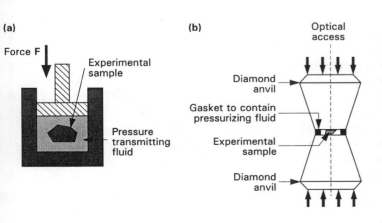

Figure 3.13 Illustration of techniques for applying high pressures to solid samples. (a) A fluid is placed around the sample to ensure that the pressure is transmitted hydrostatically. (b) For the application of the very highest pressures, diamond (the hardest material known) is used as a pressure anvil. The transparency of diamond allows relatively easy optical access. However, I draw the reader's attention to the fact that the expense of large unflawed diamonds limits the size of the apparatus illustrated in (b), and all such experiments represent a considerable challenge to the experimenter.

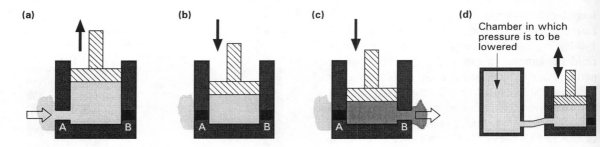

Figure 3.14 Illustration of the principle of operation of a mechanical vacuum pump. (a) A piston/chamber is enlarged and "sucks in" gas through A from the region in which the pressure is to be lowered. (b) The connection A to the area in which the pressure is to be lowered is closed, and the piston lowered to compress the gas until the pressure exceeds atmospheric pressure (c) when a valve B opens to permit the compressed gas to be expelled. This process (d) can be used to lower the pressure in an experimental enclosure. (Note: real vacuum pumps operating on this principle rarely look anything like this diagram might intimate!)

and expense, one may achieve *ultra high vacuum* pressures of around 10^{-10} Pa.

3.4.3 Magnetic field

When an electrically charged particle moves, a magnetic field is created around it. The most common way of arranging this is to use the magnetic field created around electrons when they move through a wire (see Exercise 3P12). The wires are usually wound into the form of a solenoid, as shown in Figure 3.15a. However, the magnetic field per ampere per turn is not large, and so this arrangement requires the use of both large currents and large numbers of turns of wire in the solenoid. This in turn leads to the problem of removing the heat generated by the electric current as it flows through the wire. There are two common solutions:

(a) For magnetic fields up to 20 T one uses superconducting wire. This is special wire which loses all its electrical resistance when cooled below a certain temperature. For the magnet to operate it must be cooled, usually by immersion in a bath of liquid helium at 4.2 K. Note that the bath of helium does not remove any heat from the coil, but merely maintains it in a state in which no heat is generated. The centre of the solenoid can be thermally isolated from the magnet and so samples can be examined in temperatures well above room temperature.

(b) For higher magnetic fields the heating problem is combated by pulsing the current through the

coil to a very high value (≈ 1000 A), but very briefly (≈ 0.1 s). This obviously makes experiments rather difficult, but there is no other way to obtain fields up to 60 T.

The expense of both these solutions means that they are found only in research laboratories. More common than either of these solutions is the *iron-cored electromagnet* (Fig. 3.15b) in which the relatively small magnetic field generated by the solenoid (≈ 0.1 T) is used to *magnetize* a ferromagnetic material, usually iron. The magnetic properties of iron are

Figure 3.15 Illustration of the methods of generating magnetic fields. (a) For the highest magnetic fields a simple solenoid is used, after taking special precautions to avoid over heating of the wire. (b) For magnetic fields up to about 2 T an iron-cored electromagnet may be used.

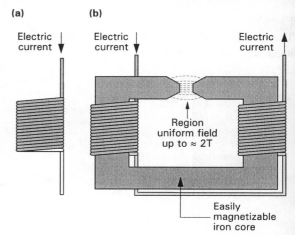

such that even a small external magnetic field is sufficient to cause magnetic fields *within* the iron of 1–2 T. By arranging to have a small gap in the piece of iron, a region is created where most of the magnetic field within the iron is made accessible.

3.5 Uncertainty

In most of the tables in the following chapters, no mention of uncertainty is made. Of course all measurements are accompanied by uncertainty, and strictly speaking the uncertainty in all the measurements in this book should be stated. The reason that statements of measurement uncertainty are lacking is that most of the sources from which I have obtained the data do not quote uncertainties. As a guide, all the figures I give in measured quantities are usually significant figures. For example, the density of the element hafnium is given in Table 7.2 as 13 276 kgm^{-3}. A realistic uncertainty for this is likely to be ±2 or ±3 in the last figure given.

Section 1.4.1 gives the sources of the data used in the tables and figures in this book and the reader is referred to these sources, and beyond them to the original measurements if they wish to establish the uncertainties of the measurements with greater certainty!

3.6 Exercises

Exercises with a P prefix are "normal" problems. Those with a C prefix are best solved numerically using a computer program or spreadsheet.

P1. What are the SI units (see Tables 3.1 & 3.3) for the specification of:
(a) Magnetic flux density?
(b) Temperature?
(c) Electrical conductance?
(d) Electrical resistance?
(e) Electrical capacitance?
(f) Amount of substance?
(g) Mass?

P2. What is the mass of (a) 1 mol of carbon atoms, (b) 1 mol of nitrogen atoms and (c) 1 mol of nitrogen molecules (N_2)?

P3. My colleagues in the geology department at Birkbeck reliably inform me that the USA and Europe are moving apart from one another at a rate of around 1 cm per year. How many metres per second does this correspond to? How many extra rows of atoms per second appear between the USA and Europe? Suggest how, given a sufficient research budget, you would attempt to confirm this result? How long would your measurement take?

P4. Verify Equation 3.7 for a Wheatstone bridge near to the balance condition. What assumption has been made about the internal resistance of the voltmeter? Compare the magnitude of the voltage change, ΔV, with the voltage change, $i_d \Delta R_d$, across R_d and state what advantages a bridge circuit offers.

P5. If the frequency standard of the National Physical Laboratory (see §3.3.1) were used as the basis of a clock, how many years would it take before the time as determined by the clock was uncertain by 1 s? If the clock lasted for 100 years before it had to be replaced, how uncertain would the time be when the clock was replaced?

P6. Suggest a practical measurement application for: (a) a slow-reacting photoresistor costing 10 pence, and (b) a fast-reacting photodiode (within a microsecond) costing £1.

P7. On a bright moonlit night it is quite possible to see one's way home from a friend's house. How many optical photons per second are striking the Earth in order to make this possible? I suggest you assume that 1 kW m^{-2} of entirely optical energy (average frequency 7×10^{14} Hz) strikes both the Earth and the Moon from the Sun. Assume that the light striking the Moon is reflected with 100% efficiency. The Moon–Earth separation is approximately 384×10^6 m and Sun–Earth or Sun–Moon separation is 150×10^9 m. The Moon's diameter is approximately 3.5×10^6 m.

P8. An apparatus similar to the one shown in Figure 3.6 is constructed. Show that the number of wavelengths of light in the reference cell of length L may be expressed as Ln_{light}/λ_0, where n_{light} is the refractive index of the gas and λ_0 is the wavelength of the light in free space. Using Equations 4.49, 5.97, 5.119 and Table 5.18, show that the refractive index of a gas is given approximately by $1 + P\alpha/2\varepsilon_0 k_B T$, where P and T are the temperature and pressure of the gas. Hence, describe how α, the molecular polarizability of an individual gas molecule, may be determined.

P9. A capacitance bridge (Fig. 3.10) can detect a change of capacitance of 1 part in 10^8. If used in conjunction with a parallel-plate capacitor ($C = \varepsilon_0 A/d$) of area $A = 1$ cm^2 and separation $d = 0.1$ mm, estimate the smallest length change that can be detected. Roughly what fraction of the diameter of an atom does that correspond to? Could this sensitivity really be achieved in practice?

P10. A mass of 10 kg is dropped through 1 m onto the blunt end of a tapered piece of steel. The blunt end of the steel has a diameter of 30 mm and the sharp end of the steel has a diameter of 0.2 mm. Estimate the maximum pressure under the tip of the taper. Could such a device be constructed on similar principles for achieving high pressures in a laboratory environment?

P11. How would a chamber of volume 100 litres be evacuated to a pressure of 10^{-3} Pa? (This requires further research outside this book.)

P12. The field at the centre of a long solenoid is $B = \mu_0 Ni$, where N is the number of turns per unit length of the coil, and i is the current through the coil. For a current of 100 A, how many turns per metre must be wound onto the solenoid? If such a magnet were constructed from wire of diameter 1 mm, estimate how many layers of wire would be needed, and hence the diameter of the magnet.

The requirement for field homogeneity around the field centre requires a coil of length 0.5 metre, and copper wire immersed in liquid nitrogen (resistivity $\rho \approx 10^{-9}\,\Omega\,\text{m}$) is used to make the coil. Estimate roughly the power dissipated in such a magnet at full field and hence show why such magnets are not commonplace.

Estimate the force per unit length ($F = iLB$) on the wires on the inner turns of the magnet by assuming that they experience the full field of the magnet. Hence produce an order of magnitude estimate of the total explosive force experienced by the coils of such a magnet.

CHAPTER 4

Gases: background theory

4.1 Introduction

In this chapter we develop one of the earliest and most successful theories of matter. This theory envisages a gas as being a collection of molecules whose average *kinetic energy* is so large that the *potential energy* of interaction between the molecules is unable to hold them together. This model will probably be familiar, but just in case Figure 4.1 shows a picture of how one imagines the motion of the molecules in a gas. The molecules are free to move around and have relatively large spaces betweeen them. A computer program which realistically simulates the dynamics of gas molecules in two dimensions is listed in Appendix 4. As we shall see in Chapter 5, quantitative explanations of the properties of gases based on this model are extraordinarily successful.

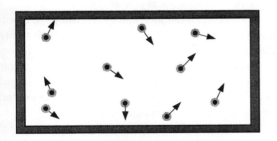

Figure 4.1 A schematic illustration of the motion of molecules in a gas. Note the large separation between the molecules and the random orientation of the velocities (arrows) of the molecules. The molecules themselves are illustrated schematically as a central darkly shaded region, where the electron charge density is high, and a peripheral lightly shaded region. Although there is no electronic charge in this peripheral region, the electric field there significantly affects the motion of any other molecules which enter that region.

4.2 The ideal gas model

The simplest model of a real gas is known as the *ideal* or *perfect* model of a gas. It consists of a list of real properties of molecules that are systematically ignored! The justification for this is that the model is relatively simple and yet has a wide range of applicability. It predicts an astoundingly simple expression for the macroscopic properties of *any gas* independent of the type of molecules that make up the gas. This universal relationship is summarized in the so called *ideal gas equation*.

$$PV = zRT \tag{4.1}$$

where

P is the pressure (Pa) of the gas,
V is the volume (m^3) of the gas,
z is the number of moles of the gas under consideration (mol),
T is the absolute temperature (K), and
R is a constant, the molar gas constant
  ($= 8.314\,JK^{-1}mol^{-1}$)

What we assume is that:
- The molecules behave as perfect point masses, i.e. they have zero volume
- The molecules do not interact with each other except instantaneously as they collide.
- The collisions between molecules are elastic.

What we neglect is that:
- The molecules have a small, but finite, volume.
- The molecules of gases do interact with each other.
- This is not appropriate is when discussing the conduction of electricity through a gas.

These will be our initial assumptions. As we proceed through the chapter we will see that, if we wish to understand the properties of real gases, we need to modify the asumptions of our simple theory.

4.3 Derivation of the ideal gas equation

In order to derive this equation we need to make two connections between the *microscopic* properties of the molecules and the *macroscopic* properties of the gas.
- First we identify the *pressure* (force per unit area) of a gas against the walls of its container as being the *average* effect of the very large number of collisions of the molecules of the gas with the wall.
- Second, we identify the *temperature* of the gas as being proportional to the *average* energy of molecules in the gas. (We define this relationship more precisely in the following sections.)

In order to do this, we will first establish a relationship between the *average momentum* and the *average energy* of a gas molecule. Then, because of the identifications above, we will be able to establish a relationship between the *pressure* of the gas and its *temperature*.

Step 1: An average molecule, S

We consider z moles of a substance, i.e. zN_A molecules of just one type, each with mass m. We assume that in the gaseous state molecular collisions are relatively infrequent, and that most of a molecule's time is spent "cruising" between collisions.

Envisage a particular molecule, S, representative of all the molecules in the box. S has mass m and velocity $\mathbf{v}$, and thus has momentum $\mathbf{p} = m\mathbf{v}$. The kinetic energy of S is given by $KE_S = \frac{1}{2}mv^2$, and if we express this in terms of the individual components of velocity v_x, v_y, and v_z, we find

$$KE_S = \tfrac{1}{2}m\mathbf{v}\cdot\mathbf{v}$$

$$= \tfrac{1}{2}m(v_x, v_y, v_z)\cdot(v_x, v_y, v_z) \qquad (4.2)$$

$$KE_S = \tfrac{1}{2}m(v_x^2 + v_y^2 + v_z^2) \qquad (4.3)$$

Step 2: hitting the wall of the box

Now, we imagine a limiting case of a low density gas with just one molecule, S, bouncing around inside a box of sides L_x, L_y and L_z and volume V (Fig. 4.2). The speed of S is chosen such that the kinetic energy KE_S is the same as the average kinetic energy of the molecules which will eventually populate the box.

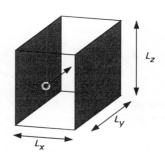

Figure 4.2 The molecule S is alone in a box and travels with a kinetic energy that is the same as the *average* kinetic energy of all the molecules that will eventually inhabit the box.

The x component of the velocity of S causes the molecule to bounce backwards and forwards between the two walls that have been shaded. The y and z components cause S to bounce between the other walls. When S hits a wall, it imparts some momentum to the wall. When S leaves the wall and re-enters the interior of the box, the wall imparts momentum to S, and so the wall is subject to a reaction force. Although these forces are tiny, we will eventually populate the box with sufficiently large numbers of molecules that the sum of these forces is appreciable. We will add up the forces that S produces as it bounces around.

We first of all imagine that the collisions of S with the walls are simple elastic collisions. If they are, then we are considering an event similar to the one depicted in Figure 4.3. More realistic assumptions concerning the interactions between the gas molecules and the molecules of the walls of the box are considered under Complication 1 below.

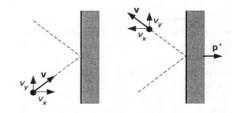

Figure 4.3 An elastic collision with the wall imparts momentum to the wall.

If momentum is conserved, then we can analyze the situation by equating momentum before and after the collision. We must consider the momentum both of the wall and the molecule.

Momentum before collision

$$mv + p_{wall} \qquad (4.4)$$

Momentum after collision

$$mv' + p'_{wall} \qquad (4.5)$$

Equating the momentum before and after the collision we have

$$mv + p_{wall} = mv' + p'_{wall} \qquad (4.6)$$

Noting that the initial momentum of the wall is zero ($p_{wall} = 0$), we have

$$p'_{wall} = mv - mv' \qquad (4.7)$$

This means that the momentum imparted to the wall is merely the difference between the initial and final momentum of S. Imagining this to be an ideal collision, we assume that v_y is unchanged by the collision and so there is no change in the y component of the momentum of S, and no force on the wall in the y direction. However, the x component of the velocity of S is exactly reversed. The momentum acquired by the wall is thus

$$p'_{wall} = mv_x\hat{x} - mv'_x\,\hat{x} \qquad (4.8)$$

$$p'_{wall} = mv_x\hat{x} - \left(-mv_x\right)\hat{x} = 2mv_x\hat{x} = \Delta p_x\hat{x} \qquad (4.9)$$

where $\hat{x}$ is a unit vector in the x direction. So the momentum imparted to the wall is *twice* the initial momentum of S. Not surprisingly, this momentum transfer will be maximized if S is travelling quickly and has a large mass.

Step 3: Bouncing around in the box

After a collision, S will head off into the box, bounce off the other wall, and eventually return to the same wall for another collision. Note that if S collides with the other walls of the box then, in the analysis above, the x component of its velocity will be unaffected, and thus we may confine our attention to the one wall we considered in Step 1. How long does it take for S to return to the same wall? If it travels with velocity v_x then S will take L_x/v_x seconds to reach the opposite wall, and a further L_x/v_x seconds to return. S thus makes collisions such as that described in Step 2 every $\Delta t = 2L_x/v_x$ seconds. Thus the *average rate* at which S imparts momentum to the wall is approximated by

$$\frac{\Delta p_x}{\Delta t} = \frac{2mv_x}{2L_x/v_x} = \frac{mv_x^2}{L_x} \qquad (4.10)$$

The rate at which momentum is transferred to the wall is nothing more than the force on the wall (Newton's second law of motion):

$$F = \frac{\Delta p_x}{\Delta t} = \frac{mv_x^2}{L_x} \qquad (4.11)$$

Thus the force per unit area on the wall, i.e. the *pressure* due to S alone, is

$$P = \frac{F}{\text{Area}} = \left(\frac{mv_x^2}{L_x}\right) \times \frac{1}{L_yL_z} = \frac{mv_x^2}{L_xL_yL_z} = \frac{mv_x^2}{V} \qquad (4.12)$$

Comparing this with the formula for the kinetic energy of S (Eq. 4.3), this is just twice the kinetic energy associated with the x component of the molecules's motion, divided by the volume.

Step 4: Pressure and kinetic energy

Now, S is not just any old molecule: its speed is chosen such that its kinetic energy is the same as the *average* kinetic energy of the real molecules with which we will fill the box in Step 5.

Neglecting the small gravitational energy term, there is no difference between the three directions in the box, and so we expect that the average speed of S will be the same in each direction, i.e.

$$\tfrac{1}{2}mv_x^2 = \tfrac{1}{2}mv_y^2 = \tfrac{1}{2}mv_z^2 \qquad (4.13)$$

Note that because S does not interact with the other molecules, it only has kinetic energy; it has no poten-

55

tial energy. So, in terms of the kinetic energy $KE_S = \frac{1}{2}m(v_x^2 + v_y^2 + v_z^2)$ we expect that

$$\frac{1}{2}mv_x^2 = \frac{1}{3}KE_S \qquad (4.14)$$

In other words, the kinetic energy associated with the motion of S in the x direction will be just one-third of its total kinetic energy. Similar expressions will hold for the energy terms associated with motion in the y and z directions. Remembering the role of S as an average molecule, we expect that the *average kinetic energy* associated with a molecule moving in the x direction will be just one-third of the *average* of its *total kinetic energy*

Now, we return to the expression (Eq. 4.12) for the pressure that S exerts on the wall perpendicular to the x direction

$$P = \frac{mv_x^2}{V} \qquad (4.15)$$

We can rewrite this as

$$P = \frac{mv_x^2}{V} = 2\frac{\frac{1}{2}mv_x^2}{V} = 2\frac{\frac{1}{3}KE_S}{V} \qquad (4.16)$$

i.e.

$$P = \frac{2}{3}\frac{KE_S}{V} \qquad (4.17)$$

Or, in words: the pressure due to a single molecule travelling with a velocity representative of the average kinetic energy of a large number of molecules, is just two-thirds of the kinetic energy per unit volume, i.e. two-thirds of the "kinetic energy density".

Step 5: Filling the box
Suppose now that the box contains not just one representative molecule, but z moles of ideal gas, i.e. zN_A molecules. The pressure on the wall will simply be zN_A times larger:

$$P = zN_A\frac{2}{3}\frac{KE_S}{V} \qquad (4.18)$$

In Complication 2 below, we consider to what extent it is really true that the pressure on the wall will sim-

ply be zN_A times larger. The total energy of the molecules in the box, called the *internal energy* of the gas, will be just

$$U = zN_A \times KE_S \qquad (4.19)$$

Step 6: Temperature
We now come to a microscopic definition of that familiar macroscopic property, temperature. Temperature is defined as being a quantity proportional to the *average energy per degree of freedom* (see §2.5.1 & Complication 3 for more details). Now S has three degrees of freedom, and so we define temperature as

$$T \propto \text{Average energy per degree of freedom} \qquad (4.20)$$

$$T \propto \frac{1}{3}KE_S \qquad (4.21)$$

This proportionality is usually written the other way around, and has a multiplying factor of 2 added (for historical reasons that we need not go into), i.e.

$$\frac{1}{3}KE_S = \frac{1}{2} \times \text{constant} \times T \qquad (4.22)$$

The constant of proportionality is called the *Boltzmann constant*, $k_B = 1.38\times10^{-23}\,\text{JK}^{-1}$. We thus have a definition of temperature as

$$\frac{1}{3}KE_S = \frac{1}{2}k_BT \qquad (4.23)$$

In words: the temperature of the gas is defined in terms of the average kinetic energy per degree of freedom of a gas molecule. If scientific understanding of matter had developed along different lines, it is quite possible that we might not today have a separate unit for temperature, but would measure it directly in joules!

Step 7: The ideal gas equation
Inserting the definition of temperature (Eq. 4.23) into the expression for pressure (Eq. 4.18), we can eliminate reference to KE_S. We begin with

$$P = zN_A\frac{2}{3}\frac{KE_S}{V} \qquad (4.24)$$

and substitute

$$\tfrac{1}{3}\text{KE}_S = \tfrac{1}{2}k_BT \qquad (4.25)$$

i.e.
$$\text{KE}_S = \tfrac{3}{2}k_BT \qquad (4.26)$$

and hence arrive at

$$P = zN_A \frac{2}{3}\frac{1}{V}\left(\frac{3}{2}k_BT\right) \qquad (4.27)$$

After cancelling terms and rearranging the equation, we arrive at

$$PV = z[N_Ak_B]T \qquad (4.28)$$

The product of the Boltzmann constant and Avogadro's number is known as the *molar gas constant, R,* which has the value $(6.023\times10^{23})\times$ (1.38×10^{-23}) = $8.314\,\text{JK}^{-1}\text{mol}^{-1}$. We have thus arrived at

$$\boxed{PV = zRT} \qquad (4.29)$$

This is a predicted relationship between the pressure, *P*, temperature, *T*, and volume, *V*, of *z* moles of ideal gas. It is an example of an *equation of state* for a substance; i.e. it links the properties that define the state of the substance. We will see that many real gases agree closely with the predictions for an ideal gas.

If you skip to Chapter 5 in which the theory is compared with experimental data you will see that the model can account for many properties of gases with a typical accuracy of around 1%. However, in order to understand some of the experimental results discussed in Chapter 5 it will be neccessary to read through Section 4.4 on calculating microscopic quantities.

Alternatively, you can read through the sections on complications and reservations below to find out about the limits of the simplifications and assumptions we have made in our seven-step derivation of $PV = zRT$.

4.3.1 Complications and reservations

Complication 1: Hitting the wall – what really happens?

The collision of a molecule with a wall is really quite a complex a process. However, we will see that in

---

Example 4.1

Let's make a rough check of *PV=zRT* on the nearest gas we have to hand, i.e. air. Air is a mixture of ≈78.1% N_2, 20.1% O_2, 0.9% Ar, 0.03% CO_2 and a variable fraction of, typically, 0.5% H_2O. Kaye & Laby give the measured value of the density of air at 20°C and 101.3 kPa (typical values) as $1.196\,\text{kg m}^{-3}$. What value does the perfect gas equation (Eq. 4.29) predict?

Rearranging *PV=zRT*, we write the number of moles per unit volume, the *molar density, z/V*, as

$$\frac{z}{V} = \frac{P}{RT}$$

Now, if we ignore the minor constituents of air, and treat it as a gas with an effective molecular weight given by the average of nitrogen (28), oxygen (16) and argon (40), the average molecular weight of air is

$$M = (78.1\% \times 28) + (20.1\% \times 16) + (0.9\% \times 40)$$
$$= 0.781 \times 28 + 0.201 \times 16 + 0.009 \times 40$$
$$= 28.66$$

So, if the molar density of air is *z/V*, then the mass density is *Mz/V*. Thus, according to perfect gas theory, the density of air should be close to

$$\rho = \frac{Mz}{V} = \frac{MP}{RT}$$

Using $M = 28.66\times10^{-3}\,\text{kg}$, $T = 293.15\,\text{K}$. $P = 101.3\,\text{kPa}$ and $R = 8.314\,\text{JK}^{-1}\,\text{mol}^{-1}$,

$$\rho = 1.191\,\text{kg m}^{-3}$$

Comparing the theoretical value with the experimental value of $1.196\,\text{kg m}^{-3}$, one can see that the perfect gas equation has predicted the density of air – a complex gas – with an accuracy of around 0.5%. This is impressive, and shows the power of the perfect gas equation.

---

terms of the exchange of momentum with the walls, the results of a more sophisticated analysis are identical with those arrived at in Step 1.

The real story is illustrated in Figure 4.4 and goes something like this. Molecules hit the wall and stick there; they interact electrically with the atoms in the wall, and then after a short time they leave the wall, and return to the gas. How can this complex process be approximated by the simple "bouncing" model outlined in Step 1? The reasoning is as follows:

Example 4.2

Suppose we have 1.3 mol of helium gas in a container of volume 1 litre at a temperature of 20°C. (a) What is the pressure of the gas? (b) If the temperature is changed to 100°C, but the gas is contained in the same volume, what is the pressure?

(a) Use $PV = zRT$:
$V = 1\,l = 10^{-3}\,m^3, z = 1.3\,mol, R = 8.31\,J\,K^{-1}\,mol^{-1}, T = 20°C$
$= 273.15 + 20 = 293.15\,K$,

so, $$P = \frac{zRT}{V} = \frac{1.3 \times 8.314 \times 293.15}{10^{-3}}$$
$$= 3.168 \times 10^6\,Pa$$

(approximately 31 times atmospheric pressure).

(b) Use $PV = zRT$:
$V = 1\,l = 10^{-3}\,m^3, z = 1.3\,mol, R = 8.31\,J\,K^{-1}\,mol^{-1}, T = 100°C = 273.15 + 100 = 373.15\,K$,

so, $$P = \frac{zRT}{V} = \frac{1.3 \times 8.314 \times 373.15}{10^{-3}}$$
$$= 4.033 \times 10^6\,Pa$$

(approximately 40 times atmospheric pressure).

- Since molecules hit the wall from random directions and leave in random directions then, on average, there is no net momentum transfer parallel to the plane of the wall. This validates our assumption that, on average, the component of the momentum of S parallel to the wall is unaffected by the collision process.
- If one looks at the initial and final situations, without asking about the details of the reflection process, then one finds that the two processes are rather similar. In each case the molecule is first approaching the wall, and then leaving the wall,

with a velocity component v_x characteristic of the temperature of the gas and wall. Thus the average momentum transfer is the same in both cases. Thus the x component of momentum imparted to the wall is unaffected by the detailed nature of the collision process.

The fact that S has spent some time on the wall instead of travelling through the box means that the average time between its collisons with the wall will be slightly greater than we assumed in Equation 4.10. This will tend to reduce the pressure of a real gas as compared to an ideal gas.

Note that the argument above is only valid when both the walls and the gas are at the same temperature. If this is not so, then S will return to the gas with kinetic energy appropriate to the temperature of the wall and not the temperature of the gas. The cooling of a gas by contact with "cold walls" is discussed in Exercise 10C1.

Complication 2: Collisions with other molecules

While S may be able to bounce around happily from side to side of the container when it is in the box alone, it will not be free to do so when the box is filled with other molecules. Does the analysis break down if S can't bounce back and hit the wall again after $2L_x/v_x$ seconds? It doesn't, but the reason it doesn't is quite subtle.

What happens is that, although S doesn't travel from side to side in the box, the *momentum* that it carries does (Fig. 4.5) In each collision with another molecule the momentum is conserved, and so the momentum does bounce from side to side of the container. In addition to this there are many other molecules whose momentum is being "bounced" across the box. In practice this makes calculations rather complicated, but in the end it amounts to exactly the

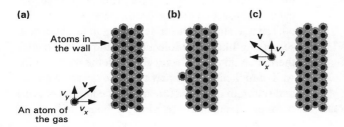

(a)

Atoms in the wall

An atom of the gas

(b)

(c)

Figure 4.4 When a molecule hits a wall what really happens is that it sticks to the wall for a short while (typically 10^{-12} s), and eventually leaves the wall and rejoins the gas with no "memory" of the trajectory with which it hit the wall. (a) The approach to the wall; (b) The adsorption on the wall; (c) The escape from the wall. The shaded grey region of each molecule indicates the region in which it interacts strongly with neighbouring molecules.

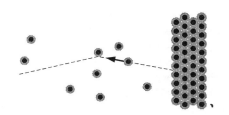

Figure 4.5 When there is more than one molecule in the box, a "representative" molecule can no longer travel freely across the box and bounce from side to side in a time $2L_x/v_x$.

same thing as our simple assumption in Step 1. The collision process may be clearly observed by running the computer simulation of a two-dimensional gas listed in Appendix 4.

Complication 3: Degrees of freedom

The idea of a degree of freedom is an important idea in statistical mechanics and has already been mentioned in Chapter 2. There are three points that we need to consider in this context:

(a) What is a degree of freedom?
(b) What is an accessible degree of freedom?
(c) The equation of energy among accessible degrees of freedom

What is a degree of freedom?

Colloquially, a degree of freedom of a molecule is a "thing it can do" or a "way in which it can possess energy". Technically, this corresponds to an independent "squared term" in the expression for the energy of a molecule. For example:

- An atom in a gas can move in three dimensions and has kinetic energy associated with each of those possibilities and so has at least three degrees of freedom

$$KE = \tfrac{1}{2}mv_x^2 + \tfrac{1}{2}mv_y^2 + \tfrac{1}{2}mv_z^2 \qquad (4.30)$$

- The same atom in a solid has more possibilities open to it. As it jiggles around it can possess kinetic energy in the same way as a molecule in a gas can, but it also has potential energy. This is stored in the deformation of its bonds with its

neighbours as it vibrates. The potential energy is

$$PE = \tfrac{1}{2} K_x\left(x - x_0\right)^2 \qquad (4.31)$$
$$+ \tfrac{1}{2} K_y\left(y - y_0\right)^2 + \tfrac{1}{2} K_z\left(z - z_0\right)^2$$

where K_x, K_y and K_z are the spring constants describing the variation of potential energy as the molecule moves away from its equilibrium position x_0, y_0, z_0. There are three independent squared terms in Equation 4.31, which correspond to an additional three degrees of freedom (see §7.4.2).

- A diatomic molecule in a gas has three degrees of freedom associated with its kinetic energy, but may also have kinetic energy of vibration and rotation, and potential energy of vibration, and so has more degrees of freedom than an atom by itself.

As a rule of thumb, the more atoms making up a molecule in a gas, then the greater will be the number of degrees of freedom (see §5.3.3).

What is an accessible degree of freedom?

The degrees of freedom refer to things that a molecule *can* do, not to whether it is doing them. Due to the importance of the quantum nature of matter on small scales, some processes cannot take place because there is an *energy gap, ΔE,* separating quantum states in which the process does not occur, from quantum states in which the process does occur. If the average energy available to a particular molecule is much less than the energy gap ΔE, then the process will only be able to take place occasionally when (rare) local fluctuations cause the particular molecule's energy to exceed the energy gap.

The average energy per degree of freedom available to a molecule in an environment at temperature T is, by definition of Equation 4.23, $\tfrac{1}{2}k_B T$. If this energy is much less than the energy gap, then the process will not take place and the degree of freedom associated with that process is said to be *inaccessible*. If the temperature increases so that $\tfrac{1}{2}k_B T$ is much greater than the energy gap, then the process becomes *accessible*.

The idea of accessibility will be extremely important when we try to understand the heat capacity of gases (§5.3).

Equipartition of energy among accessible degrees of freedom

It is a fundamental assumption of statistical mechanics that, on average, energy is stored equally in all *accessible degrees of freedom*. This is a sound assumption, and one which is extremely important.

Consider the example of a molecule that has just the three degrees of freedom of translational motion. It is perhaps not surprising that, on average, the kinetic energy associated with each degree of freedom will be equal. The *principle of equipartition of energy* allows us to go one step further by stating that, *on average*, the kinetic energy associated with each degree of freedom will *definitely* be equal.

Consider now a molecule with extra accessible degrees of freedom associated, for example, with internal vibration. It is now not at all obvious that, on average, the energy associated with each translational degree of freedom will be equal to the energy associated with each vibrational degree of feedom. The *principle of equipartition of energy* allows us to state that, on average, the energy associated with each accessible degree of freedom will be equal, independently of what type of motion each accessible degree of freedom corresponds to.

What the principle implicitly assumes is that there exists some mechanism for coupling one degree of freedom with another. A molecule travelling through space which never collided or interacted with another molecule would not exchange energy between its degrees of freedom: if it was not vibrating initially it would not spontaneously slow down and start vibrating. However, if it interacts with other molecules then it can exchange energy between its own degrees of freedom and with other degrees of freedom on other molecules. We rely on the randomizing effect of these interactions to ensure that energy is, on average, equally distributed among all the accessible degrees of freedom. Thus, during a collision, a molecule that was not vibrating could start vibrating at the expense of energy in another accessible degree of freedom.

Finally, we consider degrees of freedom associated with different molecules. Again, collisions or interactions of some kind are an essential assumption of the principle of equipartition of energy. Molecular collisions act to share out energy equally between different accessible degrees of freedom. If one degree of freedom of one molecule has much more than the average energy associated with a degree of freedom ($\frac{1}{2}k_B T$), then in interactions with other molecules the energy associated with that degree of freedom will tend to be lost to other degrees of freedom that have less than the average energy per degree of freedom.

4.4 Calculating microscopic quantities

In order to understand some propeties of gases we need go no further than the ideal gas theory outlined above. However, in order to understand, for example, the thermal conductivity of gases, we need to develop the theory of an ideal gas further. In particular, we need to find out about:

- the *size* of a gas molecule;
- the *root-mean-square speed* of a gas molecule;
- the *mean speed* of a gas molecule; and
- the *mean free path* of a gas molecule.

These are quantities that are important for understanding the behaviour of gases. However, unlike P, V and T, the quantities are not directly measurable in a straightforward way.

4.4.1 The size of a gas molecule

In the discussion of the complications of our derivation of the ideal gas equation, we mentioned the idea of molecular collisions as being important in establishing the averaging of energy among the degrees of freedom of the gas molecules. It is clear that molecules must have a finite size in order to collide. However, what do we mean by "size" exactly, and how do we represent the "size" of a molecule that may have a complex shape?

There are many definitions of the "size" of atoms, each of which differs slightly from the others. In general, the definitions refer to a radius within which, say, 90% of the electronic charge distribution is found. For molecules that are non-spherical one could specify complex shapes to describe the same criteria. In this chapter, however, we shall use the idea of an "effective" cross-sectional area of a molecule. We will specify an "effective diameter", a, such that the chance of collision with a spherical molecule of diameter a is the same as that of colliding with the molecule in question. We note that it is possible for a to change with temperature as, for example, rotational degrees of freedom are made accessible at high temperatures, or because the average speed of molecules has become so high that the region of electrical interaction around a molecule no longer affects the trajectory of nearby molecules.

4.4.2 The distribution of molecular speeds: $\bar{v}$ and $\overline{v^2}$

Clearly, not all molecules in a gas move at the same speed. Even if by some clever contrivance we arranged that all the molecules did have the same speed at some time, things would not stay that way for long. Collisions between the molecules would cause one molecule to speed up and another to slow down. It might seem that it would be impossible to say anything about the range of molecular speeds present in a gas. Surprisingly, we can in fact say rather a lot, but only if the gas is in equilibrium at a temperature T. If this is the case, then using an analysis outlined in Appendix 1, one can show that the probability that a molecule has a speed between v and $v + dv$ is given by the *Maxwell speed distribution*, $P_M(v)$:

$$P_M(v)dv = \frac{4}{\sqrt{\pi}}\left(\frac{m}{2k_BT}\right)^{\frac{3}{2}} v^2 \exp\left[\left(-mv^2\right)/(2k_BT)\right]dv$$

(4.32)

The general form of the Maxwell speed distribution curve is illustrated in Figure 4.6, and three specific examples are shown in Figure 4.7.

Given the Maxwell speed distribution function, one can define three important speeds, all of which are of a similar order of magnitude but which characterize the speed distribution curve in slightly different ways. These speeds are (Fig. 4.8):

- the average speed, $\bar{v}$;
- the root-mean-square speed, $v_{RMS} = (\overline{v^2})^{\frac{1}{2}}$; and
- the most probable speed, v_{prob}

In general, the average speed $\bar{v}$ is the relevant average, and is used when discussing the *average speed at which molecules move*. This may sound obvious, but it is not! The root-mean-square speed,

---

Example 4.3

What is the probability that a molecule of nitrogen in nitrogen gas at 1000 K has a speed between 750 m s⁻¹ and 751 m s⁻¹?

We need to evaluate Equation 4.32 to find $P_M(v)dv$ with $v = 750\,\text{m s}^{-1}$ and $dv = 1\,\text{m s}^{-1}$:

$$P_M(v)dv = \frac{4}{\sqrt{\pi}}\left(\frac{m}{2k_BT}\right)^{\frac{3}{2}} v^2 \exp\left[\left(-mv^2\right)/(2k_BT)\right]dv$$

From Table 7.2 we find that the mass of a molecule of N_2, m, is just $2 \times 14u = 28u$ where $u = 1.661 \times 10^{-27}\,\text{kg}$. Substituting this into Equation 4.32 along with $T = 1000\,\text{K}$, and the Boltzmann constant $k_B = 1.38 \times 10^{-23}\,\text{J K}^{-1}$, we arrive at

$$P_M(750) \times 1 = \frac{4}{\sqrt{\pi}}\left(\frac{28 \times 1.661 \times 10^{-27}}{2 \times 1.38 \times 10^{-23} \times 1000}\right)^{\frac{3}{2}} \times (750)^2$$

$$\times \exp\left[\frac{(750)^2 \times 28 \times 1.661 \times 10^{-27}}{2 \times 1.38 \times 10^{-23} \times 1000}\right]$$

$$P_M(750) = \frac{4}{\sqrt{\pi}}\left(1.685 \times 10^{-6}\right)^{\frac{3}{2}} \times 562500 \times e^{-0.948}$$

$$= \frac{4}{\sqrt{\pi}} \times 2.187 \times 10^{-9} \times 562500 \times e^{-0.948}$$

$$= 1.076 \times 10^{-3}$$

This point is near the peak of the 1000 K curve for nitrogen in Figure 4.8.

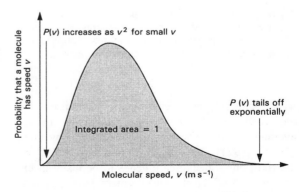

Figure 4.6 The general form of the Maxwell speed distribution curve for a gas (Equation 4.32)

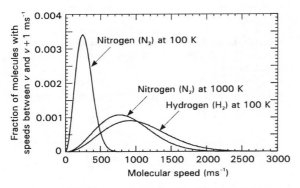

Figure 4.7 Three graphs showing the Maxwell distribution of molecular speeds in nitrogen gas and hydrogen. The vertical axis is the probability $P(v)$ that a molecule has a speed between v and $v + 1\,ms^{-1}$. Two curves show the distribution $P(v)$ at temperatures of $100\,K$ and $1000\,K$ for nitrogen. We see that the peak of the curves shifts to higher speeds at higher temperature. Note that the curve for hydrogen at $100\,K$ is similar to the curve for nitrogen at $1000\,K$.

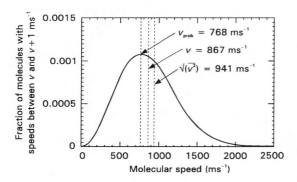

Figure 4.8 Illustration of the position the special velocities relevant to the Maxwell velocity distribution curve at a temperature of $1000\,K$ (approximately $730°C$). This curve has been drawn for nitrogen (N_2) molecules.

v_{RMS}, is the relevant average to use when discussing the *average kinetic energy of the molecules*, i.e. the average of $\frac{1}{2}mv^2$. These two averages differ by about 8%, $\bar{v} = 0.921 v_{RMS}$, and you should take care to use the right one. Special care needs to be taken in expressions in which the speed enters as a "squared" term to distinguish between cases where one has $\bar{v}^2$ and $\overline{v^2}$: the two are *not* the same! The most probable

Example 4.4

In nitrogen gas at $1000\,K$, what fraction of molecules have speeds greater than $867\,ms^{-1}$ given that

$$\int_{x=1.266}^{x=\infty} x^{\frac{1}{2}} e^{-x} dx = 0.4164 \quad ?$$

The required fraction is the integral of $P(v)dv$ (Eq. 4.32) over the required speed range, which in this case is from $v = 867\,ms^{-1}$ to $v = \infty$:

$$\int_{v=867}^{v=\infty} P(v)dv = \int_{v=867}^{v=\infty} \sqrt{\frac{2}{\pi}} \left(\frac{m}{k_B T}\right)^{\frac{3}{2}} v^2 \exp\left[\left(-mv^2\right)/\left(2k_B T\right)\right] dv$$

If we substitute $x = mv^2/2k_B T$, then we find

$$dx = dv \sqrt{\frac{2m}{k_B T}}$$

and the integral becomes

$$\text{Fraction} = \sqrt{\frac{2}{\pi}} \left(\frac{m}{k_B T}\right)^{\frac{3}{2}} \int_{x=x_1}^{x=\infty} \frac{2k_B Tx}{m} e^{-x} \frac{dx}{\sqrt{x}} \sqrt{\frac{k_B T}{2m}}$$

and, rearranging,

$$\text{Fraction} = 2\sqrt{\frac{1}{\pi}} \left(\frac{m}{k_B T}\right)^{\frac{3}{2}} \left(\frac{k_B T}{m}\right)^{\frac{3}{2}} \int_{x=x_1}^{x=\infty} x^{\frac{1}{2}} e^{-x} dx$$

Cancelling, we find

$$\text{Fraction} = 2\sqrt{\frac{1}{\pi}} \int_{x=x_1}^{x=\infty} x^{\frac{1}{2}} e^{-x} dx \,.$$

We are now able to recognize the numerical integral given in the question. We need to evaluate the lower limit of the integral using $x = mv^2/2k_B T$. We have

$$x_1 = \frac{28 \times 1.66 \times 10^{-27} \times (867)^2}{2 \times 1.38 \times 10^{-23} \times 1000} = 1.266$$

where we have used the fact that a nitrogen molecule has two atoms of nitrogen, each of mass $14u$. Substituting for the lower limit of the integral,

$$\text{Fraction} = 2\sqrt{\frac{1}{\pi}} \int_{x=1.266}^{x=\infty} x^{\frac{1}{2}} e^{-x} dx$$

We can now recognize the standard integral given at the start of the question. Hence,

$$\text{Fraction} = 2\sqrt{\frac{1}{\pi}} \times 0.4161 = 0.4695$$

Thus around 47% of molecules have speeds greater than $867\,ms^{-1}$.

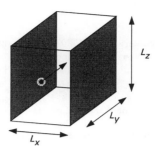

Figure 4.9 The molecule S is alone in a box travelling with a kinetic energy that is the same as the average kinetic energy of all the molecules that will eventually inhabit the box.

4.4.3 The number of molecules hitting a unit area per second

There will be several occasions, for example in considering the thermal conductivity of a gas, when it will be useful to know how many molecules are crossing a unit area within the gas per second. We can get an estimate of this if we return to the situation depicted in Figure 4.2 in our derivation of the ideal gas equation.

The molecule S is alone in a box travelling with a kinetic energy that is the same as the *average* kinetic energy of all the molecules which will eventually inhabit the box (Fig. 4.9). We consider the collisions of S with the wall of area $A = L_y L_z$. S collides with this wall, bounces off into the box, bounces off the opposite face of the box, and eventually returns to the same wall for another collision. It takes S a time L_x/v_x seconds to reach the opposite wall, and a further L_x/v_x seconds to return: it thus makes collisions with the wall in question every $\Delta t = 2L_x/v_x$ seconds. Thus

No. of collisions with the wall per unit area per sec.

$$= \frac{1}{A \Delta t} = \frac{1}{[2L_x/v_x]A} = \frac{1}{[2L_x/v_x]L_y L_z} = \frac{v_x}{2L_x L_y L_z} \quad (4.33)$$

Note that $L_x L_y L_z = V$, the volume of the box, and so Equation 4.33 becomes

No. of collisions with the wall per unit area per sec.

$$= \frac{v_x}{2V} \quad (4.34)$$

If there are N molecules in the box with an x velocity similar to that of S, then we have

No. of collisions with the wall per unit area per sec.

$$= \frac{1}{2}\frac{Nv_x}{V} = \frac{1}{2}nv_x \quad (4.35)$$

where n is the number density of molecules. Next we

notice that in this argument v_x was chosen such that the kinetic energy of S represented the average kinetic energy of the molecules in the box. Thus in terms of the discussion of the distribution of molecular speeds in the previous section we can write that

$$KE_S = \frac{1}{2}m\overline{v^2} \quad (4.36)$$

Since

$$\overline{v^2} = v_x^2 + v_y^2 + v_z^2 \quad (4.37)$$

we see that $v_x^2 = \frac{1}{3}\overline{v^2}$, and so v_x in Equation 4.35 is given by $v_x = (\frac{1}{3}\overline{v^2})^{1/2}$. Substituting this value into Equation 4.35 yields

No. of collisions with the wall per unit area per sec.

$$= \frac{1}{2}n\sqrt{\frac{1}{3}\overline{v^2}} = \frac{1}{2\sqrt{3}}n\sqrt{\overline{v^2}} = \frac{1}{2\sqrt{3}}nv_{RMS} \quad (4.38)$$

This remarkably simple expression is nearly, but not quite, correct. It is an underestimate of the correct value because of our simplified assumption that all molecules move with the same speed. Using a considerably more complicated analysis, it can be shown that

No. of collisions with the wall per unit area per sec.

$$= \frac{1}{4}n\overline{v} \quad (4.39)$$

an answer that exceeds our simple estimate by about 25%.

Example 4.5 shows that, under typical conditions, the number of collisions per second with the wall is extremely large. This allows us to understand how the tiny forces of molecular collisions can add up to the substantial forces exerted by a gas on the walls of its container.

speed, v_{prob}, is the speed marking the peak of the speed distribution curve, but is of no special physical significance; you may now forget about it.

Example 4.5

0.3 mol of gas is held at a temperature of 1000 K inside a container of volume 1 litre. What is the number of collisions per second with each wall of a container of size 10 cm × 10 cm × 10 cm?

Equation 4.39 tells us that the number of collisions per unit area per second is $\frac{1}{4}n\bar{v}$. The molecular density n can be estimated from the *molar density*, z/V. We thus find that $n = N_A z/V$. The value of $\bar{v}$ can be found using Equation 4.37, but we take our value in this case directly from Figure 4.9. We thus have:

$z = 0.3$
$V = 1 \, \text{l} = 10^{-3} \, \text{m}^3$
$\bar{v} = 867 \, \text{m s}^{-1}$
$N_A = 6.022 \times 10^{23} \, \text{mol}^{-1}$

$$\frac{1}{4}n\bar{v} = \frac{1}{4}\frac{N_A z}{V}\bar{v}$$

$$= \frac{1}{4}\frac{6.022 \times 10^{23} \times 0.3}{10^{-3}} \times 867 \, \text{m}^{-2}\text{s}^{-1}$$

$$= 3.92 \times 10^{28} \, \text{m}^{-2}\text{s}^{-1}$$

Thus the number of collisions per second with the wall of area 10 cm × 10 cm is given by

$$3.92 \times 10^{28} \times 10^{-2} = 3.92 \times 10^{26} \, \text{s}^{-1}$$

4.4.4 The mean free path of a gas molecule

Collisions between molecules play an important part in the transport of both heat and electricity through a gas. In particular, the distance that molecules travel before colliding with other molecules is an important quantity. However, on average, a slowly-moving molecule travels a considerably shorter distance before colliding than does a fast-moving molecule. For this reason we define the *mean free path* of a gas molecule, λ_{mfp}, as the *average* distance that a gas molecule travels before colliding with another molecule.

We can fairly easily estimate the average free path, λ, for molecules with effective diameter, a, that either move much slower or much faster than the mean speed of a gas molecule, $\bar{v}$. Recall that if the molecules are not spherical, then a will represent some average of its dimensions related to the *average cross-sectional area* that the molecule presents to other molecules. In this analysis we treat the molecules as hard spheres, and ignore the region of elec-

trical interaction sketched in grey around the molecules in some of the previous figures. We will consider the effect of these interactions briefly at the end of the section.

The average free path of molecules moving much faster than $\bar{v}$

Imagine taking a snapshot of the molecules of gas at some particular time, t, and concentrating our attention on a particularly fast-moving molecule, M. The situation might look like the one depicted in Figure 4.10.

On average, M travels a distance λ before colliding with another molecule. In the limit of M moving extremely quickly, it will be as if the other molecules did not move at all. In this case, there must, on average, be no other molecules whose centres lie within a cylinder of volume $\pi a^2\lambda$ where a is the effective diameter of an atom. Since there is just one molecule (M) in a volume $\pi a^2\lambda$, the number density of molecules, n, must be given by

$$n = \frac{1}{\pi a^2 \lambda} \qquad (4.40)$$

Rearranging this gives an expression for λ

$$\lambda = \frac{1}{n\pi a^2} \qquad (4.41)$$

This equation is an estimate for the average free path appropriate to fast-moving molecules. The effect of

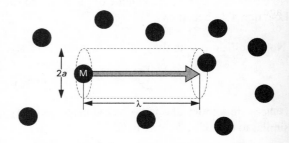

Figure 4.10 A method for making an approximation for the mean free path for fast-moving molecules. A molecule travels, on average, a distance λ before colliding with another molecule. If the other molecules are effectively stationary while M travels through the gas then there must, on average, be no other molecules whose centres lie within a volume $\pi a^2\lambda$.

the motion of the other molecules, which we have neglected entirely above, is difficult to account for quantitatively. However, it can be seen that, in general, other molecules will be likely to encroach on the "free space" of molecule M. Thus Equation 4.41 represents an overestimate of the average free path appropriate to more typical molecules.

The free path of molecules moving much slower than $\bar{v}$

Imagine taking a snapshot of the molecules of a gas at some particular time, t, and concentrating our attention on a particularly slow-moving molecule, A. The situation might look like the one depicted in Figure 4.11.

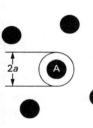

Figure 4.11 A method for making an approximation for the mean free path for slow moving molecules. A stationary molecule presents a cross-sectional area $\pi(a/2)^2$ to other moving molecules. If the centres of other molecules pass within an area πa^2 around the centre of A there will be a collision.

If A were essentially stationary in comparison to the speeds of the other molecules then, as outlined in Figure 4.11, other molecules passing within an area πa^2 around the centre of A would collide. Considering a surface through the centre of A we recall from Section 4.4.3 that the average number of molecules crossing a unit area per second is $\frac{1}{4}n\bar{v}$. Thus the average number of molecules colliding with A from each of the two sides of the surface is

Number of collisions per second
$$= 2 \times \tfrac{1}{4} n\bar{v} \times \pi a^2 \tag{4.42}$$

and so the average time Δt between collsions is

$$\Delta t = \frac{1}{2 \times \frac{1}{4} n\bar{v} \times \pi a^2} = \frac{2}{\bar{v} n\pi a^2} \tag{4.43}$$

Now if A is not quite stationary but moves with a speed v much less than $\bar{v}$, then the average distance travelled by A – the average free path – will be given by $v\Delta t$ or

$$\lambda = v\Delta t = \frac{2}{n\pi a^2}\left(\frac{v}{\bar{v}}\right) \tag{4.44}$$

The mean free path of all molecules

We have calculated that for slow-moving molecules the average free path is given by Equation 4.44:

$$\lambda = \frac{2}{n\pi a^2}\left(\frac{v}{\bar{v}}\right) \tag{4.44*}$$

which corresponds to a molecule being struck by other molecules at a constant average rate. Thus the faster it moves, the further it travels, on average, between collisions. This increase of free path with speed cannot continue indefinitely; a molecule moving so fast that other molecules appear stationary cannot travel an indefinite distance before colliding. As shown above, the limiting average free path is

$$\lambda = \frac{1}{n\pi a^2} \tag{4.41*}$$

Establishing these low and high speed limits has not been too difficult. However, finding the correct expression for the mean free path of *all* molecules, λ_{mfp}, is rather complex. We will, however, not be surpised to find – in common with Equations 4.41 and 4.45 – that λ_{mfp} is proportional to $1/(n\pi a^2)$. It is perhaps rather surprising to find that the result of the complex calculation is simply that this factor is multiplied by $1/\sqrt{2} \approx 0.71$ to give

$$\lambda_{\text{mfp}} = \frac{1}{\sqrt{2}n\pi a^2} \tag{4.45}$$

The slow molecule, fast molecule and average behaviours are summarized in Figure 4.12.

4.5 What next?

In this chapter we have developed a simple model of a gas. The extent to which this has been worthwhile depends on how well the model actually describes the properties of real gases. In Chapter 5 we will con-

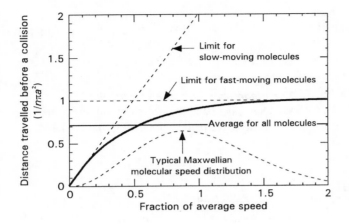

Figure 4.12 The variation of the mean free path λ of molecules as a function of the speed of the molecules expressed as a fraction of the mean speed $\bar{c}$. The figure indicates the limiting behaviour of λ for low-speed molecules and high-speed molecules, and the average value for all molecules, λ_{mfp}. Also indicated is a qualitative indication of the distribution of molecular speeds.

Example 4.6

Let us evaluate λ_{mfp} for some typical situations. But how can we estimate the number density of molecules, n, and the diameter of a molecule, a?

We can estimate n by using the perfect gas equation for z mol of gas (Eq. 4.29), $PV = zRT$. Rearranging this, we find an expression for the molar density, the number of moles of gas per unit volume, $z/V = P/RT$.

In z mol of gas there are zN_A molecules. Therefore, the molecular density, n, is given by

$$n = \frac{zN_A}{V} = \frac{N_A P}{RT}$$

Substituting values for pressure (10^5 Pa) and room temperature (20°C = 293 K), we find that n is given by

$$n = \frac{N_A P}{RT} = \frac{6.02 \times 10^{23} \times 10^5}{8.31 \times 293}$$
$$= 2.5 \times 10^{25} \text{ molecules m}^{-3}$$

Using a typical value of the spacing between atoms in a solid as an estimate for the molecular diameter $a \approx 0.3$ nm, we have

$$\lambda_{mfp} = \frac{1}{\sqrt{2}n\pi a^2} = \frac{1}{\sqrt{2} \times 2.5 \times 10^{25} \times 3.145 \times \left(0.3 \times 10^{-9}\right)^2}$$
$$\approx 10^{-7} \text{m, i.e. 0.1 } \mu\text{m}$$

Summary: In the air around us there are approximately 2.5×10^{25} molecules per cubic metre. Thus, the average separation between molecules will be approximately

$$\text{average spacing} \approx \frac{1}{\sqrt[3]{2.5 \times 10^{25}}} = 3.4 \times 10^{-9} \text{ m, i.e. 0.003 } \mu\text{m}$$

i.e. the average spacing between molecules is approximately 10 times the diameter of a molecule. The molecules travel, on average, about 10^{-7} m before colliding with another molecule, which is about 300 times the diameter of a molecule, or 10 times the average separation between molecules. Thus our initial assumptions (Eq. 4.2) about our model of a gas are at least consistent with the predictions of the ideal gas theory.

---

sider a variety of experimental data on gases and see to what extent we can understand the real behaviour of gases in terms of the theory developed above. We will find that the answer is, broadly, that the ideal gas model describes real gases well under a wide range of experimental circumstances.

4.5.1 Summary of key equations

The ideal gas equation for z moles of gas is

$$PV = zRT \tag{4.46}$$

In terms of the molar density of the gas, z/V, this may be written as

$$P = \frac{z}{V} RT \tag{4.47}$$

Since 1 mole of gas has a N_A molecules, the number density of molecules n is zN_A/V which may be written as

$$n = \frac{N_A P}{RT} \qquad (4.48)$$

Recalling that the gas constant $R = N_A k_B$ where k_B is the Boltzmann constant, we can also express the number density as

$$n = \frac{P}{k_B T} \qquad (4.49)$$

The probability that a molecule of mass m in a gas at temperature T has a speed between v and $v + dv$ is given by the Maxwell speed distribution curve

$$P_M(v)dv = \frac{4}{\sqrt{\pi}}\left(\frac{m}{2k_B T}\right)^{\frac{3}{2}} v^2 \exp\left[\left(-mv^2\right)/(2k_B T)\right]dv \qquad (4.50)$$

The number of molecules colliding with unit area of wall per second is

$$\tfrac{1}{4}n\bar{v} \qquad (4.51)$$

The mean free path of a molecule of effective diameter, a, is given by

$$\lambda_{\text{mfp}} = \frac{1}{\sqrt{2}n\pi a^2} \qquad (4.52)$$

4.5.2 Beyond the ideal gas model

Before moving on to compare the predictions of the ideal gas model with experiment, it is wise to review briefly how a more advanced theory might deal with a gas, and the kind of effects that would have to be taken into account. The two key features of molecules that have been ignored (§4.2) are the finite size of the molecules and the interactions between the molecules, which though weak do have a finite range.

The Van der Waals equation

The Van der Waals equation of state is an interesting modification of the ideal gas equation. It considers a gas composed of hard spheres, each with volume v,

which have a binding energy of Δu. Since the molecules now have a finite volume, we can plausibly argue that the actual volume available for molecules to move in is reduced by a volume equivalent to that which would be occupied by the hard spheres. Thus we can write for 1 mol of gas, i.e. N_A molecules,

$$P(V_m - b) = RT \qquad (4.53)$$

where b is approximately the volume of N_A hard sphere molecules, i.e. $b \approx N_A v$. A more sophisticated analysis indicates that the volume is in fact reduced by more than just the volume of the molecules, and normally one estimates b as

$$b = 4N_A v \qquad (4.54)$$

However, Equation 4.53 has taken no account of the attraction between the molecules. This effect is not modelled quite so realistically in the Van der Waals model of a gas, but some account may be taken by introducing a term of the form

$$\left(P + \frac{a}{V_m^2}\right)(V_m - b) = RT \qquad (4.55)$$

The Van der Waals equation

where the correction coefficient, a, is given by

$$a = \tfrac{2}{3}N_A b\Delta u \qquad (4.56)$$

Commonly, Equation 4.55 is used to model real gases with the parameters a and b determined experimentally from the small deviations of gases from $PV = RT$.

The virial equation

Modern theories of gases represent the equation of state of a gas as

$$\frac{PV_m}{RT} = 1 + \frac{B(T)}{V_m} + \frac{C(T)}{V_m^2} + \cdots \qquad (4.58)$$

The virial equation

The coefficients $B(T)$, $C(T)$, etc., are known as the *second virial coefficient, third virial coefficient*, etc., and are functions of temperature only. The second virial coefficients of many gases are tabulated in Kaye & Laby (see §1.4). Note that if the virial coefficients are taken as zero, then Equation 4.57 is the ideal gas equation. The virial coefficients model deviations from perfect gas behaviour. In general the virial coefficients are determined experimentally, and their values tabulated in reference books such as Kaye & Laby. However, in certain theories of gases, the coefficient $B(T)$ can be understood as arising from interactions between pairs of molecules, and $C(T)$ can be understood as arising from interactions between clusters of three molecules.

Molecular collisions

Finally, we consider the effect of molecular interactions on molecular collisions. The interactions between molecules are generally weakly attractive at long range (represented by the grey area around the molecules in Figure 4.13), but repulsive at short range (represented by the black area around the molecules in Figure 4.13). Note that assuming hard-sphere instantaneous-contact collisions will model the collision process correctly in most cases. However, some types of collision, such as those illustrated in Figure 4.13b,c, are not modelled at all well. The importance of this is that the "effectiveness" of such collisions is reduced when the average kinetic energy of molecules is much larger than the attractive potential energy between molecules. This leads to a temperature dependence of the apparent average cross-sectional area used to evaluate the mean free path of a molecule. The non-hard-sphere nature of molecular collisions can be clearly seen in the computer simulation listed in Appendix 4.

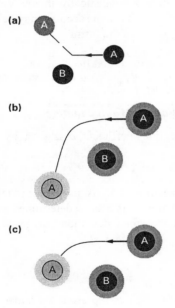

Figure 4.13 Molecular collisions are sometimes not well modelled by the hard sphere approach used in determining the mean free path of a molecule, λ_{mfp}. (a–c) A molecule A colliding with a stationary molecule B. (a) The hard sphere approach used in determining λ_{mfp}. (b) The effect of the attractive interactions (grey region) if A is moving relatively slowly; collisions such as this are quite different from those envisioned in (a). (c) The same collision as in (b) might be affected if molecule A is moving relatively quickly; note that the deviation in the trajectory of A is less than that in (b).

4.6 Exercises

Exercises with a P prefix are "normal" problems. Those with a C prefix are best solved numerically using a computer program or spreadsheet.

P1. If 4 mol of gas are contained in a chamber of volume 1 litre at around room temperature, what is the pressure inside the container (Eq. 4.47)? Is this above or below atmospheric pressure? To what temperature would you have to heat/cool the chamber in order to increase/decrease the pressure to atmospheric pressure?

P2. What is the number density of molecules in air at room temperature and atmospheric pressure (Eq. 4.49)?

P3. What is the number density of helium atoms in helium gas at room temperature and atmospheric pressure (Eq. 4.49)?

C4. The probability that a molecule of mass m in a gas at temperature T, with a speed between v and $v + dv$, is given by the Maxwell speed distribution curve. Program the formula for Equation 4.32 into a spreadsheet and so reproduce Figure 4.7.

P5. The average number of molecules colliding with unit area of wall per second is $\tfrac{1}{4}n\bar{v}$ (Eq. 4.39). Produce an order of magnitude estimate for this quantity in air at atmospheric pressure and room temperature.

 If an atom in the wall of chamber has an area of approximately 0.3 nm × 0.3 nm exposed to the chamber, estimate the frequency with which this atom is struck by atoms of the gas. To what pressure must the gas be lowered in order to reduce this frequency to one collision in 1000 s?

P6. By equating the average thermal energy per molecule,

$3/2k_B T$, with the kinetic energy of a molecule (Fig. 4.8), estimate the average speed of a molecule of argon at atmospheric pressure and room temperature. Assuming that argon molecules have a typical "effective diameter" of around 0.3 nm, estimate the mean free path of an argon molecule in argon at atmospheric pressure and room temperature (Eq. 4.45). Estimate the mean time between collisions and the average number of collisions per second.

P7. Review Appendix 1 which outlines how the Maxwell speed distribution curve may be derived. Write an essay summarizing the key stages of the calculation and suggest an experimental method by which the the speed distribution curve might be determined.

C8. Enter the text of the computer program in Appendix 4 into a QBASIC editor and investigate the behaviour of a gas composed of four molecules (enter $n = 2$). Look for the non-ideal orbiting collisions mentioned in Figure 4.13. Decrease the variable *factor* in the SETUP subprogram to increase the average energy of the gas molecules. Run the program with $n = 3$ or more and leave the program running for a long time to see if the pressure on all four walls averages out to the same value.

P9. What is adiabatic? During an isothermal expansion or contraction, the product PV stays constant. In this question you will show that during an adiabatic volume change the quantity PV^γ stays constant where γ (§5.3.2) is the ratio of the heat capacity at constant pressure C_P (§5.3.1) to the heat capacity at constant volume C_V. Showing this involves three steps.

(i) In an adiabatic expansion no heat energy flows into or out of the gas ($\Delta Q = 0$). Using expressions for the work done during a small expansion ($\Delta W = P \Delta V$) and the internal energy change in the gas ($\Delta U = -z C_V \Delta T$), the first law of thermodynamics (Eq. 2.65; $\Delta U = \Delta Q + \Delta W$) can be used to show that

$$\frac{dV}{dT} = \frac{-z C_V}{P}$$

(ii) Differentiating $PV = zRT$ with respect to T yields

$$P\frac{dV}{dT} + V\frac{dP}{dT} = zR$$

and using the chain rule of differentiation to write

$$\frac{dP}{dT} = \frac{dP}{dV} \times \frac{dV}{dT}$$

the above equation can be rewritten as

$$\frac{dV}{dT}\left[1 + \frac{V}{P}\frac{dP}{dV}\right] = \frac{zR}{P}$$

(iii) Substitution from step (i) and using $C_P - C_V = zR$ (Eqs 5.39 & 5.40) then yields

$$\frac{V}{P}\frac{dP}{dV} = -\gamma$$

(a) Write out the derivation described above, verifying all the steps involved.

(b) Show that the final differential equation

$$\frac{V}{P}\frac{dP}{dV} = -\gamma$$

is satisfied by $PV^\gamma = $ constant.

(c) One mole of gas, initially contained in a volume of 10 litres at 20°C is compressed to a volume of 5 litres. What is the initial pressure, the final pressure and the final temperature of the gas if it is compressed (i) isothermally and (ii) adiabatically?

P10. Your colleague states that the ideal gas law doesn't agree well with experiment. Think of an experiment to demonstrate that the gas law is in fact obeyed rather well by most gases. (Hint: try looking at the next chapter!)

P11. Another colleague states that the ideal gas law agrees well with experiment. Think of an experiment to demonstrate that under some circumstances the gas law breaks down completely.

CHAPTER 5

Gases: comparison with experiment

5.1 Introduction

In Chapter 4 we described the theory of a hypothetical ideal gas. The extent to which this endeavour has been worthwhile will become apparent as we compare the actual behaviour of real gases with the predictions of the theory. In the sections that follow, we consider a wide range of properties of real gases by first familiarizing ourselves with their experimental behaviour, and then developing ideal gas theory to see how good its predictions are. By juxtaposing predictions and experimental results we will see both the successes and failures of ideal gas theory, and see how the theory can be developed to model more closely the properties of real gases.

5.2 Mechanical properties

5.2.1 Data on the density of gases
The most striking property of gases is their ability to expand to fill any container, i.e. a fixed number of gas molecules has no fixed volume. For this reason the conditions under which the properties of gases are determined must be specified precisely. Table 5.1 contains data on the density of gases at a temperature of 0°C (273.15 K) and a pressure 101 325 Pa, conditions referred to as *standard temperature and pressure* (STP). Note that this pressure is usually fairly close, but rarely equal, to the average atmospheric pressure close to sea level on Earth.

Air is a mixture of several gases whose relative abundances vary slightly. The main relatively constant components of the mixture are given in Table 5.2. The variability of both the composition and the

Table 5.1 The density of various gases at STP.

Gas	Density ($kg\,m^{-3}$)
Helium	0.1786
Neon	0.9003
Argon	1.782
Krypton	3.739
Xenon	5.858
Hydrogen	0.08995
Nitrogen	1.250
Oxygen	1.428
Chlorine	3.164
Methane	0.7158
Ethane	1.342
Propane	1.968

(Calculated from data in *CRC handbook*; see §1.4.1)

Table 5.2 The major components of atmospheric air.

Gas	Molecular mass	% by volume
Nitrogen, N_2	28.01	78.09
Oxygen, O_2	32.00	20.95
Argon, Ar	39.95	0.93
Carbon dioxide, CO_2	44.00	0.03

(Data from Kaye & Laby; see §1.4.1)

density of air derive mainly from variations in water content. The density of *dry* air (i.e. with zero water content) and containing no carbon dioxide at STP is 1.293 $kg\,m^{-3}$. The typical density of air at around 20°C is 1.200 $kg\,m^{-3}$ with fluctuations due to water content that would only exceptionally exceed ±0.003 $kg\,m^{-3}$.

Following on from Example 5.1, Table 5.3 shows the volume of 1 mole of each of the gases listed in Table 5.1. The results in Table 5.3 indicate strongly that 1 mole, i.e. Avogadro's number of molecules, of any gas occupies a volume of roughly $22.41 \times 10^{-3}\,m^3$

Example 5.1

One mole of helium gas (i.e. Avogadro's number (6.022×10^{23}) of helium atoms), is held in a container at STP. As illustrated below, the container is free to expand or contract, allowing the gas to assume its equilibrium volume for STP. What is the equilibrium volume, V_m?

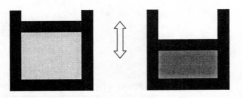

1 mole of helium gas has a mass of 4.003×10^{-3} kg. From Table 4.1, the density of helium gas at STP is 0.1786 kg m^{-3}. The equilibrium volume is thus found from

$$\text{density} = \frac{\text{mass}}{\text{volume}}$$

$$\text{so volume} = \frac{\text{mass}}{\text{density}}$$

$$V_m = \frac{4.003 \times 10^{-3}}{0.1786}$$

$$V_m = 22.413 \times 10^{-3} \, \text{m}^3$$

i.e. approximately 22.4 litres.

Table 5.3 The volume of 1 mole of various gases at STP in units of m^{-3}.

Gas	Molecular weight (amu)	Mass of 1 mol ($\times 10^{-3}$ kg)	Volume at STP ($\times 10^{-3}$ m^3)
Helium, He	4.0030	4.0030	22.4136
Neon, Ne	20.180	20.180	22.4139
Argon, Ar	39.948	39.948	22.4134
Krypton, Kr	83.800	83.800	22.4131
Xenon, Xe	131.29	131.29	22.4128
Hydrogen, H$_2$	2.0160	2.0160	22.4135
Nitrogen, N$_2$	28.014	28.014	22.4131
Oxygen, O$_2$	31.998	31.998	22.4134
Chlorine, Cl$_2$	70.906	70.906	22.4129
Methane, CH$_4$	16.043	16.043	22.4130
Ethane, C$_2$H$_6$	30.070	30.070	22.4126
Propane, C$_3$H$_{10}$	44.097	44.097	22.4124

at STP – independent of the mass or complexity of the molecules.

The questions raised by our initial examination of the experimental data on the density of real gases are:

- Why is the volume of 1 mole of any of the gases in Table 5.3 approximately 22.41×10^{-3} m^3, independent of the substance of which the gas is composed?
- Why are the densities of all gases at STP low compared with those of solids (Tables 7.1 & 7.2) and liquids (Tables 9.1 & 9.2)?

5.2.2 Understanding the density data

There are two questions raised in §5.2.1. We can answer the first question by rearranging the perfect gas equation

$$PV = zRT \tag{5.1}$$

into an expression for the volume of z moles of gas at pressure P and temperature T

$$V = \frac{zRT}{P} \tag{5.2}$$

Substituting $z = 1$ mole, and $P = 101325$ Pa and $T = 273.15$ K, corresponding to STP we have

$$V = \frac{1 \times 8.314 \times 273.15}{101325} \tag{5.3}$$

which evaluates to

$$V = 22.4127 \times 10^{-3} \, \text{m}^3 . \tag{5.4}$$

Consulting Table 5.3 reveals that the prediction of Equation 5.4 is strikingly accurate. It is clear that the perfect gas equation realistically describes at least some properties of real gases.

Considering the second point raised in Section 5.2.1, the relatively low value of the densities as compared with those of solids and liquids is a natural feature of the model illustrated in Figure 4.1. In this model we see that, compared with the size of the molecules, a gas has large spaces between the molecules. Comparing this picture with the simple pictures of solids (see Fig. 6.1) and liquids (see Fig. 8.1) shows that a given number of molecules occupies a much larger volume in the gaseous state than in either of the "condensed" states of matter.

5.2.3 Expansivity data

Gases, like most things, expand when heated. At least, they expand if they are free to do so. If they are constrained to stay at a constant volume, then the pressure they exert on their container increases.

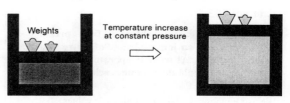

Figure 5.1 Heating a gas at constant pressure.

Heating at constant pressure

The change in volume ΔV of a gas held at *constant pressure* (Fig. 5.1) in an initial volume V_0 due to a temperature change ΔT can be expressed as

$$\Delta V = V_0 \beta_V \Delta T \qquad (5.5)$$

and so the volume V of the gas is given by

$$V = V_0 (1 + \beta_V \Delta T). \qquad (5.6)$$

The constant β_V is a called the *volume coefficient of thermal expansivity*. If β_V is small then the change in the volume of a gas due to a temperature change will be small; if β_V is large then the change in the volume of a gas due to a temperature change will be large.

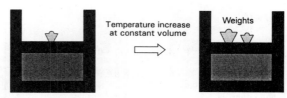

Figure 5.2 Heating a gas as constant volume.

Heating at constant volume

Similarly, when the temperature changes, the increase in pressure of a gas held at *constant volume* (Fig. 5.2) at initial pressure P_0 can be expressed as

$$\Delta P = P_0 \beta_P \Delta T \qquad (5.7)$$

and so the pressure P of the gas is given by

$$P = P_0 (1 + \beta_P \Delta T) \qquad (5.8)$$

where ΔT is the change in temperature. The constant β_P is called the *pressure coefficient of thermal expansivity*. If β_P is small then the change in the pressure of a gas due to a temperature change will be small; if β_P is large then the change in the pressure of a gas due to a temperature change will be large.

Table 5.4 shows experimental values for β_P and β_V valid at around room temperature and pressure. We see that both the volume and the pressure coefficients of expansivity of all the gases shown are close to $3.66 \times 10^{-3}\,°\text{C}^{-1}$, independent of the type of molecules of which the gas is composed. Looking ahead to Tables 7.4 and 9.4, the data given there show that this figure is also 10–100 times larger than the volume expansivity of many liquids, and around 100–1000 times larger than the expansivity of most solids.

Table 5.4 Values of the expansivity coefficients β_V and β_P for gases whose initial pressure is 0.1333 MPa at 0°C, valid in the temperature range 0°C to 100°C. The pressure 0.1333 MPa is a little greater than normal atmospheric pressure.

Gas	$\beta_V(°\text{C}^{-1})$	$\beta_P(°\text{C}^{-1})$
Helium, He	3.6580×10^{-3}	3.6605×10^{-3}
Hydrogen, H_2	3.6588×10^{-3}	3.6620×10^{-3}
Nitrogen, N_2	3.6735×10^{-3}	3.6744×10^{-3}
Air	3.6728×10^{-3}	3.6744×10^{-3}
Neon, Ne	3.6600×10^{-3}	3.6617×10^{-3}

(Data from Kaye & Laby; see §1.4.1)

Example 5.2

Some helium gas is held in a fixed volume of 1 litre at a pressure of 0.1333 MPa and a temperature of 0°C. What is the expected pressure at 100°C ?

We use Equation 5.7 with the following values:
$P_0 = 0.1333\,\text{MPa} = 0.1333 \times 10^6\,\text{Pa}$
$\beta = 3.6605 \times 10^{-3}\,°\text{C}^{-1}$ or K^{-1}
$\Delta T = 100°\text{C} - 0°\text{C} = 100\,\text{K}$

Using $\quad P = P_0(1 + \beta \Delta T)$,

we find, $\quad P = 0.1333 \times 10^6 (1 + 3.6605 \times 10^{-3} \times 100)$

$\qquad P = 0.1821 \times 10^6\,\text{Pa}$

$\qquad P = 0.1821\,\text{MPa}$

Example 5.3

Some nitrogen gas is held at a fixed pressure of 0.1333 MPa and an initial volume of 1 litre at a temperature of 0°C. At what temperature (according to Equation 4.2) would the volume reach zero?

We use Equation 5.6 with the following values:
$V_0 = 1 \, l = 1 \times 10^{-3} \, m^3$
$P_0 = 0.1333 \, MPa = 0.1333 \times 10^6 \, Pa$
$V = 0$
$\alpha = 3.6735 \times 10^{-3} \, °C^{-1}$ or K^{-1}
$\Delta T = ?$

Using, $V = V_0(1 + \alpha \Delta T) = 0$

we find, $(1 + \alpha \Delta T) = 0$

and hence, $\alpha \Delta T = -1$

and so, $\Delta T = \dfrac{-1}{\alpha} = \dfrac{-1}{3.6735 \times 10^{-3}}$

$\Delta T = -272.2° \, C$

i.e the volume is predicted by Equation 5.6 to reach zero at a temperature of $-272.2°C \, (\approx 0.95 \, K)$. If this experiment were performed it would be found that before this temperature was reached the nitrogen would first have liquefied at about $-196°C \, (77 \, K)$ and later solidified at about 63 K. In these states its expansion and contraction would be much smaller than in the gas phase. See Chapter 11 for data on the boiling and freezing points of the elements.

So the main questions raised by our initial examination of the experimental data on the expansivity of gases are:
- Why are the expansivities of the five gases listed in Table 5.4 so similar, independent of the substance of which the gas is composed?
- Why are the expansivities of gases much greater than those shown by solids (see Table 7.4) and liquids (see Table 9.4)?

5.2.4 Understanding the expansivity data

We can answer the questions raised by our examination of the experimental data by rearranging the ideal gas equation $PV = zRT$. Equation 5.6 expresses the expansivity of a gas, β_V in a form that predicts the volume of a given amount of gas in terms of its volume V_0 at temperature $T_0 = 273.15 \, K$

$$V = V_0 [1 + \beta_V \Delta T] \qquad (5.9)$$

where $\Delta T = T - T_0$. In order to compare the ideal gas theory with experiment, we need to rearrange $PV = zRT$ into a form similar to Equation 5.9. At temperatures T and T_0 the volumes of z moles of gas are V and V_0, respectively, i.e.

$$V = \frac{zR}{P_0} T \qquad V_0 = \frac{zR}{P_0} T_0 \qquad (5.10)$$

Note that both measurements are made at the same pressure P_0. Taking the ratio of the two volumes

$$\frac{V}{V_0} = \frac{(zR/P_0)T}{(zR/P_0)T_0} = \frac{T}{T_0} \qquad (5.11)$$

and simplifying, we predict that

$$V = V_0 \frac{T}{T_0} = \left(\frac{V_0}{T_0}\right) T \qquad (5.12)$$

i.e. the volume is directly proportional to the absolute temperature T with a constant of proportionality V_0/T_0. Comparing the right-hand sides of Equations 5.6 and 5.12, we have a prediction that

$$1 + \beta_V (T - T_0) = \frac{T}{T_0} \qquad (5.13)$$

Rearranging this to find β_V

$$\beta_V (T - T_0) = \frac{T}{T_0} - 1 = \frac{T - T_0}{T_0} \qquad (5.14)$$

and cancelling $(T - T_0)$ yields

$$\boxed{\beta_V = \frac{1}{T_0}} \qquad (5.15)$$

So the ideal gas theory predicts that all gases should have the same value of β_V. Since for the data in Table 5.4 $T_0 = 0 \, °C = 273.15 \, K$ we expect that

$$\beta_V = \frac{1}{273.15} = 3.6610 \times 10^{-3} \, K^{-1} \qquad (5.16)$$

Table 5.5 Comparison of experimental and theoretical expansivities of gases. See also Table 5.4.

Gas	β_V (K^{-1})	% difference between theory and experiment
Helium, He	3.6580×10^{-3}	-0.082
Hydrogen, H$_2$	3.6588×10^{-3}	-0.060
Nitrogen, N$_2$	3.6735×10^{-3}	$+0.342$
Air	3.6728×10^{-3}	$+0.323$
Neon, Ne	3.6600×10^{-3}	-0.027

which gives values that compare well with the experimental values (see Table 5.5).

As Table 5.5 shows, the prediction obtained with Equation 5.16 is accurate at a level better than 0.1% for some gases, and better than 1% even for complex mixed gases such as air, the values being impressively close for such a simple theory.

An alternative interpretation of these data is to see them as a way of *defining* the reference temperature T_0 in Equation 5.15. Given the "size" of a degree of temperature, the expansivity of gases provides a way of determining the absolute value of any chosen reference temperature.

Whichever interpretation of the results is chosen, the consistency of the results between different gases is striking. This indicates the general validity of the assumptions of the ideal gas theory outlined in the statements below Equation 4.1:

- the molecules behave as perfect point masses, i.e. they have negligible volume;
- the molecules do not interact with each other except momentarily as they collide; and
- the collisions between molecules are elastic.

Turning to the second question raised by examination of the data, let us consider why the expansivities of gases are considerably greater than those of solids and liquids. We can understand this by noting the significance of the second assumption, i.e. that the molecules interact with each other only instantaneously as they collide. This amounts to assuming that the internal energy of a gas is held entirely by the kinetic energy of its molecules, neglecting the effect of the potential energy of interaction between molecules. The extent of the agreement between ideal gas theory and experiment arises because this neglect of the potential energy is valid.

However, the average kinetic energy of a molecule is proportional to the temperature (the average

KE $= 1.5 k_B T$, Eq. 4.24) so the neglect of potential energy must break down at low enough temperatures. As the temperature is lowered, the molecules will tend to appear "sticky" and form short-lived clusters. Eventually, at low enough temperatures, they coalesce into a liquid or solid. What constitues a "low enough" temperature depends on the strength of the interactions between the gas molecules.

From the fact that solids and liquids form at low temperatures we can see that the interactions between molecules generally act to reduce the separation between molecules, and hence reduce the volume of the gas. The high expansivity of gases as compared with solids and liquids can now be understood; it is the high temperature limit of the expansivity of all materials. When the temperature of the gas is high enough that we can neglect the molecular interactions that tend to restrict the volume of the gas, then the expansivity of real gases approaches that predicted for an ideal gas.

Analysis of the pressure coefficient of expansivity is left to the reader as Exercise P4 (see §5.9).

5.3 Heat capacity

The heat capacity of a gas is defined as the limiting value of the ratio of the heat energy input ΔQ to the resulting temperature rise ΔT

$$C = \frac{\Delta Q}{\Delta T}_{\lim \Delta T \to 0} = \frac{dQ}{dT} \text{ J K}^{-1} \qquad (5.17)$$

The value obtained for this ratio depends on the conditions under which the heat input is made. The two principal measurement conditions are:

- *constant pressure*, in which the heat input can cause an expansion of the gas; and
- *constant volume*, in which the heat input can cause a rise in pressure of the gas.

The heat capacity associated with measurements at constant pressure is referred to as C_P and that associated with measurements at constant volume is referred to as C_V. The heat capacity of a substance is usually quoted either for a given *mass* of material, the *specific heat capacity* (e.g. J K^{-1}kg^{-1}), or per *mole*, the *molar heat capacity* (J K^{-1}mol^{-1}).

Instead of examining data on C_V and C_P independently, we will look first at data for C_P in §5.3.1 and then at data for γ, (the ratio of C_P to C_V) in §5.3.2.

5.3.1 The heat capacity at constant pressure C_P data

Tables 5.6 and 5.7 show the molar heat capacity C_P at several temperatures for a variety of monatomic and diatomic gases, respectively. Figures 5.3 and 5.4 show graphs of the data given in Tables 5.6 and 5.7.

Perhaps the most striking feature of the data is the constancy of C_P at a value of $20.8\,\mathrm{J\,K^{-1}\,mol^{-1}}$ for all the monatomic gases shown, independent of temperature. In contrast, the diatomic gases have a rather

larger heat capacity at low temperatures, which increases still further as the temperature is increased, tending to a maximum of around $37\,\mathrm{J\,K^{-1}\,mol^{-1}}$.

Example 5.4

What is the specific heat capacity of helium gas?

One mole of helium gas has a mass of $4 \times 10^{-3}\,\mathrm{kg}$ and the value of its the molar heat capacity at constant pressure C_P is $20.8\,\mathrm{J\,K\,mol^{-1}}$ (Table 5.6 below). Thus it takes $20.8\,\mathrm{J}$ to raise the temperature of $4 \times 10^{-3}\,\mathrm{kg}$ of helium by $1°C$.

$1\,\mathrm{kg}$ of helium $= 1/(4 \times 10^{-3}) = 250\,\mathrm{mol}$ of helium.
The *specific* heat capacity at constant pressure is therefore
$$250 \times 20.8 = 5200\,\mathrm{J\,K^{-1}\,kg^{-1}}.$$

Figure 5.3 Graph of the heat capacity of monatomic gases C_P versus absolute temperature (see Table 5.6). Data corresponding to temperatures less than the boiling point have not been plotted. Note that at the resolution of the measurement, all the gases have the same heat capacity across a temperature variation of over one order of magnitude.

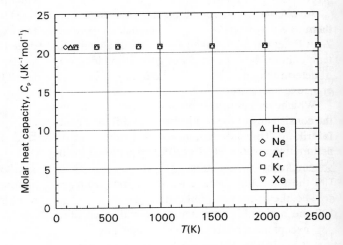

Figure 5.4 Graph of the heat capacity C_P of diatomic gases versus absolute temperature (see Table 5.7). Data points from below the boiling point have not been plotted. Note that the heat capacities increase with temperature. The lines through the data points are drawn to guide the eye.

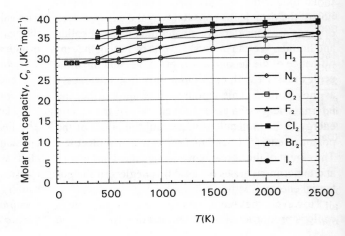

Table 5.6 The molar heat capacities at constant pressure C_p, ($JK^{-1}mol^{-1}$) for the monatomic noble gases (see also Fig. 5.3).

T (K)	He	Ne	Ar	Kr	Xe
50	–	–	24.8[a]	25.1	25.1
100	–	–	20.8	31.6[a]	28.2
150	–	–	20.8	20.8	33.6[a]
200	–	–	20.8	20.8	20.8
298.15[b]	20.786[b]	20.786[b]	20.786[b]	20.786[b]	20.786[b]
400	20.8	20.8	20.8	20.8	20.8
600	20.8	20.8	20.8	20.8	20.8
800	20.8	20.8	20.8	20.8	20.8
1000	20.8	20.8	20.8	20.8	20.8
1500	20.8	20.8	20.8	20.8	20.8
2000	20.8	20.8	20.8	20.8	20.8
2500	20.8	20.8	20.8	20.8	20.8

a. Approximate temperature of the melting and boiling points. For each gas the boiling temperature and melting temperature are separated by less than 5 K. (See Chapter 11 for data on the boiling and melting points of the elements.)
b. These data are from a different source from the rest of the table. Note that the extra measurement resolution still shows agreement between the heat capacities of the different gases.
(Data from Kaye & Laby; see §1.4.1)

Table 5.7 The molar heat capacities at constant pressure C_p, ($JK^{-1}mol^{-1}$) for some diatomic gases (see also Fig. 5.4).

T(K)	H_2	O_2	N_2	F_2	Cl_2	Br_2	I_2
50	–	46.1[a]	41.5[a]	–	29.2	33.3	35.8
100	–	29.1	29.1	–	42.3	43.6	45.6
150	–	29.1	29.1	–	51.0	49.2	49.6
200	–	29.1	29.1	–	54.2[a]	53.8[a]	51.5
400	29.2	30.1	29.2	33.0	35.3	36.7	80.3[a]
600	29.3	32.1	30.1	35.2	36.6	37.3	37.6
800	29.6	33.7	31.4	36.3	37.2	37.5	37.8
1000	30.2	34.9	32.7	37.0	37.5	37.7	37.9
1500	32.3	36.6	34.9	37.9	38.0	38.0	38.2
2000	34.3	37.8	36.0	38.4	38.3	38.2	38.5
2500	36.0	38.9	36.0	38.8	38.6	38.5	38.8

a. The approximate temperatures of the melting and boiling points. (See Chapter 11 for data on the boiling and melting points of the elements.)
(Data from Kaye & Laby; see §1.4.1)

Thus for either monatomic or diatomic gases, the molar capacity lies in the range $30 \pm 10\,JK^{-1}mol^{-1}$ at all temperatures, which forms a useful ballpark estimate for the heat capacity of a gas.

The main questions raised by our initial examination of the experimental data on C_p are:

- Why are the constant pressure heat capacities of monatomic or diatomic gases usually in the range $30 \pm 10\,JK^{-1}mol^{-1}$?
- Why, for monatomic gases, does C_p equal $20.8\,JK^{-1}mol^{-1}$ with almost no dependence of the heat capacity on temperature or on the type of molecule from which the gas is formed?
- Why do some diatomic gases have a low-temperature value of C_p which is somewhat greater than the value for monatomic gases and which increases with increasing temperature?

We will consider these questions in §5.3.3 after we have considered data on the ratio of the heat capacities $\gamma = C_p/C_V$.

5.3.2 The ratio of heat capacities $\gamma = C_p/C_V$ data

The data given in Tables 5.6 and 5.7 above refer to the heat capacity at constant pressure, C_p, i.e. the gas is allowed to expand as the heat input is made. C_p differs significantly from the heat capacity at constant volume, C_V. It is found experimentally that the ratio of C_p to C_V – known as γ (pronounced "gamma") – varies with the number of atoms in a molecule. Values of γ classified according to the number of atoms in a molecule of the gas are listed in Tables 5.8a–e.

The tables are useful for reference, but the histogram shown in Figure 5.5 of all the data from Table 5.8 indicates more clearly which values correspond to which type of molecule. It is clear from the histogram that the data cluster according to the number of atoms in a molecule of the gas. Monatomic and diatomic gases have γ values of around 1.66 and 1.4, respectively, and tri- and polyatomic gases have widely scattered smaller values, but always greater than 1. Interestingly, it seems that for air at high pressures, the values for γ are strongly increased, even beyond the values for a monatomic gas.

So the main questions raised by our initial examination of the experimental data for γ are

- Why is the heat capacity at constant pressure always greater than the heat capacity at constant volume, i.e. $\gamma > 1$.
- Why does the value of γ depend on whether the gas is monatomic, diatomic or polyatomic? In

Example 5.5

Samples of four gases are heated from room temperature (20°C or 293 K) to 100°C (373 K) in a piston device at a constant pressure of ≈ 0.1 MPa. Sample A consists of 1 kg of helium; B, 1 kg of xenon; C, 1 kg of nitrogen; D, 1 kg of an unknown monatomic gas.
(a) Work out how much heat energy is required to raise samples A, B and C to the required temperature.
(b) Sample D requires 41.65×10^3 J to raise it to 100°C. From the data given in Table 5.6, work out which gas is being heated.

The heating takes place at constant pressure therefore one uses C_P to relate the heat input to the temperature rise.

(a)
Sample A:
The heat required to raise 1 mol of gas from T_1 to T_2 is given by

$$Q = \int_{T_1}^{T_2} C_P dT$$

Consulting Table 5.6 and Figure 5.3, we see that for helium C_P is constant over the range of heating, and therefore the energy required to heat 1 mol of gas from T_1 to T_2 is given by

$$Q = C_P \int_{T_1}^{T_2} dT$$

Similarly, for z moles of gas we have

$$Q_{total} = z C_P \int_{T_1}^{T_2} dT = z C_P [T_2 - T_1]$$

1 kg of helium, molecular mass 4 contains $1/0.004 = 250$ mol of helium. Substituting for z, C_P T_2 and T_1, we have

$$Q_{total} = 250 \times 20.8 [373 - 293]$$
$$= 4.16 \times 10^5 \text{ J}$$

Sample B:
Similarly to sample A, we find $Q_{total} = z C_P (T_2 - T_1)$. Now, however, we have 1 kg of xenon, which has a molecular mass of 131.3, and hence we have only

$$z = \frac{1}{0.1313} = 7.62 \text{ mol}$$

Hence it takes only

$$Q_{total} = 7.62 \times 20.8 [373 - 293]$$
$$= 12.67 \times 10^3 \text{ J}$$

to heat the same *mass* of gas through the same temperature

rise. The reason for this somewhat counter-intuitive result is related to the fact that we have many fewer molecules in sample B than in sample A.

Sample C:
We proceed exactly as for sample A, except that now there is a possibility that C_P will change over the range of heating because nitrogen gas consists of *diatomic* (N_2) molecules. However, detailed consultation of Table 5.7 shows that the variation from 200 K to 400 K is less than 1% and so may be assumed constant over our more restricted range.

We have 1 kg of N_2, which has a molecular mass of $2 \times 14 = 28$. Note that it is the mass of the *molecule* that is important. Hence we have

$$z = \frac{1}{0.028} = 35.71 \text{ mol}$$

which takes

$$Q_{total} = 35.71 \times 20.8 [373 - 293]$$
$$= 59.42 \times 10^3 \text{ J}$$

(b)

Sample *D* requires 41.65×10^3 J to raise it to 100°C. We can use the formula used for sample A to work out the number of moles of gas present:

$$Q_{total} = z C_P (T_2 - T_1)$$

We know C_P because for all monatomic gases $C_P = 20.8$ J K^{-1} mol^{-1} and hence we can work out how many moles are contained in 1 kg of the unknown gas. Thus we can work out the relative molecular mass of gas D and use this to identify it. We find

$$z = \frac{Q_{total}}{C_P [T_2 - T_1]} = 25.03 \text{ mol}$$

Thus the relative molecular mass is given by

$$M = \frac{1}{25.03 \times 10^{-3}} = 39.95$$

which identifies the monatomic gas unambiguously as argon.

Table 5.8 The ratio of the principal heat capacities ($\gamma = C_P/C_V$).

Gas	T (°C)	T (K)	γ	Gas	T (°C)	T (K)	γ
A Some monatomic gases				*D Some triatomic gases*			
He	0.0	273.20	1.630	O_3	–	–	1.290
Ar	0.0	273.20	1.667	H_2O	100.0	373.20	1.334
Ne	19.0	292.20	1.642	CO_2	10.0	283.20	1.300
Kr	19.0	292.20	1.689	CO_2	300.0	573.20	1.220
Xe	19.0	292.20	1.666	CO_2	500.0	773.20	1.200
Hg	310.0	583.20	1.666	NH_3	–	–	1.336
(All γ close to 1.66)				N_2O	–	–	1.324
				H_2S	–	–	1.340
				CS_2	–	–	1.239
B Some diatomic gases				SO_2	20.0	293.20	1.260
H_2	10.0	283.20	1.407	SO_2	500.0	773.20	1.200
N_2	20.0	293.20	1.401	(All γ close to 1.30; γ tends to be smaller for measure-			
O_2	10.0	283.20	1.400	ments at higher temperatures (see CO_2 & SO_2))			
CO	1800.0	2073.2	1.297				
NO	–	–	1.394				
(All γ close to 1.40, except for CO, which was taken				*E Some polyatomic gases*			
at very high temperature)				CH_4	–	–	1.313
				C_2H_6	–	–	1.220
				C_3H_8	–	–	1.130
C Air (γ as a function of temperature and pressure)				C_2H_2	–	–	1.260
Air	–79.3	193.90	1.405	C_2H_4	–	–	1.264
Air	10.0	283.20	1.401	C_6H_6	20.0	293.20	1.400
Air	500.0	773.20	1.357	C_6H_6	99.7	372.90	1.105
Air	900.0	1173.2	1.320	$CHCl_3$	30.0	303.20	1.110
Air[a]	0.0	273.20	1.828	$CHCl_3$	99.8	373.00	1.150
Air[a,b]	–79.3	193.90	2.333	CCl_4	–	–	1.130
(All γ close to 1.40 as for the diatomic gases, but γ				(Spread of values between 1.1 and 1.4; γ tends to be			
falls at higher temperatures and increases at higher				smaller for measurements at higher temperatures but the			
pressures)				$CHCl_3$ results represents an exception to this general			
				rule)			

a. measurements made at a pressure of 200 atmos (20 MPa); b. this value represents a state where the gas is compressed close to the point of becoming a liquid.
(Data from Kaye & Laby; see §1.4.1)

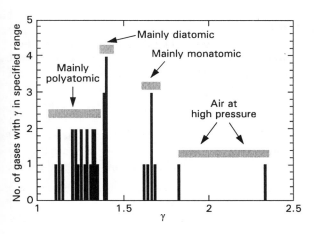

Figure 5.5 Histogram showing the distribution of values of γ in Tables 5.8a–e. The height of the columns indicates how common it is for a gas to have γ within ±0.01 of the value specified on the x axis. Note how the data are clustered into groups that can be identified as either monatomic, diatomic or polyatomic molecules, or data taken at high pressures.

Example 5.6

For helium gas, Table 5.8 gives $\gamma = 1.63$ and Table 5.6 gives $C_P = 20.8\,\mathrm{J\,K^{-1}\,mol^{-1}}$ over a wide temperature range. What is C_V for helium?

Since $\gamma = C_P/C_V$ we can calculate C_V as

$$C_V = \frac{C_P}{\gamma} = \frac{20.8}{1.63} = 12.8\,\mathrm{J\,K^{-1}\,mol^{-1}}$$

---

particular, we would like to understand why:

(i) monatomic gases have values for γ close to 1.66 ($\pm\,0.02$);

(ii) the diatomic gases listed in Table 5.8b have values for γ close to 1.4 (± 0.1), except at high pressure; and

(iii) polyatomic gases have lower values of γ than either monatomic or diatomic gases.

- Why is γ for air at extremely high pressures

Example 5.7

What is *difference* between C_P and C_V for (a) helium and (b) nitrogen, i.e. how much extra energy is required to raise 1 mole of gas through 1 K if the process takes place at *constant pressure*, as compared with the same process at *constant volume*?

From the example above we find C_V as

$$C_V = \frac{C_P}{\gamma}$$

So the difference between C_P and C_V is $(C_P - C_V)$, which is given by

$$(C_P - C_V) = C_P - \frac{C_P}{\gamma} = C_P\left[1 - \frac{1}{\gamma}\right]$$

(a) Using Tables 5.6 and 5.8, $C_P = 20.8\,\mathrm{J\,K^{-1}\,mol^{-1}}$ and $\gamma = 1.63$ for helium, so

$$(C_P - C_V) = 20.8\left[1 - \frac{1}{1.63}\right] = 8.04\,\mathrm{J\,K^{-1}\,mol^{-1}}$$

(b) Using Tables 4.6 and 4.7, $C_P = 29.2\,\mathrm{J\,K^{-1}\,mol^{-1}}$ at around room temperature and $\gamma = 1.401$ for nitrogen, so

$$(C_P - C_V) = 29.2\left[1 - \frac{1}{1.401}\right] = 8.36\,\mathrm{J\,K^{-1}\,mol^{-1}}$$

---

increased to values above those for monatomic gases?

5.3.3 Understanding the data on the heat capacity of gases

In this section we will attempt to understand the questions raised by the data from §5.3.1 on C_P and from §5.3.2 on γ. In order to do this we will develop the theory of an ideal gas further to derive expressions first for C_V, then for C_P and hence for their ratio, γ.

Heat capacity: background theory

Rearranging the First Law of Thermodynamics, $\Delta U = \Delta Q + \Delta W$ (Eq. 2.55), to

$$\Delta Q = \Delta U - \Delta W \qquad (5.18)$$

tells us that when heat ΔQ is supplied to a substance, it may either increase the internal energy ΔU of the substance, or cause the substance to do ΔW of work on its environment. Note that ΔW is defined in Equations 2.65 and 5.18 as the work done *on* the gas by its environment. Thus if the gas does net work on its environment ΔW will be negative. We will develop our theory of the heat capacity of a gas by

- deriving an expression for the work done by an expanding gas,
- deriving an expression for the heat capacity when no work is done by the gas, and
- combining the first two steps to derive an expression for the heat capacity of a gas when it does work on its environment.

Work done by an expanding gas

First of all we recall that, in general, work is done when a force moves its point of action. Thus, in order to do work, a gas must change its volume. Figure 5.6 illustrates a hypothetical experiment that will allow us to calculate the work done during a volume change.

If the piston apparatus shown in Figure 5.6 has an area A and the pressure of the gas is P, then the total force on the piston is $F = PA$. Now, suppose the piston moves an infinitessimal amount dx, then the work done is

$$\mathrm{d}W = F\,\mathrm{d}x = PA\,\mathrm{d}x$$

$$(5.19)$$

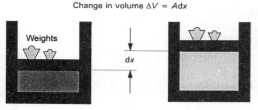

Change in volume $\Delta V = A\,dx$

Weights

dx

Before heat input   After heat input

Figure 5.6 The input of heat into a gas under conditions of constant pressure. The figure shows gas trapped in a thermally insulated piston device. The piston is free to move so that it will finally stop when the net force on the piston is zero.

increase dV and so the work done is given by

$$dW = P\,dV \qquad (5.20)$$

Integrating to yield the work done in a finite expansion yields

$$\Delta W = P \int_{V}^{V+\Delta V} dV \qquad (5.21)$$

In general, P will change as the gas expands, but if we consider the case when heat is added in order that the pressure remains constant during the expansion, then P can slip outside the integral to yield

$$\Delta W = P \int_{V}^{V+\Delta V} dV = P[V]_{V}^{V+\Delta V} = P\Delta V \qquad (5.22)$$

Using the ideal gas equation $PV = zRT$, we may eliminate $P\Delta V$ in favour of $zR\Delta T$, where ΔT is the temperature change required to keep the pressure constant during the expansion. Hence, if z mole of ideal gas expand at constant pressure, the work done is

$$\boxed{\Delta W = P\Delta V = zR\Delta T} \qquad (5.23)$$

work done *by* gas

Note that if the volume of the gas is kept constant, no work is done. Also, take care with the sign of ΔW when used in Equation 5.18 where ΔW is defined as the work done *on* the gas by its environment, i.e. the negative of Equation 5.23.

Heat capacity at constant volume

If the heat is supplied to a gas at constant volume then from Equation 5.23, the gas does no work on its environment. Thus the first law of thermodynamics tells us that

$$\Delta Q = \Delta U \qquad (5.24)$$

Under these conditions the heat energy input goes entirely into increasing the average energy of the molecules and hence, by Equation 4.19, the internal energy, U, of the gas. Recall that the internal energy of an ideal gas is just the sum of the kinetic energies of the individual molecules, each of which has just three degrees of freedom (§4.3, Step 6 and Complication 3). Generalizing our calculation to a gas whose molecules have p degrees of freedom each with $\frac{1}{2}k_B T$ of energy, the internal energy of z moles of gas, i.e. zN_A molecules, will be

$$U = zN_A \times p\tfrac{1}{2}k_B T = \tfrac{1}{2}zpN_A k_B T \qquad (5.25)$$

Using $N_A k_B = R$, this reduces to

$$U_T = \tfrac{1}{2}zpRT \qquad (5.26)$$

so the internal energy of the gas at $T + \Delta T$ will be given by

$$U_{T+\Delta T} = \tfrac{1}{2}zpR(T + \Delta T) \qquad (5.27)$$

and so the change in internal energy ΔU on heating from T to $T + \Delta T$ is

$$\Delta U = U_{T+\Delta T} - U_T = \tfrac{1}{2}zpR\Delta T \qquad (5.28)$$

By Equation 5.24 this internal energy must be supplied by the heat ΔQ, and so we may write

$$\Delta Q = \tfrac{1}{2}zpR\Delta T \qquad (5.29)$$

and so the heat capacity at constant volume $C_V = \Delta Q/\Delta T$ of z moles of gas is given by

$$C_V = \frac{\Delta Q}{\Delta T} = \tfrac{1}{2}zpR \qquad (5.30)$$

We thus predict that the molar (i.e. $z = 1$) constant volume heat capacity of a gas will be

$$C_V = 4.157p \ \mathrm{J\,K^{-1}\,mol^{-1}} \qquad (5.31)$$

in general, and

$$C_V = 12.472 \text{ J K}^{-1} \text{ mol}^{-1} \qquad (5.32)$$

for $p = 3$.

These are simple results. They indicate that the measurement of C_V allows the determination of the number of internal degrees of freedom of the molecules of the gas.

Heat capacity at constant pressure

In the previous two sections, we have developed an expression for the change in internal energy of a gas whose molecules have p degrees of freedom

$$\Delta U = \tfrac{1}{2} z p R \, \Delta T \qquad (5.33)$$

and for the work done by such a gas in expanding at constant pressure

$$\Delta W = P \Delta V = z R \, \Delta T \qquad (5.34)$$

Taking note of the sign of ΔW as discussed following Equation 5.18, we may substitute these expressions into the first law of thermodynamics, $\Delta Q = \Delta U - \Delta W$, to yield

$$\begin{aligned} \Delta Q &= \tfrac{1}{2} z p R \, \Delta T - \left(-z R \, \Delta T\right) \\ &= \left(\tfrac{1}{2} z p R + z R\right) \Delta T \end{aligned} \qquad (5.35)$$

Remembering that $C_P = \Delta Q / \Delta T$, we find

$$C_P = z R \left(1 + \frac{p}{2}\right) \qquad (5.36)$$

We thus predict that the molar (i.e. $z = 1$) constant pressure heat capacity of a gas will be

$$C_P = 8.314 \times \left(1 + \frac{p}{2}\right) \text{J K}^{-1} \text{ mol}^{-1} \qquad (5.37)$$

in general, and

$$C_P = 20.786 \text{ J K}^{-1} \text{ mol}^{-1} \qquad (5.38)$$

for $p = 3$.

Gamma (γ)

Combining Equation 5.30 for C_V

$$\boxed{C_V = \tfrac{1}{2} z p R} \qquad (5.39)$$

and Equation 5.36 for C_P

$$\boxed{C_P = z R \left(1 + \frac{p}{2}\right)} \qquad (5.40)$$

yields an expression for $\gamma = C_P / C_V$

$$\gamma = \frac{z R \left(1 + p/2\right)}{\tfrac{1}{2} z p R} \qquad (5.41)$$

which simplifies to

$$\boxed{\gamma = 1 + \frac{2}{p}} \qquad (5.42)$$

Comparison with experiment

We are now in a position to compare the theoretical expressions (Eqs 5.39–5.42) with experiment. We will do this by considering in turn the points raised at the end of §5.3.2 and §5.3.3.

Why are the constant pressure heat capacities of monatomic or diatomic gases usually in the range $30 \pm 10 \, J K^{-1} mol^{-1}$?

We can see from Equation 5.40 for C_P that we expect heat capacity at constant pressure to vary with the number of degrees of freedom of the molecules of the gas. We can rearrange Equation 5.40 as an expression for p

$$p = 2\left(\frac{C_P}{R} - 1\right) \qquad (5.44)$$

in terms of the molar heat capacity. Now we interpret the range of values of C_P as being due to different numbers of internal degrees of freedom of the molecules of the gas. Substituting $C_P \approx 30 \pm 10 \, \text{J K}^{-1} \text{mol}^{-1}$ yields an estimate for $p \approx 5.2 \pm 2.4$ (or, more plausibly, $3 < p < 8$) for the typical number of degrees of freedom of the molecules of gases examined in §5.3.1.

Why, for monatomic gases, does C_P equal 20.8 J K^{-1} mol^{-1} with almost no dependence of the heat capacity on temperature or on the type of molecule from which the gas is formed?

For monatomic gases we expect molecules to have no internal degrees of freedom, and so there should be just three degrees of freedom per molecule corresponding to the kinetic energy of molecule. Using $p = 3$ in Equation 5.40 predicts $C_P = 20.786$ JK^{-1} mol^{-1} and this value compares favourably (Fig. 5.7) with the data for monatomic gases given in Table 5.6 and Figure 5.3. We may take this as an indication that the internal energy of a monatomic gas really is held in the kinetic energy of its constituent atoms, *and in no other way*.

Why do some diatomic gases have a low-temperature value of C_P which is rather greater than the value for monatomic gases and which increases with increasing temperature?

Using Equation 5.43

$$p = 2\left(\frac{C_P}{R} - 1\right) \qquad (5.44)$$

we see that we can understand these results by supposing that diatomic molecules possess extra degrees of freedom. Furthermore, it appears that not only are there more degrees of freedom, but also the number of degrees of freedom required to explain the data becomes larger as the temperature increases. The types of *internal molecular motion* associated with these extra degrees of freedom are discussed below. Here we note that the diatomic gases in our sample appear to have a heat capacity corresponding to around five degrees of freedom at low temperatures, increasing to around seven or eight at higher temperatures.

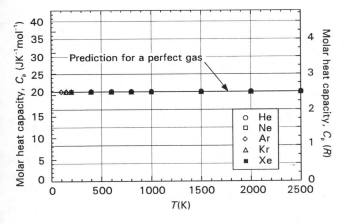

Figure 5.7 Comparison of the prediction of perfect gas theory with experimental results for monatomic gases. R is the universal gas constant.

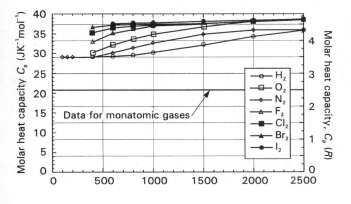

Figure 5.8 The experimental results for diatomic gases compared with the predictions for monatomic gases (see Fig. 5.7). The data exceed the prediction for monatomic gases by at least R (the universal gas constant) at low temperatures, increasing to around $2R$ to $2.5R$ at high temperatures.

We will discuss the diatomic gas data more completely in the following section on γ.

Gamma (γ)

Why is the heat capacity at constant pressure always greater than the heat capacity at constant volume, i.e. $\gamma > 1$?

We can see immediately why this must be so. The first law of thermodynamics tells us that

$$\Delta Q = \Delta U + \Delta W \qquad (5.45)$$

In order to achieve a given change in temperature, ΔT, we need to increase the internal energy of the gas by $\Delta U = \frac{1}{2}pR\,\Delta T$. So in order to raise the temperature of the gas by ΔT, the minimum energy required is $\Delta Q = \Delta U$, which occurs when $\Delta W = 0$, i.e. when the gas is held at fixed volume. If the gas is allowed to expand during heating, then additional energy must be supplied to do the work, ΔW, involved in the expansion. Thus C_P must always be greater than C_V, and hence γ must always be greater than 1.

Why does the value of γ depend on whether the gas is monatomic, diatomic or polyatomic?

In particular we would like to understand why:
- monatomic gases have values for γ close to 1.66 (± 0.02);
- all the diatomic gases listed in Table 5.8b have values for γ close to 1.4 (± 0.1), except at high pressure; and
- polyatomic gases have lower values of γ than either monatomic or diatomic gases.

From Equation 5.42, we see that we expect γ to depend directly on the number of degrees of freedom per molecule, p

$$\gamma = 1 + \frac{2}{p} \qquad (5.42)*$$

Rearranging, we obtain an expression for the number of degrees of freedom per molecule in terms of γ

Table 5.9 Experimental values of γ for gases. The number of degrees of freedom p is predicted from the measured value of γ according to Equation 5.47 (see text and Table 5.8).

Gas	γ	T (K)	p	Gas	γ	T (K)	p
A Monatomic gases; the expected value for a monatomic gas is 3.00				*D Triatomic gases*			
				O_3	1.290	–	6.90
He	1.630	273.20	3.17	H_2O	1.334	373.20	5.99
Ar	1.667	273.20	3.00	CO_2	1.300	283.20	6.67
Ne	1.642	292.20	3.12	CO_2	1.220	573.20	9.09
Kr	1.689	292.20	2.90	CO_2	1.200	773.20	10.00
Xe	1.666	292.20	3.00	NH_3	1.336	–	5.95
Hg	1.666	583.20	3.00	N_2O	1.324	–	6.17
				H_2S	1.340	–	5.88
B Diatomic gases				CS_2	1.239	–	8.37
H_2	1.407	283.20	4.91	SO_2	1.260	293.20	7.69
N_2	1.401	293.20	4.99	SO_2	1.200	773.20	10.00
O_2	1.400	283.20	5.00				
CO	1.297	2073.2	6.73	*E Polyatomic gases*			
NO	1.394	–	5.08	CH_4	1.313	–	6.39
				C_2H_6	1.220	–	9.09
C Air; whose molecules are mostly diatomic				C_3H_8	1.130	–	15.4
				C_2H_2	1.260	–	7.69
Air	1.405	193.90	4.94	C_2H_4	1.264	–	7.58
Air	1.401	283.20	4.99	C_6H_6	1.400	293.20	5.00
Air	1.357	773.20	5.60	C_6H_6	1.105	372.90	19.0
Air	1.320	1173.2	6.25	$CHCl_3$	1.110	303.20	18.2
Air[a]	1.828	273.20	2.42	$CHCl_3$	1.150	373.00	13.3
Air[a]	2.333	193.90	1.50	CCl_4	1.130	–	15.4

a. Pressure = 200 atm.

$$p = \frac{2}{\gamma - 1} \qquad (5.46)$$

Table 5.8 shows that, experimentally, monatomic gases have values of $\gamma = 1.66 \pm 0.02$, which, using Equation 5.46 corresponds to $p \approx 3$ (within ± 0.1), as expected. Using Equation 5.46 to interpret the γ values of all the gases listed in Table 5.8 yields the results tabulated in Table 5.9.

We see that we can plausibly explain the data for γ in terms of quite a reasonable number of degrees of freedom per molecule. We now need to understand the way in which the values of p vary in Table 5.9. In particular, we need to understand why p varies with:

- the different molecules constituting the gas;
- temperature; and
- pressure.

All molecules in a gas possess the three degrees of freedom associated with their kinetic energy in each of the x, y and z directions (Fig. 5.9).

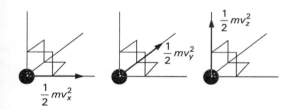

Figure 5.9 The three degrees of freedom of a gas molecule associated with the kinetic energy in the x, y and z directions.

Rotational degrees of freedom

More complicated molecules also possess degrees of freedom associated with *rotation* about each of the x, y and z axes. The energy associated with rotations differs depending on the axis of rotation. The rotation of a diatomic molecule is illustrated in Figure 5.10. Rotation about the y and z axes is equally likely, but rotation about the x axis, the axis through the centres of the two atoms, is much less likely. This is because the energy separation ΔE between the non-rotating quantum state of the molecule and the rotating quantum state is much greater for rotations about the x axis than for rotations about the y or z axes. The idea that the energy ΔE separating non-rotating states

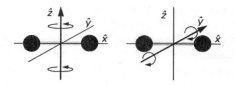

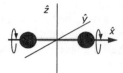

Figure 5.10 The three degrees of freedom of a gas molecule associated with rotation about the x, y and z axes.

from rotating states is anything other than zero can only be understood through consideration of the quantum mechanics of molecular motion. Classically, we could put as little energy as we chose into the rotation of a molecule; there was no lower limit on the rotational speed of a molecule. Quantum mechanically, there is a lower limit to the energy tied up in rotation given by $\Delta E = \hbar \omega$, where ω is the minimum angular frequency of rotation.

Vibrational degrees of freedom

The rotational degrees of freedom refer to *rigid* rotation of the molecule i.e. motion in which there is no relative motion of the component atoms of the molecule. In general however, the atoms within molecules have several *modes of vibration*. For example, the relative separation of the two atoms in a diatomic molecule can vary, and the atoms will then oscillate about their average positions (Fig. 5.11). The atoms then posses both *kinetic energy of vibration* and *potential energy of vibration* corresponding to a further two degrees of freedom.

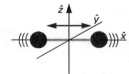

Figure 5.11 The degree of freedom of a gas molecule associated with vibration.

Diatomic molecules

For diatomic molecules we can understand how a molecule can have:

- three translational degrees of freedom,

- three rotational degrees of freedom, and
- two vibrational degrees of freedom,

making a total of eight degrees of freedom in all. Using Equation 5.40, $C_P = nR(1+p/2)$ with $p = 8$ predicts a *maximum* value of C_P of $5R = 41.6 \, \text{JK}^{-1}\text{mol}^{-1}$, which is just greater than the largest value in Figure 5.4. However, C_P takes this value only at high temperatures. We can understand this as being due to the fact that some types of motion do not take place at low temperatures. Or, in the language of statistical mechanics, some *degrees of freedom* are *inaccessible* at low temperatures (see §2.5 and Table 5.10).

Tri- and polyatomic molecules

In general, we expect the number of degrees of freedom to increase as the number of atoms in a molecule increases, since more types of motion will become available. This tendency is clearly borne out by the data given in Table 5.9. Diatomic molecules have

more degrees of freedom than monatomic molecules, and polyatomic molecules have more degrees of freedom than diatomic molecules. As discussed below, the actual number observed depends on the way in which the atoms within the molecule are bonded together, and on the temperature.

Variation of γ with temperature

As the temperature of the gas is increased then, by definition, the average energy per degree of freedom of the gas molecules is increased. In order to assess whether a particular degree of freedom is *accessible* or not, we need to compare $k_B T$ with ΔE, the energy difference between the quantum states of the molecule. There are three general situations that are summarized in Table 5.10 below.

The data for oxygen (O_2) (Figs 5.4 and 5.12) most clearly indicate the gradual transition in which quantum states are inaccessible at low temperature, but

Table 5.10 An example of the use of the term "accessibility" with regard to quantum states.

Inaccessible	Marginal accessiblity	Fully accessible
$k_B T \ll \Delta E$	$k_B T \approx \Delta E$	$k_B T \gg \Delta E$
e.g. $k_B T < 0.1 \Delta E$	e.g $0.1 \Delta E < k_B T < 1.5 \Delta E$	e.g $2k_B T > 1.5 \Delta E$
In this case only occasionally do molecules make transitions to the higher quantum state, and we can consider the degrees of freedom associated with these transitions to be inaccessible	In this case molecules make transitions to the higher quantum state, but detailed calculations are required to assess the extent to which the quantum state can be considered accessible	In this case molecules frequently make transitions to the higher quantum state, and we can consider the degrees of freedom associated with these transitions to be fully accessible
In colloquial terms, the process associated with transitions between quantum states occurs so infrequently that it may generally be ignored		In colloquial terms, the process associated with transitions between quantum states occurs so frequently that the quantum nature of the states may generallay be ignored

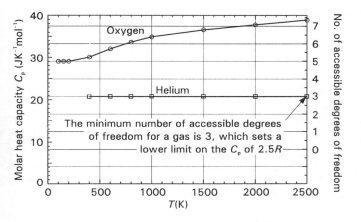

Figure 5.12 The experimental C_P results for helium and oxygen as a function of temperature (data extracted from Figs 5.3 and 5.4, respectively). The left-hand axis gives C_P in units of joules per kelvin while the right-hand axis shows the number of accessible degrees of freedom inferred from Equation 5.44.

become accessible at higher temperature. These data indicate that at around 2000 K oxygen molecules have acquired an extra two degrees of freedom as compared with the situation at 100 K. Based on the discussion above, we can tentatively assign these degrees of freedom to *either* the two easier modes of rotation, *or* the potential and kinetic energies of internal molecular vibration. Detailed calculations (beyond the scope of this book) indicate that the two extra degrees of freedom already accessible at low temperatures are due to molecular *rotation*. We thus infer that the two or three extra degrees of freedom becoming available over the temperature range up to 2000 K are associated with internal vibration of the molecule. From these data we can infer that the separation of the energy levels involved in these vibrational degrees of freedom is

$$\Delta E \approx k_B T \approx 1.38 \times 10^{-23} \times 2000 \, J$$
$$= 2.76 \times 10^{-20} \, J = 0.173 \, eV.$$

Using the formula for the frequency of vibration, f_0, of a simple harmonic oscillator, $\Delta E = h f_0$ where h is the Planck constant ($6.62 \times 10^{-34} \, Js$), this implies a vibrational frequency of $\approx 4.2 \times 10^{13} \, Hz$, a frequency in the infrared part of the electromagnetic spectrum.

So as more degrees of freedom become available, we expect that γ (given by Eq. 5.42 as $\gamma = 1 + 2/p$) should fall at higher temperatures, which is in broad agreement with the data.

Variation of γ with pressure
The data for air given in Table 5.9c indicate that γ increases with increasing pressure. Comparing the

datum for air at around atmospheric pressure and $T = 283.2 \, K$ and the datum at around 200 atm ($\approx 20 \, MPa$) and $T = 273.2 \, K$, we see that increasing the pressure has reduced the number of accessible degrees of freedom from a plausible 4.99 (i.e. 5), to a value of 2.42. This latter value is less than the three degrees of freedom that must be possessed by a gas whose molecules are free to move in three dimensions. How can we understand this?

Increasing the pressure of a gas at a fixed temperature requires increasing the number density of molecules and hence reducing the average separation between molecules. This increases the frequency of molecular collisions and interactions and these interactions restrict the motion of molecules, and so the associated degrees of freedom become inaccessible.

Consider the degrees of freedom associated with rotation. Because of its inertia, an oxygen molecule requires a certain time to rotate, typically $10^{-10} \, s$, and if the molecules, on average, collide before this rotation can be completed then the degree of freedom is restricted. Calculations of the extent of this restriction are complicated, and we will not go through them here. However, we note that if the pressure is increased sufficiently, the gas will condense into a liquid, or enter a state where the molecules are so close together that it is not clear whether it should be called a liquid or a gas. Gases compressed to this extent begin to show significant deviations from perfect gas behaviour, and the simple interpretation of γ according to Equation 5.42 is no longer realistic. Gases compressed sufficiently that they cannot be distinguished from a liquid are discussed in §10.7.1 on the critcal point.

Summary of heat capacity

Broadly speaking, we have been able to explain plausibly the behaviour of the heat capacity of gases on the basis of a relatively simple theory. The key feature of the theory is the assumption that the internal energy of the gas is "held" in the individual degrees of freedom of the gas molecules. In order to understand the temperature dependence, it was necessary to develop the idea of the *accessibility* of degrees of freedom. This requires acknowledgement of the quantum mechanical nature of the vibrations and rotations of gas molecules.

5.4 Thermal conductivity

Heat flow through fluids can be extremely complex. This is because, in addition to the normal flow of heat that takes place through any material, it is also possible for the fluid itself to move from one place to another. This process, known as *convection*, allows fluid to "carry" heat with it in the form of its heat capacity. Thus, for example, a fluid flow of z moles per second from part of a container that is ΔT hotter than a cold part, will result in a delivery of heat energy to the colder part of the container of $zC_P\Delta T$ joules per second. This effect is illustrated in Figure 5.13.

Convection arises from the interaction of two effects:

- The density of a gas changes significantly on heating (see §5.1.2 on the expansivity of gases).
- Most experiments we perform on gases take place in a gravitational field. If this were not so there would be no tendency for the hotter, less dense gas to "rise" or "fall".

Both factors are necessary for convection.

When heat flow by convection occurs in a fluid it can easily overwhelm any other thermal conduction processes occuring through the "still" fluid. Furthermore, heat flow by convection varies dramatically from one measurement apparatus to another and so is difficult to quantify. For this reason, even though

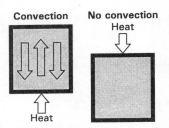

Figure 5.13 Convection occurs when heating produces a region of low density gas (hot) which is beneath a region of high density gas (cold). This arrangement is unstable and the low density (hot) gas rises and moves to the colder part of the chamber. When it reaches the colder part of the chamber it cools and gives up thermal energy $zC_P\Delta T$. When convection occurs the heat flow does not depend on the intrinsic thermal conductivity of the gas, but on its heat capacity and the rate of flow of the gas.

convection is important in many practical situations, more fundamental information can be extracted from experiments on fluids where the fluid itself does not flow. The measurements presented below are made in the absence of convection on "still gas".

The thermal conductivity, κ, is defined by the equation

$$\frac{dQ}{dt} = -\kappa A \frac{dT}{dx} \qquad (5.47)$$

where

$\dfrac{dQ}{dt}$ is the rate of heat flow (W);

κ is the thermal conductivity ($Wm^{-1}K^{-1}$);

A is the cross-sectional area across which heat is flowing (m^2); and

$\dfrac{dT}{dx}$ is the temperature gradient (Km^{-1}).

The minus sign indicates that heat flows down a temperature gradient, from hot to cold.

5.4.1 Data on the thermal conductivity of gases

The data given in Table 5.11 show the still-gas values of the thermal conductivity of a selection of monatomic, diatomic and triatomic gases at several temperatures. All the data refer to gases at pressures around one atmosphere. A typical value of κ for gases

Table 5.11 A selection of measured values of the thermal conductivities of gases.(The units are $10^{-2}\,Wm^{-1}K^{-1}$, i.e. multiply the number in the table by 10^{-2} to convert the value to $Wm^{-1}K^{-1}$).

| Gas | Temperature (K) | | | | |
---	73.2	173.2	273.2	373.2	1273
A Some monatomic gases					
Helium, He	5.95	10.45	14.22	17.77	41.90
Neon, Ne	1.74	3.37	4.65	5.66	12.80
Argon, Ar	–	1.09	1.63	2.12	5.00
Krypton, Kr	–	0.57	0.87	1.15	2.90
Xenon, Xe	–	0.34	0.52	0.70	1.90
Radon, Ra	–	–	0.33	0.45	–
B Some diatomic gases					
Hydrogen, H_2	5.09	11.24	16.82	21.18	–
Fluorine, Fl_2	–	1.56	2.54	3.47	–
Chlorine, Cl_2	–	–	0.79	1.15	–
Bromine, Br_2	–	–	0.40	0.60	–
Nitrogen, N_2	–	1.59	2.40	3.09	7.40
Oxygen, O_2	–	1.59	2.45	3.23	8.60
Carbon monoxide, CO	–	1.51	2.32	3.04	–
Air, N_2/O_2	–	1.58	2.41	3.17	7.60
C Some polyatomic gases					
Ammonia, NH_4	–	–	2.18	3.38	–
Carbon dioxide, CO_2	–	–	1.45	2.23	7.90
Ethane, C_2H_6	–	1.80	–	–	–
Ethene, C_2H_4	–	1.40	–	–	–
Methane, CH_4	–	1.88	3.02	–	–
Sulphur dioxide, SO_2	–	–	0.77	–	–
Water/steam, H_2O	–	–	1.58	2.35	–

(Data from Kaye & Laby; see §1.4.1)

is $\approx 0.1\,Wm^{-1}K^{-1}$, which is much less than the value for a metal such as copper ($\approx 400\,Wm^{-1}K^{-1}$) or an electrical insulator such as quartz (≈ 1 to $10\,Wm^{-1}K^{-1}$) (see Tables 7.15 and 7.16, respectively).

The data given in Table 5.11 are shown graphically in Figure 5.14, which shows that there is clear qualitative similarity in the behaviour of the thermal conductivity of all the gases. The thermal conductivity increases as the temperature is raised, but the increase is not linear: the *rate* of increase becomes less at higher temperatures.

More surprisingly, Figure 5.15 shows that the thermal conductivity of a gas is independent of pressure across a wide range of pressures around atmospheric pressure (note the logarithmic pressure range on the *x* axis). Thus removing 99% of the molecules from a gas at atmospheric pressure (i.e. reducing the pressure by a factor 100) produces no change in the thermal conductivity of the gas!

The main questions raised by our preliminary examination of the experimental data on the thermal conductivity of gases are:

(a) Why does the thermal conductivity of gases increase at high temperatures?

(b) Why is the thermal conductivity of gases independent of pressure across a wide range of pressures around atmospheric pressure?

(c) Why are the thermal conductivities of gases as low as they are?

5.4.2 Understanding the data on the thermal conductivity of gases

Before considering the theory of thermal conductivity, we note that the thermal conductivity of a gas is a different type of property from the heat capacity of a

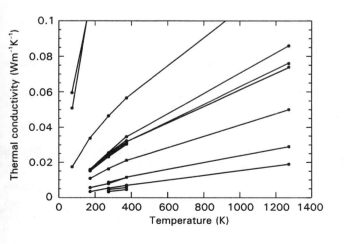

Figure 5.14 Variation in the thermal conductivity of the gases listed in Tables 5.11A–C with absolute temperature. All the curves are broadly similar in shape, i.e. monatomic and diatomic gases behave broadly similarly, but the absolute magnitude of conductivity varies significantly from one gas to another. The lines joining the data points are only a guide for the eye.

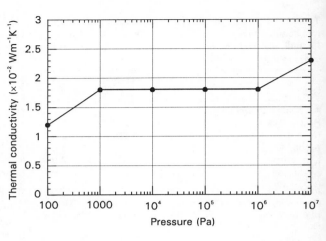

Figure 5.15 Variation in the thermal conductivity of argon with pressure. The lines join the data points and are only a guide for the eye. Note that the pressure scale is logarithmic and represents a pressure range of five orders of magnitude. Atmospheric pressure is around 10^5 Pa. (Data from B. H. Flowers & E. Menduza 1978, *Properties of matter*, New York, John Wiley.)

gas. The heat capacity relates the temperature at the beginning and end of a heating process to the heat added to the gas. In each state (before and after heating) the gas is in thermal equilibrium. In contrast, the thermal conductivity relates to the process of heat flow, which necessarily indicates that the gas is not in equilibrium. It is a general feature of physics that equilibrium properties are easier to understand than transport processes.

Background theory

In order to understand thermal conductivity of gases, we need to use the kinetic theory ideas developed in Section 4.3 (Fig. 5.16). In particular, we will use the following results:

(a) The *mean free path*, λ_{mfp}, of a gas molecule, is given by

$$\lambda_{mfp} = \frac{1}{\sqrt{2}n\pi a^2} \tag{5.48}$$

where n is number density of molecules and a is the effective diameter of a gas molecule.

(b) The *average* energy of a molecule with p degrees of freedom is given by

$$\bar{u} = \tfrac{1}{2} p k_B T \tag{5.49}$$

(c) The number of molecular collisions with an area A per second is given by

$$\tfrac{1}{4} n \bar{v} A \tag{5.50}$$

where n is number density of molecules and $\bar{v}$ is the mean speed of a gas molecule.

Example 5.8

A container of cross-sectional area 25 cm² and length 30 cm is heated from the top such that air at the bottom is at approximately 100°C and air at the top is approxiately 10°C warmer. What is the rate at which energy is transported down the tube?

We first note that the situation is stable against convection, and so we may use the intrinsic values of the thermal conductivity listed in Table 5.11. We then note that the pressure is not mentioned in the question, so we assume that it is within an order of magnitude of atmospheric pressure and will thus be the same as the figures listed in Table 5.31. We are now in a position to use Equation 5.48 in a fairly straightforward manner. We have:

$$\frac{dQ}{dt} = -\kappa A \frac{dT}{dx}$$

with k = 3.17×10^{-2} W m⁻¹ K⁻¹ and $A = 25 \times 10^{-4}$ m². The temperature difference across the length of the tube is 10°C and the length of tube is 30×10^{-2} m, so we can estimate the temperature gradient within the tube as

$$\frac{dT}{dx} \approx \frac{\Delta T}{\Delta X} = \frac{10}{0.3} = 33°\text{C m}^{-1}$$

Thus the rate of heat flow down the tube is given by

$$\frac{dQ}{dt} = -3.17 \times 10^{-2} \times 25 \times 10^{-4} \times 33 = 2.6 \times 10^{-3} \text{W}$$

Thus the maintenance of a 10°C temperature difference across the length of the tube requires only a heat input of around 2.6 mW.

110°C

100°C

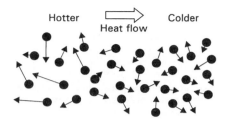

Figure 5.16 The basic process by which we imagine heat flow to take place. The arrows represent the directions in which the molecules are moving. Note that the *average* speed of molecules (represented by the length of the arrows) in the hotter region is greater than the *average* speed in the colder region.

Figure 5.17 illustrates a gas in which there is a temperature gradient, and highlights an area A perpendicular to the temperature gradient. Molecules strike area A from both the hotter side and the colder side. On average, molecules striking A have travelled a distance λ_{mfp} before striking A, and so have come from within a sphere of radius λ_{mfp}. In order to work out the net energy transported across A per second we make a simplifying assumption that the molecules have come not from a sphere, but from two planes a distance λ_{mfp} from A in either direction along the temperature gradient. The energy transported from these planes will be in the form of the kinetic energy of the molecules, plus the energy associated with any internal degrees of freedom.

The number of molecules crossing a unit area per second is given by Equation 5.50 as $\frac{1}{4}n\bar{v}$, and each molecule brings with it, on average, $\frac{1}{2}k_BT$ of energy per accessible degree of freedom. Thus if each molecule has p accessible degrees of freedom, then, on average, the energy crossing area A per second is

$$\tfrac{1}{4}n\bar{v} \times A \times p \times \tfrac{1}{2}k_BT \qquad (5.51)$$

Now at $x + \Delta x$, the temperature is approximately

$$T + \frac{\partial T}{\partial x}\lambda_{mfp} \qquad (5.52)$$

and at $x - \Delta x$, the temperature is approximately

$$T - \frac{\partial T}{\partial x}\lambda_{mfp} \qquad (5.53)$$

This small difference in temperature between

Equations 5.52 and 5.53 means that molecules striking area A from $x + \Delta x$ carry, on average, slightly more energy than molecules striking area A from $x - \Delta x$. The difference between the two energy fluxes gives the net energy transported across A per second. Note that in Figure 5.17 the temperature increases in the positive x direction, and so the heat flow will be in the negative x direction.

Energy per second across area A in the positive x direction is given by

$$\tfrac{1}{4}n\bar{v}A\,\tfrac{1}{2}\,pk_B\left(T - \frac{\partial T}{\partial x}\lambda_{mfp}\right) \qquad (5.54)$$

Energy per second across area A in the negative x direction is given by

$$\tfrac{1}{4}n\bar{v}A\,\tfrac{1}{2}\,pk_B\left(T + \frac{\partial T}{\partial x}\lambda_{mfp}\right) \qquad (5.55)$$

We now subtract Equation 5.55 from Equation 5.54. Thus, the net energy per second in the positive x direction is given by

$$\frac{dQ}{dt} = \tfrac{1}{4}n\bar{v}A\,\tfrac{1}{2}\,pk_B\left(T - \frac{\partial T}{\partial x}\lambda_{mfp} - T - \frac{\partial T}{\partial x}\lambda_{mfp}\right) \qquad (5.56)$$

$$\frac{dQ}{dt} = \tfrac{1}{4}n\bar{v}A\,\tfrac{1}{2}\,pk_B\left(-2\frac{\partial T}{\partial x}\lambda_{mfp}\right) \qquad (5.57)$$

Figure 5.17 Simplified illustration of the calculation of the thermal conductivity of a gas. The figure shows three planes in the gas perpendicular to a temperature gradient. The separation of the planes is λ_{mfp}, the mean free path of the molecules. Thus molecules that are travelling in the appropriate direction in either the top or the bottom plane will (probably) cross the central plane before colliding. Energy is carried across A in both directions but, on average, more energy flows across A from the plane in the hotter region of the gas. The analysis of the thermal conductivity centres on the problem of evaluating the net energy flow across area A .

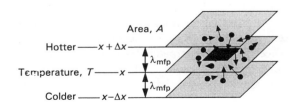

$$\frac{dQ}{dt} = \left(\tfrac{1}{4} n\bar{v}pk_B\lambda_{mfp}\right)A\frac{\partial T}{\partial x} \qquad (5.58)$$

Comparing Equation 5.58 with the standard form of the heat flow (Eq. 5.47), we see that the expression within the brackets is just κ

$$\kappa = \tfrac{1}{4} n\bar{v}pk_B\lambda_{mfp} \qquad (5.59)$$

This expression requires only refinement and re-arrangement before we can compare it with experimental results. If we had been more careful in our averaging of the heat flows across area A, we would have found an expression identical to Equation 5.59, but with the "4" in the denominator replaced by a "6". This modified expression is the one we will now work with, namely

$$\boxed{\kappa = \tfrac{1}{6} n\bar{v}pk_B\lambda_{mfp}} \qquad (5.60)$$

If we substitute for λ_{mfp} using Equation 5.48, we arrive at an expression that eliminates n completely

$$\kappa = \tfrac{1}{6} n\bar{v}pk_B \times \frac{1}{\sqrt{2}n\pi a^2} \qquad (5.61)$$

which simplifies to

$$\boxed{\kappa = \frac{\bar{v}pk_B}{6\sqrt{2}\pi a^2}} \qquad (5.62)$$

We are now in a position to compare our theoretical prediction with the experimental results given in §5.4.1.

Comparison with experiment

Let us consider the questions raised at the end of §5.4.1 in turn.

Why do the thermal conductivities of gases increase at high temperatures?

Example 5.9 shows how we may use Equation 5.62 to estimate the thermal conductivity of a gas. The example shows that we expect to find a $\sqrt{T}$ dependence for κ, which broadly describes the behaviour of all the gases in Figure 5.14. However, we shall see that the theory as we have developed it is not able to

---

Example 5.9

Evaluate Equation 5.62 for argon gas at temperature T.

$$\kappa = \frac{\bar{v}pk_B}{6\sqrt{2}\pi a^2}$$

There are four quantities in Equation 5.62 that we need to estimate: the number of accessible degrees of freedom, p; the average molecular speed, $\bar{v}$; the mean free path λ_{mfp}; and the effective molecular diameter, a.

From the analysis of heat capacity measurements (§5.3.3) we can be sure that for a monatomic gas there are exactly three degrees of freedom i.e. $p = 3$

We can make an estimate of $\bar{v}$ in terms of the *root mean square velocity* $\sqrt{\overline{v^2}}$. We remember that for any molecule there are three degrees of freedom associated with kinetic energy. Because of the equipartition of energy between the degrees of freedom, we can write the *average* kinetic energy

$$\tfrac{1}{2}m\overline{v^2} = 3 \times \tfrac{1}{2}k_BT$$

where m is the mass of a molecule. As mentioned in §4.4.2, $\sqrt{\overline{v^2}} = 1.085\bar{v}$, so we can estimate $\bar{v}$ as

$$\bar{v} = \frac{1}{1.085}\sqrt{\frac{3k_BT}{m}}$$

Finally, we estimate a from a typical value for the separation between atoms in the solid state, which is $a \approx 0.3 \times 10^{-9}$ m. This is not a very accurate estimate, and does not allow us quantitatively to compare results for different gases. However, it should be a constant for each type of gas and be of the correct order of magnitude.

Putting these estimates together we find that

$$\kappa = \frac{pk_B\bar{v}}{6\sqrt{2}\pi a^2}$$

becomes

$$\kappa \approx \frac{3k_B}{6\sqrt{2}\pi a^2}\frac{1}{1.085}\sqrt{\frac{3k_BT}{m}} \approx \frac{k_B}{2.17\pi a^2}\sqrt{\frac{3k_BT}{m}}$$

Or, collecting terms,

$$\boxed{\kappa \approx \left(\frac{k_B}{2.17\pi a^2}\sqrt{\frac{3k_B}{m}}\right)\sqrt{T}}$$

Substituting data for argon (relative atomic mass 39.95)
$k_B = 1.38 \times 10^{-23}$ J K^{-1}
$a \approx 0.3 \times 10^{-9}$ m
$m = 39.95 \times 1.66 \times 10^{-27}$ kg
we find that

$$\kappa \approx 3.98 \times 10^{-4}\sqrt{T}\ \mathrm{W\,m^{-1}\,K^{-1}}$$

Table 5.12 Calculated and experimental values of κ for argon at various temperatures.

	Temperature (K)		
173.2	273.2	373.2	1273
Experimental value (W m⁻¹ K⁻¹)			
1.09×10^{-2}	1.63×10^{-2}	2.12×10^{-2}	5.00×10^{-2}
Predicted value (W m⁻¹ K⁻¹)			
5.23×10^{-3}	6.57×10^{-3}	7.68×10^{-3}	14.19×10^{-3}
Ratio			
2.08	2.48	2.76	3.52
Argon "diameter" (nm)			
0.21	0.19	0.18	0.16

explain fully the experimental data in more than this qualitative manner.

We can look at the extent of detailed agreement between the calculation of κ in Example 5.9 and the experimental data. Table 5.12 shows this calculation repeated for each of the temperatures for which we have κ data. The predictions are clearly of the correct order of magnitude, but differ by a factor of about 2–3 from the experimental data. Noting that our estimate of the diameter of an argon atom in Example 5.9 was just a ballpark guess, we could improve the overall level of agreement between theory and experiment if we revised our estimate for the diameter of an argon atom. The diameter required to obtain a match between theory and experiment is given in the bottom row of Table 5.12. The magnitude of the diameters is entirely plausible; however, in order to explain the data we would have to suppose that the

molecules become effectively smaller at higher temperatures.

A more quantitative approach

We can quantitatively examine the deviations of the data from the $\sqrt{T}$ dependence if we replot some of the data from Figure 5.14 in a different form. Example 5.9 predicts that $\kappa = AT^{\frac{1}{2}}$ and so a plot of $\log \kappa$ versus $\log T$ should have slope of 0.5 and an intercept of $\log A$ where

$$A = \frac{k_B}{2.17\pi a^2} \sqrt{\frac{3k_B}{2m}}$$

Figure 5.18 shows the data for the monatomic gases plotted in this way.

The slopes and intercepts of the best fit lines of the form AT^x are recorded in Table 5.13. The intercepts A behave qualitatively as expected, becoming smaller for the gases with the larger molecules. In the right-hand column of Table 5.13 we have used the experi-

Table 5.13 Results from an analysis of the thermal conductivity data assuming the data has the form $\kappa = AT^x$ (Fig. 5.18). The significance of a is discussed in the text.

	A	x	a (nm)
He	30.91×10^{-4}	0.685	0.108
Ne	9.14×10^{-4}	0.695	0.198
Ar	2.34×10^{-4}	0.754	0.391
Kr	0.93×10^{-4}	0.806	0.620
Xe	0.42×10^{-4}	0.857	0.923
Ra	0.12×10^{-4}	0.994	1.73

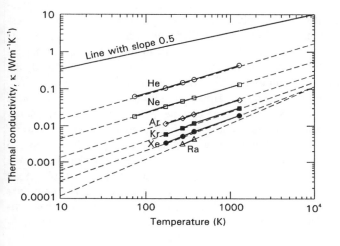

Figure 5.18 A log–log plot of the thermal conductivity κ of the monatomic gases versus temperature. The thick lines connect data points. The thin dashed lines are the best lines of the form AT^x fitted to the data. The thin solid line is drawn to illustrate what a line of slope 0.5 looks like. The fitted results for the gases are given in Table 5.13.

mental values for A to extract a value for the low-temperature limit of the effective diameter a of the noble gas atoms from Equation 5.62. We see that the data indicate a low-temperature effective diameter of an argon atom of 0.391 nm compared with a value of 0.21 nm at 173.2 K. We see that the low temperature limit is considerably greater than the effective diameter at high temperature.

The origin of this effect is illustrated in Figure 4.13 and concerns the detailed nature of molecular collisions. At low temperatures, the average speed of molecules is relatively slow, and molecules may be strongly affected by the weak attraction between molecules. At higher temperatures, the higher average speed of molecules causes their trajectories to be only relatively weakly affected. The dynamics of molecular collisions between molecules interacting realistically may be observed using the computer simulation listed in Appendix 4.

Further evidence for the idea that molecular interactions are the sources of the non-$\sqrt{T}$ dependence of the data can be obtained by examination of the slopes x in Figure 5.18 (recorded in Table 5.13). We note that the experimental values are all higher than the predicted value of 0.5, and that the difference from 0.5 becomes larger for the larger atoms.

We have not yet discussed the detailed nature of the interactions between molecules, but we will do so in §6.2 when we consider the properties of solids made from gas molecules. There we will see that the strength of the interaction between molecules is related to the electrical polarizability of a molecule (§5.6), i.e. the ease with which the electron distribution around an atom may be distorted. The large noble gas atoms have relatively large and easily distorted (polarizable) electron distributions and hence relatively strong interatomic interactions. In contrast, helium has a small, very rigid electron distribution and hence relatively weak interatomic interactions. This picture corresponds well with the data given in Table 5.13.

Before moving on we note that we have come to an explanation of the non-$\sqrt{T}$ dependence of the thermal conductivity of gases. However, in order to achieve this we have had to consider details of the interatomic interactions in the gas. In contrast, we could understand the heat capacity data on monatomic gases without any consideration of the details of inter-

atomic interactions. This is typical of the difference between an equilibrium property of a substance and a non-equilibrium transport property. Equilibrium properties are relatively straightforward to interpret compared with transport properties. However, by studying the details of the non-$\sqrt{T}$ dependence one could, in principle, compare particular models of molecular interactions with experiment.

Why are the thermal conductivities of gases independent of pressure across a wide range of pressures around atmospheric pressure?

Figure 5.15 shows that, for argon at least, thermal conductivity is independent of pressure over a range of at least three orders of magnitude. We can understand this immediately from the form of Equation 5.62:

$$\kappa = \frac{\bar{v} p k_B}{6\sqrt{2}\pi a^2} \qquad (5.62)^*$$

This equation does not mention pressure, and so the theory predicts, correctly, that κ is independent of pressure. But how does this come about? If one examines the precursor equations to 5.62, such as Equation 5.59, one can see that the formula includes the *product* of n and λ_{mfp}. This makes sense, since the thermal conductivity will be maximized if:

- there are more molecules to carry the thermal energy (i.e. n is large); and
- each molecule can travel unimpeded through the gas (i.e. λ_{mfp} is large).

However, λ_{mfp} is *inversely proportional* to n, and so increasing n reduces λ_{mfp} by an exactly compensating factor.

Figure 5.15 shows that κ is independent of pressure only over a certain (large) range. At lower pressures κ gets smaller, and at higher pressures κ gets larger. We can understand the low pressure behaviour quite simply. As the pressure is reduced, the number density of molecules, n, falls, and so λ_{mfp} gets larger. However, a point will be reached where λ_{mfp} is so large that the molecules are simply bouncing from side to side in their container and only rarely colliding with each other. Then λ_{mfp} will be unable to increase and κ will simply be proportional to n. We can estimate the pressure at which this occurs: if we fix at $\lambda_{mfp} \approx L$, the dimension of the gas container which might be, say, 1 cm $= 10^{-2}$ m, then we can estimate the value of n for which this occurs. From Equation 5.48 we have

$$n = \frac{1}{\sqrt{2}\lambda_{mfp}\pi a^2} \qquad (5.63)$$

Using our usual estimate of a molecular diameter of $a = 0.3\,nm$ we find $n = 2.5 \times 10^{20}\,m^{-3}$. Using our estimate of n in terms of pressure and temperature from §4.5.1 (Eq. 4.49)

$$n = \frac{P}{k_B T} \qquad (5.64)$$

Assuming a temperature of around 300 K, we have

$$P = nk_B T = 2.5 \times 10^{20} \times 1.38 \times 10^{-23} \times 300$$

$$P \approx 1.04\,Pa \; \left(= 1.04 \times 10^{-2}\,mbar\right) \qquad (5.65)$$

Atmospheric pressure is approximately $10^5\,Pa$ (1000 mbar), so this phenomenon does not occur until relatively low pressure – a regime known as the *high vacuum* regime. For example, the pressure inside the wall of a "vacuum flask" is typically $10^{-5}\,mbar$ (approximately $10^{-8}\,Pa$) and so the thermal conductivity of the gas in the flask is well into the high vacuum regime. Note that lowering the pressure through five orders of magnitude from atmospheric pressure would have no effect on the thermal conductivity! Only when the pressure is reduced below the critical pressure indicated by Equation 5.65 is the thermal conductivity affected.

The rise in thermal conductivity at higher pressures is due to the fact that the molecules are becoming so densely packed that the assumptions about molecular trajectories are becoming invalid and the mean free path, λ_{mfp} is again becoming fixed, but now at a distance of approximately an atomic diameter.

Why are the thermal conductivities of gases as low as they are?

The rise in thermal conductivity at high pressures gives a clue to the origin of the small absolute magnitude of the thermal conductivity in gases as compared with solids or liquids. The smallness of κ arises, at least in part, because the number density of molecules, n, is so small for a gas. If we consider a highly compressed gas in which λ_{mfp} is estimated as a small fraction of a molecular diameter, a, and n as approximately $1/a^3$, then estimates of the order of magnitude of κ obtained from Equation 5.60 produce answers similar to those found for liquids. We will consider this regime further when analyzing the thermal conductivity of liquids in §9.7.3.

5.5 Speed of sound

5.5.1 Data on the speed of sound in gases

Sound travels considerably slower through gases than through solids or liquids. As shown in Table 5.14, the maximum speed at 0°C is approximately $1000\,m\,s^{-1}$ for the gases with the lightest molecules (helium and hydrogen). This may be compared with

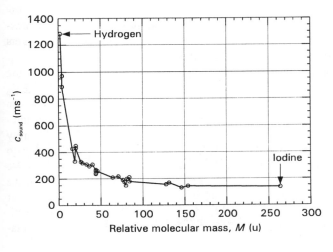

Figure 5.19 A graph of c_{sound} versus the relative molecular mass, M (u). The graph seems to show that the molecular weight of the molecules of the gas plays a significant role in determining the velocity of sound through the gas.

Table 5.14 The speed of sound in a selection of gases. The gases are listed in order of increasing molecular mass M in atomic mass units, (u $= 1.66 \times 10^{-27}$ kg) and the absolute temperature T at which the measurement was made. The shaded entries in the table are gases that have a "partner" gas in the table with the same molecular mass (see text). Out of general interest, we also record that a high precision value for the speed of sound in dry air with 0.03% CO_2 by volume at 1.0kHz is 331.45±0.05ms^{-1} (T = 273.15°C and P = 101325Pa).

Gas	M	T (K)	c_{sound} (m s^{-1})	Gas	M	T (K)	c_{sound} (m s^{-1})
Hydrogen, H_2	2.0	273.2	1286	Hydrogen sulphide, H_2S	33.1	273.2	310
Helium, He	4.0	273.2	971.9	Hydrogen chloride, HCl	36.5	273.2	296
Deuterium, D_2	4.0	273.2	890	Argon, Ar	40.0	273.2	307.8
Methane, CH_4	16.0	273.2	430	Nitrous oxide, N_2O	44.0	298.2	268
Ammonia, NH_3	17.0	273.2	415	Propane, C_3H_8	44.0	273.2	238
Water (steam), H_2O	18.0	373.2	473	Carbon dioxide, CO_2	44.0	273.2	259
Water (steam), H_2O	18.0	407.2	494	Ethanol, $C_2H\backslash OH$	46.0	326.2	258
Fluorine, F_2	19.0	373.2	332	Sulphur dioxide, SO_2	64.0	273.2	211
Heavy water (steam), D_2O	20.0	373.2	451	Chlorine, Cl_2	70.9	293.2	219
Neon, Ne	20.2	273.2	434	Carbon disulphide, CS_2	76.0	273.2	192
Acetylene, C_2H_2	26.0	273.2	329	Benzene, C_6H_6	78.0	273.2	177
Nitrogen, N_2	28.0	273.2	337	Bromine, Br_2	79.9	331.2	149
Carbon monoxide, CO	28.0	273.2	337	Hydrogen bromide, HBr	80.9	273.2	200
Ethylene, C_2H_4	28.0	273.2	318	Krypton, Kr	83.8	273.2	213
Ethane, C_2H_6	30.0	283.2	308	Cyclohexane, C_6H1_2	84.0	303.2	181
Ethane, C_2H_6	30.0	304.2	316	Hydrogen iodide, HI	127.9	273.2	157
Nitric oxide, NO	30.0	283.2	324	Xenon, Xe	131.3	273.2	170
Nitric oxide, NO	30.0	289.2	334	Sulphur hexafluoride, SF_6	146.0	284.2	133
Oxygen, O_2	32.0	303.2	332	Carbon tetrachloride, CCl_4	153.8	370.2	145
Methanol, CH_3OH	32.0	370.2	335	Iodine, I_2	263.8	453.2	138

the values in Tables 7.6 (for solids) and 9.5 (for liquids) that have typical values of around 3000ms^{-1}. The speed of sound at frequency 1.0kHz in dry air with 0.03% CO_2 is 331.45 ± 0.05ms^{-1} at a temperature of 273.15°C and a pressure of 101325 Pa.

Figure 5.19 shows a graph of c_{sound} versus M, the relative molecular mass of the molecules of the gas. It is fairly clear that the mass of the molecules of the gas plays a significant role in determining the speed of sound. Table 5.14 does not contain enough data to allow a systematic study of the effects of temperature and other factors on the speed of sound. However, we can obtain some clues about which other factors affect the speed of sound by looking in detail at the shaded entries in table. These have at least one other "partner entry", either a gas with the same molecular mass or the same gas at a different temperature. We will now examine each of the shaded entries in turn.

Table A

Gas	M	T (K)	c_{sound} (m s^{-1})
Helium, He	4.0	273.2	971.9
Deuterium, D_2	4.0	273.2	890

The data for helium and deuterium are taken at the same temperature, and yet the sound velocities differ by about 9%. So there must be a factor other than temperature and molecular weight that affects c_{sound}.

Table B

Gas	M	T (K)	c_{sound} (m s^{-1})
Water (steam), H_2O	18.0	373.2	473
Water (steam), H_2O	18.0	407.2	494

The data for steam show an increase in the speed of sound with increasing temperature.

Table C

Gas	M	T (K)	c_{sound} (m s^{-1})
Nitrogen, N_2	28.0	273.2	337
Carbon monoxide, CO	28.0	273.2	337
Ethylene, C_2H_4	28.0	273.2	318

The data for nitrogen and carbon monoxide are taken at the same temperature, and have the same speed of sound: the level of agreement is striking. However, ethylene, with the same molecular mass and at the same temperature, has a sound velocity about 6% lower. Let us hypothesize that the *number* of atoms in the gas molecule is also a factor affecting c_{sound}.

Reviewing the data so far we have some evidence that:

- c_{sound} is increased at higher temperatures, and
- c_{sound} is decreased if a molecule has many atoms.

These hypotheses are consistent with all the data so far. We will now examine the rest of the data.

Table D

Gas	M	T (K)	c_{sound} (m s^{-1})
Ethane, C_2H_6	30.0	283.2	308
Ethane, C_2H_6	30.0	304.2	316
Nitric oxide, NO	30.0	283.2	324
Nitric oxide, NO	30.0	289.2	334

We now have data that allow comparison of the effects of both temperature (over a limited range) and molecular complexity. We see that these data are again consistent with the hypotheses above: the more complex molecule has the lower c_{sound}; the higher temperature data have higher values of c_{sound}.

Table E

Gas	M	T (K)	c_{sound} (m s^{-1})
Oxygen, O_2	32.0	303.2	332
Methanol, CH_3OH	32.0	370.2	335

These data correspond both to different numbers of atoms per molecule and to different temperatures.

The data are not inconsistent with the hypotheses because the effects of temperature and molecular complexity could be compensating one another.

Table F

Gas	M	T (K)	c_{sound} (m s^{-1})
Nitrous oxide, N_2O	44.0	298.2	268
Propane, C_3H_8	44.0	273.2	238
Carbon dioxide, CO_2	44.0	273.2	259

Again the hypotheses are confirmed. Comparing propane and carbon dioxide we see that the molecule with more atoms has the lower c_{sound}. Comparing nitrous oxide and carbon dioxide, both triatomic molecules, we see that the substance at the higher temperature has the greater c_{sound}.

Thus, the hypotheses set out above seem to be confirmed by all the data in Table 5.14. There are a few data available on the dependence of the speed of sound on pressure, because – perhaps surprisingly – the speed of sound is broadly independent of the pressure.

So the questions raised by our preliminary examination of the experimental data on the speed of sound in gases are:

- Why is the speed of sound greater at higher temperatures, but roughly independent of pressure?
- Why is the speed of sound reduced by molecular complexity, i.e. why for different gases of the same molecular mass does sound travel faster in the gas with the least complex molecules?
- Why is the speed of sound higher in gases with low molecular mass?
- Why is the speed of sound in gases at around atmospheric pressure around a few hundered metres per second?

5.5.2 Understanding the data on the speed of sound

Sound is a *displacement wave* or *pressure wave* which propagates through a gas. The wave consists of layers of *compressed gas* at a pressure slightly higher than ambient pressure, and *rarefied gas* at a pressure slightly lower than ambient pressure. Figure 5.20 illustrates the variation of the pressure/

Figure 5.20 Two illustrations of the variation of pressure within a sound wave at a given time. (a) A perspective drawing of planes of constant pressure: shaded planes represent compressed regions of the gas, unshaded planes represent rarefied regions of gas. (b) Plot of the pressure as a function of position in a sound wave; the pressure amplitude of a sound wave rarely exceeds 1% of the ambient pressure.

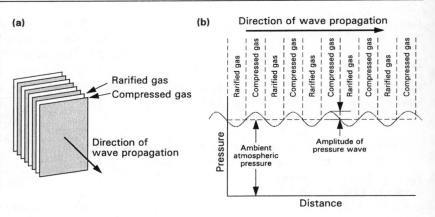

density within the wave. Further pictures and a discussion of the formula for the speed of sound are given in Appendix 2.

There we predict that the speed of sound, c_{sound}, in a gas is given in terms of its mass density ρ, and either the *compressibility*, K, or the *bulk modulus*, B, by

$$c_{sound} = \sqrt{\frac{1}{K\rho}} = \sqrt{\frac{B}{\rho}} \qquad (5.66)$$

We can see that this prediction makes physical sense by considering a region of compressed gas: the ease with which the gas will "spring back" to try to restore ambient pressure depends on the quantities in Equation 5.66:

- It depends on the *compressibility* of the gas, K, because if the gas is easily compressed then the restoring force will be small, i.e. the gas is not very "springy". Thus highly compressible gases will tend to have a low speed of sound.
- It depends on the *density* of the gas, ρ, because if the density of the gas is high then the *mass per unit volume* will be high and a given restoring force will produce less acceleration of the gas. Thus dense gases will "spring back" slower and hence tend to have lower speed of sound.

The compressibility of a gas is defined as

$$K = \frac{1}{V}\frac{\partial V}{\partial P} \qquad (5.67)$$

and its inverse ($1/K$) is called the *bulk modulus*. The compressibility has different values, depending on

whether the gas is compressed at constant temperature, or whether the gas is isolated during the compression, in which case the temperature may rise on compression. In the former case the compressions are said to be *isothermal*; in the latter the compressions are said to be *adiabatic*. The value of K corresponding to each of these types of compression is discussed below.

Adiabatic or isothermal compressions?

It is not obvious which type of compression takes place in a sound wave. Since we are not aware of temperature oscillations when we hear a sound, we might at first conclude that the compressions in the wave are isothermal. However, it is also possible that if the temperature oscillations are small, and the heat capacity of our thermometers (or our skin) large, then we might not be aware of them. In order to decide which type of compression takes place in a sound wave we will work out the predicted speed of sound in air (Eq. 5.66) assuming each of the two cases in turn. Then, by comparing the predictions with experiment, we will see that it is possible to deduce which type of compression actually occurs.

Isothermal compressions

By assuming that the compressions are isothermal, we implicitly assume that the compressions take place slowly enough that heat can travel from the compressed regions of the gas to the rarefied regions of the gas. During isothermal expansions and compressions the gas obeys $PV = zRT$, with T held constant.

Differentiating $PV = zRT$ with respect to P

$$\frac{\partial}{\partial P} PV = \frac{\partial}{\partial P} zRT \qquad (5.68)$$

we see that the right-hand side is zero because T stays constant during the compression. Using the product rule on the left-hand side we find

$$P\frac{\partial V}{\partial P} + V = 0 \qquad (5.69)$$

Dividing through by V and rearranging yields

$$\frac{P}{V}\frac{\partial V}{\partial P} + 1 = 0 \qquad (5.70)$$

and recalling the definition of K (Eq. 5.67), we find

$$\boxed{K_{iso} = -\frac{1}{V}\frac{\partial V}{\partial P} = \frac{1}{P}} \qquad (5.71)$$

where the subscript "iso" indicates that the compression is isothermal.

Adiabatic compressions

By assuming that the compressions are adiabatic, we assume that the compressions and rarefactions take place so quickly that there is no time for heat to be exchanged from the compressed (hotter) regions of the gas to the rarefied (colder) regions of the gas. During adiabatic expansions and compressions the gas obeys $PV^\gamma = $ constant. (see Exercise P9 in section 4.6). Thus, differentiating with respect to pressure

$$\frac{\partial}{\partial P} PV^\gamma = \frac{\partial}{\partial P} \times \text{constant} \qquad (5.72)$$

we find $\qquad \gamma P V^{\gamma-1}\frac{\partial V}{\partial P} + V^\gamma = 0 \qquad (5.73)$

Dividing through by V^γ and rearranging yields

$$\frac{\gamma P}{V}\frac{\partial V}{\partial P} + 1 = 0 \qquad (5.74)$$

and, recalling the definition of K (Eq. 5.67), we find

$$\boxed{K_{ad} = -\frac{1}{V}\frac{\partial V}{\partial P} = \frac{1}{\gamma P}} \qquad (5.75)$$

where the subscript "ad" indicates that the compression is adiabatic.

Comparison with experiment

Taking Equations 5.71 and 5.75 for K_{iso} and K_{ad}, we can compare predictions for the speed of sound with experimental values. Taking the case of air under the standard conditions of 0°C (273.15 K) and pressure 1.013×10^5 Pa, we find (§5.1) that the density of air is 1.293 kg m^{-3}. Taking a value for γ of 1.4 (Table 5.8C), we can predict values for the speed of sound assuming either isothermal or adiabatic compressions in the sound wave:

$$c_{sound} = \sqrt{\frac{1}{K\rho}} \qquad (5.76)$$

Isothermal prediction

$$c_{sound} = \sqrt{\frac{1}{K\rho}} = \sqrt{\frac{P}{\rho}}$$

$$c_{sound} = \sqrt{\frac{1.013 \times 10^5}{1.293}}$$

$$c_{sound} = 279.9 \text{ m s}^{-1}$$

Adiabatic prediction

$$c_{sound} = \sqrt{\frac{\gamma P}{\rho}}$$

$$c_{sound} = \sqrt{\frac{1.4 \times 1.013 \times 10^5}{1.293}}$$

$$c_{sound} = 331.2 \text{ m s}^{-1}$$

$$(5.77)$$

Comparing these predictions with the experimental value of 331.45 ± 0.05 m s^{-1} (see legend to Table 5.14) clearly favours the theory in which the compressions of the gas are adiabatic. In retrospect, one can understand this fairly easily. The wavelength of a sound wave is, for example, ≈ 1 m at a frequency of ≈ 1 kHz. In order for the compressions of the wave to be isothermal the very small temperature differences in the wave would have to equalize over a distance of ≈ 0.5 m in ≈ 0.5 ms. For a poor thermal conductor such as a gas this is unrealistic. Thus for all sound waves in gases over all practical frequencies, the compressions and rarefactions are adiabatic, and the speed of sound is given by

$$c_{sound} = \sqrt{\frac{\gamma P}{\rho}} \qquad (5.78)$$

The temperature oscillations associated with sound waves are considered in Exercise P29 in §5.9.

Comparison with experiment

The pressure and temperature dependence of the speed of sound

Equation 5.78 allows us to understand immediately why the speed of sound is not significantly dependent on pressure: the density of a gas is proportional to its pressure. Thus changing the pressure of a gas affects the compressibility of a gas, but changes its density by an exactly compensating amount. This lack of pressure dependence becomes clearer if we rearrange Equation 5.78 by substituting for P and ρ to allow a direct comparison with the results of Table 5.14. We use:

- the perfect gas equation to substitute for $P = zRT/V$; and
- $\rho = zM/V$ (where z/V is the number of moles per unit volume, and M is the mass in kilograms of 1 mole of gas) to substitute for ρ.

We find that the speed of sound may thus be expressed as

$$c_{\text{sound}} = \sqrt{\frac{\gamma zRT}{V} \times \frac{V}{zM}} \qquad (5.79)$$

$$c_{\text{sound}} = \sqrt{\frac{\gamma RT}{M}} \qquad (5.80)$$

Equation 5.80 makes it clear that if the temperature of the gas is kept constant, then we expect c_{sound} to show no variation with pressure.

It is actually rather hard to compare the predictions of Equation 5.80 with experiment. One might

---

think it fairly straighforward to look up the values of γ and M for the gas, and then note the temperature T at which c_{sound} was measured. However, the tables of γ are commonly compiled by *assuming* that Equation 5.80 holds true and then deducing γ from measurements of c_{sound} and T. Hence one cannot then use the γ values to check the values of c_{sound}! However, below we can look at questions raised by examination of the data with the aim of ensuring the overall consistency of Equation 5.46 in describing gases with different types of molecule.

The effect of molecular complexity

We saw in §5.3.2 that the number of degrees of freedom of a molecule can be determined from measurements of γ using Equation 5.46:

$$p = \frac{2}{\gamma - 1} \qquad (5.46)*$$

We also saw that more complex molecules have more degrees of freedom (p) and hence have smaller values of γ. Thus through Equation 5.80

$$c_{\text{sound}} = \sqrt{\frac{\gamma RT}{M}} \qquad (5.80)*$$

we expect gases of more complex molecules to have a lower speed of sound.

We can examine the dependence on p and γ by looking for gases that have the same molecular mass M, but different numbers of molecules, and hence different values of γ. Searching through Table 5.14 we see that we have several examples, of which I have chosen just two, whose details are set out in Table 5.15.

For the monatomic or diatomic molecules, the agreement is excellent if we assume that at around room temperature there are two accessible rotational degrees of freedom, but no vibrational degrees of freedom. However, for more complex molecules such as ethylene (CH_2CH_2) it is difficult to arrive at an estimate for γ which is anything more than a guess. In order to improve on guesswork one would need to calculate carefully the molecular dynamics of an ethylene molecule, a task well beyond the scope of this book.

Example 5.10

What is the predicted speed of sound in helium at 0°C?

Helium:
$\gamma = 1.63$ (from Table 5.8A)
$R = 8.31\,\text{JK}^{-1}\text{mol}^{-1}$
$T = 273.1\,\text{K}$
$M = 4.0\,\text{g} = 4.0 \times 10^{-3}\,\text{kg}$
So

$$c_{\text{sound}} = \sqrt{\frac{1.63 \times 8.31 \times 273.1}{4.0 \times 10^{-3}}} = 961.3\,\text{ms}^{-1}$$

This compares with the experimental value (Table 5.14) of $971.9\,\text{ms}^{-1}$, i.e. the error is around 1%.

Table 5.15 Details of gases whose molecules have relative molecular mass of 4 and 28. The table allows a detailed comparison of theoretical expectations and experimental results for the dependence of the speed of sound upon molecular complexity. The question marks indicate uncertainty in the expected values of p and γ.

Gas	M	Number of atoms per molecule	Expected γ $(1+2/p)$	T (K)	c_{sound} Theoretical $\sqrt{(\gamma RT/M)}$	Experimental (Table 5.14)
He	4.0	1	1.667 ($p = 3$)	273.1	972.8	971.9
D_2	4.0	2	1.400 ($p = 5$)	273.1	891.5	890.0
N_2	28	2	1.400 ($p = 5$)	298.1	336.9	337.0
CO	28	2	1.400 ($p = 5$)	273.1	336.9	337.0
CH_2CH_2	28	6	1.2 ($p = 10$)??	273.1	$\approx$312	318.0

Why γ?

It is interesting to ask why γ, the ratio of the heat capacities C_P/C_V, is involved in an expression for the speed of sound. We can see why if we consider what happens in the compression of the gas that occurs in the wave. Since the compressions are adiabatic, the compression heats the gas (slightly) (see Exercise 5P29). By the *Equipartition principle* (see §4.3.1, Complication 3), the increased internal energy of the gas caused by the compression is shared equally amongst the p accessible degrees of freedom of the molecules. So if the average energy of each molecule is increased by Δu, the temperature rise ΔT is such that $\Delta u = 0.5pk_B\Delta T$. So ΔT is given by

$$\Delta T = \frac{\Delta u}{0.5pk_B} \qquad (5.81)$$

Thus ΔT is inversely proportional to p. So, for example, if $p = 3$ then the temperature rise, ΔT, will be larger than if $p = 5$, because the energy of compression is shared amongst fewer degrees of freedom. This temperature rise causes the gas pressure to increase more quickly as its volume is reduced. That is, the gas is made less compressible, and the larger the temperature rise the greater the reduction in compressibility. Thus if two types of gas have molecules of the same mass, the gas whose molecules have fewer internal degrees of freedom will be "springier".

Dependence of the speed of sound on M

The strong dependence of c_{sound} on the molecular mass M is clearly shown in Figure 5.19. With hindsight, we can now see how this dependence arises, and by plotting c_{soind} against $M^{-\frac{1}{2}}$ we should find that the data lie close to a straight line through the origin. Further, if we make this plot using a set of gases for which γ is the same we should be able to reduce the apparently random scatter on Figure 5.19. Figure 5.21 shows such a plot for the monatomic gases for which

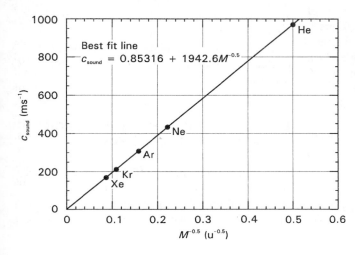

Figure 5.21 Graph of the speed of sound versus $1/\sqrt{A}$, where A is the molecular mass in atomic mass units u. The data show a clear linear relationship, as indicated by the least-squares fit shown as a line on the graph. Where appropriate, the data have been "corrected" to 273.1 K by multiplying by $\sqrt{(273.1/289.1)}$.

we are confident from analysis of heat capacity data that there are exactly 3 degrees of freedom, i.e. $p = 3$.

If we substitute the expression for γ (Equation 5.48) into Equation 5.82, and then separate out the dependence on molecular mass, we find

$$c_{sound} = \sqrt{\left(1 + \tfrac{2}{3}\right)RT} \times M^{-\frac{1}{2}} \qquad (5.82)$$

We now note that M is the mass of 1 mole of gas, i.e. N_A times the mass of a molecule (in atomic units). Evaluating this yields

$$\text{Slope} = \left[\left(1 + \tfrac{2}{3}\right)\frac{8.314 \times 273.1}{6.022 \times 10^{23} \times 1.661 \times 10^{-27}}\right]^{0.5} \qquad (5.83)$$

$$= 1945.1$$

which differs from the experimental slope (1942.6) by $\approx 0.13\%$ i.e. the agreement is better than two parts in 1000. Clearly we understand all the factors at play in determining the speed of sound, in monatomic gases at least.

Order of magnitude of the speed of sound

Why does c_{sound} have the value it has for a gas? One answer is to refer to Equation 5.80 and its derivation and say "it just does". However, there is another type of answer. It is reasonable to suppose that *a sound wave cannot travel through a gas more quickly than the molecules of the gas are moving*. We can see this by noting that, if we have a compressed region of gas, the pressure changes that constitute the sound wave clearly cannot move faster than the fastest molecules, and will probably move at a speed related to the average speed of molecules. Since the average speed of molecules in air at room temperature is $\approx 500\,\mathrm{m\,s^{-1}}$, and the speed of sound is $\approx 340\,\mathrm{m\,s^{-1}}$ this certainly does bear out this rough upper limit theory. We can in fact do better than this and convert our Equation 5.80 for c_{sound} into a relationship between c_{sound} and $\overline{v^2}$.

First of all, we recall that T is defined by the fact that the average energy per accesible degree of freedom is $0.5k_B T$. So for the three degrees of freedom of the kinetic energy of the gas we have (§4.3. Eq. 4.24)

$$\tfrac{1}{2}m\overline{v^2} = \tfrac{3}{2}k_B T \qquad (5.84)$$

Substituting for T in the expression for c_{sound} yields

$$c_{sound} = \sqrt{\frac{\gamma RT}{M}} = \sqrt{\frac{\gamma Rm\overline{v^2}}{3k_B M}} \qquad (5.85)$$

Now M is the mass in kilograms of 1 mole of the gas, i.e. $M = N_A m$, where m is the mass of one molecule. Remembering that, by definition, $R = k_B N_A$, we have

$$c_{sound} = \sqrt{\frac{\gamma Rm\overline{v^2}}{3k_B M}} = \sqrt{\frac{\gamma Rm\overline{v^2}}{3k_B N_A m}} = \sqrt{\frac{\gamma \overline{v^2}}{3}} \qquad (5.86)$$

$$c_{sound} = \sqrt{\frac{\gamma}{3}}\sqrt{\overline{v^2}} = \left(\sqrt{\frac{\gamma}{3}}\right)v_{RMS} \qquad (5.87)$$

Since γ ranges from ≈ 1.7 to 1, c_{sound} should vary between 58% and 75% of the root-mean-square speed of the molecules.

5.6 Electrical properties

The electrical properties of gases are complex, and are thus difficult to describe succinctly. However, we can simplify our discussion by dividing the properties of the gas into the behaviour observed in weak electric fields and that observed in strong electric fields.

In this section we will look mainly at the properties of gases under weak, *static* electric fields. The term "static" in this context refers to electric fields varying at frequencies much lower than approximately $10^9\,\mathrm{Hz}$. The response to electric fields varying at infrared frequencies or higher is discussed in §5.7 on the optical properties of gases.

In §5.6.3 we will look briefly at the phenomenon of electrical breakdown, which occurs in strong electrical fields.

5.6.1 Data on the electrical properties of gases in weak electric fields

The application of a weak electric field to a gas does not produce any dramatic effects. A "weak" electric field may be defined as one that is less than a characteristic field strength known as the *dielectric strength* of the gas. The dielectric strength varies with temperature and pressure, but at around room temperature and normal atmospheric pressure it is of the order of 10^6 volts per metre (Vm^{-1}).

For electric fields less than the dielectric strength of the gas it is not a trivial matter to detect the effects of an electric field on the gas. To a first approximation, one may ignore the presence of a gas in a region of weak electric field and treat the situation as if there were no matter present, i.e. a vacuum. However, careful experiments have shown that there is a small effect that may be described in terms of a *dielectric constant*, ε. The dielectric constant has the value unity for a vacuum and is often tabulated as ($\varepsilon-1$) in order to indicate the difference between the presence and absence of molecules of the substance. For gases, ε usually differs only slightly from unity with typical values (Table 5.16) being in the region of 1.00001.

Since a gas weakens the interaction between electric charges, or lowers the electric field around a charge (which amounts to the same thing), we find $\varepsilon \geq 1$. The quantity ($\varepsilon-1$) is tabulated for various gases around room temperature and atmospheric pressure in Table 5.16.

Table 5.16 and Figure 5.22 both show a good deal of apparently random variability. However, despite this, it is possible to discern some systematic trends in the data. From Figure 5.22 it appears that, with striking exceptions, there is a weak trend towards larger dielectric constants in gases with large molecules. Furthermore, in the two cases where Table 5.16 has data for a gas at several temperatures (carbon dioxide and ammonia), the data show a systematic decrease in ε with increasing temperature.

Figure 5.23 shows the variation of ($\varepsilon-1$) for carbon dioxide as a function of pressure at a temperature of 100°C. We see that ($\varepsilon-1$) depends linearly on the pressure, with only a slight sign of curvature at the highest pressures.

The main questions raised by our preliminary examination of the experimental data on the dielectric permittivities of gases are:

- Why does the value of ε depend linearly on pressure?
- Why does the value of ε decrease with increasing temperature?
- Why does the value of ε depend on the type of molecule of the gas? There is some indication that ε may be larger for larger molecules, but there are some molecules that display particularly large dielectric constants.

Table 5.16 The relative dielectric permittivity, ($\varepsilon-1$), of various gases at atmospheric presure (1.013×10^5 Pa): for pressures below atmospheric pressure, ε, varies linearly with pressure. Different experimenters find different values of ε and the data for $10^4(\varepsilon-1)$ should all be treated as accurate to only about 10% The entry for ethanol has two alternative values to indicate two particularly divergent values for $10^4(\varepsilon-1)$. For other entries I have taken averages of tabulated results, or ignored entries in tables that were clearly in error. The data refer to values obtained with electric fields oscillating at radio frequencies, ($\approx 10^6$ Hz). The shaded entries in the table i.e. helium, hydrogen, argon, oxygen, nitrogen and air, are typical results for ε valid from DC up to optical frequencies $\approx 10^{15}$ Hz. The variation over that range is within ±2 of the least significant figure in the table.

Gas	M	T (°C)	$10^4(\varepsilon-1)$	Gas	M	T (°C)	$10^4(\varepsilon-1)$
Monatomic gases				*Triatomic gases*			
Helium, He	4.0	20	0.65	Carbon dioxide, CO_2	44.0	0	9.88
Neon, Ne	20.2	0	1.3	Carbon dioxide, CO_2	44.0	20	9.22
Argon, Ar	40.0	20	5.16	Carbon dioxide, CO_2	44.0	100	7.23
Mercury, Hg	200.6	180	7.4	Nitrous oxide, N_2O	44.0	25	11
Mercury, Hg	200.6	180	7.4	Water (steam), H_2O	18.0	100	60
Diatomic gases				*Polyatomic gases*			
Hydrogen, H_2	2.0	0	2.72	Ethane, C_2H_6	30.0	0	15
Hydrogen, H_2	2.0	20	2.54	Benzene, C_6H_6	65.0	100	32.7
Nitrogen, N_2	28.0	20	5.47	Methanol, CH_3OH	32.0	100	57
Oxygen, O_2	32.0	20	4.94	Ethanol, C_2H_5OH	44.0	100	61 or 78
Air (dry, no CO_2)	28.8	20	5.36	Ammonia, NH_3	18.0	0	8.34
Carbon monoxide, CO	28.0	23	6.92	Ammonia, NH_3	18.0	100	4.87

(Data from Kaye & Laby, *CRC handbook* and Bleaney & Bleaney; see §1.4.1)

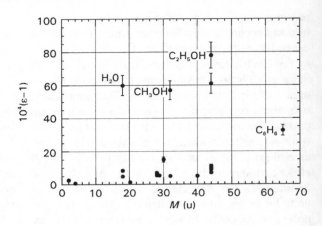

Figure 5.22 Plot of the data in Table 5.16 showing $10^4(\varepsilon-1)$ as a function of the mass of the molecules of the gas. Uncertainty indications represent ±10% of the values in Table 5.16. The value for mercury (Hg) has not been plotted because its large molecular mass distorts the scale of the graph.

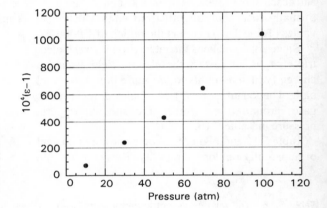

Figure 5.23 The variation in the dielectric constant plotted as $10^4(\varepsilon-1)$ for carbon dioxide as a function of pressure at a temperature of 100°C. We see that $(\varepsilon-1)$ depends linearly on the pressure, with a slight sign of curvature at the highest pressures. (Data from *CRC handbook*; see §1.4.1)

- Why, under the action of weak electric fields, is the relative dielectric permittivity, ε, so close to unity?

5.6.2 Understanding the electrical properties of gases in weak electric fields

In order to understand the phenomena described above, we must first develop a theory of the dielectric constant of a gas. We do this in the following sections and then proceed to compare the predictions of the theory with experiment.

Background theory

The theory that we develop below models a gas as a collection of molecules that do not interact with one another. Thus the properties of the gas will depend directly on the response of individual molecules to an applied electric field. It is found that the response of molecules falls into two distinct classes:

- Atoms or molecules that, in the absence of an applied electric field, possess no *electric dipole moment*. Such atoms and molecules are called *non-polar*. All atoms are non-polar, but only a few molecules – those with a high degree of symmetry such as N_2 or O_2 – fall into this category.

- Molecules that, in the absence of an applied electric field possess a finite *electric dipole moment*. Such atoms and molecules are called *polar*. Most molecules fall into this category, although the magnitude of the dipole moment varies considerably from one type of molecule to another.

Let us examine the cases of non-polar and polar molecules in turn.

Non-polar molecules

The situation when a weak field is applied to an atom (or molecule) with no intrinsic electric dipole moment is illustrated in Figure 5.24. The electric field perturbs the electronic charge distribution around the nucleus of each atom, drawing the electrons slightly to one side of the nucleus. Thus an electric dipole moment is *induced* on each atom and the atoms are said to be *polarized*. However, the atoms are still electrically neutral and so there is no net force on them in the applied electric field. The magnitude of the induced dipole moment, p_i, is given by the product of the total charge on the nucleus, Ze, multiplied by Δx, the distance between the centres of symmetry of the electronic and nuclear charge distributions. In general, Δx is extremely small, and atomic dipole moments are sometimes measured in units of $10^{-30}\,\mathrm{Cm}$, called *debye* units.

Polar molecules

The situation when a weak applied field is applied to a molecule which has a permanent electric dipole moment, p_p, is illustrated in Figure 5.25. Commonly,

Figure 5.25 (a) Schematic illustration of a charge distribution that possesses a dipole moment. (b) How the actual molecule is modelled as a permanent dipole moment. (c) The forces acting on a dipole moment form a torque that twists the molecule until **p** lies parallel to **E**.

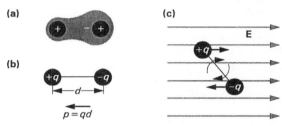

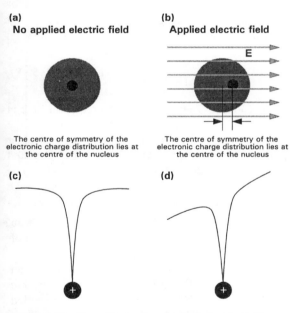

(a) No applied electric field

The centre of symmetry of the electronic charge distribution lies at the centre of the nucleus

(b) Applied electric field

The centre of symmetry of the electronic charge distribution lies at the centre of the nucleus

(c)

(d)

Figure 5.24 The effect of an applied electric field on a non-polar molecule. (a, c) The situation when no electric field is applied. The charge distribution of the electrons is arranged symmetrically around the nucleus (a), because the Coulomb potential energy of an electron (which varies approximately as $1/r$, as shown in (c)) is symmetrical.(b, d) The situation when there is an external electric field applied. The charge distribution of the electrons is arranged asymmetrically around the nucleus (b), because the Coulomb potential energy of an electron (d) is asymmetrical. The extent of the disturbance of the charge distribution depends on the ratio of the applied electric field to the internal field. Usually this ratio is extremely small (see Example 2.1).

Example 5.11

In an applied electric field, the centre of electronic symmetry of an argon atom moves a distance of about $10^{-13}\,\mathrm{m}$ away from the nucleus. What is the magnitude of the electric dipole moment induced on the atom?

An argon nucleus has a charge $+18e$ and the electrons have a total charge $-18e$. If we assume that the entire charge distribution moves rigidly by a distance $10^{-13}\,\mathrm{m}$, then the induced electric dipole moment is (see §2.3.3)

$$p = qd$$
$$p = 18 \times 1.6 \times 10^{-19} \times 10^{-13} = 2.88 \times 10^{-31}\,\mathrm{Cm}$$
$$p = 0.288\,\mathrm{debye}$$

In fact it is unlikely that the entire charge distribution would shift rigidly in the applied field. The electric field of the nucleus falls off as $\approx 1/r^2$, and so is much weaker in the outer regions of the atom than near the nucleus (Figure 5.24). The electrons respond to the *total electric field* due to both the nucleus and the applied electric field. Near the nucleus, the applied electric field is a small fraction of the total electric field, but for electrons in the outer regions of the atom the applied electric field may be a significant fraction of the total field. Thus the outer (valence) electrons tend to move more than the inner electrons in an applied field. These considerations would reduce the actual dipole moment below that predicted above.

the permanent electric dipole moment, p_p, is much larger than the induced electric dipole moment, p_i. In this case the main effect of the electric field is to rotate the molecule. Recall (see §2.3.3) that the *energy* of an electric dipole **p** in a field **E** is $-\mathbf{p}.\mathbf{E}$ and that the *torque* on an electric dipole **p** in a field **E** is $\Gamma = \mathbf{p} \times \mathbf{E}$.

Dielectric constant of a non-polar gas

We start by assuming that the electric field **E** around each molecule induces a electric dipole moment $\mathbf{p}_i$ on the molecule given by

$$\mathbf{p}_i = \alpha \mathbf{E} \tag{5.88}$$

where α is called the *molecular polarizability*. If there are n such molecules per unit volume, then the total polarization per unit volume, **P**, is given by

$$\mathbf{P} = n\mathbf{p}_i = n\alpha\mathbf{E} \tag{5.89}$$

Now the electric field **E** around each molecule is slightly reduced from the applied electric field $\mathbf{E}_{app}$ according to

$$\mathbf{E} = \frac{\mathbf{E}_{app}}{\varepsilon} \tag{5.90}$$

which defines the dielectric constant ε. However, the electric field **E** is also given by

$$\mathbf{E} = \mathbf{E}_{app} - \frac{\mathbf{P}}{\varepsilon_0} \tag{5.91}$$

by the definition of polarization (see Bleaney & Bleaney, §1.4.1). The minus sign arises because the electric field at a molecule due to all the other molecules ($\mathbf{P}/\varepsilon_0$) opposes the applied field. Combining Equation 5.90 with Equation 5.91 we have

$$\frac{\mathbf{E}_{app}}{\varepsilon} = \mathbf{E}_{app} - \frac{\mathbf{P}}{\varepsilon_0} \tag{5.92}$$

We now substitute for **P** using Equation 5.89 to yield

$$\frac{\mathbf{E}_{app}}{\varepsilon} = \mathbf{E}_{app} - \frac{n\alpha\mathbf{E}}{\varepsilon_0} \tag{5.93}$$

Neglecting the small difference between the applied field $\mathbf{E}_{app}$ and field around each molecule **E**, we arrive at

$$\frac{1}{\varepsilon} = 1 - \frac{n\alpha}{\varepsilon_0} \tag{5.94}$$

Solving for ε,

$$1 - \frac{1}{\varepsilon} = \frac{n\alpha}{\varepsilon_0} \tag{5.95}$$

and rearranging

$$\frac{\varepsilon - 1}{\varepsilon} = \frac{n\alpha}{\varepsilon_0} \tag{5.96}$$

Noting that ε is very close to unity, Equation 5.96 becomes

$$\boxed{\varepsilon - 1 = \frac{n\alpha}{\varepsilon_0}} \tag{5.97}$$

This may be compared with the exact expression for ε obtained after considerably more trouble by Clausius & Mossotti (Bleaney & Bleaney: p. 298),

$$\frac{\varepsilon - 1}{\varepsilon + 2} = \frac{n\alpha}{3\varepsilon_0} \tag{5.98}$$

Expressions 5.97 and 5.98 yield a prediction for ε in terms of the number density of molecules, n, the permitivitty of free space, ε_0, and the molecular polarizability, α, a quantity which may be calculated (with some difficulty) for each type of molecule. The magnitude of α depends on the ease with which the electronic charge around a molecule may be deformed by an applied electric field.

Dielectric constant of a polar gas

For a gas of polar molecules we have to take account of two different effects of the applied electric field.

- The molecules will have an electric dipole moment $\mathbf{p}_i$ *induced* on them.
- The orientation of molecules will be affected by the torque on their *permanent* electric dipole moment $\mathbf{p}_p$.
- The first of these effects has been discussed in the previous section, and so we consider here only the second effect. In the absence of an applied electric field the permanent electric dipole moments are randomly oriented resulting in zero net polarization of the gas. In the presence of an applied field there will be a tendency for more molecules to orient their per-

manent dipole moments $\mathbf{p_p}$ parallel to the applied field, resulting in a net polarization $\mathbf{P}$. We thus expect the polarization to be

$$\mathbf{P} \propto \text{Fraction} \times n \times \mathbf{p_p} \qquad (5.99)$$

where n is the number density of molecules and the "Fraction" (the fractional excess of molecules oriented parallel to $\mathbf{E}$) varies between 0 and 1. The fraction depends on the ratio of two energies: the random kinetic energy of the molecules, which is of the order of $k_B T$; and the orientational energy of the molecule in the applied field, which is of the order of $\mathbf{p_p} . \mathbf{E} \approx p_p E$. We thus expect the *fraction* in Equation 5.99 to be of the form

$$\text{Fraction} \propto \frac{p_p E}{k_B T} \qquad (5.100)$$

at least for small electric fields. Hence we expect the polarization of the gas to be given by

$$\mathbf{P} \propto \frac{n p_p E}{k_B T} \times \mathbf{p_p} \qquad (5.101)$$

or dropping the vector notation by

$$P \propto \frac{n p_p^2 E}{k_B T} \qquad (5.102)$$

An exact calculation (see Bleaney & Bleaney, §1.4.1) shows the constant of proportionality in Equation 5.102 to be 1/3, and so the polarization is given by

$$P = \frac{n p_p^2 E}{3 k_B T} \qquad (5.103)$$

This expression has exactly the same form as the expression known as the *Curie's law* for the *magnetization* of a substance containing freely rotating permanent *magnetic* dipole moments (see §7.8.4). Following the argument of Equations 5.92–5.97 in the previous section, we can convert this expression for the polarization into an expression for the dielectric constant due to the reorientation of the permanent dipole moments

$$\varepsilon - 1 = \frac{n p_p^2}{3 \varepsilon_0 k_B T} \qquad (5.104)$$

Comparison with experiment

We are now in a position to compare the data on the dielectric constants of gases with the predictions of the preceding sections. We expect that for gases of non-polar molecules ε should be given by

$$\varepsilon - 1 = \frac{n \alpha}{\varepsilon_0} \qquad (5.105)$$

where the molecular polarizability, α, is an intrinsic property of a molecule. For gases of polar molecules we expect that in addition to any induced dipole moment there should be an additional term given by

$$\varepsilon - 1 = \frac{n p_p^2}{3 \varepsilon_0 k_B T} \qquad (5.106)$$

where p_p is an intrinsic property of a molecule.

Let us now turn to the questions raised by the data.

Pressure dependence

Note that both the expressions for $\varepsilon - 1$ (Eqs. 5.105 and 5.106) depend linearly on n, the number density of molecules. Hence at a fixed temperature we expect $\varepsilon - 1$ to vary linearly with pressure, in accordance with the experimental behaviour of CO_2 shown in Figure 5.23.

Temperature dependence of non-polar molecules

The temperature dependence of $\varepsilon - 1$ is slightly harder to understand than the pressure dependence. From a wider understanding of the structure of molecules, we expect that the CO_2 molecule should have no permanent dipole moment and so it should obey Equation 5.97 for non-polar molecules. If this is so, then we would expect $\varepsilon - 1$ to show no temperature dependence. However, the data tabulated below (extracted from Table 5.16) show a clear dependence on temperature.

Gas	T (°C)	T (K)	$10^4(\varepsilon - 1)$
Carbon dioxide, CO_2	0	273	9.88
Carbon dioxide, CO_2	20	293	9.22
Carbon dioxide, CO_2	100	373	7.23

In fact the temperature dependence is "hidden" in the expression for the number density, n. The data in Table 5.16 are taken at constant pressure, and increasing the temperature at constant pressure implies a reduction in the number density, n (see §4.5, Eq. 4.49):

$$n = \frac{P}{k_B T} \qquad (5.107)$$

showing that, at constant pressure, n is proportional to $1/T$. Substitution into Equation 5.107 predicts that

$$\varepsilon - 1 = \frac{P\alpha}{\varepsilon_0 k_B T} \qquad (5.108)$$

If this is the origin of the temperature dependence, then we should expect the value of $\varepsilon-1$ at 373 K to be a factor of $273/373 = 0.732$ less than its value at 273 K. Comparing this with the data in Table 5.16 predicts that the value at 100°C should be $0.732 \times 9.88 \times 10^{-4} = 7.23 \times 10^{-4}$, which is in agreement with the experimental value.

Temperature dependence of polar molecules

Using this knowledge of the "hidden" temperature dependence of n we now expect the value of $\varepsilon-1$ for polar molecules to vary as

$$\varepsilon - 1 = \frac{p_p^2}{3\varepsilon_0 k_B T} \times \frac{P}{k_B T} = \frac{P p_p^2}{3\varepsilon_0 k_B^2 T^2} \qquad (5.109)$$

i.e. with a $1/T^2$ temperature dependence. Examining the data for the polar molecule ammonia in Table 5.16, we expect the value of $\varepsilon-1$ at 373 K to be a factor of $(273/373)^2 = 0.536$ less than its value at 273 K. Comparing this with the data in Table 5.16 predicts that the value at 100°C should be $0.536 \times 8.34 \times 10^{-4} = 4.46 \times 10^{-4}$, which is in rough agreement with the experimental value of 4.87.

Gas	T (°C)	T (K)	$10^4(\varepsilon-1)$
Ammonia, NH_3	0	273	8.34
Ammonia, NH_3	100	373	4.87

The discrepancy between experiment and theory can be understood as arising from the fact that polar molecules have contributions to $\varepsilon-1$, both from the reorientation of their permanent electric dipole moments and from their induced dipole moment. Thus we assign the difference between the observed and predicted value to the induced electric dipole

moment. (See also §5.7.2 on the optical properties of gases.)

Dependence of ε−1 on the type of molecule

The final points we need to understand are the general magnitude of $\varepsilon-1$ and the occasional occurrence of some extremely large values (Figure 5.22). We can understand this by considering as examples the distribution of electric charge within two simple diatomic molecules: oxygen and carbon monoxide.

Even without knowing the details of the charge distribution within an oxygen molecule, we can say that we would expect the charge to be distributed symmetrically between the two atoms. We say this because each oxygen atom is identical and has an equal attraction (or *affinity*) for electrons. If we think now about carbon monoxide (CO), we would expect the charge to be distributed asymmetrically between the two atoms. We say this because each atom is different and so has a different characteristic *electron affinity*. In this case, the electrons are more attracted to the oxygen atom than the carbon atom and so the oxygen atom becomes negatively charged with respect to the carbon atom. In other words a CO molecule has permanent electric dipole moment.

We can estimate the magnitude of the permanent dipole moment, p_p, by considering the CO datum in Table 5.16: $10^4(\varepsilon-1) = 6.92$ at 23°C (296 K). Rearranging Equation 5.109, into an expression for p_p we obtain

$$p_p = \sqrt{\frac{3 k_B^2 T^2 (\varepsilon-1) \varepsilon_0}{P}} \qquad (5.110)$$

Evaluating this at atmospheric pressure ($P = 1.013 \times 10^5$ Pa) we find, $p_p =$

$$\sqrt{\frac{3 \times \left(1.38 \times 10^{-23} \times 296\right)^2 \times 6.92 \times 10^{-4} \times 8.85 \times 10^{-12}}{1.013 \times 10^5}}$$

$$(5.111)$$

$$p_p = 1.74 \times 10^{-30} \text{ C m} = 1.74 \text{ debye} \qquad (5.112)$$

Does this make sense? Imagine that some fraction, f, of an entire electronic charge, e, is transferred from the carbon to the oxygen atom. The separation, r_0, of

108

the atoms in a CO molecule is given in Kaye & Laby (see §1.4.1) as 0.1131 nm. We thus expect that the permanent dipole moment will be given by

$$p_p = 1.74 \times 10^{-30}$$
$$= r_0 fe = 0.113 \times 10^{-9} \times f \times 1.6 \times 10^{-19} \quad (5.113)$$

which predicts $f = 0.096 \approx 0.1$, which seems reasonable. Note that the magnitude of p_p yields the *product* of fe and r_0, and so we may interpret the magnitude of p_p as being:

• the transfer of a fractional charge, fe, across the entire length of the molecule; or
• the transfer of an entire electronic charge, e, across a fraction, f, of the entire length of the molecule; or
• any combination of the above two effects.

The above discussion allows us to understand why larger molecules tend to possess larger dipole moments. In larger molecules it is possible for an amount of charge of the order of e to be transferred across relatively large distances, and hence give rise to relatively large dipole moments. This charge transfer arises as part of the processes of chemical interaction between the atoms within a molecule. The larger physical extent of larger molecules allows charge to be spread across larger distances and hence give rise to larger *permanent* electric dipole moments.

Considering the non-polar gases listed in Table 5.16, we see that $\varepsilon - 1$, and hence the *induced* dipole moment, tends to increase with increasing atomic number, i.e. the molecular polarizabilty, α, is larger for larger molecules. We can understand this by considering the electric field that acts on the valence electrons in an atom (Fig. 5.24c,d). For an atom such as helium, the two electrons are extremely close to the electric charge on the nucleus, $+2e$. In an atom such as xenon, however, the outer electrons are further away from the nucleus and experience a much weaker electric field. As a consequence the electric field around the valence electrons in xenon is much less than that around the valence electrons in helium. Thus an applied electric field of a given strength affects the valence electrons around a xenon atom more strongly than the electrons around a helium atom. This amounts to making a xenon atom more deformable or technically more *polarizable* than a helium atom.

We note that in Figure 5.22, all three of the substances with anomalously large dipole moments contain an OH group of atoms. A wide range of chemical experiments indicate that the OH group, known as a *hydroxyl* group, has a particularly strong affinity for electrons, thus causing a large amount of charge to be transported away from the centre of the molecule.

5.6.3 Data on the electrical properties of gases in a strong electric field

Remember that, at low electric fields, gases form effective electrical insulators with low values of electrical conductivity. At high electric fields however electric charge can move through a gas to produce a variety of interesting phenomena. The value of electric field at which this change in behaviour takes place is known as the *breakdown electric field* or the *dielectric strength* of the gas.

Taking air as a typical gas, the dielectric strength of air at 25°C and normal atmospheric pressure (0.1013 MPa) is $3.13 \times 10^6 \text{ V m}^{-1}$. The variation of the dielectric strength of air with temperature and pressure is shown in Figure 5.26.

The value of the electric field at which breakdown occurs varies between gases, and also depends critically on several "minor" properties of the gas. In particular, the presence of ions, i.e. atoms that are not electrically neutral, significantly lowers the dielectric strength of a gas. In practice, one must also consider the humidity of the gas: humid air can deposit a microscopic layer of water on solid surfaces that it contacts. Although this extremely thin layer is only a weak conductor of electricity, it can "short circuit" the even more highly resistive air. In different experiments this can lead to anomalous apparent increases or decreases in the dielectric strength of humid air.

In the nineteenth century, great efforts were put into understanding the nature of the flow of electric current through a gas, and a considerable understanding of the phenomena associated with "cathode rays" was achieved. These experiments advanced our understanding of the nature of gases and of the internal structure of atoms enormously. A selection of the phenomena we have to understand are succinctly described by one of the foremost scientists of that age, J. J. Thomson, in his book *Conduction of electricity through gases*. In the following extract he

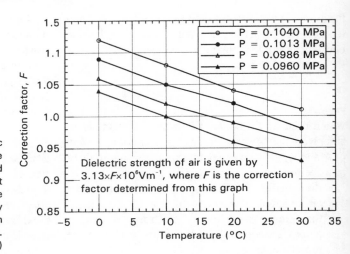

Figure 5.26 The variation in the dielectric strength of air with temperature and pressure in the region of ambient temperature and pressure. The dielectric strength is increased at low temperatures and high pressures. The dielectric strength of air is given by $3.13 \times F \times 10^6 \, V \, m^{-1}$, where F is the correction factor determined from the graph shown. (Data from *CRC handbook*; see §1.4.1)

Example 5.12

An air-cored parallel-plate capacitor has a plate separation of $d = 1 \, mm$. If the air pressure is 0.1 MPa and the temperature is 5°C, what is the maximum voltage that can be applied between the plates before sparking occurs?

From the text, the dielectric strength of air at 25°C and normal atmospheric pressure (0.1013 MPa) is $3.13 \times 10^6 \, V \, m^{-1}$. We use the graph (Figure 5.21) to find a corrected value for our temperature and pressure. Looking up the 5°C line on the graph reveals that the correction factor will be between 1.04 (which would be appropriate at a pressure of 0.0986 MPa) and 1.07 (which would be appropriate at a pressure of 0.1013 MPa). Our pressure lies roughly midway between these values, and so we estimate a value of 1.055. The dielectric strength of air at 5°C and 0.100 MPa is $1.055 \times 3.13 \times 10^6 \, V \, m^{-1} = 3.30 \times 10^6 \, V \, m^{-1}$.

The electric field between the plates of a capacitor is given approximately by $E = V/d$. If we work out the voltage which will yield the maximum sustainable field we have

$$V = Ed = 3.30 \times 10^6 \times 1 \times 10^{-3} = 3.30 \times 10^3 \, V$$

Thus the capacitor can, in principle, be used up to 3.3 kV. However, the electric field around the edge of a capacitor plate can be significantly greater than the field between the plates, and breakdown would occur there before it would in between the plates. A more likely value of maximum working voltage is $\approx 1 \, kV$. Note that for very small gaps of the order of $1 \, \mu m$ ($10^{-6} \, m$) this breakdown voltage can be only a few volts.

describes the general nature of the phenomena that occur when an electric current is passed through a glass tube containing gas at low pressure:

When the electric discharge passes through a gas at low pressure, differences in the appearance of the gas at various points in its path become very clearly marked. The discharge, as illustrated [see Fig. 5.27], presents the following features: starting from the cathode (negative terminal) there is a thin layer of luminosity (1) spread over its surface; next to this there is a comparatively dark space called "Crookes dark space" (2), the width of which depends on the pressure of the gas, increasing as the pressure diminishes – it also depends, under some conditions, on the intensity of the current. The boundary of the dark space is approximately the surface traced out by normals of constant length drawn to the surface of the cathode. Beyond the dark space there is a luminous region (3) called the "negative glow". Beyond this again is another comparatively dark region (4) called by some writers the "second negative dark space" and by others the "Faraday dark space". Its length is very variable, even when the pressure is constant. Beyond this again there is a luminous column (5) reaching right up to the anode and called the "positive column". When the current and pressure are within certain limits this column exhibits remarkable alternations of dark and bright spaces: these are called striations. In long tubes the positive column constitutes by far the greater part of the discharge, for the Crookes space, negative glow and Faraday dark space do not depend markedly on the length of the tube. So that when the length of the dis-

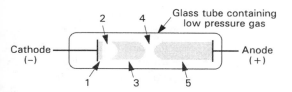

Figure 5.27 An illustration of the pattern of light emitted when an electric current flows through a gas, called a *discharge*. The details of the pattern (1–5) are discussed in the text. The shaded areas in the figure represent luminous regions of the gas, and the unshaded regions represent regions from which no light is emitted.

charge is increased, the increase is practically only in the length of the positive column. Thus for example in a tube about 15 metres used by one of us, the positive column occupied the whole of the tube with the exception of two or three centimetres close to the cathode.

J. J. Thomson, *Conduction of electricity through gases*, Cambridge University Press: Cambridge, 1933, p.292)

The colour of the "luminosity", particularly within the positive column (feature 5 in Fig. 5.27), depends on the type of gas. Hydrogen, for instance, has a red glow, as does neon, but argon has a bluish tint. If one examines the spectrum of the light then one finds that the spectrum is characteristic of the type of gas in the tube. More details of these *emission spectra* are given in §5.7.5 on the optical properties of gases.

The main questions raised by our preliminary examination of the experimental data on the properties of gases under strong electric fields are:

(a) Why is the dielectric strength largest at low tem-

peratures and high pressures, i.e. when the gas is most dense?

(b) Why, below a critical electric field, the dielectric strength, do gases behave as electrical insulators, but above it they behave as electrical conductors, displaying a variety of optical effects?

5.6.4 Understanding the data on the electrical properties of gases in strong electric fields

When the electric field exceeds the *dielectric strength*, the gas changes from behaving like an insulator to become an electrical conductor. In this section we will propose a formula for the temperature and pressure dependence of the dielectric strength of a gas, and then use the ideas developed for this formula to discuss qualitatively the phenomena observed in a discharge tube by J. J. Thompson.

Background theory

In order to understand how the breakdown of the insulating properties of the gas occurs we need to consider the situation where a single charged particle, an ion, is present in the gas while an electric field **E** is applied. The ion might be the result of an interaction between an atom and a fast-moving particle from a cosmic ray shower. In the absence of an electric field, the electron and ion created by the ionization will quickly recombine (Fig. 5.28). In the presence of an electric field, the electron and ion created by the ionization may be prevented from recombining and can

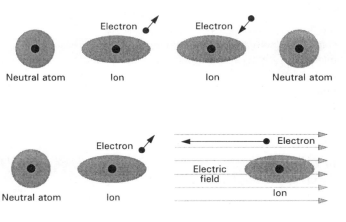

Figure 5.28 (a) Initially we have a neutral atom, then (b) one of the events described in the text causes an ionization event. (c) In the absence of an applied field the negatively charged electron and the positively charged ion quickly recombine. (d) Finally, we return to the situation we started with. Thus the gas is stable against the formation of ions.

Figure 5.29 (a) Initially we have a neutral atom, then (b) one of the events described in the text causes an ionization event. (c) In contrast with Figure 5.28, due to the presence of an applied electric field, the negatively charged electron and the positively charged ion are drawn apart, and recombination is prevented.

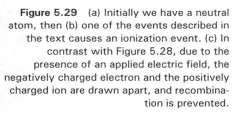

be accelerated through the gas by the applied electric field (Fig. 5.29).

Given that a free ion, separated from "its" electron, has been created, it will experience force $q\mathbf{E}$ which will accelerate it. Hence its speed will increase and it will acquire more kinetic energy the further it travels in the field. If the *mean free path*, λ_{mfp}, of the ion in the gas is long enough, or the electric field strength is great enough, then the ion may acquire enough kinetic energy to ionize the atom with which it next collides. The consequences of this are enormous. There will now be two ions instead of just one. What happened to the first ion may also happen to the second one and, if it does, the two ions will cause further ions to be created Thus, starting with just a single ion, the number of ions in a gas can increase in an effect called an *avalanche*. Also, under the influence of the applied electric field the second ion will move along in the same direction as the first, whereas the electron released from the second ion will travel in the opposite direction (Fig. 5.29). After many collisions an electric current flows through the gas, with electrons travelling in one direction, and ions travelling in the other.

Thus the presence of a single ion has caused the gas to become an electrical conductor instead of an electrical insulator. The electric field that must be applied to a gas to cause it to become a conductor is called the *breakdown electric field*, or the *dielectric strength* of the gas.

Pressure and temperature dependence of the dielectric strength

We can consider the ionization process that initiates and maintains the density of ions responsible for conduction in a gas as a two-step process. In the first stage, an electron or ion (charge q) accelerates for a distance, λ, under the action of the electric field and gains energy u (typically of the order of few electron-volts), given by

$$u = \int_0^\lambda q\mathbf{E}.\mathbf{ds} = eE\lambda \qquad (5.119)$$

In the second stage, some of this energy (typically of the order of a few hundredths of an electron-volt) is given up in collisions with molecules that lead, on average, to a further ionization process.

However, although this two-stage process is easy

to describe at this level, working out λ, the average path length of an ion or an electron between collisions, is rather difficult. Note that because the ions and electrons are electrically charged, ion–molecule and electron–molecule interactions are considerably stronger than molecule–molecule interactions. Calculating the effective diameters or cross-sectional areas of molecules for ionic collisions is not an easy problem. However, we can say one thing for sure: whatever the value of λ is, it will be inversely proportional to n, the number density of molecules in the gas.

So from Equation 5.114 we expect that the break down field E_B will be defined when the work done in accelerating an ion or electron is sufficient to ionize a molecule i.e. when

$$u_0 = eE_B\lambda \qquad (5.115)$$

where u_0 is the ionization energy. Since u_0 is characteristic of the molecules of the gas we expect to find that E_B is proportional to $1/\lambda$. Hence, based on the arguments in the preceding paragraph, we expect quite generally to find

$$E_B \propto \frac{1}{\lambda} \propto n \qquad (5.116)$$

Recalling that $n = P/k_B T$ (see §4.5, Eq. 4.49), we therefore expect that

$$E_B = A\frac{P}{T} \qquad (5.117)$$

where A is an expression involving λ and u_0. Thus we expect the dielectric strength to be proportional to pressure and inversely proportional to absolute temperature.

Examining Figure 5.26, we see that at 0°C the correction factor increases from 1.04 at $P = 0.096$ MPa to 1.12 at $P = 0.104$ MPa. Thus the dielectric strength increases by a factor of $(1.12/1.04) = 1.08$ as a result of a pressure increase of $(0.104/0.960) = 1.08$. At limited resolution, this is a fairly good agreement.

Similarly, at a pressure of 0.1013 MPa, the correction factor of 1.09 at $T = 273.2$ K falls to a correction factor of 0.98 at a $T = 273.2 + 30$ K. Thus the dielectric strength decreases by a factor of $(0.98/1.09) = 0.90$ as a result of an increase in temperature by a factor of $(303.2/273.2) = 1.11$. Since $1/1.11 = 0.90$ this is a fairly good agreement, albeit at limited resolution.

This result is somewhat counterintuitive. It might be thought that the best way to stop a gas from conducting would be to use a low gas pressure i.e. an approximation to a vacuum which is an excellent insulator. It is not too difficult to reduce the pressure of a gas by a factor of about 10^9 from atmospheric pressure $(10^{-6}\,\text{mbar} \approx 10^{-4}\,\text{Pa})$. However, even here there are still $\approx 10^{16}$ molecules m^{-3}. In these circumstances, if a single ion is created it will travel straight across the vacuum chamber with virtually no chance of collision with another molecule. Thus the mean free path becomes fixed at the dimension of the container and doesn't get any longer as the pressure is

lowered. This situation is similar to the behaviour of the pressure dependence of the thermal conductivity of a gas (see §5.4). The relevant ionization energy is then that of the atoms in the walls of the container. If an ion hits the walls and creates more ions, then ions will simply bounce between the walls of the container. Lowering the gas pressure will make no difference until there are no atoms at all – presently an unachievable situation.

Figure 5.30 An illustration of the pattern of light emitted when an electric current flows through a gas, called a *discharge*. The details of the pattern (1–5) are discussed in the text. The shaded areas in the figure represent luminous regions of the gas and the unshaded regions represent regions from which no light is emitted.

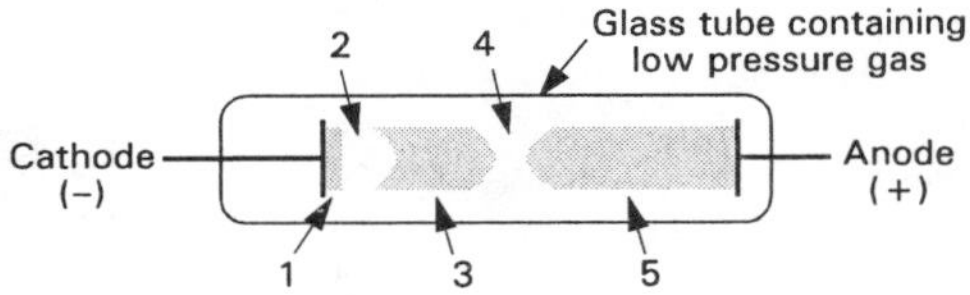

The phenomena within a discharge tube

In the light of our discussions above, let us re-examine the description given by J. J. Thomson, in his book *Conduction of electricity through gases*, which was reproduced in §5.6.3.

> When the electric discharge passes through a gas at low pressure, differences in the appearance of the gas at various points in its path become very clearly marked. The discharge [as illustrated in Fig. 5.30] presents the following features: [Table 5.17]

Table 5.17

J. J. Thompson's description	What's happening
Starting from the cathode (negative terminal) there is a thin layer of luminosity (1) spread over its surface.	The luminosity is caused by positive ions striking the surface of the cathode.
Next to this there is a comparatively dark space called "Crookes dark space" (2), the width of which depends on the pressure of the gas, increasing as the pressure diminishes – it also depends, under some conditions, on the intensity of the current. The boundary of the dark space is approximately the surface traced out by normals of constant length drawn to the surface of the cathode.	In this region electrons liberated from the cathode (by the impact of positive ions) are being accelerated by the electric field in the tube. They are colliding with atoms and ions in this region but give off no light because they do not have sufficient energy to excite the atoms. The boundary of the region has this form because electrons have travelled in straight lines for a distance of around one ionic mean free path from the cathode.
Beyond the dark space there is a luminous region (3) called the "negative glow".	Now electrons do have sufficient energy to excite the atoms.
Beyond this again is another comparatively dark region (4) called by some writers the "second negative dark space" and by others the "Faraday dark space". Its length is very variable, even when the pressure is constant.	Having collided inelastically with atoms of the gas in region (3), the electrons lost kinetic energy and are now being accelerated again.
Beyond this again there is a luminous column (5) reaching right up to the anode and called the "positive column". When the current and pressure are within certain limits this column exhibits remarkable alternations of dark and bright spaces: these are called "striations". In long tubes the positive column constitutes by far the greater part of the discharge, for the Crookes space, negative glow and Faraday dark space do not depend markedly on the length of the tube. So that when the length of the discharge is increased, the increase is practically only in the length of the positive column. Thus for example in a tube about 15 metres used by one of us, the positive column occupied the whole of the tube with the exception of two or three centimetres close to the cathode.	The striations are caused by alternating regions of the gas in which: • electrons and ions are accelerated and do not yet have sufficient energy to ionize/excite the atoms of the gas: these are the non-luminous dark regions, • electrons and ions accelerated in the above "dark" regions now have sufficient energy to ionize/excite the atoms of the gas: these are the luminous regions. In some circumstances where the geometry of the acceleration is not very well defined, or where the cathode is heated to create a spread of initial electron velocities, the dark and light striations become blurred and overlap one another.

The spectra of the "luminosity" is discussed below in §5.7 on the optical properties of gases. The state in which a substantial fraction of gas molecules are ionized is generally referred to as a *plasma*. By far the majority of the matter in the universe exists in this state, but on Earth it is relatively rare, and we pass it by without further mention.

5.7 Optical properties

"Light" is the name given to oscillations of the electromagnetic field that take place with frequencies in the range 400 THz (red) to 750 THz (blue) (1 THz = 10^{12} Hz). Below we concentrate entirely on the effect of the oscillations of the *electric* component of the electromagnetic field.

5.7.1 Data on the speed of light in gases: refractive index

Light travels at a slightly slower speed, c_g, through a gas than its speed through vacuum c. An apparatus illustrating how such changes may be measured is illustrated in Figure 3.6. The ratio c/c_g is known as the *refractive index* of the gas, and normally has the symbol n. In order to avoid confusion with n used to signify number density, the refractive index in this text has the symbol n_{light}. The refractive indices of various gases are shown in Table 5.18 and plotted as a function of relative molecular mass in Figure 5.31. Because n_{light} for a gas is close to unity, the table

Table 5.18 The refractive index of various gases as $10^6(n_{light} - 1)$ and the molecular weight, M, of the molecules of the gas*.

Gas	M	$(n_{light} - 1) \times 10^6$
Hydrogen, H_2	2	132
Helium, He	4	36
Methanol, CH_3OH	32	586
Methane, CH_4	18	444
Water vapour, H_2O	18	254
Ammonia, NH_4	18	376
Neon, Ne	20	67
Nitrogen, N_2	28	297
Carbon monoxide, CO	28	338
Air	29	293
Nitric oxide, NO	30	297
Oxygen, O_2	32	271
Hydrogen sulphide, H_2S	34	633
Hydrogen chloride, HCl	36	447
Fluorine, F_2	38	195
Argon, Ar	40	281
Nitrous oxide, N_2O	44	516
Carbon dioxide, CO_2	44	451
Ethanol, C_2H_5OH	46	878
Sulphur dioxide, SO_2	64	686
Chlorine, Cl_2	71	773
Carbon disulphide, CS_2	76	481
Benzene, C_6H_6	78	1762
Hydrogen bromide, HBr	81	570
Krypton, Kr	84	427
Hydrogen iodide, HI	128	906
Xenon, Xe	131	702
Bromine, Br_2	160	1132

*The data refer to gases at standard temperature and pressure (P = 0.1013 MPa, T = 0°C). The refractive index is that appropriate to the bright yellow "D" lines in the emission spectrum of sodium vapour and varies slightly with frequency.
(Data from Kaye & Laby; see §1.4.1)

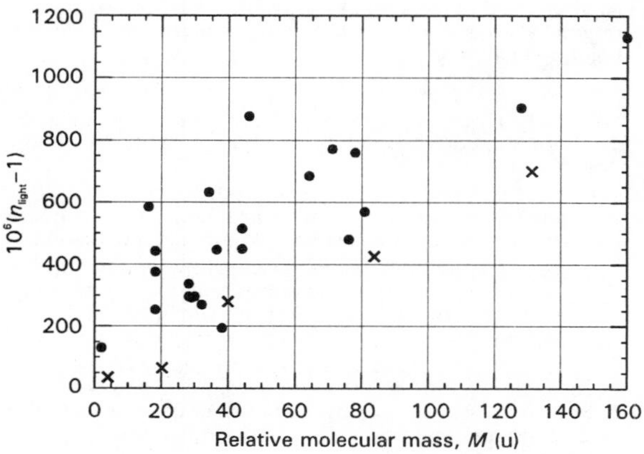

Figure 5.31 The refractive index, n, of the gases listed in Table 5.18 plotted as a function of the relative molecular weight of the gas molecules. The refractive indices of the gases are all within 0.2% of unity and so the quantity $10^6(n - 1)$ has been plotted in order to make visible the differences in the data. The crosses mark the data points for the noble gases helium, neon, argon, krypton and xenon.

Example 5.13

It is required to split an optical signal in two, and to delay one beam with respect to the other travelling through a vacuum by 4.7 ps (4.7×10^{-12} s). It is suggested that the light be passed through a length, L, of gas in order to achieve this delay and, for reasons of its inertness and cost, argon is thought suitable. What length of tube is required if the gas is at STP? Is this method really practical?

We require a tube which will take a light signal 4.7 ps longer to traverse than an equivalent tube filled with vacuum. If the tube were evacuated the transit time would be $t = L/c$, but when filled with gas it will be $t_g = L/c_g$.

We need a device in which $t_g - t = 4.7$ ps, i.e.

$$t_g - t = \frac{L}{c_g} - \frac{L}{c} = 4.7 \times 10^{-12} \, \text{s}$$

We now note that Table 5.18 tells us that for argon at STP the refractive index is given by $n-1 = 281 \times 10^{-6}$ i.e. $n = 1.000281$. From the simple definition of the refractive index this implies that $c_g = c/1.000281$. Rearranging our equation we have

$$L\left(\frac{1}{c/1.000281} - \frac{1}{c} \right) = 4.7 \times 10^{-12}$$

which simplifies to

$$\frac{L}{c}(1.000281 - 1) = 4.7 \times 10^{-12} \quad \text{so} \quad L = \frac{4.7 \times 10^{-12} \, c}{281 \times 10^{-6}}$$

which evaluates to

$$L = \frac{4.7 \times 10^{-12} \times 2.998 \times 10^8}{281 \times 10^{-6}} = 5.01 \, \text{m}$$

This is a rather long tube for practical use, given that it needs to be kept temperature stabilized. However, the technique might be appropriate for much shorter delays.

records data in the form $10^6(n_{light} - 1)$. Thus the refractive index of air recorded as 293 is actually 1.000293, so the speed of light in air is $c/1.000293$, i.e. 99.97% of the speed of light in vacuum.

Figure 5.31 shows that there is a general trend towards large molecular mass molecules displaying a higher refractive index, but the wide variation in n_{light} shows that there are significant factors other than molecular mass. The data shown for the monatomic noble gases helium, neon, argon, krypton and xenon in Figure 5.31 show that amongst similar molecules the link between n_{light} and M is relatively direct.

The main questions raised by our preliminary examination of the experimental data on the refractive indices of gases are

(a) Why do the refractive indices of a wide variety of gases at STP all lie between 1.000 and 1.002?

(b) Why, within this range, is there a significant dependence of n upon the molecular mass?

5.7.2 Understanding data on the refractive index of gases

Our explanation of the interaction of light with gases follows on from the discussion of the dielectric constant data (see §5.6.2). We assume that each molecule responds independently to the oscillating electric field of the light wave, and that in response to this field the molecule either:

(a) acquires an induced dipole moment p_i; or

(b) orients its own permanent dipole moment p_p (if it has one) parallel to the instantaneous value of the electric field.

As we shall see, in general the second process is not as important as the first because (Example 5.14) the time taken for a molecule to rotate is typically of the order 10^{-10} s which is much longer than the period of the electric field oscillation in a light wave (of the order of 10^{-15} s) (Fig. 5.32).

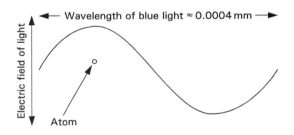

Figure 5.32 Comparison of the wavelength of a blue light wave with the size of an atom. Note that in order to be clearly printed the atom has been shown at approximately 10 times its correct scale size.

So the main effect of a light wave is to induce an oscillating electric dipole moment on the molecules of the gas (Fig. 5.33). In general, the magnitude of the induced dipole moment varies with frequency, and generally grows resonantly large at a frequency in the ultraviolet region of the spectrum. The *frequency dependence* of the dipole moments induced on atoms

Example 5.14

Consider a diatomic polar molecule (i.e. one with a "built-in" electric dipole moment) that has a permanent electric dipole moment of approximately 0.1×10^{-30} C m.
(a) Estimate the torque on the molecule in an electric field of $1000\,V\,m^{-1}$.
(b) Calculate the time taken for the molecule to rotate through 90°. This is tricky if you are unfamiliar with dealing with rotational calculations.

(a) Assuming the moment is oriented perpendicular to the field, then $\Gamma = pE = 0.1 \times 10^{-30} \times 10^3 = 10^{-28}$ Nm.

(b) We recall that we use the rotational analogues to the equations of linear motion. Analogous to $F = ma$ we have $\Gamma = I(d^2\theta/dt^2)$ where Γ is the torque described above, I is the *moment of interia* defined below, and $d^2\theta/dt^2$ is the *angular* acceleration.

Similarly, analogous to $s = ut + \frac{1}{2}at^2$ or, with $u=0$, $s = \frac{1}{2}at^2$, we have $\theta = \frac{1}{2}(d^2\theta/dt^2)t^2$

The moment of inertia of the molecule is given by $I = m_1 \times d_1^2 + m_2 \times d_2^2$, where d_1 and d_2 are the distances to the two atoms from the centre of mass of the molecule. This amounts to approximately,

$$I \approx \text{Average mass of two atoms} \times (\tfrac{1}{2}r)^2$$

where r is the distance between the two atoms, typically $\approx 1.3 \times 10^{-10}$ m (*CRC handbook*). Assuming average masses of the two atoms of $\approx 14u$ (i.e. like CO), the moment of inertia is approximately

$$I \approx 14u \times \left(\tfrac{1}{2} \times 1.3 \times 10^{-10}\right)^2$$

$$\approx \left(14 \times 1.66 \times 10^{-27}\right) \times \left(0.65 \times 10^{-10}\right)^2 = 9.8 \times 10^{-47}\,\text{kg m}^2$$

Substituting this into the analogue of $F = ma$, we find an angular acceleration of

$$\frac{d^2\theta}{dt^2} = \frac{\Gamma}{I} \approx \frac{10^{-28}}{10^{-46}} = 10^{18}\,\text{rad s}^{-2}$$

So we can now find out how long it takes the molecule to rotate. Using $\theta = \frac{1}{2}(d^2\theta/dt^2)t^2$ and rearranging to solve for t

$$t = \sqrt{\frac{2\theta}{d^2\theta/dt^2}} \quad \text{so} \quad t = \sqrt{\frac{2(\pi/2)}{10^{18}}} \approx \sqrt{\pi \times 10^{-18}} \approx 1.8 \times 10^{-9}\,\text{s}$$

So we find that it takes a molecule a couple of nanoseconds to rotate in a weak electric field. This may seem like a short time, but in fact it can in certain circumstances amount to a long time. For example, the electric field of a light wave oscillates at around 10^{15} times per second. In a field oscillating this fast the molecule will have no time to rotate, whereas in a DC field the molecule will have plenty of time to rotate. This causes a difference between the electrical properties of gases at DC and optical frequencies.

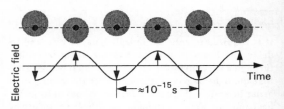

Figure 5.33 Illustration of the origin of the oscillating dipole moment on a non-polar molecule subject to an oscillating electric field. Note that in the very short time between successive periods of the light field, only the low-mass electrons can move a significant distance. The heavy nucleus moves relatively little. It is an analogous inertia that prevents polar molecules from rotating significantly during a single period of the light wave.

and molecules is considered more fully in the analysis of the optical properties of solids (see §7.7). Here we concentrate on the effect of these induced dipole moments on the speed of light through the gas.

The speed of light is related to the relative permittivity ε and relative permeability, μ, of a medium by (see §2.3.5, Eq. 2.23)

$$c_g = \frac{1}{\sqrt{\varepsilon\varepsilon_0\mu\mu_0}} \tag{5.118}$$

Since $\mu = \varepsilon = 1$ for a vacuum, we can write down an expression for the refractive index in a gas as

$$n_{light} = \frac{c}{c_g} = \frac{\sqrt{\varepsilon\varepsilon_0\mu\mu_0}}{\sqrt{\varepsilon_0\mu_0}} = \sqrt{\varepsilon\mu} \approx \sqrt{\varepsilon} \tag{5.119}$$

where we have assumed (see §7.8) that the magnetic permeability of a gas is close to 1.

We can check Equation 5.119 by using the values of $\varepsilon - 1$ given in Table 5.16 to predict values for $n_{light} - 1$ (Table 5.19). The non-polar gases (helium, neon and argon) show excellent agreement. Recalling that ε is determined at low frequencies (i.e. $<10^9$ Hz), this agreement indicates that the molecular polarizability of these atoms is relatively constant from radio frequencies up to optical frequencies.

When we consider the polar molecules listed in Table 5.19 we find that this agreement is not as good: knowing the value of $\varepsilon - 1$ does not enable us to predict the value of $n_{light} - 1$. This disagreement is slight in the case of ammonia, but for water vapour the difference is dramatic. The reason is that the value of $\varepsilon - 1$ for polar molecules includes the effect of

Table 5.19 Comparison of the experimental values of the refractive index of gases with the values predicted using Equation 5.119*.

Gas	$10^4(\varepsilon-1)$	T	Correction factor	$10^4(\varepsilon-1)$ (STP)	Predicted $10^6(\sqrt{\varepsilon}-1)$	Experimental $10^6(n_{light}-1)$
He	0.65	20	293/273	0.70	35	36
Ne	1.3	0	1	1.3	65	67
Ar	5.16	20	293/273	5.54	277	281
NH_3	8.34	0	1	8.34	416	376
H_2O	60	100	$(293/273)^2$	69.1	3449	254

*The dielectric constant data have been corrected to STP using factors discussed in §5.6.2. The first three entries in the table are for non-polar gases, and the last two are for polar gases. Note the good agreement between theory and experiment for the non-polar gases, and the massive disagreement for water vapour.

molecular rotation and, as mentioned above, the oscillations of electric field at optical frequencies are too rapid to allow molecules to rotate. Thus the refractive index measurement is sensitive only to the induced component of the electric dipole moment of a molecule.

So turning to the questions raised at the end of §5.7.1, we can understand both these phenomena in terms of the connection outlined in Equation 5.119 between n_{light} and ε. The small magnitude of $(n_{light}-1)$ results from the small magnitude of $\varepsilon-1$ (Eq. 5.105)

$$\varepsilon - 1 = \frac{n\alpha}{\varepsilon_0} \qquad (5.120)$$

which in turn is small because of the intrinsic magnitude of molecular polarizability, α and the low number density of molecules in the gas. The dependence of n_{light} upon molecular mass arises because larger molecules tend to have larger values of molecular polarizability.

5.7.3 The scattering of light by gases

At optical frequencies, most gases are extremely transparent. However, they are not 100% transparent. This can be easily seen by looking through a few kilometres of our most readily available gas: air. Objects viewed at a distance can appear "hazy" or "coloured". Also, looking upwards into the air, we see mainly blue light coming apparently from nowhere (we say "the sky is blue"). Furthermore, the sun – which appears to be yellow during the day – appears red as it rises or sets. All these effects are due to the *scattering*

of light by the molecules of the gases in the atmosphere. The main questions raised by our qualitative observations of the scattering of light are:

- Why are gases highly, but not completely, transparent?
- Why are some colours scattered more strongly than others?

What is scattering?

The scattering of light caused by atoms in a gas is known as *Rayleigh scattering* (Figure 5.34). The process may be considered in two stages:

- An incoming electromagnetic wave polarizes molecules of the gas and induces an oscillating electric dipole moment.
- An oscillating electric dipole moment radiates energy in a complicated pattern. The radiated – or scattered – wave is at the same frequency as the incoming wave, but travels radially away from the scattering centre.

Figure 5.34 Rayleigh scattering. An incoming plane wave causes an oscillating electric dipole moment (see Fig. 5.33). The oscillating dipole moment then re-radiates some of the incoming wave as a spherical wave emanating from the molecule. The figure shows a two-dimensional analogue of the scattering, such as might occur with water waves on the surface of a pond.

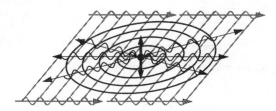

The "scattering power" of an individual molecule depends on the magnitude of the dipole moment induced on it by the electric field of the light wave. Clearly, if no dipole moment is induced there will be no scattering. Recall that, as we saw in examining the data on the refractive index of gases (see §5.7.2), at optical frequencies only the induced dipole moment needs to be considered – polar molecules have no time to rotate in a single cycle of the oscillating electric field of the light wave. The theory of an oscillating dipole radiator (Bleaney & Bleaney, see §1.4.1) indicates that the power radiated by a dipole whose magnitude varies as

$$p = p_0 \cos(2\pi f t) \tag{5.121}$$

is given by

$$\text{Power} = \frac{4\mu_0 \pi^3 f^4 p_0^2}{3c} \tag{5.122}$$

Thus the scattered power depends on the fourth power of the frequency i.e. f^4. Since blue light has a frequency of $f \approx 7 \times 10^{14}\,\text{Hz}$ and red light has a frequency $f \approx 4 \times 10^{14}\,\text{Hz}$ then, according to Equation 5.122 blue light should be scattered at least $(7/4)^4 \approx 2.5^4 \approx 39$ times more strongly than red light.

***Blue skies, sunsets, and the difficulty
of seeing stars in the daytime***
We are familiar with the fact that when standing on earth looking out into space we cannot see the stars if the atmosphere through which we are looking is illuminated by sunlight (Fig. 5.35). This is because the small amount of scattered light from the sun is much brighter than the feeble light from the stars. The scattered light is primarily blue because, as we noted above, blue light is more strongly scattered than red light.

Furthermore, the light from the sun at sunset, has passed through a larger than normal thickness of the Earth's atmosphere. Thus at sunset the sun appears considerably less bright than it does at midday, indicating that some scattering must be taking place. In addition, because the blue light has been scattered more strongly than the red (in order to make blue skies for other observers) the light that reaches us has a stronger yellow/red component than the original spectrum of sunlight.

5.7.4 Data on the emission spectra of gases

As outlined in §5.6.3 on the electrical properties of gases, when an electric current is passed through a gas, the gas gives off light from its *positive column*. The colour of the light is indicative of the type of gas through which the current is passed. However, "colour" is a subjective description of the light which can be made quantitative by examination of the *spectrum* of the light.

A *spectrometer* is a device that separates the different frequencies of electromagnetic wave present in

Figure 5.35 At the observer's position on the Earth it is sunset. The light that the observer sees is red because the blue light has been scattered out of it to make the "sky" appear blue to an observer in the daylight.

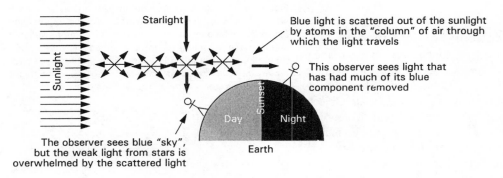

Starlight

Blue light is scattered out of the sunlight by atoms in the "column" of air through which the light travels

This observer sees light that has had much of its blue component removed

The observer sees blue "sky", but the weak light from stars is overwhelmed by the scattered light

Earth

light by means of a prism or, more usually, a diffraction grating (Fig. 5.36). If one examines the spectrum of a narrow slit illuminated by light from the gas, one sees striking differences between white light from a heated filament lamp or the sun, and light from the gas.

The *emission spectrum* of a gas is discrete: only certain frequencies (colours) characteristic of the molecules of the gas are present. For example, the presence of two very bright "lines" close together with an average frequency of 509×10^{12} Hz (i.e. a wavelength of 589 nm) (yellow) indicates the presence of sodium atoms in the gas. The details of the spectra of sodium (vapour) and neon are shown in Figure 5.37.

Spectroscopic investigations can be used to assess the composition of an unknown gas by comparing the spectrum of the gas with reference spectra for known substances. This technique is used to determine the composition of the outer layers of stars, and as an alternative to "wet" chemical analysis in the examination of unknown materials. This is the technique used to obtain the data on the composition of the outer part of the sun detailed in Figure 2.1.

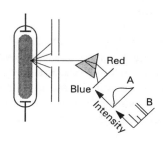

Figure 5.36 A prism spectrometer. Light is passed through slits, and then through a prism. The different velocities (refractive indices) of the different frequencies of light through the glass of the prism (Figure 7.48) causes the different colours to take different paths through the prism. The light is then projected onto a screen. If the light is from an incandescent lamp, the intensity varies across the screen in a way (A) similar to the familiar "rainbow" spectrum. However, light from the positive column of a discharge tube produces a line spectrum (B), the precise positions and relative intensities of the lines being characteristic of the type of gas within the discharge tube.

Figure 5.37 The intensity of the light emitted as function of wavelength for sodium vapour and neon gas. On each spectrum the dotted curve represents the average sensitivity of the human eye. Note that:
(a) The vertical axis is logarithmic and is plotted in arbitrary units.
(b) Only about half the lines (those with intensities greater than 300 a.u.) are plotted.
(c) The positive column of a sodium vapour discharge lamp appears bright yellow because of the two closely spaced intense lines (the "D" lines) at around 589 nm, which is close to the peak of the spectral sensitivity of the eye.
(d) The positive column of a neon discharge lamp appears red because of the relative scarcity of *any* intense lines near the peak of the eye's sensitivity. The intense cluster of peaks at the red end of the spectrum thus dominates the appearance of the spectrum, even though the eye is very insensitive in this region.

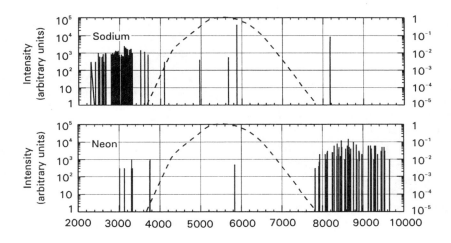

Historically, it was from studying the emission spectra of gases that our understanding of atomic structure was derived and checked. Thus the emission spectra of gases have played a key role in the development of the theory of quantum mechanics used to describe the electronic structure of atoms.

So the main question raised by this preliminary discussion of the experimental data on the emission spectra of gases is: why are the emission spectra of gases in the positive column of a discharge tube discrete and characteristic of the atoms of the gas?

5.7.5 Understanding the data on the emission spectra of gases

The emission spectra of gases is more properly the subject of a book on atomic physics, but we can see roughly why the emission spectra are characteristic, without trying to calculate why the spectra have the particular form they have.

First, we note that within the positive column of a discharge tube we have molecules in several different states: electrically neutral molecules, singly ionized molecules, doubly ionized molecules and, possibly, higher ionization states of the molecule. The internal electronic structure of each of these states of the molecule is similar, but slightly different. Let us assume for the moment that the molecules within the positive column are predominantly electrically neutral. These molecules are being continually bombarded by (a) other molecules as in a non-conducting gas and (b) by ions and electrons with energies ranging up to the ionization energy of the molecules.

Taking the analogy of a mechanical object, this high-energy bombardment sets a molecule "ringing" at its natural resonant frequencies in the same way that striking a bell sets it oscillating in characteristic patterns, the precise frequencies depending on the structure of the bell. So the molecules oscillate in a characteristic manner, with characteristic frequencies depending on the internal electronic structure of the molecules. The oscillation of charge density within the molecule causes radiation of energy according to Equation 5.122, the radiated power being proportional to the magnitude of the oscillating dipole moment, p. If the resonant vibration of the charge within the atom is large, then the oscillating dipole moment p also becomes resonantly large and

the radiated power at that frequency is intense.

As mentioned in §2.5 (and later in §7.7.3 on the optical properties of solids), quantum mechanically the phenomenon of resonance is described as a transition between quantum states. Thus in the correct quantum mechanical description a resonance of frequency f_0 is described in terms of the emission or absorption of quanta of radiation with energy $E = hf_0$. Thus the frequencies of light (the colours) emitted from the positive column of a discharge tube are the characteristic frequencies of the oscillation of electric charge within the molecules of the gas. These frequencies are in turn related to the energy differences between quantum states.

5.8 Magnetic properties

Gases in general have subtle magnetic properties that may for many purposes be ignored, i.e. the magnetic permeability of the gas is very close to unity. However, in detailed studies of the emission spectra (see above) of gases in magnetic fields it was observed that under the action of strong applied magnetic field some of the lines in the spectrum of light from a gas can either change their position or split into several lines. The observation and explanation of this *Zeeman effect* proved important historically in attempts to understand atomic structure.

5.9 Exercises

Exercises with a P prefix are "normal" problems. Those with a C prefix are best solved numerically using a computer program or spreadsheet.

P1. Of the properties of gases listed below, which do not depend strongly on the microscopic properties of gas molecules other than their mass?
(a) Thermal conductivity (§5.4).
(b) Heat capacity at constant volume (§5.3).
(c) Heat capacity at constant pressure (§5.3).
(d) Density (§5.2).
(e) Thermal expansivity (§5.2).
(f) Speed of sound (§5.5).
(h) Dielectric constant (§5.6).
(i) Refractive index (§5.7).
(h) Emission spectrum (§5.7).

Density and expansivity

P2. What is the approximate volume of 1 mol of any gas (a) at STP and (b) at room temperature and pressure? Work out the volume of the room you are in *now*: Considering air to be made of 80% nitrogen and 20% oxygen (Table 5.2), approximately how many moles of oxygen and nitrogen are in the room? What is the mass of the air? If all the air were condensed into a solid, what volume would it occupy (Tables 5.2 and 7.1)?

P3. Air held in a fixed volume of 1 l has a pressure of 0.1333 MPa and a temperature of 0°C. What is the mass of air in the container and what is the expected pressure at 100°C (Section 5.2.2)?

P4. Compare the experimental results on the pressure coefficient, β_P (Table 5.4), with those of a theory of your own devising. What is the relationship between the deviations from theory of the pressure and volume coefficients (Table 5.5)?

P5. The cylinder of a petrol engine initially contains approximately 10^{-3} mol of gas in a volume of 60 cm^3 at a temperature of around 200°C. During the subsequent explosion, the gas pressure peaks at approximately 3×10^5 Pa before the cylinder expands adiabatically to a volume of 240 cm^3.

(a) What is the initial pressure of the gas (§5.2.3)?

(b) What is the peak force on the piston if its diameter is 5 cm?

(c) What is the peak temperature of the gas (§5.2.3)?

(d) Assuming the gas obeys PV^γ = constant throughout the expansion, with l = 1.3, estimate the pressure and temperature at the end of the expansion.

P6. Objects float if their average density is less than the density of the fluid in which they are immersed. A balloon of neglible wall thickness and mass is filled with helium gas at STP until it just floats in air at STP. What volume must the balloon have if it is to lift the weight of an adult human being, say 70 kg (§5.2.3)?

If the balloon just floats at STP, will it float or sink if the temperature is increased to room temperature? (Assume the volume of the balloon stays constant but that the pressure increases as for an ideal gas.)

P7. A solid object of density 2000 kg m^{-3} is weighed on a simple balance against weights of density 8000 kg m^{-3}. Calculate the buoyancy (uplift) force on both the object and the weights if the measurement is carried out in air at room temperature and normal atmospheric pressure (§5.2 and Tables 5.1 and 5.2). By considering your results show that no balance can be read with an accuracy greater than around 1 part in 10^4 unless the relative densities of the weights and the object being weighed are considered.

P8. Approximately how many moles of gas are there in the Earth's atmosphere? What is the mass of this gas? Assume that (a) the atmosphere exists at a uniform pressure of around half the normal atmospheric pressure at sea level, (b) the average temperature is around 0°C, and (c) that the atmosphere exists only up to a height of 10 km. The radius of the Earth is approximately 6400 km.

Assuming that the atmosphere has the same composition as given in Table 5.2, estimate the mass of CO_2 in the atmosphere. If the concentration of CO_2 increases by 1% each year, what mass of carbon joins the atmosphere each year (§5.2.3)? (Note: the concentration of CO_2 in the atmosphere reflects the balance between processes which produce CO_2 (e.g. animals breathing, fuel-burning) and processes which remove CO_2 (e.g. rain, plant–growth). Understanding the actual concentrations is rather complex.)

Heat capacity

P9. What is the molar heat capacity at constant pressure of (a) argon at 2000 K, (b) oxygen at −73.15°C, (c) oxygen at 1000 K? (Tables 5.6 and 5.7)?

P10. What is the the molar heat capacity at constant pressure of the major components of air (Table 5.2) at around room temperature. (Tables 5.6 and 5.7)?

P11. At around room temperature and pressure, what is the value of γ, the ratio of the heat capacity at constant pressure to the heat capacity at constant volume, for (a) oxygen, (b) nitrogen, (c) argon, (d) carbon dioxide and (e) air (Table 5.8)?

For each gas, work out the number of degrees of freedom per molecule according to Equation 5.43 and compare your calculated values with mine in Table 5.8. To what extent is it fair to assume that, with regard to thermal properties, air behaves similarly to nitrogen?

P12. What is the ratio of the principal heat capacities, γ, for (a) helium at 0°C, (b) mercury vapour at 310°C, (c) krypton at 20°C, (d) air at 500°C, (e) steam at 100°C, (f) methane and (g) ethane (Table 5.8)?

For each gas estimate the number of degrees of freedom of molecules of the gas according to Equation 5.43 and write a paragraph suggesting the molecular motions to which these degrees of freedom might correspond (Tables 5.8 and 5.9)

P13.

(a) What is the heat capacity per kilogram (at constant pressure) of argon gas at STP? Is it greater or less than that of helium?

(b) What is the heat capacity per cubic metre (at constant pressure) of argon gas at STP? Is it greater or less than that of helium?

P14. The room in which I am writing this question is approximately 3 m × 4 m × 5 m in size. Assuming that no heat is lost through the walls or my newly double-glazed windows, estimate how long I will have to run a 1 kW fan heater if I am to raise the air temperature in the room from 8°C to 25°C (Tables 5.2 and 5.8 and Exercise P2 in this section).

Do this experiment in your own room and compare the results of the experiment with a similar calculation.

Based on the results of your experiment, state to what extent the heat in a room is stored in the air, and to what extent it is stored in a thin layer of plaster on the inside of the walls?

P15. From analysis of the variation with temperature of the heat capacity of oxygen, the text estmates that oxygen molecules vibrate with a frequency of around $f_0 = 4.2 \times 10^{13}$ Hz. Following a similar method, estimate the vibrational frequencies of (a) hydrogen (H_2) and (b) chlorine (Cl_2) (Table 5.7 and Fig. 5.4). By considering the relative mass of hydrogen, oxygen and chlorine atoms, estimate the effective "spring constant", K, for each type of molecular bond (Eq. 2.31 and Example 2.5).

Note: as discussed following Exercise P9 in Chapter 2, the correct mass to use in Equation 2.31 is actually the reduced mass, but since this calculation is "order of magnitude" only, this does not affect the general conclusions drawn from the calculation.)

P16. Following on from Exercise P7, estimate the heat capacity at constant pressure of the Earth's atmosphere.

Thermal conductivity

C17. Plot the thermal conductivity at 100°C of the noble gases (Table 5.11A) as a function of molecular mass, m. By replotting the data as a function of $m^{-1/2}$, discuss to what the extent the data agree with the theory summarized in Example 5.9. Repeat the analysis for diatomic gases in Table 5.11B at both 0°C and 100°C.

C18. Enter the text of the computer program in Appendix 4 into a QBASIC editor and investigate the behaviour of a gas composed of four molecules (enter $n = 2$). Examine the non-ideal orbiting collisions mentioned in Figure 4.13 and discussed in the text surrounding Figure 5.18 and Table 5.13. Decrease the variable *factor* in the SETUP subprogram to increase the average energy of the gas molecules and see if the frequency of such collsions is reduced. Be sure to take account of the warnings in Appendix 4 regarding the length of the time step dt in the program.

P19. Estimate the mean free path, λ_{mfp}, of helium molecules at a pressure of 1 Pa based on the molecular size data given in Table 5.13. If the gas is contained in a space of thickness 1 mm, will the gas still be in the plateau region of thermal conductivity, or will the thermal conductivity have begun to decrease (Fig. 5.15)?

P20. A *Pirani gauge* is a device used to measure the pressure of a gas at low pressures. The device consists of a heated element which is cooled by the gas. In the pressure range where the thermal conductivity of the gas changes with pressure, the temperature of the element varies with gas pressure. Thus, after calibration, measurement of the temperature of the element by a thermocouple or resistance thermometer (Chap. 2) allows determination of the pressure. By considering Figure 5.15 and Equation 5.63, estimate the size of the container in which the element is suspended if the device works only at pressures below 100 Pa.

P21. A cubic container of volume 1 m^{-3} is heated from the bottom such that air at the bottom is approximately 10°C hotter than air at the top (Fig. 5.13). (a) If the box is at a temperature around room temperature and at normal atmospheric pressure, approximately how many moles of gas are in the box? (b) Convection lifts 0.1 mol s^{-1} of air from the bottom to the top. Estimate the heat flow across the container due to convection. Is it greater or less than would be expected due to the "still air" thermal conductivity listed in Table 5.8B alone?

C22. Table 5.13 shows the reduction of the effective diameter, a, of monatomic gas molecules as the temperature is increased. Plot a graph of a^2 versus $1/T$. You should see straight-line behaviour – a phenomenon noted by Sutherland. From the $1/T = 0$ intercept (i.e. corresponding to $T = \infty$), estimate the high-temperature "hard core" diameter of an argon atom.

Analyze the data for the thermal conductivity of a diatomic gas (e.g. N_2) in same way as the text analyzes the thermal conductivity of the monatomic gases. Does this analysis support or contradict the conclusions that the text draws from the analysis of the monatomic gas data?

Speed of sound

P23. What is the speed of sound in (a) helium and (b) deuterium at STP? What conclusion can be drawn from the difference between these results (Table 5.14)?

P24. What is the speed of sound in dry air at STP (Table 5.14)? What value would you expect for the speed of sound in air at normal room temperature (approximately 20°C)? (Eq. 5.80)?

P25. A lightning flash and the start of a peel of thunder are separated by t seconds. By considering the likely speed of sound in the atmosphere (Table 5.14 and Eq. 5.80), derive a rule of thumb for estimating the distance of a lightning strike (in kilometres), based on a measurement of t.

P26. A siren in a flat region of England emits sound waves upwards and outwards in a hemispherical pattern. If the speed of sound were constant, the wave fronts would form concentric hemispheres centred on the siren. However, the speed of sound falls with decreasing temperature, and since the atmosphere becomes progressively colder with increasing altitude, the wave fronts become distorted. Sketch the wave fronts for (a) the normal situation and (b) the rare occasion (known as a *temperature inversion*) when warm air is trapped above colder air. In the second case, show graphically how sound can be "focused" back onto the ground at large distances from the siren.

P27. If a volcano explosion near Australia were loud enough to be heard in England, roughly how long after the explosion would it be heard (Table 5.14)? If, instead, seismometers were used to detect the sound waves that travelled directly through the Earth, estimate very roughly (Table 7.6) when the explosion would be detected. The radius of the Earth is 6400 km.

P28. Could a device be built which would measure the temperature of the air by measuring the speed of sound through it (§5.5)? If you did build such a device, what problems would you expect to have? What frequency of sound would you choose if the device were to be portable?

P29. The amplitude of the pressure oscillations in a sound wave is usually expressed in terms of the ratio of the root-mean-square (rms) value of the pressure oscillations to a reference rms value of P_0 of 2×10^{-5} Pa. The ratio is usually expressed logarithmically in units of decibel (dB). Thus the sound pressure level defined in decibels is given by

$$\text{sound pressure level} = 20 \log(P_1/P_0)$$

Thus, for example, if a sound wave has an amplitude of 0.2 Pa then the sound pressure level in decibels is 80 dB. At a frequency of around 1 kHz, sound waves of this amplitude would be subjectively experienced as very loud – equivalent to standing near a road as a lorry passes by. By recalling that (a) the pressure and volume of a gas during adiabatic compressions obey $PV^\gamma =$ constant, and (b) the pressure oscillations are much smaller in amplitude than the total pressure, establish a relationship between P and T in adiabatic compression and simplify it to show that the magnitude of the temperature oscillations is

$$\frac{\Delta T}{T} = \left(1 - \frac{1}{\gamma}\right)\frac{\Delta t}{P}$$

Show that the magnitude of the temperature oscillations associated with an 80 dB sound wave in air at STP is around 0.16 mK. How would you detect such oscillations?

P30. Based on the speed of sound in air at STP, estimate the mean speed of molecules in air at STP according to Equation 5.87. How does this estimate compare with one derived from the Maxwell speed distribution for nitrogen at 0°C (Tables 5.8 and Table 5.14, and Exercise C4 in Chap. 4).

P31. Describe with the aid of labelled sketches, an apparatus that could be used to determine γ for gases at a range of temperatures, by measuring the speed of sound in the gas. In an analysis of your results, show how would you separate out the $T^{1/2}$ dependence of the speed of sound from the variations of γ with temperature (Eq. 5.80).

Electrical properties

P32. A parallel-plate capacitor of plate area A and plate separation d has a capacitance of $C = \varepsilon \varepsilon_0 A/d$. Work out the capacitance if there is a vacuum between the two plates. What would be the percentage change in the capacitance if the space between the two plates was filled with (a) air at STP (b) CO_2 at STP (c) CO_2 at 1 atm and 100°C and (d) CO_2 at 1000 atm and 100°C. (Table 5.16 and Fig. 5.23)?

P33. What is the dielectric constant of (a) helium, (b) neon (c) argon and (d) nitrogen at STP (Table 5.16)? Use Equation 5.108 to estimate the polarizability of a single molecule of the each of the gases (a) to (d). State which molecule is the least polarizable, and which the most, and briefly suggest the features of each kind of atom which determine its polarizability (Fig. 5.24).

P34. What is the dielectric constant of water (steam) at 100°C (Table 5.16)? Use Equation 5.106 to estimate the size of the permanant dipole moment on a water molecule in Debye units (Example 5.11).

P35. What is the approximate breakdown field of air around room temperature and atmospheric pressure (Fig. 5.26)?

P36. Around high voltage apparatus such as van der Graaf generators, electrical breakdown may be prevented by enclosing the entire apparatus in an atmosphere of sulphur hexafluoride (SF_6) gas. Given that the device must operate at around room temperature, should the gas be at high or low pressure (Fig. 5.26)? What is the maximum conceivable breakdown field for the gas?

Optical properties

P37. What are (i) the refractive index and (ii) the speed of light in: (a) a vacuum, (b) air at STP (c) helium at STP, and (d) xenon at STP (Table 5.18)?

P38. What is the ratio of the wavelength of light to the approximate diameter of an argon atom (Table 5.13 and Fig. 5.32)?

P39. Equation 5.112 predicts that, at constant temperature, the difference of the dielectric constant of a gas from unity $(\varepsilon - 1)$ is proportional to pressure. By considering Equation 5.119, show that as long as $\varepsilon \ll 1$, the difference between the refractive index of a gas and unity is also proportional to pressure.

P40. What is the wavelength of (yellow) light with a frequency of 5.09×10^{14} Hz (a) in a vacuum and (b) in air at STP (0°C and 1.013×10^5 Pa). As you may have shown in Question P39, the difference of the refractive index of a gas from unity is proportional to pressure. Use the answers to part (a) and (b) to estimate the wavelength of the light in air at 0°C and 1.023×10^5 Pa (Table 5.18).

Before it reaches us, light from stars passes through the atmosphere which is not quite uniform. The pressure of the air fluctuates slightly over a time-scale of a few seconds, and differs slightly above neighbouring points on the ground. Show with the aid of a sketch how this effect can lead to distortions of the initially parallel wave fronts from a distant star. Explain how this leads to the "twinkling" of stars when observed from the ground.

P41. Figure 3.6 shows the use of an optical interferometer to determine the refractive index of light in a gas. Initially the sample arm of the apparatus is at vacuum. For a path length in the gas of 50cm and yellow light with a frequency of 5.09×10^{14} Hz, calculate the pressure of helium gas required to produce a single wavelength phase shift between the two beams. (See also problem P8 in Chapter 3.)

P42. The penetration of light through the atmosphere obeys a law of the form $I = I_0 \exp(-x/\lambda)$. Derive an order of magnitude estimate of λ for air under typical condi-

tions based on your own observations of the atmosphere.

P43. Figure 5.37 shows the emission spectra of sodium and neon. Both spectra have lines at around 5800×10^{-10} m to 5900×10^{-10} m, close to the peak sensitivity of the human eye. Taking note of the logarithmic scale, explain why an electrical discharge through sodium vapour appears yellow and that through neon appears red.

P44. Estimate the refractive index of water vapour at STP by extrapolation from the refractive index of liquid water. How accurate is your estimate (Tables 5.18 and 9.20, and §9.8)?

CHAPTER 6

Solids: background theory

6.1 Introduction

We envisage a solid as a collection of atoms whose *average* positions are fixed with respect to one another. When we studied the properties of gases we essentially ignored the potential energy of interaction between atoms, but in solids we cannot do this: solids exist *because* of the potential energy of interaction between atoms. The atoms of a solid can vibrate about their average position, but can only exceptionally change their position with respect to their neighbours. This will probably be familiar to you, but in case it isn't, Figure 6.1 illustrates how one imagines the situation of the atoms in a solid. A computer program that simulates the dynamics of atoms in a simple two-dimensional solid is listed in Appendix 4, or can be downloaded from my world wide web home page at

http://www.bbk.ac.uk/Departments/Physics

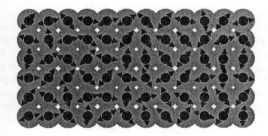

Figure 6.1 A schematic illustration of the motion of atoms in a solid; the arrows represent the directions in which the molecules are moving. Note the small separation between the molecules and the random orientation of the vibrations of the molecules. The atoms themselves are illustrated schematically as a central darkly shaded region, where the electron charge density is high, and a peripheral lightly shaded region. The electric field in this peripheral region significantly affects the motion of neighbouring atoms, and disturbs the electronic charge density of neighbouring atoms.

When we discussed the properties of gases we were able to arrive at the theory of a "perfect gas" which for many purposes is a good approximation to the properties of real gases. Solids, however, have many fewer properties that can be explained in terms of a theory of a "perfect solid". The diversity of properties exhibited by solids calls for us to make several simple models to serve as starting points for attempts to understand the behaviour of real solids. Despite the diversity in the properties of solids, it is important to realize that in all the materials the only force acting between atoms is the electrostatic coulomb force. The coulomb force, coupled with the different configurations of electrons in the outer parts of the 100 or so different types of atoms, is sufficient to produce solids with the diversity that you find around you.

In this chapter we will discuss four simplified model solids representing idealized categories. Most real solids do not fall neatly into one category or another. Our hope is that by looking at a few (rather rare) "simple solids" that do fit this categorization scheme, we will be able to shed light on what is happening in more common, but more complex, solids.

6.1.1 Four simple models of solids

The four simple models of solids which we will discuss in the following sections are illustrated in Figure 6.2. As Figure 6.2 makes clear, it is the arrangement of the outer (valence) electrons on atoms in a solid that is the basis of our categorization. The electrostatic interaction of valence electrons with neighbouring atoms is responsible for *bonding* atoms together into a solid, and the four models below correspond to four relatively distinct bonding mechanisms between atoms.

6.2 Molecular solids

6.2.1 General description

In molecular solids, the electronic structure of the atoms or molecules that make up the solid is the same as (or similar to) the electronic structure of the atoms or molecules that moved around independently in the gas. The solid is held together by the weak interactions that take place between all molecules – exactly the same interactions that we ignored when we considered such molecules in the gaseous state. This kind of solid is formed when the molecules that make up the gas are chemically stable enough that the outer electrons of each molecule stay with their "parent" molecule when they form a solid.

The types of substance in which molecular bonding plays a key role are:

- the noble "gases", i.e. the same gases that were such fine exemplars of the perfect gas law also make fine molecular solids;
- the halogen molecules, i.e. F_2, Cl_2, Br_2 and I_2; and
- many long-chain polymer molecules such as polyethylene or poly vinyl chloride.

Two key factors affect the properties of solids made from these molecules: the *shape* of the molecules, and the *strength* of the interaction between the molecules. In this section we will look quantitatively at the effect of the strength of the interaction on the properties of simply shaped molecules. The effects of molecular shape are discussed further in §7.2.3 on the thermal expansivity of plastics and §8.2.3 and §8.3.3 on the properties of organic liquids.

6.2.2 Attractive and repulsive forces between neutral molecules

The attractive force between neutral molecules

The origin of the attraction

The distribution of electric charge around molecules in a molecular solid is broadly unchanged from the distribution when the molecules were in the gaseous state. In particular, each molecule is electrically neutral. So what holds the molecules together? The force that holds the molecules together is the coulomb force, but the action of the force is subtle. The attraction between molecules arises not because each molecule has a net charge, but because the distribution of electrons around each molecule fluctuates. Consider a series of imagined snapshots of the electronic distribution around a single atom such as an argon atom (Fig. 6.3).

Averaged over a time greater than a *few* $\times 10^{-16}$s, the charge distribution will be symmetrical (Fig. 6.3d). But on a short time-scale, the charge distribution fluctuates and is generally slightly "imbalanced" which means the atom behaves as a tiny *electric dipole*. An electric dipole has a weak electric field

Figure 6.2 (a) Molecular solids, in which the entities that make up the solid (atoms or molecules) are essentially the same as the entities that make up the gas. (b) Ionic solids, in which the entities that make up the solid are ions rather than atoms or molecules. Electrons from the outer part of one atom move *wholly* to another atom. (c) Covalent solids, in which the entities that make up the solid (atoms or molecules) are greatly altered from their state in the gas. Electrons from the outer part of one atom change their "orbits" so that they move round more than one atom. This leads to a high electronic charge density in regions *in between* the mean positions of the atoms. (d) Metallic solids, in which the electrons from the outer parts of the atoms can move anywhere within the solid and are not attached to any individual atom.

(a)   **(b)**   **(c)**   **(d)**

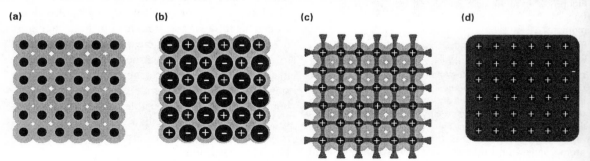

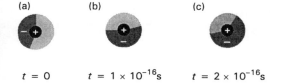

$t = 0$ $t = 1 \times 10^{-16}$s $t = 2 \times 10^{-16}$s

(d)   Time averaged charge distribution

Figure 6.3 A representation of the fluctuations of charge density around a neutral atom. The figure shows the charge distribution at three times separated by $\approx 10^{-16}$ s.

around it that can affect other atoms nearby (§2.3.3), and it is through the action of this field that bonding takes place.

Rough calculation

Suppose that at a particular instant the charge distribution is asymmetrical such that an atom has an electric dipole, p. The electric field due to p is given approximately by

$$E(r) \approx \frac{p}{4\pi\varepsilon_0 r^3} \qquad (6.1)$$

A second atom placed in this field will become electrically polarized by an amount that is proportional to the magnitude of the electric field (see §5.6.2 on gases). The electric dipole moment *induced* on the second atom, p_2, will be given by

$$p_2 = \alpha E(r) \qquad (6.2)$$

Figure 6.4 Fluctuations of the charge density distribution on atom 1 lead to a temporary electric dipole moment. The electric field from the electric dipole moment *induces* a dipole moment on neighbouring atoms, causing the two atoms to be attracted to one another.

Atom 1   Atom 2

Dipole caused   Dipole induced
by fluctuating   by electric field
charge density   from atom 1

where α is the *molecular polarizability* of atom 2 (Fig. 6.4). The second atom will have an energy of interaction u with the first atom given by

$$u = -p_2 E(r) \qquad (6.3)$$

We now substitute for p_2 using Equation 6.2 and $E(r)$ using Equation 6.1 to yield

$$u \approx -\alpha E(r)E(r) \sim -\frac{\alpha}{r^6} \qquad (6.4)$$

which predicts that the energy of interaction between the two atoms therefore varies as $1/r^6$, and is proportional to the molecular polarizability, α, of the atoms involved. The attractive energy varies with distance more rapidly than the $1/r$ variation that occurs between uncompensated charges, is weak in magnitude, and is generally known as *Van der Waals* or *molecular* bonding. The polarizability of atoms and molecules is discussed extensively in §5.6.2 on the electrical properties of gases.

The repulsive force between neutral molecules

The attractive force described above cannot be the only term involved in the interaction between two neutral atoms since the energy would become more negative without limit as the atoms came closer. There must be another term that increases the energy as the separation between the atoms gets smaller.

The origin of the repulsion

The repulsion arises because as the atoms get closer the outer electrons on each atom come closer together. The processes that take place as atoms approach each other closely are complex, but one can identify two separate ways in which bringing atoms closer together costs energy:

- The outer electrons of one atom have to attempt to occupy orbits around the other atom. The lowest energy orbitals are all occupied and so the *Pauli exclusion principle* forces electrons to occupy higher energy orbitals.
- When the electrons on the outer part of each atom get very close they begin to repel each other directly through their coulomb interaction.

There is no simple direct way to derive the variation of the repulsion with distance, but two approximate forms are commonly used that capture some of the behaviour of real solids. It is sometimes assumed that the repulsive energy varies as either

$$u \approx +\frac{\text{Constant}}{r^{12}} \qquad (6.5\text{a})$$

or

$$u \approx +\text{Constant} \times \exp(-r/\rho) \qquad (6.5\text{b})$$

where ρ is a "range" parameter. We will choose the first of these alternative forms because it is mathematically easier to manipulate.

6.2.3 Pair potential:
The Lennard–Jones potential

We can combine the attractive and repulsive terms (Eq. 6.4 & 6.5a) into an expression for the potential energy of a pair of interacting neutral atoms

$$u_{\text{pair}}(r) = \frac{A}{r^{12}} - \frac{B}{r^6} \qquad (6.6)$$

where A and B are constants that determine the relative size of the attractive and repulsive terms. Because they describe the interaction of just two atoms, expressions such as Equation 6.6 are known as *pair potentials*. This particular form of pair potential (with powers of 12 and 6) was discovered by Lennard and Jones to be particularly appropriate to molecular solids and so is called a Lennard–Jones pair potential.

The Lennard–Jones potential is often rearranged in the form

$$u_{\text{pair}}(r) = -4\varepsilon\left\{\left[\frac{\sigma}{r}\right]^6 - \left[\frac{\sigma}{r}\right]^{12}\right\} \qquad (6.7)$$

where

$$\sigma = \left[\frac{B}{A}\right]^{\frac{1}{6}} \quad \text{and} \quad \varepsilon = \left[\frac{A^2}{4B}\right] \qquad (6.8)$$

Although Equation 6.7 looks more complicated than Equation 6.6, it has the advantage that the parameters σ and ε have a clear physical interpretation in terms of the form of $u_{\text{pair}}(r)$ (see Fig. 6.5):

(a)

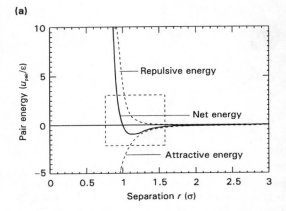

(b)

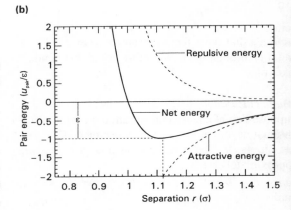

(c)

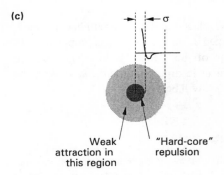

Figure 6.5 Variation of potential energy with the separation of two atoms interacting via a Lennard–Jones potential. The energy scale is drawn in units of ε and the separation scale is drawn in units of σ, where ε and σ are defined according to Equation 6.8. (a) The general shape of the curve, which highlights the shallowness of the minimum in potential energy. (b) The region of (a) indicated by the rectangle in detail, showing that the minimum occurs at $r = 1.12\sigma$ and $U = -\varepsilon$ on these arbitrary scales. (c) The relationship between σ and the representation of a molecule shown in Figure 6.1.

- σ is a *range* parameter and indicates the approximate "size" of an atom. At $r = \sigma$ the value of u_{pair} is zero.
- ε is an *energy* parameter and indicates the *strength* of the interaction between atoms. At the minimum of the $u_{\text{pair}}(r)$ curve u_{pair} has the value $-\varepsilon$.

It is important to note that the interactions between molecules described above take place in both the solid state *and* the gaseous state. In the gaseous state, the most important feature is the strong repulsion that makes atoms which collide with one another behave as if they have a "hard core" – in this case with radius of around σ (Fig. 6.5). However, in the solid state the key feature of the pair potential is the shallow minimum. Molecules try to achieve positions relative to other molecules that allow them to sit as close to this minimum as possible.

6.2.4 Calculation of cohesive energy

In this section we will calculate the electrostatic energy of 1 mole (i.e. Avogadro's number, N_A), of molecules which interact through the Lennard–Jones pair potential. This quantity is known as the *cohesive energy*, U, of a substance and is generally specified in units of kilojoules per mole ($\text{kJ}\,\text{mol}^{-1}$).

We will assume that the molecules are arranged in a regular crystal structure, and in §6.2.5 we shall work out which crystal structure the molecules would choose if they were to maximize their cohesive energy.

Stage 1

To begin, we consider a particular molecule, i, surrounded by all the other molecules, j, in the crystal. We write its energy u_i as

$$u_i = \sum_{j \neq i}^{N_A} u_{\text{pair}}(r_{ij}) \tag{6.9}$$

where r_{ij} is the distance between the ith molecule and each other molecule, j, in the crystal. The energy for N_A molecules in 1 mole of the solid is then given by

$$U = \frac{N_A u_i}{2} \tag{6.10}$$

Note the factor of ½ which arises because the energy u_i is the sum of the electrostatic energy of *pairs* of molecules, and we wish to calculate the energy per molecule. Remembering the form of the Lennard–Jones potential (Eq. 6.7) we can then write

$$U = \frac{N_A}{2} \sum_{j \neq i} 4\varepsilon \left\{ \left[\frac{\sigma}{r_{ij}} \right]^{12} - \left[\frac{\sigma}{r_{ij}} \right]^{6} \right\} \tag{6.11}$$

Stage 2: a trick

Now we use a mathematical trick. We write the distance between the ith and jth atoms as a multiple of the nearest-neighbour distance r_0 (as yet unknown), i.e.

$$r_{ij} = \alpha_{ij} r_0 \tag{6.12}$$

Note that α_{ij} is just a number, not a distance. We can now rewrite our expression for U using this substitution

$$U = 2\varepsilon N_A \sum_{j \neq i} \left\{ \left[\frac{1}{\alpha_{ij}} \right]^{12} \left[\frac{\sigma}{r_0} \right]^{12} - \left[\frac{1}{\alpha_{ij}} \right]^{6} \left[\frac{\sigma}{r_0} \right]^{6} \right\} \tag{6.13}$$

and then separate the summation into two terms

$$U = 2\varepsilon N_A \sum_{j \neq i} \left[\frac{1}{\alpha_{ij}} \right]^{12} \left[\frac{\sigma}{r_0} \right]^{12} - 2\varepsilon N_A \sum_{j \neq i} \left[\frac{1}{\alpha_{ij}} \right]^{6} \left[\frac{\sigma}{r_0} \right]^{6} \tag{6.14}$$

Note that the fraction σ/r_0 does not depend on the index j in either summation, and hence may be brought outside the summation sign since it is a common factor to each term in the sum:

$$U = 2\varepsilon N_A \underbrace{\left\{ \sum_{j \neq i} \left[\frac{1}{\alpha_{ij}} \right]^{12} \right\}}_{A_{12}} \left[\frac{\sigma}{r_0} \right]^{12} - 2\varepsilon N_A \underbrace{\left\{ \sum_{j \neq i} \left[\frac{1}{\alpha_{ij}} \right]^{6} \right\}}_{A_6} \left[\frac{\sigma}{r_0} \right]^{6} \tag{6.15}$$

Each of the two terms in Equation 6.15 now has two parts: the term in σ/r_0, and the quantity in square brackets known as the *lattice sum*, and referred to as A_{12} and A_6, respectively. On substituting for A_6 and A_{12}, the expression for the cohesive energy begins to look a little more tractable:

$$U = 2\varepsilon N_A A_{12}\left[\frac{\sigma}{r_0}\right]^{12} - 2\varepsilon N_A A_6\left[\frac{\sigma}{r_0}\right]^6 \quad (6.16)$$

Lattice sums

Evaluating lattice sums is a straightforward mathematical exercise (Example 6.1). The lattice sums A_6 and A_{12} depend only on the *form* of the lattice, i.e. on the *type* of crystal structure (e.g. face-centred cubic or body-centred cubic) and not on the particular separation between the molecules in the crystal of σ or r_0, i.e. A_6 and A_{12} are just numbers. Each different crystal structure has characteristic values of A_6 and A_{12}.

The cohesive energy of an assembly of N_A atoms is now conveniently expressed in terms of the lattice sums as

$$U = 2\varepsilon N_A\left\{A_{12}\left[\frac{\sigma}{r_0}\right]^{12} - A_6\left[\frac{\sigma}{r_0}\right]^6\right\} \quad (6.17)$$

Minimizing the cohesive energy

Now we need to find the value of (σ/r_0) that minimizes U. To do this we differentiate Equation 6.17 for U with respect to (σ/r_0) and set the result equal to zero. Differentiating we find

$$\frac{dU}{d(\sigma/r_0)} = 2\varepsilon N_A\left\{12A_{12}\left[\frac{\sigma}{r_0}\right]^{11} - 6A_6\left[\frac{\sigma}{r_0}\right]^5\right\} \quad (6.18)$$

Example 6.1

Calculate the lattice sum A_{12} for a face-centred cubic lattice.

The lattice sum A_{12} is defined according to Equation 6.15 as

$$A_{12} = \sum_{j\neq i}\alpha_{ij}^{\frac{1}{12}}$$

where α is the "multiplier" of the nearest-neighbour distance required to reach a particular atom j from any chosen atom i.

Each nearest neighbour has a multiplier distance of 1 ($\times r_0$) and there are 12 such atoms (labelled *nn* in the illustration).

Each next-nearest neighbour has a multiplier distance of $\sqrt{2}$ ($\times r_0$) and there are six such atoms (labelled *nnn* in the illustration, though not all the atoms are shown).

Each next-next-nearest neighbour has a multiplier distance of $\sqrt{3}$ ($\times r_0$) and there are 24 such atoms (labelled *nnnn* in the illustration, though not all the atoms are shown).

Proceeding to successively more distant atoms, we see that the expression for the lattice sum is given by

$$A_{12} = 12\times\frac{1}{1^{12}} + 6\times\frac{1}{\left(\sqrt{2}\right)^{12}} + 24\times\frac{1}{\left(\sqrt{3}\right)^{12}} + 24\times\frac{1}{\left(\sqrt{5}\right)^{12}} +\ldots$$

Evaluating the terms we find

$$A_{12} = \frac{12}{1} + \frac{6}{64} + \frac{24}{729} + \frac{24}{15625} +\ldots$$

$$= 12 + 0.09375 + 0.03292 + 0.00154+\ldots$$

$$A_{12} \approx 12.128$$

This sum converges very rapidly and only a few terms are required to estimate A_{12} with good precision. The A_6 summation converges more slowly, and one must consider more distant neighbours than those shown in the illustration.

---

which is equal to zero when

$$12 A_{12} \left[\frac{\sigma}{r_0} \right]^{11} = 6 A_6 \left[\frac{\sigma}{r_0} \right]^5 \qquad (6.19)$$

Simplifying and rearranging we find

$$\left[\frac{\sigma}{r_0} \right]^6 = \frac{A_6}{2 A_{12}} \qquad (6.20)$$

$$\frac{\sigma}{r_0} = \left[\frac{A_6}{2 A_{12}} \right]^{\frac{1}{6}} \qquad (6.21)$$

$$r_0 = \sigma \left[\frac{2 A_{12}}{A_6} \right]^{\frac{1}{6}} \qquad (6.22)$$

This is the value of the nearest-neighbour distance that minimizes the potential energy of the lattice. Note that r_0 is a multiple of σ, the "range" parameter in Equation 6.7. In order to evaluate the cohesive energy of the lattice, we can substitute the value for r_0 or r_0/σ from Equation 6.22 into Equation 6.12. This calculation is performed in Example 6.2.

According to Example 6.2, the cohesive energy per mole is given by

$$U = - \left[\frac{A_6^2}{2 A_{12}} \right] N_A \varepsilon \qquad (6.23)$$

and so the cohesive energy per molecule, $u = U/N_A$, is just

$$u = - \left[\frac{A_6^2}{2 A_{12}} \right] \varepsilon \qquad (6.24)$$

Summary of cohesive energy calculation
The calculation culminating in Equations 6.23 and 6.24 involved the use of technical tricks such as the use of lattice sums. Because of this it is worth recalling that these tricks have been used in order simply to add up the electrostatic interactions of a large number of molecules.

Example 6.2

Calculate the cohesive energy per mole for a substance whose atoms interact via a Lennard–Jones potential.

According to Equation 6.17 the cohesive energy per mole is given by

$$U = 2 N_A \varepsilon \left[A_{12} \left(\frac{\sigma}{r_0} \right)^{12} - A_6 \left(\frac{\sigma}{r_0} \right)^6 \right]$$

and at equilibrium r_0 is given by Equation 6.22 as

$$r_0 = \sigma \left(\frac{2 A_{12}}{A_6} \right)^{\frac{1}{6}}$$

Substituting the equilibrium value of r_0 into the expression for U should therefore yield the equilibrium value of cohesive energy per mole. The expression for U has two brackets in it: $(\sigma/r_0)^{12}$ and $(\sigma/r_0)^6$. These brackets simplify to

$$\left(\frac{\sigma}{r_0} \right)^{12} = \left[\frac{\sigma}{\sigma(2 A_{12}/A_6)^{\frac{1}{6}}} \right]^{12} = \left(\frac{A_6}{2 A_{12}} \right)^2$$

and

$$\left(\frac{\sigma}{r_0} \right)^6 = \left[\frac{\sigma}{\sigma(2 A_{12}/A_6)^{\frac{1}{6}}} \right]^6 = \frac{A_6}{2 A_{12}}$$

Substituting into the expression for U we find

$$U = 2 N_A \varepsilon \left[A_{12} \left(\frac{A_6}{2 A_{12}} \right)^2 - A_6 \frac{A_6}{2 A_{12}} \right]$$

On cancelling terms we find

$$U = 2 N_A \varepsilon \left(\frac{A_6^2}{4 A_{12}} - \frac{A_6^2}{2 A_{12}} \right)$$

which simplifies to

$$U = - \left(\frac{A_6^2}{2 A_{12}} \right) N_A \varepsilon$$

6.2.5 Equilibrium crystal structure

Equation 6.24 succinctly expresses the cohesive energy of a crystal in terms of the lattice sums A_6 and A_{12}. Note that A_6 and A_{12} could refer to *any* crystal structure. When a crystal forms from a melt, the crystal structure adopted is generally the one with the largest (i.e. most negative) cohesive energy. (This matter is discussed further in Chapter 10.)

In order to predict which crystal structure actually forms, we need only to calculate u according to Equation 6.24 for all likely crystal structures, and work out which crystal structure has the largest value of $A_6/2A_{12}$. Table 6.1 shows the lattice sums and calculated cohesive energy for three different crystal structures in which a molecular solid might conceivably crystallize. The face-centred cubic structure is illustrated in Example 6.1 and the body-centred cubic and simple cubic structures are illustrated in Figure 6.6.

Table 6.1 Values of the lattice sums A_6 and A_{12} for three crystal structures.

	Simple cubic	Body-centred cubic	Face-centred cubic
A_6	8.4	12.25	14.45
A_{12}	6.2	9.11	12.13
Cohesive energy*, u	-5.69ε	-8.24ε	-8.61ε

*Calculated in terms of A_6 and A_{12} according to Equation 6.24.

All the crystal structures listed in Table 6.1 have a negative cohesive energy, i.e. they all represent arrangements of molecules that are energetically favourable compared to molecules being separated from one another by large distances. However, the crystal structure naturally chosen will be the one with the *largest* cohesive energy (i.e. the most negative value), which in this case is the face-centred cubic (*f.c.c.*) structure. This expectation is indeed borne out and all the noble gases (neon, argon, krypton and xenon) form solids with a f.c.c. structure at low temperature.

Comparison with experiment for noble gases

Once the crystal structure is known to be f.c.c., the appropriate values of A_6 and A_{12} can be substituted into Equation 6.22 for r_0. For f.c.c., the predicted values of r_0 and u are $r_0 = 1.09\sigma$ and $u = -8.6\varepsilon$. Values of σ and r_0 can be deduced from analysis of the deviations of gases from perfect gas behaviour (e.g. §5.4), i.e. σ and ε are estimated from the behaviour of the substance *in the gaseous phase*. Using these values of σ and ε derived from these experiments allows us to predict values of r_0 and u which may be compared with experimental values deduced from analysis of data on X-ray scattering, and latent heat of vaporization of the substances. Table 6.2 compares predicted and experimental values of the cohesive energy per molecule and the lattice spacing in the f.c.c. crystal structure. We see that there is fair agreement between theory and experiment.

The computer program listed in Appendix 4 shows the dynamics of a two-dimensional solid made from molecules interacting via a Lennard–Jones 6–12 potential with $A = B = 1$ (Eq. 6.6). Initially, the molecules are positioned close to the lattice sites of a simple cubic lattice. However, it is clear that the molecules quickly adopt a more favourable crystal structure with six nearest neighbours.

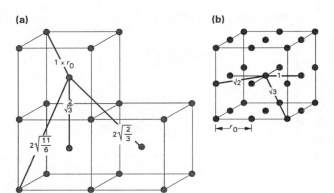

(a)   **(b)**

Figure 6.6 When a crystal forms it could conceivably form any one of several different structures. The actual structure formed depends strongly on the cohesive energy, U, of the structure. For atoms that interact via the Lennard–Jones potential (Eq. 6.7), the cohesive energy can be expressed in terms of the lattice sums A_6 and A_{12} described in the text. The figures illustrate two crystal structures showing that (i) the *number* of nearest-neighbour atoms differs from crystal structure to crystal structure and (ii) the relative distances to next-nearest neighbours also differs from one structure to another. The calculation of the lattice sums for face-centred cubic crystals is considered in Example 6.1

Table 6.2 Values of the lattice constant, r_0, and the cohesive energy per atom, u, calculated according to Equations 6.22 and 6.24 compared with experimental values for neon, argon, krypton and xenon.*

	Substance			
	Ne	Ar	Kr	Xe
r_0(exp.)$\times10^{-10}$ m	3.13	3.75	3.99	4.33
σ	2.74	3.44	3.65	3.98
$r_0 (=1.09\sigma)$	2.99	3.71	3.98	4.34
u(exp.)$\times10^{-3}$ eV	−20	−80	−110	−170
ε	3.1	10.3	14.0	20.0
$u = -8.6\varepsilon$	−27	−89	−120	−172

*The values of ε and σ for each substance were deduced from measurements in the gas phase of each substance by observing the deviations from perfect-gas behaviour. See §5.4.2.

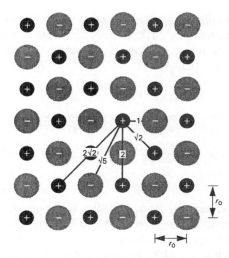

Figure 6.7 A section from an infinite two-dimensional "net" of charges. The numbers on the lines indicate the number of times r_0 must be multiplied to reach between the lattice sites indicated.

6.3 Ionic solids

6.3.1 General description

In ionic solids the "entities" that make up the solid are essentially *ions*, i.e. atoms stripped of one or more of their electrons, or with one or two more added.

As with molecular solids, it is the coulomb force that holds the substance together, but in ionic solids not all the forces are attractive. There are equal numbers of positively and negatively charged ions and, although the positively and negatively charged ions attract one another, the negative ions repel the other negative ions and the positive ions repel the other positive ions. The fact that ionic solids actually exist tells us that the attractive forces will eventually outweigh the repulsive forces, but we need to make a detailed calculation in order to see why.

6.3.2 Calculation of the cohesive energy

The calculation of the cohesive energy of an ionic solid is similar to the calculation given in §6.2 for the cohesive energy of a molecular solid. We have to add all the pair interactions between ions in the solid. Before we do the calculation for a real three-dimensional crystal, we can get the flavour for what the calculation involves by looking at a two-dimensional problem.

A two-dimensional problem
What is the potential energy of a single ion in a large "net" of charges, a section of which is illustrated in Figure 6.7?

In order to answer this question, consider one particular ion, a positive ion say, and add up the energy of interaction of all the other ions with the chosen ion. One must be careful to consider the ions systematically in order to make sure that all of them are counted (but only once!). One starts by adding up the coulomb terms as follows

$$\text{Energy} = \frac{-e^2}{4\pi\varepsilon_0 r_0} + \frac{-e^2}{4\pi\varepsilon_0 r_0} + \frac{-e^2}{4\pi\varepsilon_0 r_0}$$
$$+ \frac{-e^2}{4\pi\varepsilon_0 r_0} + \frac{+e^2}{4\pi\varepsilon_0\left(\sqrt{2}r_0\right)} + \dots \quad (6.25)$$

Note that the energy in Equation 6.25 is the sum of energies of ion *pairs*. The energy associated with the central ion is just half this quantity. We therefore write the coulomb energy per ion, u, as

$$u = \frac{1}{2}\left[\frac{-e^2}{4\pi\varepsilon_0 r_0} + \frac{-e^2}{4\pi\varepsilon_0 r_0} + \frac{-e^2}{4\pi\varepsilon_0 r_0} \right.$$
$$\left. + \frac{-e^2}{4\pi\varepsilon_0 r_0} + \frac{+e^2}{4\pi\varepsilon_0\left(\sqrt{2}r_0\right)} + \dots \right] \quad (6.26)$$

If one collects together the terms that correspond to ions at the same distance from the origin, and takes out a common factor, one finds a series, the first four terms of which are

$$u = \frac{e^2}{8\pi\varepsilon_0 r_0}\left[4\times\frac{-1}{1}+4\times\frac{+1}{\sqrt{1^2+1^2}}\right.$$
$$\left.+4\times\frac{+1}{2}+8\times\frac{-1}{\sqrt{1^2+2^2}}+...\right] \qquad (6.27)$$

This series is similar to the lattice sums considered in the previous section on molecular bonding (Example 6.1), except that in this case the sign of each term alternates. In each term within the brackets in Equation 6.27:

- the first number (4 or 8 for the terms shown) is the number of ions at that distance;
- the sign in the numerator of the fraction is determined by whether the charges are of the same or opposite sign; and
- the denominator of the fraction is the distance from the origin in units of r_0.

Evaluating the terms, one finds that the first eight terms are

$$u = \frac{e^2}{8\pi\varepsilon_0 r_0}\left[-4+\frac{4}{\sqrt{2}}+2-\frac{8}{\sqrt{5}}+\frac{4}{2\sqrt{2}}-\frac{4}{3}\right.$$
$$\left.+\frac{8}{\sqrt{10}}-\frac{8}{\sqrt{13}}...\right] \qquad (6.28)$$

The problem with this summation may be becoming apparent to you. The terms are getting smaller, but only very slowly. Using a "brute force" computer program to whip through the first few thousand terms in the sum yields an approximate answer of –1.57; this answer is likely to be close, but not quite right. And that's the end of the two-dimensional problem. What this result means is that an ion in a net such as the one described will be bound into the net with an energy equal to approximately 1.57 times the energy of interaction of the ion with one other ion of the opposite charge, separated by distance r_0.

The three-dimensional problem

Now we proceed to the three-dimensional problem (Fig. 6.8) actually encountered in ionic crystals. We approach the calculation in a similar way to the calculation for the two dimensional case, but the geometry is now a little more taxing.

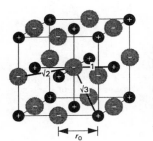

Figure 6.8 The arrangement of ions in a crystal with the same structure as sodium chloride (NaCl). The numbers on the lines are the distance between the ions in units of r_0.

The summation for the energy of an ion now looks like

$$u = \frac{e^2}{4\pi\varepsilon_0 r}\times\frac{1}{2}\left[6\times\frac{-1}{1}+12\times\frac{+1}{\sqrt{1^2+1^2}}\right. \qquad (6.29)$$
$$\left.+8\times\frac{-1}{\sqrt{1^2+1^2+1^2}}+6\times\frac{+1}{2}+...\right]$$

$$u = \frac{e^2}{8\pi\varepsilon_0 r_0}\left[-6+\frac{12}{\sqrt{2}}-\frac{8}{\sqrt{3}}+\frac{6}{2}+...\right] \qquad (6.30)$$

where we have again taken care to calculate the energy per ion and not per ion pair. Using a "brute force" computer program to evaluate the sum, (Eq. 6.30) yields a sum that does not clearly converge but oscillates at a value of around – 1.5. Series of this type are called *Madelung sums* after the person who first showed that for the three-dimensional problem the sum evaluates to –1.748 for the sodium chloride (NaCl) type crystal structure and to similar values for other common ionic crystal structures, i.e. the energy of an ion in an ionic crystal of the NaCl type is

$$u = \frac{e^2}{8\pi\varepsilon_0 r_0}\left[-1.748\right] \qquad (6.31)$$

However this seems to indicate that if r_0 is reduced the energy of the ion will become very large and negative. Of course there must be some repulsive term between atoms that stops the crystal "collapsing". This repulsive term is due to the Pauli exclusion principle effect discussed in §6.2.2. It may be modelled once again as

$$u = \frac{c}{r_0^{12}} \qquad (6.32)$$

Adding the repulsive effect between the ion and its six nearest neighbours we find its energy is

$$u = \frac{-1.748e^2}{8\pi\varepsilon_0 r_0} + \frac{6c}{r_0^{12}} \qquad (6.33)$$

It is clear that reducing r_0 makes the attractive term more negative, but it also rapidly increases the Pauli repulsion between the outer orbitals of each ion.

Finding the equilibrium separation

If we differentiate Equation 6.33 for u with respect to r_0 we can find the value of r_0 that minimizes the cohesive energy per ion:

$$\frac{du}{dr_0} = \frac{+1.748e^2}{8\pi\varepsilon_0 r_0^2} - \frac{12 \times 6c}{r_0^{13}} \qquad (6.34)$$

Equating this to zero yields

$$\frac{1.748e^2}{8\pi\varepsilon_0 r_0^2} = \frac{72c}{r_0^{13}} \qquad (6.35)$$

Solving for r_0 yields

$$r_0^{11} = \frac{576c\pi\varepsilon_0}{1.748e^2} \qquad (6.36)$$

$$\boxed{r_0 = \left[\frac{576c\pi\varepsilon_0}{1.748e^2}\right]^{\frac{1}{11}}} \qquad (6.37)$$

This formula predicts the nearest-neighbour ionic separation in terms of the Madelung sum (−1.748) and the constant, c, which determines the magnitude of the repulsive forces.

6.3.3 Crystal structure

Determining the crystal structure of an ionically bonded substance from first principles is considerably more difficult than the equivalent task for a molecularly bonded substance (§6.2.3). This is because in the molecularly bonded solid that we considered, all the molecules were identical. For ionically bonded substances there are always at least two different types of ion. This introduces a new consideration: the relative sizes of the ions.

However, we are able to say some general things about the types of crystal structure we may expect ionically bonded substances to form. The presence of both positive and negative ions rules out densely packed crystal structures with large numbers of nearest neighbours.

Consider the two-dimensional situation illustrated in Figure 6.9. Suppose a single ion A surrounded itself with 6 nearest neighbours of opposite charge to itself (Figure 6.9a). This would lower A's electrostatic energy considerably, but the 6 neighbours would all have the same sign of charge and thus their energy would be *increased* considerably. Further, the structure could not be extended throughout the crystal while keeping the crystal electrically neutral overall. Similar arguments apply to the equivalent three-dimensional close packed structures. For this reason ionically bonded substances form crystal structures that are more "open". Note that in Figure 6.9b the nearest neighbours always have opposite charges: only next-nearest neighbours have the same charge.

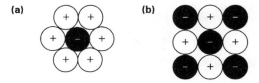

Figure 6.9 Possible two-dimensional structures: ion A is surrounded by (a) six and (b) four nearest neighbours of opposite charge.

6.3.4 Cohesive energy

We have already gone a substantial way to calculating the cohesive energy of an ionic solid by calculating the separation between ions in equilibrium,

(Eq. 6.37). In this section we will complete the calculation by substituting the expression for the equilibrium separation (Eq. 6.37) back into the expression for the cohesive energy.

The expression for the cohesive energy per ion is

$$u = \frac{-1.748e^2}{8\pi\varepsilon_0 r_0} + \frac{6c}{r_0^{12}} \qquad (6.33)*$$

We can make the substitution for r_0 much simpler if we first take out a factor $1/r_0$ from each term to yield

$$u = \frac{1}{r_0}\left(\frac{-1.748e^2}{8\pi\varepsilon_0} + \frac{6c}{r_0^{11}}\right) \qquad (6.38)$$

Substituting for r_0 inside the bracket only yields

$$u = \frac{1}{r_0}\left[\frac{-1.748e^2}{8\pi\varepsilon_0} + \frac{6c}{\left(576c\pi\varepsilon_0/1.748e^2\right)^{\frac{11}{11}}}\right] \qquad (6.39)$$

Simplifying, and cancelling terms

$$u = \frac{1}{r_0}\left(\frac{-1.748e^2}{8\pi\varepsilon_0} + \frac{1.748e^2}{96\pi\varepsilon_0}\right) \qquad (6.40)$$

we arrive at

$$u = \frac{-1.748e^2}{8\pi\varepsilon_0 r_0}\left(1 - \frac{1}{12}\right) \qquad (6.41)$$

which evaluates to

$$u = \frac{-1.602e^2}{8\pi\varepsilon_0 r_0} \qquad (6.42)$$

Note that the repulsive component in Equation 6.41 is one-twelfth of the attractive component. The expression yields the cohesive energy per ion in a simple-cubic ionic solid. In order to convert this into a cohesive energy per mole, one must multiply this figure by the number of formula units in a mole. Note that in the simple two-component ionic solids we have been considering, each chemical formula unit contains *two* ions (e.g. Na^+ and Cl^-). We therefore expect to find that the cohesive energy per mole for

Example 6.3

Let us evaluate the cohesive energy per ion (Eq. 6.41) for NaCl (common salt).

We can evaluate r_0 from: the density (see Table 7.2); the atomic masses of Na and Cl (see Table 7.2); and the crystal structure which is known from X-ray diffraction patterns to be of the simple type described in Figure 6.8. We have:

1 mol of Na has a mass of 22.99×10^{-3} kg
1 mol of Cl has a mass of 35.45×10^{-3} kg

So 1 mol of NaCl has a mass of 58.44×10^{-3} kg and one chemical formula unit of NaCl has a mass of

$$m = 58.44\times10^{-3}/N_A = 9.708\times10^{-26}\text{kg}.$$

Consideration of Figure 6.8 indicates that one chemical formula unit of NaCl occupies a volume of $2r_0^3$ and hence a density $\rho = m/2r_0^3$.

The experimentally determined density of NaCl is $r = 2165$ kg m^{-3}, and so we deduce that

$$r_0 = \left(\frac{m}{2\rho}\right)^{\frac{1}{2}} = \left(\frac{9.708\times10^{-26}}{2\times2165}\right)^{\frac{1}{3}} = 2.82\times10^{-10}\text{m}$$

Using this value in Equation 6.42 for the cohesive energy per ion we have

$$u = \frac{1.602\times\left(1.6\times10^{-19}\right)^2}{8\pi\times8.85\times10^{-12}\times2.82\times10^{-10}} = -6.55\times10^{-19}\text{J/ion}$$

(We may also express this as 4.09 eV/ion.)

such a substance is

$$U = 2N_A u \quad \text{per mole} \qquad (6.43)$$

Using the result from Example 6.3 we conclude that the molar cohesive energy of NaCl is given by $U = 2\times6.02\times10^{23}\times6.55\times10^{-19} = 789$ kJ mol^{-1}. This is the energy required to separate one mole of NaCl into one mole of Na$^+$ ions and one mole of Cl$^-$ ions. The value of U inferred from experiments is given in Kittel, *Introduction to solid state physics*, Table 7 (see §1.4.1) as 774 kJ mol^{-1}. We may take this agreement at the level of a few per cent as validation of the theory we have developed so far.

Cohesive energy per atom

The cohesive energy determined in Equation 6.42 is the energy required to separate an ionic solid into its constituent *ions*. It is interesting to consider what energy would be required to separate an ionic solid into its constituent *atoms*. In order to calculate this we need to know:

(a) The energy u_1 (known as the *ionization energy*) required to remove an electron from a sodium atom

$$Na \xrightarrow{+u_1} Na^+ + e^- \qquad (6.44)$$

(b) The energy u_2 (known as the *electron affinity*) required to add an electron from a chlorine atom

$$Cl + e^- \xrightarrow{+u_2} Cl^- \qquad (6.45)$$

For NaCl, $u_1 = +5.139\,eV$ per Na ion and $u_2 = -3.617\,eV$ per Cl ion (Emsley, see §1.4.1). Thus to take two neutral atoms of Na and Cl and form an Na^+ and a Cl^- ion requires $u_1 + u_2 = +5.139 - 3.617 = 1.522\,eV$ per NaCl formula unit (or $146.9\,kJ\,mol^{-1}$ of NaCl). Note that it costs energy to transfer an electron from the sodium atom to the chlorine atom. However, if we take account of the cohesive electrostatic energy of the ions when they are arranged in a crystal, there is an additional energy gain u given by Equation 6.43 and evaluated in Example 6.3 as $-789\,kJ\,mol^{-1}$. Thus the cohesive energy of NaCl relative to neutral atoms is $+146.9 - 789 \approx -642\,kJ\,mol^{-1}$ (or $-6.65\,eV$ per NaCl formula unit).

Summary

We have seen in the above sections how we are able to understand the magnitude of the cohesive energy of ionic solids in terms of the coulomb interaction of ions distributed on a crystal lattice.

6.4 Covalent solids

6.4.1 General description

In covalent solids, the electrons that in isolated atoms occupy quantum states localized around a single atom, occupy orbitals that are distributed around at least one other atom in addition to their "parent" atom – i.e. they are literally co-valent. The materials that tend to form covalent solids are

- elements from the centre of the periodic table

(e.g. carbon, silicon and germanium);
- compounds of elements from the centre of the periodic table;
- many compounds of oxygen (oxides), sulphur (sulphides) and nitrogen (nitrides).

Covalency often occurs when the outer valence electrons have orbitals that are not spherical, but have "lobes" that point in different directions (Fig. 6.10). This gives rise to a high degree of directionality in the attraction between atoms, i.e. atoms will be attracted to one another if their orbitals point in a certain direction (perhaps towards each other), but if the orbitals point in a slightly different direction the atoms may well repel one another.

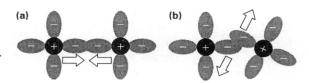

Figure 6.10 A simple model of a covalent bond between atoms illustrating the directional nature of the bond. (a) Two similar atoms with overlapping electron orbits. The increased charge density in between the atoms pulls the atoms together (arrows). (b) A misalignment of the atoms causes a repulsion (arrows) between the atoms.

Because of the directionality of the attraction/repulsion, one cannot simply write down a law of attraction that depends only on the separation between atoms, such as the Lennard–Jones equation (Eq. 6.7). Thus calculating the properties of covalent solids is considerably more complex than calculating the properties of molecular solids. However, we will attempt a simple calculation for one of the simplest (but rarest) covalent solids: carbon in the form of diamond.

6.4.2 Calculation of covalent bond length or strength

In what follows, we use the examples of §6.2 & 6.3 and calculate only the electrostatic potential energy of a covalent bond. The kinetic energy of electrons in a covalent bond is considered in Exercise P5 in §6.7.

A "point charge" model of a covalent bond

We can make a simple model of a covalent bond as follows. We consider a "solid" consisting of just carbon atoms, and imagine the charge in the region between the atoms to be distributed as shown in Figure 6.11.

Figure 6.11 A simple model of the charge distribution in a C–C bond.

The C–C bond shown in Figure 6.11 is electrically neutral overall, but electrons "originally" from each atom spend some time orbiting both atoms, and some time in the bond region. We will assume that each electron spends some fraction, f, of its time orbiting the atoms at each end of the bond, and therefore spends a fraction $(1 - 2f)$ of its time in the region between the atoms. We see that if $f = 0.5$ then the electrons spend essentially no time in the bond region, and if $f = 0$ the electron spends all its time in the bond region. If we recall that one electron from each atom will take part in the bond then, on average, we can make a "point charge" model of the charge distribution in the bond like that illustrated in Figure 6.12.

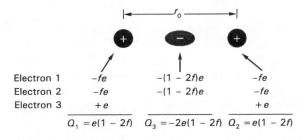

Electron 1 $-fe$ $-(1 - 2f)e$ $-fe$
Electron 2 $-fe$ $-(1 - 2f)e$ $-fe$
Electron 3 $+e$ $+e$

$$Q_1 = e(1 - 2f) \qquad Q_3 = -2e(1 - 2f) \qquad Q_2 = e(1 - 2f)$$

Figure 6.12 A point charge model of the charge distribution in a covalent bond.

Note that when an electron leaves an atom, the atom is left positively charged due to the balance of nuclear charges and core electrons. If we imagine the charges Q_1, Q_2 and Q_3 to be point charges, we can work out the approximate energy of this charge distribution using Coulomb's law. We have

$$u = \frac{Q_1 Q_2}{4\pi\varepsilon_0 r_0} + \frac{Q_1 Q_3}{4\pi\varepsilon_0 (r_0/2)} + \frac{Q_2 Q_3}{4\pi\varepsilon_0 (r_0/2)} \quad (6.46)$$

Substituting for Q_1, Q_2 and Q_3 from Figure 6.12

$$u = \frac{e^2(1-2f)^2}{4\pi\varepsilon_0 r_0} + \frac{-2e^2(1-2f)^2}{4\pi\varepsilon_0 (r_0/2)} + \frac{-2e^2(1-2f)^2}{4\pi\varepsilon_0 (r_0/2)} \quad (6.47)$$

and taking out a common factor, we have

$$u = \frac{e^2(1-2f)^2}{4\pi\varepsilon_0 r_0} \left(\frac{1}{1} + \frac{-2}{1/2} + \frac{-2}{1/2} \right) \quad (6.48)$$

which simplifies to

$$\boxed{u = \frac{e^2(1-2f)^2}{4\pi\varepsilon_0 r_0}(-7)} \quad (6.49)$$

Equation 6.49 indicates that if an electron spends no time in the bond region ($f = 0.5$) then the binding energy u is zero. The maximum binding energy occurs for $f = 0$ which corresponds to an electron spending all its time in the bond region and has a value

$$u_{max} = -7 \times \frac{e^2}{4\pi\varepsilon_0 r_0} \quad (6.50)$$

Comparison with experiment: a plausibility test

Kaye & Laby (see §1.4.1) give tabulated values of the C–C bond length and binding energy in different compounds and molecules. Taking the C–C bond length in a typical diatomic molecule, we have $r_0 = 0.1312\,\text{nm}$ and a cohesive energy per bond (actually tabulated as a *dissociation energy D*) of $D = 603\,\text{kJ mol}^{-1}$. Converting this into a cohesive energy per bond $u = D/N_A = 603\times10^3/6.02\times10^{23} = 1.00\times10^{-18}\,\text{J} = 6.25\,\text{eV}$.

We can use these *experimental* values to deduce an estimate for f. Rearranging Equation 6.49 and solving for f we have

$$(1-2f)^2 = -\frac{4\pi\varepsilon_0 r_0 u}{7e^2} \quad (6.51)$$

$$f = 0.5\left(1 - \sqrt{-\frac{4\pi\varepsilon_0 r_0 u}{7e^2}}\right) \quad (6.52)$$

Substituting the values discussed above

$$f = 0.5\left(1 - \sqrt{-\frac{4\pi \times 8.85 \times 10^{-12} \times 0.1312 \times 10^{-9} \times -1.00 \times 10^{-18}}{7 \times \left(1.6 \times 10^{-19}\right)^2}}\right)$$

(6.53)

$$f = 0.5\left(1 - \sqrt{0.08146}\right) = 0.357 \qquad (6.54)$$

Equation 6.54 indicates that an electron in a C–C covalent bond spends roughly 35% of its time orbiting the atoms at either end of the bond, and around 30% of its time in the central bond region. This estimate for f is unlikely to be accurate. However, it does show at least that our assumptions are self-consistent.

The above calculation is extremely crude: we have ignored, for example, the quantum mechanical rules that govern the distribution of charge around the atom and in the bond region. However, the purpose of the calculation was *not* to calculate a realistic answer for the charge distribution in a covalent bond. If we wanted to do this we would have to calculate quantum mechanically (or determine experimentally) the charge density around, and in between, each atom. The purpose of this calculation is to show that, given the existence of the bond structure (i.e. its length and strength) we can understand the connection between the two using only Coulomb's law.

We have also ignored the fact that in a solid each carbon atom takes part in other covalent bonds with other atoms. There will therefore be other contributions to the binding energy of a solid because the neighbouring electrically negative bond regions will repel one another. The consequences of this are discussed in the next section on crystal structure and illustrated in Figures 6.13–6.15.

6.4.3 Crystal structure

The directionality of the covalent bond means that solids with covalent bonds tend to form crystals with quite different structures from those formed by molecular solids. In molecular solids, the maximum cohesive force was achieved by atoms packing themselves so as to achieve the largest possible number of nearest–neighbours. In covalent solids, the structures

are dictated by the geometry of the electron orbitals around atoms; adding extra nearest-neighbours results, in general, in no extra bonding. For this reason covalent solids often form rather "open" structures with relatively low numbers of nearest neighbours.

For example, a carbon atom has four valence electrons and so can take part in, at most, four covalent bonds. It will therefore be of no benefit to form a crystal structure with more than four nearest neighbours. Each lobed orbital on the atom points in a different direction, and since each orbital has an excess of negative charge in it, it will repel the other orbitals in the same atom. The lobes therefore point away from one another, in directions towards the corners of a regular tetrahedron (Fig. 6.13a). In order to benefit from a covalent bond, other atoms must orient their orbitals with respect to the first atom in order to create overlapping orbital regions (Fig. 6.13b). The crystal structure formed is rather difficult to describe in words but is sketched in Figures 6.14 & 6.15.

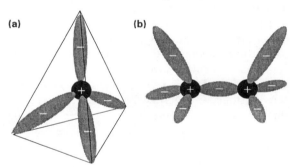

Figure 6.13 (a) An illustration of the arrangement of electric charge around a carbon atom. The four valence electrons orient their orbitals towards the points of a regular tetrahedron in order to minimize their repulsive Coulomb interaction. (b) An illustration of the arrangement of orbitals in two carbon atoms covalently bonded together.

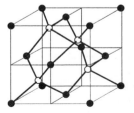

Figure 6.14 The crystal structure of diamond as deduced from X-ray scattering. The separation between atoms is 0.154 nm. The lines indicate the bonding regions and the spheres indicate the location of the atoms. The different shading is to aid the clarity of the picture, and allow identification of the four atoms entirely contained within the outer cube.

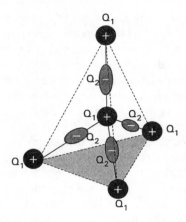

Figure 6.15 A simplified model of a small region of a diamond lattice. The model uses point charges to represent the continuous charge distribution that exists in the real lattice. It is a three-dimensional extension of the two-atom point-charge model discussed in the text.

6.4.4 Cohesive energy calculations

The cohesive energy of a covalently-bonded solid is rather difficult to calculate, and we will not proceed to a full calculation here. However, we can indicate how such a calculation would be made. We will consider the case of diamond which has the crystal structure illustrated in Figures 6.14 & 6.15. The complex three-dimensional nature of the charge distribution throughout the lattice makes a systematic evaluation of the electrostatic energy of the structure more complicated than it was for ionic substances (see §6.3). In particular, the situation of the positively charged atoms and the negatively charged bond regions are not equivalent. However, one may reasonably imagine how the cohesive energy of such a lattice can be determined in a way broadly similar to that discussed in §6.3.2.

6.5 Metals

6.5.1 General description

In metals the entities that make up the solid are essentially positive ions, i.e. atoms stripped of one or more of their electrons, and electrons free to move in between the ions. One can think of this as a kind of generalized covalent bond. However, in order to understand even the simplest properties of metals we need to look at one factor that has been ignored in the previous three models of solids: *quantum mechanics*.

The "particle in a box" problem

In this section we will develop a strikingly simple model of a metal. We envisage that the electrons are unable to leave the metal, but that they are free to move around within it. This is a description of a metal as a "box" and the determination of the quantum states of electrons in a metal is then reduced to the "particle in a box" problem (see §2.4.3). The way in which the coulomb potentials around single ions can add up to form a box-like potential is illustrated in Figure 6.16.

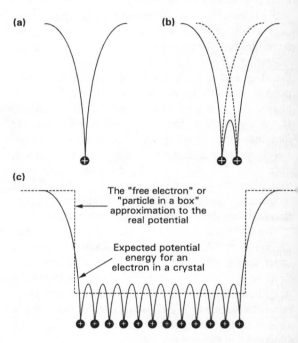

The "free electron" or "particle in a box" approximation to the real potential

Expected potential energy for an electron in a crystal

Figure 6.16 The variation of the potential energy of a single electron in the vicinity of (a) a single ion, (b) two ions close together, and (c) a collection of ions close together. In (a) the potential will vary approximately like the Coulomb law ($\approx 1/r$). In (b) the two potentials add together to make the region in between the ions a particularly attractive place to be for an electron. In (c) the ions have formed a box-like region capable of containing an electron within its walls, but allowing it to move within the box if its energy is not so low as to get trapped into the "corrugations" at the bottom of the box. The dotted line shows the essence of the "free electron" approximation discussed in the text.

The approximation that the potential energy experienced by an electron is box-like is known as the *free electron approximation*, and amounts to neglecting the corrugated details on the bottom of the expected potential in Figure 6.16c. The main justification of this approximation is that – as we shall see – the electrical properties of metals are generally rather similar. This being so, their properties might be expected to arise from some general feature of their situation, and not from details particular to certain types of atoms.

Quantum mechanics tells us that the energy of an electron in a cubic box of side L is given by (Eq. 2.59)

$$E(n_x, n_y, n_z) = \frac{h^2}{8mL^2}\left(n_x^2 + n_y^2 + n_z^2\right) \quad (6.55)$$

where n_x, n_y and n_z are quantum numbers allowed to take only integer values 1, 2, 3, etc. The quantum state with lowest energy is the state with $n_x = n_y = n_z = 1$, which we denote by $E(1,1,1)$. The quantum state with the next lowest energy is $E(1,1,2)$. There are three states with this same energy since from Equation 6.55, $E(1,1,2) = E(1,2,1) = E(2,1,1)$. The properties of a few of the low energy states are listed in Table 6.3.

The Pauli exclusion principle allows only one electron in each quantum state. In order to apply the exclusion principle correctly we need to recall that because an electron in a state (n_x, n_y, n_z) can have two possible spin states (see §2.4) there are *two* electron quantum states associated with each unique (n_x, n_y, n_z) combination. Note that if we place several electrons in the same box, then not all electrons will be able to occupy the lowest energy state, and so some electrons are forced to occupy higher energy states.

An example: an eight-atom metal

In order to see how these quantum mechanical results affect our understanding of metals, we will consider a specific example of a cluster of eight atoms. We consider that each atom has one valence electron which could be relatively easily detached from its parent atom. We thus have eight electrons to be accommodated in one of two ways, let's call them options A and B:

- In option A we model each atom as a box of side a, with each box containing a single electron. The solid is then just a collection of these eight separate boxes.
- In option B we model the solid as a single large box of side $2a$ that contains all eight electrons.

Note that both options refer to electrons with the same overall number density, $n = 1/a^3$.

Table 6.3 The first few energy levels for particles trapped in a box.*

Quantum numbers, n_x, n_y, n_z	Energy in units of $h^2/8mL^2$	No. of states	No. of electrons (including spin) able to be accommodated with this energy	No. of electrons able to be accommodated with energy less than the current energy
1,1,1	3	1	2	2
1,1,2 1,2,1 2,1,1	5	3	6	2 + 6 = 8
1,2,2 2,1,2 2,2,1	9	3	6	2 + 6 + 6 = 14
1,1,3 1,3,1 3,1,1	11	3	6	2 + 6 + 6 + 6 = 20
2,2,2	12	1	2	2 + 6 + 6 + 6 + 2 = 22

*The columns show, in order: (i) the quantum numbers of the states, (ii) the energy of the states with particular quantum numbers, in units of $h^2/8mL^2$. For example, $E(1,1,3) = A(1^2 + 1^2 + 3^2) = 11A$ where $A = h^2/8mL^2$. (iii) The number of quantum states with the same energy level. (iv) The number of electrons that can be accommodated at that energy. (v) The running total of the number of electrons able to be accommodated with energy equal to or less than the current energy, i.e. the running total of column (iv).

Option A

Each electron is confined to its own separate box (atom) with side a.

Each electron enters the ground state in its own box, so the total energy will be

$$E = 8 \times \frac{h^2}{8ma^2} \times 3$$

And so the total energy of option A is

$$\boxed{E_{\text{option A}} = 24 \times \frac{h^2}{8ma^2}}$$

Option B

←—2a—→

Each electron is contained in the same box with side $2a$.

The electrons enter the lowest quantum states available. Two electrons can occupy the ground state and so the total energy of these two electrons will be

$$E = 2 \times \frac{1}{4} \times \frac{h^2}{8ma^2} \times 3$$

The factor 1/4 arises from Equation 6.55 because the box has side $L = 2a$ (as opposed to $L = a$ in option A). The next six electrons can occupy the states (1,1,2), (1,2,1) and (2,1,1), so their total energy will be

$$E = 6 \times \frac{1}{4} \times \frac{h^2}{8ma^2} \times 6 .$$

Adding these energies we find that the total energy of option B is

$$E_{\text{option B}} = \left(\frac{2 \times 3}{4} + \frac{6 \times 6}{4} \right) \frac{h^2}{8ma^2}$$

$$\boxed{E_{\text{option B}} = 10.5 \times \frac{h^2}{8ma^2}}$$

Figure 6.17 Options A and B for accommodating eight electrons.

The energy required for option B is dramatically lower than the energy of option A. This means that electrons prefer to be in one big box rather than eight small boxes. Although this calculation is specific, the result is quite general: *electrons like to spread out if they can.*

Of course what we are really interested in is not the properties of an eight-atom piece of metal, but the limit of this problem for large numbers of atoms, i.e. the comparison between the energy of N electrons trapped in boxes of side a and N electrons in a single box of side $(N)^{1/3}a$. We consider this problem in §6.5.4, but surprisingly, the result does not change much.

The importance of quantum mechanics

Considering once again the eight-atom metal of the previous section, we see that is not difficult to imagine how changes in electron density could result in different total energies. However, in both options A and B *the electron density was the same*, $n = 1/a^3$, so the change in energy does not arise from a change in electron density. The lowered energy is a result of the change in the nature of the wave functions. Thus an arrangement of atoms can lower their energy by allowing their electrons to become *delocalized*, i.e. to move freely through the crystal, but be trapped within the crystal and not allowed to leave. This is nothing more than a gigantic version of option B, the big box. In other words, the cohesive energy of metals derives from this delocalization energy.

The importance of coulomb interactions

Notice that no mention has been made of the coulomb interaction either

(a) between electrons and ions, or
(b) between electrons and the other electrons with which they share a "box".

Since we have seen the importance of coulomb interactions in the other three types of bonding, we can assume that coulomb interactions are important and that a full theory must take them into account. The theory outlined above describing many particles in boxes is known as the theory of a *free electron gas*. The neglect of coulomb interactions is dignified by giving it two special names:

- The *free electron approximation* assumes one can neglect the interactions between electrons and ions, *except that occasionally electrons are scattered by vibrations of the lattice.*

- The *independent electron approximation* assumes one can neglect the interactions between electrons and other electrons *except that electrons know about each other sufficiently to obey the exclusion principle.*

Overcoming either of these approximations is beyond the scope of this book.

6.5.2 Counting quantum states

In a metal we have to deal with more than "just a few" electrons in a box, and counting the occupied quantum states individually becomes impossible. In order to cope with the large number of electrons, two pictures have been developed to allow calculations to be made about metals. The pictures correspond to rather sophisticated ways of counting quantum states the energies of which lie in particular energy ranges, as opposed to the rather laborious method of "just counting" adopted for Table 6.3.

Standing waves

The first method of counting is derived from Equation 6.55 for the energies of allowed quantum states

$$E(n_x, n_y, n_z) = \frac{h^2}{8mL^2}\left(n_x^2 + n_y^2 + n_z^2\right) \quad (6.56)$$

If we represent an allowed quantum state by a point on a three-dimensional graph the axes of which are n_x, n_y and n_z, respectively, then the energy of a particular

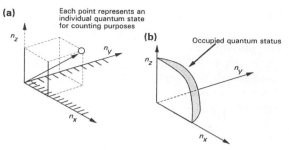

Figure 6.18 A scheme for counting quantum states in the particle in a box problem when many particles are present. (a) Each quantum state is represented by a point on an n_x, n_y, n_z graph. The energy of the state is proportional to the square of its distance from the origin on this graph. For this reason, the occupied quantum states cluster in an octant of a sphere around the origin (b) in order to minimize their energy.

state will be given by $h^2/8mL^2$ multiplied by the square of the length of the vector from the origin to the point representing that quantum state (Fig. 6.18).

Imagine filling the quantum states one by one until we have filled sufficient states to accommodate all the electrons in a particular piece of metal. If one fills the lowest energy states first and then proceeds to higher and higher energies, one can see that eventually one will fill states in an octant of a sphere centred on the origin. If we were dealing with only a few electrons, then the surface of the sphere would be irregular because the difference between the quantum numbers of neighbouring states on the graph ($\Delta n_x = 1$ or $\Delta n_y = 1$) is comparable to the value of the quantum numbers (n_x or n_y) themselves. However, for large n the surface of the sphere is defined with good precision, as shown (in two-dimensions) in Figure 6.19.

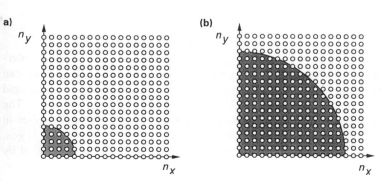

Figure 6.19 Details of the x,y plane on the n_x, n_y, n_z diagram shown in Figure 6.18. If one considers just a few occupied states then the occupied states form only an ill-defined octant of a sphere around the origin. However, when one considers many electrons in the box (i.e. a large box with many electrons) the states occupied form a relatively well-defined octant of a sphere around the origin.

143

Because of this, by measuring "volumes" on a three-dimensional graph one can, with a fair degree of approximation, also count quantum states. We can do this because the quantum states are distributed uniformly on the graph. This aspect is developed further if we go over to a second view of the quantum states of a particle in a box.

The quantum states: travelling waves

The second method of counting arises from a reconsideration of the particle in a box problem. We will arrive at similar, but distinctly different, solutions to the Shrödinger equation: travelling wave solutions.

In §2.4.3 we required the wave functions to be zero at the edge of the box because we imagined the edge of the box to represent the walls of the box (or, in the case of a metal, the surface of the metal). Now, however, we consider a more complex situation: we imagine a volume that is *representative* of the situation within the box as a whole and we imagine it to be surrounded by *identical volumes*. The differences between this situation and the standing wave situation are illustrated in Figure 6.20. The reader may consider this to be "trickery", and it is. It is a mathematical trick, but it will allow us to see more clearly the physics of what is happening in the metal.

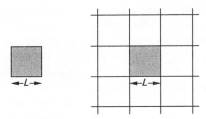

Figure 6.20 Illustration of two approaches to the particle in a box problem. The first approach uses the idea of an isolated box representing the entire crystal. The second approach uses the idea of a box of representative material surrounded by identical copies of itself. The edges of the actual crystal are imagined to be far enough away such that they do not significantly affect the electrons deep inside the box.

The key effect of this change in representation is that, now, instead of requiring that the wave functions be zero at the edges of the box, i.e. $\Psi(-L/2) = \Psi(+L/2) = 0$, we require that the wave functions be identical in neighbouring boxes, i.e.

$$\Psi(x) = \Psi(x + L_x) \tag{6.57}$$

where L_x is the length of the box under consideration. If the solutions are $\sin(kx)$ or $\cos(kx)$ waves this is equivalent to requiring that

$$\cos(k_x x) = \cos[k_x(x + L_x)] \tag{6.58a}$$

$$\sin(k_x x) = \sin[k_x(x + L_x)] \tag{6.58b}$$

The requirements given by Equations 6.58a,b are known as *Born–von Karmen boundary conditions*, and lead to only discrete values of k being allowed solutions. Let's follow one of the equations (Eq. 6.58a) through explicitly to find these conditions for k_x.

The boundary conditions require that

$$\cos(k_x x) = \cos[k_x(x + L_x)] = \cos(k_x x + k_x L_x). \tag{6.59}$$

Using the trigonometric identity

$$\cos(A+B) = \cos A \cos B - \sin A \sin B \tag{6.60}$$

Equation 6.59 can be rewritten as

$$\cos(k_x x) = \cos(k_x x)\cos(k_x L_x) - \sin(k_x x)\sin(k_x L_x) \tag{6.61}$$

Clearly, in general this equation is not true! It is only true when $\cos(k_x L_x) = 1$ and when the second term is zero. Since x can vary, this will only be true when $\sin(k_x L_x) = 0$. Now $\cos(k_x L_x) = 1$ when $k_x L_x = 0, \pm 2\pi, \pm 4\pi$, etc. and these values of $k_x L_x$ also cause $\sin(k_x L_x)$ to be zero. So the Born–von Karmen boundary conditions are satisfied when

$$k_x L_x = 0, \pm 2\pi, \pm 4\pi, \ldots \tag{6.62}$$

i.e. when

$$k = 0, \frac{\pm 2\pi}{L_x}, \frac{\pm 4\pi}{L_x}, \ldots \tag{6.63}$$

or, in general, when

$$k_x = \frac{2m_x \pi}{L_x} \quad \text{where } m_x = 0, \pm 1, \pm 2, \ldots \tag{6.64}$$

It is left to the reader to show that following Equation 6.58b through in a similar way (using $\sin(A + B) = \sin A \cos B + \cos A \sin B$) results in exactly the same conclusion.

k-space

We can make sketches analogous to those shown in Figure 6.18 for this new situation. In these sketches we plot points representing the k_x, k_y and k_z components of a wave. The k_x, k_y, k_z graph is often referred to as "k-space". The main difference between the k-space description of the quantum states of a particle in a box and the (n_x, n_y, n_z) representation discussed previously, is that each allowed point in space now represents a travelling wave rather a standing wave. As a result of this k can also take negative values and so the low energy states cluster uniformly around $k = 0$ rather than being confined to the positive $|k|$ octant. If we occupy the states starting at low energies and work outwards, then we fill a sphere centred on the origin (Fig. 6.21).

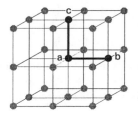

Figure 6.22 Close-up view of k-space. The spheres represent allowed values of k. If the central point a represents a solution of the Schrödinger equation with a particular value of k_x, k_y, k_z, then point b represents a solution with the k_x component increased by $2\pi/L_x$. Similarly, point c represents a solution with the k_z component increased by $2\pi/L_z$.

From the Figure 6.22 we can see that each state has around it a "volume of k space", $\Delta\Omega$, given by

$$\Delta\Omega = \frac{2\pi}{L_x} \times \frac{2\pi}{L_y} \times \frac{2\pi}{L_z} \qquad (6.65)$$

i.e.

$$\Delta\Omega = \frac{8\pi^3}{L_x L_y L_z} = \frac{8\pi^3}{V} \qquad (6.66)$$

We must also remember that each allowed k value can accommodate two electrons with opposite spin. Therefore each quantum state has associated with it a "volume of k-space" given by

$$\boxed{\Delta\Omega = \frac{1}{2} \times \frac{8\pi^3}{V} = \frac{4\pi^3}{V}} \qquad (6.67)$$

We are now in a position to calculate some useful quantities such as the number of quantum states in any specified energy range. The simple result embodied in Equation 6.67 is crucial to these calculations.

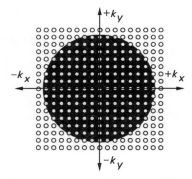

Figure 6.21 A cross-section through k-space. Each small circle represents an allowed travelling wave solution to the Shrödinger equation. The filled circles represent occupied quantum states and the unfilled circles represent empty quantum states. The configuration of occupied and empty states shown is similar to that expected in a metal at absolute zero. Only the lowest energy states (low k = long wavelength = low energy) are occupied. In three-dimensions the occupied states form a sphere in k-space known as the *Fermi sphere*.

Counting states by measuring volumes in k-space

We know that in k-space the points representing the occupied quantum states form a sphere centred around $k = 0$. We also know further that the allowed states are distributed uniformly through k space so that if we consider a "volume" of k-space we can work out how many points representing allowed quantum states it contains. A close-up view of k-space is shown in Figure 6.22.

The number of quantum states with k between k and k + dk

We will use Equation 6.67 to work out how many quantum states of the particle in a box problem have k values in the range k to k + dk. All the quantum states with k in this range have representative points on a k-space graph that lie *within a thin shell of inner radius k and thickness* dk. The "volume" occupied by this shell on the k-space diagram is

$$d\Omega = 4\pi k^2 dk \qquad (6.68)$$

Equation 6.67 tells us that each quantum state occupies a volume $\Delta\Omega = 4\pi^3/V$ so the number of quantum states dN in this shell must be given by

$$dN = \frac{4\pi k^2 dk}{4\pi^3/V} = \frac{Vk^2}{\pi^2}dk \qquad (6.69)$$

The number of quantum states with E between E and E + dE

We can now use Equation 6.69 to work out how many quantum states there are between energy E and $E + dE$, i.e. the density of states function, $g(E)$, discussed in Section 2.5. Our plan will be to use the relationship between E and k for an electron ($E = \hbar^2 k^2/2m$) to replace references to k in Equation 6.69 with references to E. Thus we substitute

$$k = \frac{1}{\hbar}\sqrt{2mE} \quad \text{and} \quad \frac{dk}{dE} = \frac{1}{2\hbar}\sqrt{\frac{2m}{E}} \qquad (6.70)$$

in Equation 6.69 to yield

$$dN = \frac{Vk^2}{\pi^2}dk = \frac{V2mE}{\pi^2\hbar^2} \times \frac{1}{2\hbar}\sqrt{\frac{2m}{E}}dE \qquad (6.71)$$

which simplifies to

$$dN = \frac{V\sqrt{2m^3E}}{\pi^2\hbar^3}dE \qquad (6.72)$$

This is the sought-after density of states function

$$\boxed{g(E) = \frac{V\sqrt{2m^3E}}{\pi^2\hbar^3}} \qquad (6.73)$$

The occupied quantum state with the highest energy

Using the density of states function we can work out answers to questions such as "What is the energy of the highest occupied quantum state in a metal?". At $T = 0\,$K this is a well-defined energy generally known as the *Fermi energy*, E_F. It is the energy possessed by

---

Example 6.4

In the particle in a box model of a piece of metal of volume $1\,cm^3$, how many quantum states are there with energy in the range 1–$1.01\,eV$?

Since the energy range dE in this question is small compared with E, we can use Equation 6.72 to answer this question without integrating. Substituting values for the fundamental constants involved we find

$$dN = \frac{10^{-6}\sqrt{2(9.1 \times 10^{-31})^3(1.6 \times 10^{-19})}}{\pi^2(1.054 \times 10^{-34})^3} \times 0.01 \times 1.6 \times 10^{-19}$$

$$dN = \frac{4.91 \times 10^{-61}}{11.56 \times 10^{102}} \times 1.6 \times 10^{-21} = 6.8 \times 10^{20} \text{ quantum states}$$

You may reflect that this is a tricky calculation. However, it is a lot easier than counting $\approx 10^{20}$ quantum states by hand! The vast numbers of states involved are the reason why we have moved to a geometrical (k-space) representation of the quantum states of the particle in a box problem.

---

those electrons occupying quantum states on the surface of the Fermi sphere (see Fig. 6.21).

We begin by noting that, as discussed in §2.5, the number of *particles* in quantum states with energies between E and $E + dE$ is given by the product $f(E,T)g(E)dE$ where $f(E,T)$ is the occupation function. Since there are N particles to be accommodated in quantum states we can write, in general, that

$$N = \int_{E=0}^{E=\infty} f(E,T)g(E)dE \qquad (6.74)$$

However, at $T = 0$ life is considerably simpler. The $f(E,T)$ is the Fermi–Dirac occupation function, and so states with energies below E_F are definitely occupied and states above this energy are empty. We can thus write

$$N = \int_{E=0}^{E=E_F} g(E)dE \qquad (6.75)$$

Substituting for the density of states function we find

$$N = \int_{E=0}^{E=E_F} \frac{V\sqrt{2m^3 E}}{\pi^2 \hbar^3} \, dE$$

$$= \frac{V\sqrt{2m^3}}{\pi^2 \hbar^3} \int_{E=0}^{E=E_F} E^{\frac{1}{2}} dE \qquad (6.76)$$

Rewriting this in terms of the number density of electrons, n, and integrating, we find

$$n = \frac{\sqrt{2m^3}}{\pi^2 \hbar^3} \left[\frac{2}{3} E^{\frac{3}{2}} \right]_{E=0}^{E=E_F} = \frac{\sqrt{2m^3}}{\pi^2 \hbar^3} \left[\frac{2}{3} E_F^{\frac{3}{2}} \right] \quad (6.77)$$

Rearranging this as an expression for the Fermi energy, we find

$$E_F = \left(\frac{3n\pi^2 \hbar^3}{2\sqrt{2}m^3} \right)^{\frac{2}{3}} = \frac{\hbar^2 \left(3n\pi^2\right)^{\frac{2}{3}}}{2m} \qquad (6.78)$$

which implies that the Fermi wave vector – the "radius" of the Fermi sphere – is

$$k_F = \left(3n\pi^2\right)^{\frac{1}{3}} \qquad (6.79)$$

So, to summarize, we have taken the k-space method of counting the quantum states of the particle in a box problem a long way. We have used the counting formalism along with some statistical mechanical ideas (see §2.5) to arrive at an expression for the energy and wave vector of the highest occupied quantum state:

$$\boxed{E_F = \frac{\hbar^2 k_F^2}{2m} \quad \text{and} \quad k_F = \left(3n\pi^2\right)^{\frac{1}{3}}} \quad (6.80)$$

6.5.3 Real simple metals

In the previous section we have given a rather abstract view of a metal. Let's apply some numbers that might be appropriate to real metals to the model developed in the previous section in order to make its consequences clearer.

A numerical example

To estimate the number density, n, of "free electrons" in a metal we can work out the number density of atoms in a metal, and then imagine that 1, 2 or 3 electrons, the *valence electrons*, from each atom can leave the "parent" atom to form the electron gas. There are several ways to do this, but perhaps the simplest is to look up the density of an elemental metal, and to divide by the mass of an atom to get the number density of atoms and then multiply by the valence to get n. From Table 7.2 we see that the density of copper is 8.933×10^3 kg m^{-3}, and the mass per atom is $63.55 \times 1.66 \times 10^{-27}$ kg. Assuming copper has a valence of one, then our estimate for n is

$$n = \frac{8.933 \times 10^3 \, \text{kg m}^{-3}}{63.55 \times 1.66 \times 10^{-27} \, \text{kg}} \approx 8.47 \times 10^{28} \, \text{m}^{-3} \quad (6.81)$$

If we use this estimate for n in Equation 6.80 for the Fermi wave vector, k_F, we find

$$k_F = \left[3\pi^2 \times 8.47 \times 10^{28}\right]^{\frac{1}{3}} = 1.36 \times 10^{10} \, \text{m}^{-1} \quad (6.82)$$

Substituting this in Equation 6.80 for the Fermi energy, E_F, we find

$$E_F = \frac{\left(1.054 \times 10^{-34}\right)^2 \times \left(1.356 \times 10^{10}\right)^2}{2 \times 9.1 \times 10^{-31}}$$

$$= 1.13 \times 10^{-18} \, \text{J}$$

$$E_F \approx 7.0 \, \text{eV} \qquad (6.83)$$

We will now look at the significance of k_F and E_F in turn.

The significance of k_F

The significance of k_F can be seen more easily if we look at the wavelength of electrons with wave vector k_F. Since k_F is the *maximum* k-vector, then electrons with this k-vector will have the *minimum wavelength* since $k = 2\pi/\lambda$. Substituting from Equation 6.82

$$\lambda_{min} = \frac{2\pi}{1.36 \times 10^{10}} = 4.63 \times 10^{-10} \, \text{m} \qquad (6.84)$$

we find a wavelength comparable to the spacing, a,

between atoms. We can see the significance of this if we consider again Figure 6.16 which was used as the basis of the free-electron approximation. We have been able to ignore the "corrugations" on the bottom of the potential well because the wavelengths of electron wave functions are not related to the spacing between ions, a. Thus the peaks in the probability density $|\psi|^2$ sometimes fall exactly on an ion and sometimes in between ions. On average, each electron experiences a smoothly averaged potential.

However, if the periodicity of the electron probability density has a periodicity that *exactly* matches the lattice periodicity, then the peaks in the probability density will either always fall on an ion, or always fall between ions. In either case, the electron–ion interaction becomes much larger, and they can no longer be treated in the free-electron approximation which we have used so far. Although this phenomenon is critical for understanding the electronic properties of all solids in detail, we will not deal with it further in this text.

The significance of E_F

The significance of E_F can be seen if we compare it with the typical amount of thermal energy an electron might expect to receive as it interacts with the crystal ($\approx 3 \times 0.5 k_B T$). Evaluating this at room temperature (≈ 290 K)

$$\text{Typical thermal energy} = 6.0 \times 10^{-21} \text{J}$$
$$= 0.038 \text{ eV} \quad (6.85)$$

This energy is smaller than E_F by a factor of about 200. Even at a temperature of 2000 K, which is well above the melting point of most metals, the thermal energy would constitute less than 10% of E_F. Thus electrons in a metal are not greatly disturbed by temperature.

The electrons with energies near E_F also travel very fast. Since E_F is the kinetic energy of the electron, $m v_F^2/2$, we can work out the Fermi speed, v_F, as

$$v_F = \sqrt{\frac{2 E_F}{m}} = \sqrt{\frac{2 \times 1.122 \times 10^{-18}}{0.1 \times 10^{-31}}}$$
$$\approx 1.6 \times 10^6 \text{ ms}^{-1} \quad (6.86)$$

which is about 1% of the speed of light.

Note that the electrons with energies near E_F travel with this speed, regardless of the temperature. This may be contrasted with the situation of molecules in a *molecular* gas whose distribution of speeds depends strongly on temperature (see Fig. 4.7).

6.5.4 Cohesive energy

The cohesive energy of a metal is rather difficult to calculate accurately. However, even at the level of this book, we can show that the cohesive energy is quite large, potentially of the order of several electron-volts per atom (several hundred kilojoules per mole).

The cohesive energy of an eight-electron metal

To determine the cohesive energy of an eight-electron metal, we can consider the calculations illustrated in Figure 6.17, which showed that the energy of eight electrons on eight isolated "atoms" of volume a^3 (option A) is significantly greater than the energy of eight electrons that are able to move around a single "box" of volume $8a^3$ (option B). Table 6.4 summarizes the total energy of the eight electrons, the *average energy* per electron, and the difference between the average energy per electron in the two options.

Now, for a metal, a has a typical value of about 0.3 nm, and so a typical energy benefit u per electron in Table 6.4 is given by

$$u = -1.69 \times \frac{h^2}{8 m a^2} \quad (6.87)$$

Substituting

$$u = -1.69 \times \frac{\left(6.626 \times 10^{-34}\right)^2}{8 \times 9.1 \times 10^{-31} \times \left(3 \times 10^{-10}\right)^2}$$

Table 6.4 The total energy of eight electrons, the average energy per electron and the difference between the average energy per electron in options A and B.

	Option A	Option B
Energy of 8 electrons	$24 \times h^2/8ma^2$	$10.5 \times h^2/8ma^2$
Average energy per electron	$3 \times h^2/8ma^2$	$1.31 \times h^2/8ma^2$
Energy *difference* per electron		$-1.69 \times h^2/8ma^2$

we find

$$u \approx -1.13 \times 10^{-18} \, \text{J / electron} \qquad (6.88)$$

$$u \approx -7.1 \, \text{eV / electron} \qquad (6.93)$$

$$u \approx -681 \, \text{kJ / mole} \qquad (6.90)$$

The cohesive energy of a "real" metal

In order to calculate the cohesive energy of real metal consisting of more than eight electrons, we need to calculate the average energy per electron. This was done in Example 2.8, where it was shown that the average energy per electron in the free electron approximation is $3/5E_F$.

Table 6.5 summarizes the calculation we need to make. Let's try to evaluate this energy difference. First of all we write E_F in terms of k_F (Eq. 6.80).

$$u = \frac{3}{5} E_F - \frac{3h^2}{8ma^2} \qquad (6.91)$$

$$u = \frac{3}{5} \times \frac{\hbar^2 k_F^2}{2m} - \frac{3h^2}{8ma^2} \qquad (6.92)$$

We now write k_F in terms of n (Eq. 6.80), the number density of electrons,

$$u = \frac{3}{5} \times \frac{\hbar^2 \left(3n\pi^2\right)^{\frac{2}{3}}}{2m} - \frac{3h^2}{8ma^2} \qquad (6.93)$$

Finally we recall that $\hbar = (h/2\pi)$ and that the number density of electrons n is just $1/a^3$:

$$u = \frac{3}{5} \times \frac{(h/2\pi)^2 \left(3\pi^2/a^3\right)^{\frac{2}{3}}}{2m} - \frac{3h^2}{8ma^2} \qquad (6.94)$$

Table 6.5 Summary of the calculation of the cohesive energy of a "real" metal.

	Option A (individual atoms)	Option B (metallic state)
Energy of N electrons	$3N(h^2/8ma^2)$	$N \times 3E_F/5$
Average energy per electron	$3(h^2/8ma^2)$	$3E_F/5$
Energy *difference* per electron		$3E_F/5 - 3(h^2/8ma^2)$

Simplifying

$$u = \frac{3}{5} \times \frac{h^2 \left(3\pi^2\right)^{\frac{2}{3}}}{4\pi^2 \times 2ma^2} - \frac{3h^2}{8ma^2} \qquad (6.95)$$

and taking out common factors

$$u = \frac{3h^2}{8ma^2} \left[\frac{3\left(3\pi^2\right)^{\frac{2}{3}}}{5\pi^2} - 1 \right] \qquad (6.96)$$

we find

$$u = -1.254 \times \frac{3h^2}{8ma^2} \qquad (6.97)$$

which differs by only $\approx 30\%$ from the cohesive energy per electron of the eight-electron metal we considered earlier. We therefore find for the cohesive energy of a typical metal a figure of

$$u = -8.394 \times 10^{-19} \, \text{J} \qquad (6.98)$$

$$u = -5.24 \, \text{eV} \qquad (6.99)$$

$$U = -505 \, \text{kJ mol}^{-1} \qquad (6.100)$$

Clearly this is a potentially strong form of bonding. Astonishingly, this figure is arrived at without any consideration of the coulomb interaction between electrons. When the coulomb interactions between electrons and other electrons, and between electrons and ions are also taken into account, the problem becomes considerably more complex.

6.6 Real solids

The four types of bonding discussed in §6.2 to 6.5 represent "ideal" situations: only rarely can solids be understood by considering just one type of bonding. The solids which can be understood are sometimes called "simple" (although they aren't!) and have been studied extensively over the first 60 years or so of modern solid-state physics. However, most solids are not "simple" in the sense of having only one type

149

of bonding: most materials have two or more types of bonding. Below we consider briefly the combinations of types of bonding that one finds in real solids.

6.6.1 Organic solids

Many of the most common solids encountered in normal life are classified as *organic*. Originally this meant that the substance was derived from a living organism, but in modern terms an organic substance is one composed primarily of carbon and hydrogen atoms. If we consider *plastics* – just one class of organic substance – we can begin to appreciate the astonishingly diverse properties of organic solids. Clearly, we cannot specify everything about bonding in organic substances, but we can make one or two generalizations. In particular, we will look at bonding between molecules and bonding within molecules (Fig. 6.23).

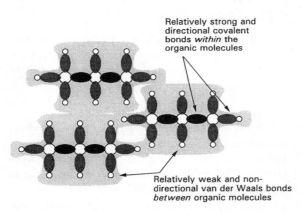

Relatively strong and directional covalent bonds *within* the organic molecules

Relatively weak and non-directional van der Waals bonds *between* organic molecules

Figure 6.23 Within an organic substance there are in general at least two types of bonding. *Within* an organic molecule the atoms are held together by relatively strong, directional covalent bonds. *Between* organic molecules the bonding is due to relatively weak, non-directional, Van der Waals bonds.

Bonding between molecules

Organic substances are composed of molecules that are bound to each other by Van der Waals bonding similar to that discussed in §6.2. The relatively weak Van der Waals force acts between *all* molecules, and is non-directional in nature.

However, some molecules have other types of bonds between them. In particular, *hydrogen bonds* are of particular importance for molecules that contain the OH chemical group. We discuss these further in Chapter 8.

Bonding within molecules

Organic molecules can consist of anything from just a few to many thousands of atoms. *Within* each organic molecule, the atoms are held together by relatively strong covalent bonding similar to those discussed in §6.4. Furthermore, the covalent bonding is directional in nature, which leads to organic molecules having relatively well-defined shapes.

6.6.2 Hydrogen bonding

Hydrogen bonding is a specific combination of ionic and covalent bonding which occurs in substances that contain oxygen and hydrogen atoms bonded together, e.g. H_2O. It is given a specially category to itself because of its importance in organic substances where it acts *between* molecules and is much stronger than the Van der Waals force. The nature of the hydrogen bond is discussed in detail in §9.2.2 where we discuss its effect on the density of liquid water.

6.6.3 The ionic–covalent continuum

In "pure" ionic bonding, an electron is transferred completely from one atom to another. Of course, once this transfer is complete the electric forces will act to try to pull this electron back to its parent ion. If these forces partially succeed then there will be some electron density in the region between the ions, which is the situation in a covalent bond. Thus pure ionic and pure covalent bonds can be seen as two extremes of a continuum. Symmetry considerations dictate that a "pure" covalent bond can only exist between two identical atoms. Any difference in the *electron affinity* of the atoms at either end of the bond will result in an asymmetric charge distribution that has both ionic and covalent character (Fig. 6.24).

The differing electron affinities of atoms are caused by the different distributions of charge around the nucleus of different atoms. Atoms with complete "shells" of electrons have essentially zero electron affinity. Atoms with one electron missing from a

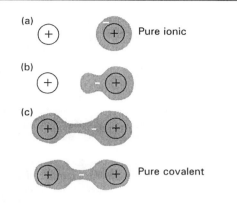

Figure 6.24 An illustration of the ionic-covalent continuum. In the text, ionic and covalent bonding are presented as distinct categories. In practice, many substances have bonding which is intermediate between the two cases. The figure illustrates four possible electronic charge distributions between two atoms. In (c), the atoms have identical electron affinity. (a) and (b) refer to situations where the atom on the right has increasing electron affinity compared with the atom on the left.

shell have a particularly high electron affinity. This arises because an atom with an electron missing from a shell allows other electrons to "see" the nuclear charge and hence be attracted to the atom.

6.6.4 Transition metals

Most elements are metallic. However, most elements are bonded in a way that is a mixture of metallic and covalent bonding. In the transition metals for example, electrons occupying d-orbitals form a mixture somewhere between a covalent bond and a metallic bond, while electrons occupying s-orbitals form metallic bonds. The situation is too complex for detailed consideration at this level.

6.6.5 What have we neglected?

In considering molecular, ionic and covalent bonding in §6.2–6.4 we concentrated entirely on the effect of coulomb interactions, and neglected to take account of quantum mechanics. Clearly this can't be quite right, since we saw when considering metals in §6.5 that quantum mechanics plays a critical role in determining, for example, the cohesive energy. Also, in metallic bonding, when we did consider the effect the of quantum mechanics, we neglected to take account of the coulomb interaction between electrons.

In reality, both coulomb interactions and quantum mechanics are important considerations in all types of bonding. Table 6.6 summarizes some of the effects that we have neglected in §6.2–6.5.

Table 6.6 Summary of what has been considered and neglected in Chapter 6.

Bonding type	What we considered	What we neglected
Molecular	We calculated the cohesive energy of an array of neutral molecules interacting through fluctuations of their electric charge distribution	We neglected to consider the origin of the charge distributions, and the origin of their fluctuations. These can only be properly calculated using quantum mechanics.
Ionic	We calculated the cohesive energy of an array of positive and negative ions interacting through the Coulomb interaction	We neglected to consider why electrons transfer themselves completely from one ion to another. This can only be understood by quantum mechanical calculations of the charge distribution around atoms and ions
Covalent	We calculated the cohesive energy of a covalent bond in terms of a simple model of the charge distribution within the bond	We neglected to consider why that particular charge distribution occurs. This can only be understood by quantum mechanical calculations of the electron wave functions near both atoms. Note that in addition to the Coulomb energy, which we calculated, there will be an additional "delocalization" term analogous to that which occurs on a larger scale in metals. In short, we calculated only the potential energy and ignored the kinetic energy.
Metallic	We calculated the cohesive energy of a metal by considering the change in energy of the wavefunctions as they were allowed to expand into a large volume	We neglected to include the effects of the strong Coulomb interactions between electrons and other electrons, and between electrons and ions

6.7 Exercises

Exercises with a P prefix are "normal" problems. Those with a C prefix are best solved numerically using a computer program or spreadsheet.

C1. Use the computer program listed in Appendix 4 to observe the dynamics of molecules in a solid. Note how the initial "square" structure spontaneously "collapses" into a "hexagonal" structure.

C2. The computer program listed in Appendix 4 simulates the dynamics of circular (spherical) molecules interacting via a potential of the form given by Equation 6.6 with $A = B = 1$. Investigate the changes in the dynamics of the solid when the relative values of A and B are changed. Use a spreadsheet program to draw the functional form of Equation 6.6 for a range of values of A and B.

P3. Given the entries in Table 6.2 for α and ε for krypton, estimate (a) the molar cohesive energy and (b) the density. Check your answers in Table 11.5 and Table 7.2, respectively.

C4. Write a computer program to evaluate the Madelung sum for an ion at the centre of a small ionic crystal with 12×12×12 ions. Re-evaluate the Madelung sum for an ion displaced from its equilibrium position by an amount Δ and plot a graph of the Madelung sums versus Δ. You should find a roughly quadratic variation with Δ

indicating that the ion will vibrate in a roughly simple harmonic potential.

P5. In Section 6.4 we considered calculations of the potential energy of electrons in a covalent bond, but did not consider their kinetic energy. The kinetic energy is difficult to calculate accurately, but one can estimate it roughly by considering the electrons in the bond region of length L to be confined to a three-dimensional box of side L. The kinetic energy of an electron in such a box may be estimated using Equation 2.39 (or Eq. 6.55). Calculate this energy.

The calculated kinetic energy for the C–C bond is large, but a more realistic calculation would compare the kinetic and potential energies of electrons on isolated atoms with the kinetic and potential energies of electrons when bonded. Unfortunately, this is a much more difficult calculation than the simple estimates we have made.

P6. Work out the separation between neighbouring atoms in (a) silicon and (b) germanium given that their crystal structure is illustrated in Figure 6.14. The densities of silicon and germanium may be found from Table 7.1. (See also Example 7.3.)

P7. Repeat the Option A/Option B (Fig. 6.17) exercise for metals with $3^3 = 27$ electrons and $4^3 = 64$ electrons. Compare the energy per electron with the value calculated for the eight-electron metal and the value for "real" metal (Tables 6.4 and 6.5).

CHAPTER 7

Solids: comparison with experiment

The diverse properties of solids make direct comparison of the simple models outlined in Chapter 6 with the properties of real solids particularly difficult. There are so many "special cases" and "exceptions" in the study of solids that it is unusual to find strict numerical agreement between the theories outlined in the previous sections and experimental data on real solids. However, although one is not able to understand the precise results of experiments, one can come to understand:

(a) the order of magnitude of the results; and
(b) the trends in results taken at different temperatures or for different materials.

With these limitations on our ambition in mind, let us examine the data on the properties of solids.

7.1 The density of solids

In contrast with gases, the most striking property of solids is that they have both a well-defined *volume* and *shape*. Furthermore, unlike gases, neither the volume or shape are particularly sensitive to changes in temperature or pressure.

7.1.1 Data on the density of solids

The densities of most solids lie in the range between 0.5 and 20 times that of water. Historically, the *gram* was defined as the mass of $1\,cm^3$ of water at 4°C and so we find that the density of water lies close to $1\,g\,cm^{-3}$, or, in the rather incongruous SI units, $1000\,kg\,m^{-3}$. We thus find solid densities in the range $\approx 500\,kg\,m^{-3}$ to $\approx 20000\,kg\,m^{-3}$. Table 7.1 gives the density of an arbitrary selection of substances. The

Table 7.1 Density of some common solids (approximate values).

Solid	ρ (kg m^{-3})	Solid	ρ (kg m^{-3})
Metals		*Natural materials*	
Aluminium/Dural	2700–2800	Amber	1100
Phosphor bronze	8900	Beeswax	950
Brass	8400–8500	Granite	2700
Gold (22 carat)	17500	Ice	920
Gold (9 carat)	11300	Coal	1400–1600
Mild steel	7900	Mica	2800
Stainless steel	7700–7800		
Wrought iron	7800	*Wood*	
Invar	8000	Balsa	200
Platinum/Iridium	21500	Pine	500
		Oak	700
		Beech	750
		Teak	850
		Ebony	1200

(Data from Kaye & Laby; see §1.4.1)

density of the elements is recorded in Table 7.2 and plotted as function of atomic number in Figure 7.1.

The first point we note is that solid densities are of a similar order of magnitude to liquid densities (see Tables 9.1 & 9.2), but dramatically greater than the densities of gases which are typically only a few kilograms per cubic metre at around standard temperature and pressure (STP) (Table 5.1).

From the data in Table 7.2 we can work out many quantities that are useful for understanding the properties of solids (Example 7.1), for example:

- the number densities of the atoms in the elements;
- the volume per atom taken up by the elements;
- the typical separation between atoms in the solid state; and
- the molar volume of the elements.

The densities of elements as recorded in Table 7.2

Table 7.2 The densities of the elements($\times 10^3\,\mathrm{kg\,m^{-3}}$) shown as varying with the atomic number and the atomic weight (A) in atomic mass units, $u = 1.66\times10^{-27}\,\mathrm{kg}$. [*]

Z	Element		A (u)	Density[†] ($10^3\,\mathrm{kg\,m^{-3}}$)	Z	Element		A (u)	Density[†] ($10^3\,\mathrm{kg\,m^{-3}}$)
1	Hydrogen	H	1.008	89	54	Xenon	Xe	131.3	3560
2	Helium	He	4.003	120	55	Caesium	Cs	132.9	1900
3	Lithium	Li	6.941	533	56	Barium	Ba	137.3	3594
4	Beryllium	Be	9.012	1846	57	Lanthanum	La	138.9	6174
5	Boron	B	10.81	2466	58	Cerium	Ce	140.1	6711
6	Carbon	C	12.01	2266	59	Praseodymium	Pr	140.9	6779
7	Nitrogen	N	14.01	1035	60	Neodymium	Ne	144.2	7000
8	Oxygen	O	16.00	1460	61	Promethium	Pm	145.0	7220
9	Fluorine	F	19.00	1140	62	Samarium	Sm	150.4	7536
10	Neon	Ne	20.18	1442	63	Europium	Eu	152.0	5248
11	Sodium	Na	22.99	966	64	Gadolinium	Gd	157.2	7870
12	Magnesium	Mg	24.31	1738	65	Terbium	Tb	158.9	8267
13	Aluminium	Al	26.98	2698	66	Dysprosium	Dy	162.5	8531
14	Silicon	Si	28.09	2329	67	Holmium	Ho	164.9	8797
15	Phosphorous	P	30.97	1820	68	Erbium	Er	167.3	9044
16	Sulphur	S	32.06	2086	69	Thulium	Th	168.9	9325
17	Chlorine	Cl	35.45	2030	70	Ytterbium	Yb	173.0	6966
18	Argon	Ar	39.95	1656	71	Lutetium	Lu	175.0	9842
19	Potassium	K	39.10	862	72	Hafnium	Hf	178.5	13276
20	Calcium	Ca	40.08	1530	73	Tantalum	Ta	180.9	16670
21	Scandium	Sc	44.96	2992	74	Tungsten	W	183.9	19254
22	Titanium	Ti	47.90	4508	75	Rhenium	Re	186.2	21023
23	Vanadium	V	50.94	6090	76	Osmium	Os	190.2	22580
24	Chromium	Cr	52.00	7194	77	Iridium	Ir	192.2	22550
25	Manganese	Mn	54.94	7473	78	Platinum	Pt	195.1	21450
26	Iron	Fe	55.85	7873	79	Gold	Au	197.0	19281
27	Cobalt	Co	58.93	8800	80	Mercury	Hg	200.6	13546
28	Nickel	Ni	58.70	8907	81	Thallium	Th	204.4	11871
29	Copper	Cu	63.55	8933	82	Lead	Pb	207.2	11343
30	Zinc	Zn	65.38	7135	83	Bismuth	Bi	209.0	9803
31	Gallium	Ga	69.72	5905	84	Polonium	Po	209.0	9400
32	Germanium	Ge	72.59	5323	85	Astatine	At	210.0	–
33	Arsenic	As	74.92	5776	86	Radon	Rn	222.0	4400
34	Selenium	Se	78.96	4808	87	Francium	Fr	223.0	–
35	Bromine	Br	79.90	3120	88	Radium	Ra	226.0	5000
36	Krypton	Kr	83.80	3000	89	Actinium	Ac	227.0	10060
37	Rubidium	Rb	85.47	1533	90	Thorium	Th	232.0	11725
38	Strontium	Sr	87.62	2583	91	Protractinium	Pa	231.0	15370
39	Yttrium	Y	88.91	4475	92	Uranium	U	238.0	19050
40	Zirconium	Zr	91.22	6507	93	Neptunium	Np	237.0	20250
41	Niobium	Nb	92.91	8578	94	Plutonium	Pu	244.0	19840
42	Molybdenum	Mo	95.94	10222	95	Americium	Am	243.0	13670
43	Technetium	Tc	97.00	11496	96	Curium	Cm	247.0	13300
44	Ruthenium	Ru	101.1	12360	97	Berkelium	Bk	247.0	14790
45	Rhodium	Rh	102.9	12420	98	Californium	Cf	251.0	15100
46	Palladium	Pd	106.4	11995	99	Einsteinium	Es	254.0	–
47	Silver	Ag	107.9	10500	100	Fermium	Fm	257.0	–
48	Cadmium	Cd	112.4	8647	101	Mendelevium	Md	258.0	–
49	Indium	In	114.8	7290	102	Nobelium	No	259.0	–
50	Tin	Sn	118.7	7285	103	Lawrencium	Lr	260.0	–
51	Antimony	Sb	121.7	6692	104	Unnilquadium	Unq	261.0	17000
52	Tellurium	Te	127.6	6247	105	Unnilpentium	Unp	262.0	21600
53	Iodine	I	126.9	4953					

[*]For example, the density of magnesium, whose atoms each contain 12 protons, is $1.738\times10^3\,\mathrm{kg\,m^{-3}}$. The mass of an atom of magnesium is $26.98\times1.66\times10^{-27}\,\mathrm{kg}$.

[†]The densities of elements that are normally gaseous at room temperature are evaluated at a temperature just below the freezing point of the element at atmospheric pressure. The exception to this is helium, which does not solidify at atmospheric pressure at any temperature. In this case the density is evaluated at 4.2 K and 25 atm ($25\times10^5\,\mathrm{Pa}$) pressure, which is sufficient to cause solidification.

(Data from Kaye & Laby; see §1.4.1)

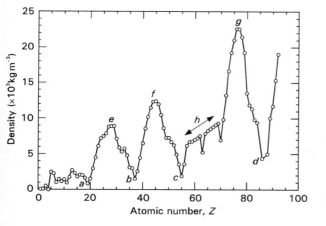

Figure 7.1 The densities of the elements plotted against atomic number i.e. the number of electrons on each atom. A great deal of structure is evident. The troughs (a–d) occur just after the completion of the main electron shells of the atoms: a, K (potassium); b, Rb (rubidium); c, Cs (caesium); d, Fr (francium). The peaks e–g are between the middle and end of the filling of the d-electron states in the three rows of transition metals; e, Zn (zinc); f, Rh (rhodium); g, Os (osmium). The regular slope (h) represents the Lanthanides.

Example 7.1

Work out each of the four quantities listed above for the element iron.

From Table 7.2 we see that iron (atomic number, $Z = 26$) has a density of $7.783 \times 10^3 \, \text{kg m}^{-3}$ and that the average mass of an iron atom is $55.85u = 55.85 \times 1.66 \times 10^{-27} \text{kg}$. Thus, dividing the mass density by the mass per atom yields the number density of atoms

$$n = \frac{7.783 \times 10^3 \, \text{kg m}^{-3}}{55.85 \times 1.66 \times 10^{-27} \, \text{kg}} = 8.39 \times 10^{28} \text{ atoms m}^{-3}$$

If there are n atoms per cubic metre, then the volume per atom $v = 1/n$, which for iron is

$$v = \frac{1}{8.39 \times 10^{28} \text{ atoms m}^{-3}} = 1.19 \times 10^{-29} \, \text{m}^3 \text{ atom}^{-1}$$

It is not possible to evaluate the exact separation between atoms without knowledge of the crystal structure of the element. However, we can make an estimate of the separation if we imagine an atom to be confined to a tiny cube of side L, then we can estimate L by using $L = (V)^{1/3}$, i.e. we have

$$L = \sqrt[3]{1.19 \times 10^{-29}} = 2.28 \times 10^{-10} \, \text{m}$$

Finally, we can work out the volume of 1 mole, i.e. N_A atoms, of iron. This is just $N_A v$, which is

$$\text{molar volume} = 6.022 \times 10^{23} \times 1.19 \times 10^{-29}$$

$$= 7.18 \times 10^{-6} \, \text{m}^{-3}$$

This corresponds to a cube of iron a little less than 2 cm on each side.

appear to vary almost randomly, but when plotted as a function of atomic number (Fig. 7.1) it becomes apparent that there is a significant amount of structure in the data. The most striking features are the periodic increases and decreases in density, and the curious linear portion between elements 57 to 71. There are also more subtle features: note for instance the "shoulders" on the high-atomic-number side of each peak (e–g) and the periodicities at small atomic number (§7.9, Ex. C2). We also note that there is a general trend towards elements with high atomic number having higher densities.

So the main questions raised by our preliminary examination of the experimental data on the density of solids are

- Why are the densities of solids much greater than those of gases?
- Why is there a general trend to increasing density for elements with increasing atomic number?
- Why are there periodic increases and decreases in density?
- Why is there a linear density increase between elements with atomic numbers 57 to 71?

7.1.2 Understanding the density of solids

Of the questions raised at the end of the previous section, we can understand the first point immediately: why the densities of solids are much greater than those of gases. All the models of solids that we discussed in Chapter 6 consist of atoms arranged close to one another, whereas a gas consists of molecules

with large spaces between them. It is not really surprising therefore that the amount of mass per unit
volume in a solid is greater than for a gas.

However, understanding the other questions is
rather more complicated because it is difficult to
make predictive calculations of the expected density
of a particular solid. The reason for this is that the
atoms of a solid can arrange themselves in any of several different ways, and each different arrangement
of atoms would have a different density, even if the
distance between neighbouring atoms were the same.
Thus in order to predict the density of a material one
needs to be able to predict both the atomic separation
and the crystal structure which the material will
adopt. Thus one needs to evaluate the energy of several different "competing" crystal structures with
high accuracy and then compare the results to find
out which crystal structure the atoms will naturally
choose. Although we managed to do this for the noble gas solids (§6.2.3), for most solids this task is still
a challenge even for the most modern and sophisticated theories.

Bearing these reservations in mind, let's look in
turn at each of the questions raised about the data.

General trend to increasing density

The general trend towards increasing density arises
because the atoms that make up solids get heavier as
their atomic number increases. The heavier atoms in
Figure 7.2 have masses over 200 times greater than
the lighter atoms on the graph, and so it would be surprising if materials made from heavier atoms were

---

Example 7.2

**What would the density of the elements be if they all
had a simple cubic crystal structure with nearest neighbour separation $a = 0.3$ nm?**

On these assumptions the number density of all elements
would be $1/a^3$. Since the mass of each atom is Au, where A
is the relative atomic mass and u is an atomic mass unit, we
conclude that the mass density, ρ, of the elements would
be

$$\rho = \frac{Au}{a^3}$$

Substituting $a = 0.3 \times 10^{-9}$ m and $u = 1.66 \times 10^{-25}$ kg yields

$$\rho = 615A$$

This line is plotted on Figure 7.2.

---

not denser.

However, we need also to consider the way in
which the spacing between atoms changes. The simplest model of a solid we can envisage assumes that
the spacing doesn't change much, and that the crystal
structures of the elements are all the same. Example
7.2 is a calculation of the density of the elements,
assuming that they all have a simple-cubic crystal
structure (which in fact none of them do) with an
atomic spacing of 0.3 nm. The results of this calculation are plotted in Figure 7.2. It seems that these simple assumptions produce density estimates that are of
the correct order of magnitude and that reproduce the
trend towards increasing density. This broad agreement implies that, as a general trend, the separation

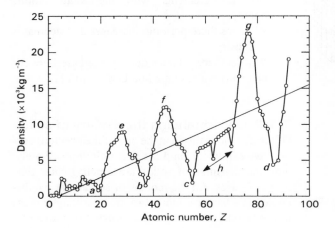

Figure 7.2 The densities of the elements
plotted against atomic number. See Figure 7.1
for details. The solid line represents the expected
density of the solid elements if they all had a
simple-cubic crystal structure with a nearest-
neighbour separation of 0.3 nm (see Example
7.2). Note that the line is not straight because A
is only approximately proportional to Z.

between atoms must stay roughly constant as the atoms get heavier. However, this approach clearly cannot explain the periodic variations in the density.

Periodic increases and decreases in density

In order to understand the periodic increases and decreases in density, we need to examine the data in rather more detail and to consider the type of bonding present in different elements. Let's look at the low density and high density elements in turn.

Low density elements

The lowest density elements (near points a–d in Fig. 7.2) occur for elements made from atoms that have either:

- filled electron shells, i.e. the molecularly bonded noble gases (Ne, Ar, Kr and Xe); or
- one electron outside a filled electron shell, i.e. the alkali metals (Na, K, Rb and Cs).

In the first case we have seen that the bonding mechanism between the atoms is the weak Van der Waals attraction (§6.2), and so we are not surprised if the atoms don't pull each other together tightly. In the second case the addition of one proton and one electron changes Ne to Na, Ar to K, Kr to Rb and Xe to Cs, and the bonding changes from molecular to metallic, but the density changes rather little. We can understand this by recalling that molecularly bonded solids choose a close-packed crystal structure (face-centred cubic) to allow as many atoms as possible to get as close together as their pair potential allows. However, metals achieve their lowest energy state by allowing their conduction electrons "extra room" to move around in. The alkali metals thus choose a crystal structure (body-centred cubic) that allows for this "extra room". Thus, although the metallic bonding is stronger, as indicated by the higher melting and boiling temperatures of the alkali metals (see Table 11.2), the density does not increase much compared with their molecularly bonded neighbours.

High density elements

The high density elements occur in the middle of rows of the periodic table. For these materials, in addition to metallic bonding there is also strong covalent bonding between atoms, and the number of covalent bonds that an atom can make reaches a maximum in the middle of a row of the periodic table. Each covalent bond pulls atoms tighter together and thus leads to higher densities. Elements from near the centre of a row in the periodic table can form more covalent bonds than can elements from near the ends of rows, because elements in the middle have both electrons that are able to form covalent bonds *and* unoccupied quantum states that can accommodate bond-forming electrons from neighbouring electrons.

The discussion above seems to indicate that there might be a correlation between the density of substances and their cohesive energy. This is in fact so, and the relationship is discussed more fully in §11.6.

Elements 57 to 71

Finally we need to understand the striking linear density increase shown by elements 57 to 71. These elements are known as the *lanthanide series* or the *rare earth elements* although, as Figure 2.1 shows, they are not particularly rare. Remember that as one moves across Figure 7.2, each step corresponds to the addition of:

(a) one electron to the atom;
(b) one proton to the nucleus; and
(c) one or two more neutrons to the nucleus.

Normally the extra electron occupies a quantum state that places it in an orbit which results in most of its charge density being around the outer part of the atom, i.e. in the valence electron states. In the lanthanides a new orbital is available (called the 4f orbital; Fig. 7.3) that can hold 14 electrons. Recall that s

Figure 7.3 A schematic diagram showing the location of the peak of the charge density associated with the 4f orbitals, and contrasting their position with that of the valence orbitals and the core orbitals.

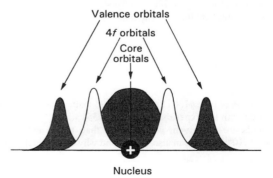

Valence orbitals

4f orbitals

Core orbitals

Nucleus

Table 7.3 The atomic number (Z), atomic mass (A) and the density (ρ) of the lanthanide series of elements extracted from Table 7.2.

	La	Ce	Pr	Nd	Pm	Sm	Eu	Gd	Tb	Dy	Ho	Er	Tm	Yb	Lu
Z	57	58	59	60	61	62	63	64	65	66	67	68	69	70	71
A	138.9	140.1	140.9	144.2	145	150.4	152.0	157.3	158.9	162.5	164.9	167.3	168.9	173.0	175.0
%A*	0	0.87	1.44	3.84	4.39	8.28	9.40	13.21	14.41	16.99	18.74	20.41	21.62	24.57	25.96
ρ	6174	6711	6779	7000	7220	7536	5248	7870	8267	8531	8797	9044	9325	6966	9842
%ρ†	0	8.7	9.8	13.4	16.9	22.1	−15.0	27.4	33.9	38.2	42.5	46.5	51.4	12.8	59.4

*The percentage density increase (compared with La) expected if the separation between atoms is unchanged and only the atomic mass changes.
†The percentage density increase (compared with La) actually found. It shows that the 59% density increase is much greater than can be explained by the 26% increase in atomic mass alone.

Example 7.3

We can work out the density of some covalently bonded materials if we know (a) the length of a covalent bond, and (b) the crystal structure. As we saw in the discussion of the covalent bond in Section 6.4, we cannot predict the length of the bond until we know the precise distribution of charge within the bond. In this section we take the length of the bond as being determined experimentally, and work out the density by looking at a representative sample of the material.

The crystal structure of diamond as deduced from X-ray scattering. The separation between atoms is 0.154 nm. The lines indicate the bonding regions and the spheres indicate the location of the atoms. The different shading is to aid the clarity of the picture, and allow identification of the four atoms entirely contained within the outer cube (grey).

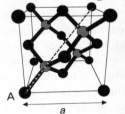

The sample of material shown here is a representative section of crystal known as a *unit cell*. If we work out the density of the unit cell we will find the same density as the entire crystal. If the cube shown has a side a, then the density will be

$$\rho = \frac{\text{number of atoms} \times \text{mass of atom}}{a^3}$$

The mass of a carbon atom is $12u$, i.e. $12 \times 1.66 \times 10^{-27}$ kg. The number of atoms is easy to work out in principle, but can be rather confusing in practice. One just counts the number of atoms in the above drawing, but one must take account of the fact that some atoms at the edge of the unit cell will also be counted in neighbouring unit cells. For cubic unit cells the rules for dealing with this are illustrated in the construction shown to the right.

(a) atoms shared between *faces* of neighbouring cells (white) count as only 1/2 an atom;
(b) atoms shared between *edges* of neighbouring cells (grey) count as only 1/4 an atom; and
(c) atoms shared between *corners* of neighbouring cells (black) count as only 1/8 an atom.

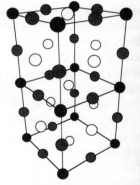

Counting carefully on the diamond structure one finds: (8×1/8 corner atoms) + (6×1/2 corner atoms) + (4 whole atoms) = 8 atoms in all. Finally, we need to work out the side of this unit cell in terms of the separation between two atoms. The *body diagonal*, AD, shown above, can be shown (by using Pythagoras' theorem) to be equal to $\sqrt{3}a$. The spacing between atoms is just 1/4 of this distance, i.e. $r_0 = \sqrt{3}a/4$, so that $a = 4r_0/\sqrt{3}$. We can now put the numbers into our calculation.

$$\rho = \frac{8 \times 12 \times 1.67 \times 10^{-27}}{\left(4r_0/\sqrt{3}\right)^3}$$

Emsley (see §1.4.1) gives the C–C distance as $r_0 = 1.54 \times 10^{-10}$ m and so we find a prediction that $\rho = 3564$ kg m^{-3} compared with the experimental value of 3513 kg m^{-3}, a disagreement of 1.5%. Since the temperatures of the various measurements are not given, we conclude that, to the extent that we understand why the C–C separation is approximately 1.54×10^{-10} m, we can also understand why carbon in the form of a diamond has the density it has.

orbitals hold two electrons, p orbitals six electrons, and d orbitals ten electrons.

This orbital results in the additional electronic charge residing *inside* the atom rather than near the outside of the atom. This has two consequences for the density of the lathanides. First, as the charge density in the outer part of the atom is not changed, the bonding to neighbouring atoms (which is both metallic and covalent) is broadly unaffected. If this was the only effect then the density of elements would increase as the mass of atoms increased, while the separation between atoms would remain roughly constant. Second, the effect of each extra nuclear charge pulls all the orbitals a little closer to the nucleus. This makes the orbitals, and hence the atom itself, become systematically smaller as one proceeds across the series – an effect known as the *lanthanide contraction*.

Table 7.3 shows a calculation of how the density of the lanthanides would be expected to vary if the separation between atoms stayed constant. It is clear that the 59% density increase that actually occurs is much greater than can be explained by the 26% increase in atomic mass alone. The dips in the density at Eu (europium) and Yb (ytterbium) occur for reasons that are connected with the detailed arrangement of electrons within the $4f$ shell.

Summary
We conclude that we can understand the density of solids rather well, even though there are few materials for which we can, a priori, predict the density. Broadly speaking, the variation in the density of the elements arises from the combination of two trends: a periodic variation connected with the type of bonding present, and a linear trend to higher densities for elements with high mass atoms.

7.2 The thermal expansivity of solids

7.2.1 Background

In general, solids expand when heated. If a solid has initial volume V_0, the increase in volume, ΔV, on raising the temperature by ΔT at constant pressure may be expressed as

$$\Delta V = V_0 \beta \Delta T \tag{7.1}$$

where β is the *coefficient of volume expansion*, sometimes called the *coefficient of cubical expansion*. Hence the volume may be written as

$$V = V_0 + \Delta V = V_0(1 + \beta \Delta T) \tag{7.2}$$

For solids it is also possible to define α, the *coefficient of linear expansion*, which describes the way the *length* of a sample of solid will change with temperature. It doesn't make much sense to describe a coefficient of linear expansion for a gas or a liquid, which doesn't have the constancy of shape possessed by solids.

If a solid has initial length L_0, the increase in length, ΔL, on raising the temperature by ΔT at constant pressure may be expressed as

$$\Delta L = L_0 \alpha \Delta T \tag{7.3}$$

where α is the *coefficient of linear expansion*. Hence the length may be written as

$$L = L_0 + \Delta L = L_0(1 + \alpha \Delta T) \tag{7.4}$$

As illustrated in Example 7.4, the coefficients of linear and volume expansivity are linked by the simple relationship

$$\beta = 3\alpha \tag{7.5}$$

Example 7.4

Consider a sphere of initial volume $4\pi r_0^3/3$. Each dimension increases in length by a factor $(1 + \alpha \Delta T)$ and so the volume increases by a factor of

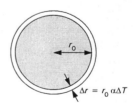

$$(1 + \alpha \Delta T)^3 = 1 + 3\alpha \Delta T + 3(\alpha \Delta T)^2 + (\alpha \Delta T)^3$$

Since $\alpha \Delta T$ is usually $\ll 1$, this is given to an excellent approximation by

$$(1 + \alpha \Delta T)^3 \approx 1 + 3\alpha \Delta T$$

Thus the volume expansivity, β (Eq. 7.2), is given by

$$\beta = 3\alpha.$$

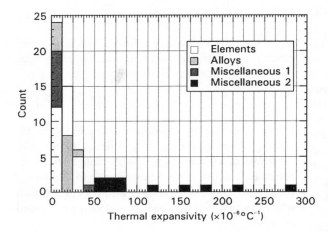

Figure 7.4 Histogram of the data given in Tables 7.4A–C. Where a range of values is indicated in the table, the midpoint of the range has been plotted.

7.2.2 Data on the thermal expansivity of solids

The coefficients of linear expansivity for a variety of solids are shown in Tables 7.4A–C, and are plotted by category on the histogram in Figure 7.4. In general these results indicate that for metals, either elements or alloys, a typical expansivity is of the order $10^{-5}\,°C^{-1}$, but with significant variability about this figure. This is considerably less than the expansivities of gases ($\approx 3 \times 10^{-3}\,°C^{-1}$ at around STP) (Table 5.4).

Table 7.4 The coefficient of linear expansivity, α, for various solids at temperatures around room temperature.

A Elemental metals		B Alloys		C Miscellaneous materials	
Elemental metal	$\alpha(\times10^{-6}°C^{-1})$	Alloy	$\alpha(\times10^{-6}°C^{-1})$	Miscellaneous	$\alpha(\times10^{-6}°C^{-1})$
Aluminium (Al)	23	Brass	18–19	Brick	3–10
Antimony (Sb)	≈11	(68% Cu, 32% Zn)		Cement and concrete	10–14
Bismuth (Bi)	≈13	Bronze	7–18	Marble	3–15
Cadmium (Cd)	≈30	(80% Cu, 20% Sn)		Lead glass (46% PbO)	≈ 8
Chromium (Cr)	≈7	Constantan	15–17	Typical glass	≈ 8–10
Cobalt (Co)	≈12	(60% Cu, 40% Ni)		Porcelain	2–6
Copper (Cu)	16.7	Duralumin	23	Silica	0.4
Gold (Au)	13	(95% Al, 4% Cu)		Typical wood (along grain)	3–5
Iridium (Ir)	6.5	Magnalium	≈23	Typical wood (across grain)	35–60
Iron (Fe)	11.7	(90% Al, 10% Mg)		Epoxy resins	45–65
Lead (Pb)	29	Nickel steel	13	Epoxy resins	45–65
Magnesium (Mg)	25	(10% Ni, 90% Fe)		Polycarbonates	66
Nickel (Ni)	12.8	Nickel steel	0–1.5	Low density	
Palladium (Pd)	≈11	(36% Ni, 64% Fe)		polyethylene	40–150
Platinum (Pt)	8.9	Nickel steel	7.9	Medium density	
Rhodium (Rh)	8.4	(43% Ni, 57% Fe)		polyethylene	80–220
Silver (Ag)	19	Nickel steel	11.4	High density	
Tantalum (Ta)	6.5	(58% Ni, 42% Fe)		polyethylene	200–360
Thallium (Tl)	≈28	Carbon steel	≈11	Natural rubber	220
Tin (Sn)	≈21	Stainless steel	29	Hard rubber	60
Titanium (Ti)	≈9	(74% Fe, 18% Cr/8% Ni)		Perspex	50–90
Tungsten (W)	4.5	Phosphor bronze	17	Nylon	80–280
Vanadium (V)	≈8	Platinum–Iridium	8.7	Polystyrene	34–210
Zinc (Zn)	≈30	(90% Pt, 10% Ir)		Polyvinyl chloride (PVC)	70–80

The volume expansivity of the elements is given by $\beta = 3\alpha$ as shown in Example 7.4.
(Data from Kaye and Laby; see §1.4.1)

Notice in particular the results for nickel–iron alloys (Table 7.4B) for a composition of 36% Ni and 64% Fe (i.e. a ratio of just under 2:1 of iron atoms to nickel atoms), the expansivity falls by a factor 10 to $\approx 10^{-6}\,°C^{-1}$. Alloys with this composition, known as *invar* alloys (short for "invariable"), are important in the construction of mechanical components in which high precision must be maintained over a range of temperatures, such as the shadow mask of a colour television.

The typical figure of $10^{-5}\,°C^{-1}$ is also a reasonable guide to the materials in the first part of the "Miscellaneous" section of Table 7.4C. However, the plastics and rubbers in the lower part of the column have expansivities that are typically 10 times larger than this – all the substances in Figure 7.4 with expansivities greater than $50 \times 10^{-6}\,°C^{-1}$ belong to this category.

So the main questions raised by our preliminary examination of the experimental data on the expansivity of solids are:

- Why is the expansivity of solids much smaller than that shown by gases?
- Why is the expansivity of an alloy not always the average of the expansivity of its components?
- Why is the expansivity of plastics much larger than that of metals?

7.2.3 Understanding the expansivity of solids

Background

First of all let's see how we can understand the expansion of solids in general terms and then turn to the questions raised above.

Consider two neighbouring atoms or ions in a crystalline solid such as any of those we considered in §6.2–6.5. We can understand the thermal expansion of the crystal only if the average separation between the two atoms or ions increases with temperature. Now, in general, heating the crystal increases the amplitude of vibration of atoms, and so the separation between the two atoms oscillates, typically at around $10^{13}\,Hz$.

The harmonic approximation

If the potential energy of interaction of the two atoms is symmetric about r_0, then the *average* separation of

---

Example 7.5

Samples of copper, perspex, and invar, each of length 2 m at 20°C, are warmed to 30°C. Calculate the change in length of each sample.

$L_0 = 2\,m$
$\alpha = 16.7 \times 10^{-6}\,°C^{-1}$ for copper
$\alpha = 50–90 \times 10^{-6}\,°C^{-1}$ for perspex
$\alpha = 0–1.5 \times 10^{-5}\,°C^{-1}$ for invar
$\Delta T = 30–20°C = 10\,K$

We start with Equation 7.4

$$L = L_0(1 + \Delta T)$$

Remembering that $L = L_0 + \Delta L$ we write this as

$$L_0 + \Delta L = L_0 + L_0 \alpha \Delta T$$
$$\Delta L = L_0 \alpha \Delta T$$

For our examples this becomes

$$\Delta L = 2\alpha \times 10 = 20\alpha$$

Thus the changes in length are:

copper	$\Delta L = 20 \times 16.7 \times 10^{-6}\,°C^{-1}$	$= 0.33\,mm$
perspex	$\Delta L = 20 \times (50–90) \times 10^{-6}\,°C^{-1}$	$= 1–1.8\,mm$
invar	$\Delta L = 20 \times (0–1.5) \times 10^{-6}\,°C^{-1}$	$= 0–0.03\,mm$

This example is intended to illustrate that, while the thermal expansion of solids is indeed small, it is still appreciable and must be taken into account in many real measurements. Note that the invar alloy produces significantly smaller length changes than either of the other two materials.

Example 7.6

In Example 7.1, we calculated the typical separation between atoms in a piece of iron at 20°C to be 2.29×10^{-10} m. How much does this average separation change if the iron is heated to 100°C?

Table 7.4A gives the coefficient of linear expansion of iron as $\alpha = 11.7 \times 10^{-6}\,°C^{-1}$. Thus each length in a piece of iron changes by a factor $(1 + \alpha \Delta T)$ when the temperature changes by ΔT. In this case $\Delta T = 100–20 = 80°C$. We thus have

$$L = L_0(1 + \alpha \Delta T) = L(1 + 11.7 \times 10^{-6} \times 80)$$
$$= L(1.000936)$$

Thus the average separation between iron atoms changes from $L_0 = 2.29 \times 10^{-10}\,m$ to $L = 2.292 \times 10^{-10}\,m$, i.e. it increases by ≈ 1 part in 1000, or 0.1%.

161

the atoms neither increases nor decreases with increased vibrational amplitude (Fig. 7.5a). It is as if the atoms are connected together by a perfect spring whose potential energy may be written as

$$u(r) = u_0 + \frac{1}{2} K(r - r_0)^2 \qquad (7.6)$$

where K is the "spring constant" of the bond. Assuming a pair potential such as Equation 7.6 is known as the *harmonic approximation*, and is quite sufficient for many purposes (see, for example, the section on the heat capacity of solids, §7.4). However, it is not possible to understand thermal expansion in the harmonic approximation.

Anharmonic vibrations

In general, the potential energy of interacting atoms or ions is not quite symmetrical. As the amplitude of atomic oscillations increases, so does the average separation between the atoms (Fig. 7.5b). We can describe the potential energy of the two atoms by an equation such as

$$u(r) = u_0 + \frac{1}{2} K(r - r_0)^2 + \frac{1}{6} K'(r - r_0)^3 \qquad (7.7)$$

where K' (usually negative) determines the magnitude of the asymmetric – or *anharmonic* – contribution. Note that if $(r - r_0)$ is small enough we may always neglect the anharmonic part of the potential energy, because the $(r - r_0)^3$ term goes to zero faster than does the $(r - r_0)^2$ term. Thus at low temperatures, where the vibrational amplitudes are small, the anharmonic term is generally negligible. At higher temperatures, the anharmonic term begins to contribute to the potential energy of the atoms. When the anharmonic term contributes to u, we see that $u(r - r_0) \neq u(r_0 - r)$ and the vibrations of the atoms in the potential become asymmetric, with atoms spending, on average, more time at separations greater than r_0, giving rise to an increase in the average separation of the atoms, i.e. thermal expansion.

Unfortunately, estimating the magnitude of the anharmonic coefficient K' and determining its effect on the average separation between atoms are both rather difficult. Thus, in general, we shall attempt to understand the data on thermal expansion in terms of the general shape of a graph of potential energy of in-

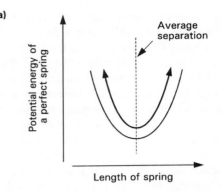

(a)

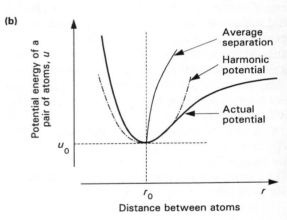

(b)

Figure 7.5 The variation with separation of the potential energy (PE) of interaction between atoms in a solid. (a) The variation in PE if the atoms are connected by hypothetical "perfect springs" — the *harmonic approximation*. (b) Typical deviations from the harmonic approximation of the real interatomic potential energy. The thick line indicates the increasing average separation between the atoms as the average energy of oscillation (i.e. the temperature) is increased.

teraction between two atoms versus their separation. If this explanation of the thermal expansion of solids is correct, we should find ourselves able to naturally understand the questions raised by the experimental data given in §7.2.2.

Why is the thermal expansivity of solids less than that of gases?

We can understand fairly directly why the thermal expansivity of solids is less than that of gases. A gas represents the high-temperature limit of the thermal expansivity of a substance. When the temperature of a gas is raised, one may usually ignore the potential energy of interaction between the molecules which would act to constrain the volume of the gas. However, in a solid the atoms are restrained from "flying away" by their potential energy of interaction with their neighbours. Thus the thermal expansivity of solids is "restricted" by the potential energy of interaction between atoms.

It is a surprising fact that the oscillation amplitudes of atoms in solids are never very large. In §11.3.2 we show that solids melt when the oscillation amplitude reaches around 5% of the atomic separation. So, well below the melting temperature, the amplitude of atomic oscillations is rather small, and the anharmonic contributions to the potential energy that give rise to thermal expansion are generally smaller than the harmonic contributions.

The thermal expansivity of alloys: invar

Given our basic thesis of the underlying origin of thermal expansion, let us now look at why the expansivity of alloys appears sometimes not to be the average of the expansivity of its components.

A one-dimensional alloy

For simplicity, we will imagine a row of atoms in a sort of one-dimensional alloy composed of two elements, A and B. Figure 7.6 shows rows of atoms representing both the pure elements and the "alloy".

From the point of view of the expansivity of the alloy, we need to concentrate not on the distribution of atoms themselves but on the distribution of bonds *between* atoms. In pure A material there are just A–A bonds, and in pure B material there are just B–B bonds. However, in the alloy there is a third type of bond, an A–B bond which in general will have different characteristics from either A–A or B–B bonds.

A three-dimensional alloy

If we envisage a real random alloy as a collection of one-dimensional alloys held parallel to one another, then we can understand the thermal expansivity of alloys as being due to the average number of A–A, A–B and B–B bonds in the alloy. Since we can measure the properties of A–A and B–B bonds by studying the pure material, alloys give us a chance to study

the behaviour of A–B bonds.

The composition of a random alloy may be specified as A_xB_{1-x} where x varies between 0 (pure B) and 1 (pure A), and, in general, x will be some arbitrary number (e.g. for $x = 0.47$ we have $A_{0.47}B_{0.53}$). The probability that the atom at a particular site in the alloy will be type A is x, and so the probability that a particular bond will be an A–A bond is x^2. Similarly, the probability that a particular atom will be type B is $1-x$, and so the probability of a B–B bond is $(1-x)^2$. Finally, the probability that a particular bond will either A–B or B–A is $2x(1-x)$. The expansivity of the alloy will be given by the average of each of the three types of bond within the material, i.e

Expansivity of alloy =

(Expansivity of A–A bond)x^2 +

(Expansivity of B–B bond)$(1-x)^2$ +

(Expansivity of A–B bond)$2x(1-x)$ (7.8)

With the ideas summarized in Equation 7.8 in mind, let us examine the alloy data given in Tables 7.5a–d. The "expected" value is the average expan-

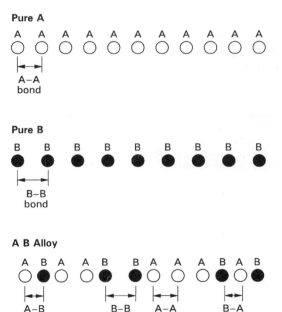

Figure 7.6 An illustration of the arrangement of bonds in one-dimensional models of pure A, pure B, and an alloy of A and B. Note that the alloy contains a type of interatomic bond that is not present in either of the pure substances.

163

sivity of the elements, weighted by the percentage of the two elements present. These values represent estimates of the thermal expansivity that neglects the final term in Equation 7.8. Thus if the expected value in Table 7.5 is close to the experimental value, then the "A–B" bonds between the alloy components are similar to the average of an A–A and B–B bond.

For the aluminium alloys (Table 7.5a), copper alloys (Table 7.5b) and platinum alloys (Table 7.5c), the expected value of the expansivity is close to the experimental value. However, the data for iron-nickel alloys shows that Fe–Ni bonds behave dramatically differently from either Fe–Fe or Ni–Ni bonds. In particular, $Fe_{0.64}Ni_{0.36}$ shows essentially zero expansivity: this seems to imply that under certain circumstances an Fe–Ni bond must have a *negative* expansivity in order to compensate for the fact that the Fe–Fe and Ni–Ni bonds have a positive expansivity. This corresponds to an interatomic potential between an iron and a nickel atom that is asymmetric in the opposite way to the normal potential (Fig. 7.7)

The origin of this unusual behaviour is connected with the ferromagnetism shown by these elements. In addition to normal contributions to the energy of an Fe–Ni bond there is an additional term in the potential energy associated with the magnetic interaction between neighbouring atoms. The dependence of this energy on the separation between iron and nickel

Table 7.5 Expected and experimentally determined values of the linear thermal expensivities of some alloys and their component elements.

Alloy (composition)	Expected (see text)	Experimental
a Aluminum alloys		
Duralumin (95% Al, 4% Cu)	22.5×10^{-6}	23×10^{-6}
Magnalium (90% Al, 10% Mg)	23.2×10^{-6}	$\approx 23 \times 10^{-6}$
Aluminium	–	23×10^{-6}
Copper	–	16.7×10^{-6}
Magnesium	–	$\approx 25 \times 10^{-6}$
b Copper alloys		
Brass (68% Cu, 32% Zn)	21×10^{-6}	$18–19 \times 10^{-6}$
Bronze (80% Cu, 20% Sn)	17.6×10^{-6}	$17–18 \times 10^{-6}$
Constantan (60% Cu, 40% Ni)	15.1×10^{-6}	$15–17 \times 10^{-6}$
Copper	–	16.7×10^{-6}
Zinc	–	$\approx 30 \times 10^{-6}$
Tin	–	$\approx 21 \times 10^{-6}$
Nickel	–	12.8×10^{-6}
c Platinum alloys		
Platinum–iridium (90% Pt, 10% Ir)	8.66×10^{-6}	8.7×10^{-6}
Platinum	–	8.9×10^{-6}
Iridium	–	6.5×10^{-6}
d Iron alloys		
Nickel steel (10% Ni, 90%Fe)	11.8×10^{-6}	13×10^{-6}
Nickel steel (36% Ni, 64%Fe)	12.1×10^{-6}	$0–1.5 \times 10^{-6}$
Nickel steel (43% Ni, 57%Fe)	12.2×10^{-6}	7.9×10^{-6}
Nickel steel (58% Ni, 42%Fe)	12.3×10^{-6}	11.4×10^{-6}
Stainless steel (74%Fe, 18% Cr,8%Ni)	10.9×10^{-6}	29×10^{-6}
Iron	–	11.7×10^{-6}
Nickel	–	12.8×10^{-6}
Chromium	–	7×10^{-6}

Figure 7.7 Schematic illustration of the variation of potential energy with separation in an Fe–Ni bond in an invar alloy (Tables 7.4 & 7.5d). Note that the asymmetry of the potential (over a certain range) is opposite to that which occurs in normal bonds (Fig. 7.5b).

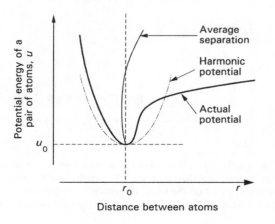

Distance between atoms

atoms is critical, and increases strongly with only a small increase in separation, r (Fig. 7.7). Over a certain temperature range this effect can give rise to a *contraction* in the average length of an Fe–Ni bond, which can compensate for the expansion of the Fe–Fe and Ni–Ni bonds. Outside this temperature range, the expansivity of the invar alloy assumes values closer to the "expected" values in Table 7.5d.

The thermal expansivity of plastics

Table 7.4C and Figure 7.4 show that the thermal expansivity of plastics is typically one order of magnitude or more greater than that of crystalline solids. Since all the plastics listed in Table 7.4C share high thermal expansivity values, we would expect the explanation to be general to the class of materials, rather than specific to particular features of each individual substance. So what general features of plastics cause them to have a high thermal expansivity?

All the plastics listed in Table 7.4C are made from organic *polymer* molecules. These molecules are extremely long, varying from a few hundred to many thousands of atomic bonds in length. The molecules are composed of repeating units of simple *monomer*, such as ethylene, linked together in a chain the length of which varies with the conditions under which the *polymerization* is carried out.

The bonding in polymers is similar to that present in other organic substances (§6.6.1). The bonding *within* the molecules is covalent and is thus relatively strong: making the molecules themselves extremely strong. However, the bonding *between* the molecules is the relatively weak Van der Waals interaction: this allows molecules to move relatively easily with respect to one another.

An extra factor in the case of polymers is the extreme length of the molecules, which causes them to become *entangled* in one another. In other words, although the molecules are not strongly bonded to each other, their physical entanglement holds them together in the solid state. In §8.3 & §9.6 on the cell model of liquids, we will see that the complex shapes of organic molecules lead to their having a relatively large viscosity, i.e. it is relatively difficult for organic molecules to move relative to one another. The entanglement of polymer molecules in a solid is an extreme limit of this "viscosity" effect.

The thermal expansivity of plastics is therefore not dominated by the variation in the average separation of bonded molecules, but by rearrangements of the *shapes* of the molecules that become possible as the level of thermal excitation is increased. The processes of molecular rearrangement are too complex to consider here, but one can fairly directly imagine that this rearrangement – in combination with weak intermolecular forces – could lead to large values of thermal expansivity

7.3 The speed of sound in solids

The passage of sound through a solid is more complicated to describe than the passage of sound through a

gas because, in addition to the longitudinal (compressive) waves supported by a gas, solids can support waves of transverse stress. Recall that in a longitudinal sound wave the material is alternately compressed and rarefied as one travels along the direction of wave propagation (see Figs 5.20 & 7.8a). In a transverse sound wave, the material is transversely stressed, or *sheared*, in alternating directions as one travels along the direction of wave propagation (Figs 7.8b,c).

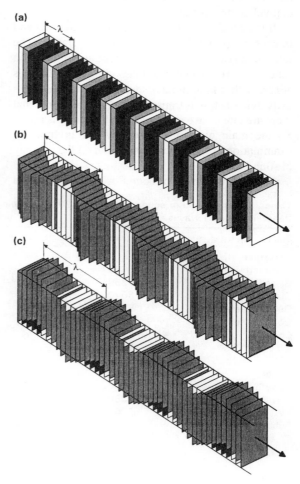

Figure 7.8 Sound waves in solids: (a) longitudinal waves; (b, c) transverse waves. The figure shows planes of material perpendicular to the direction of propagation of a sound wave. In the longitudinal wave (a) the intensity of shading indicates the degree of compression or rarefaction. (b, c) A *transverse* wave displaces layers of material perpendicular to the direction of propagation.

7.3.1 Data on the speed of sound in solids

The speed of longitudinal and transverse sound waves in various bulk media are shown in Table 7.6. The data for the elements are plotted as a function of relative atomic mass in Figure 7.9 which also shows the speed of sound in gases at STP for comparison (see Fig. 5.19). It is apparent that sound travels considerably faster through solids than gases at STP, typically by a factor of ≈ 10. Also, the data for solids show considerably more variability than was seen in the equivalent data for gases.

It is also clear that the speeds of longitudinal and transverse sound waves differ considerably. The ratio of the two speeds is plotted for the elements as a function of relative atomic mass in Figure 7.10. It is apparent that longitudinal sound waves travel typically twice as fast as transverse sound waves, but that there are one or two striking exceptions to this rule. So the main questions raised by our preliminary examination of the experimental data on the velocity of sound in solids are:

- Why is the speed of sound in solids typically a few thousand metres per second, i.e. about 10 times faster than in gases at STP?
- Why do longitudinal sound waves travel about twice as quickly as transverse sound waves?
- Why does the speed of sound tend to be less in elements with large atomic mass?

7.3.2 Understanding the speed of sound in solids

All these questions can be answered fairly well with reference to the formulae for the velocities of sound in a solid. These are derived in Appendix 2 and discussed below.

Compressive and shear strain

Sound is a *strain wave* that propagates through a solid. The nature of a sound wave in a solid can be considerably more complex than in a gas, because in addition to *compressive* strains (such as those in gases), a solid can sustain *shear* strain. Shear strain is a little more complicated than compressive strain, but

Table 7.6 The speed of sound in solids (at 20°C) showing c_L the speed of longitudinal waves, and c_T the speed of transverse (shear) waves.

Solid	c_L (ms⁻¹)	c_T (ms⁻¹)	c_T (ms⁻¹)	Solid		c_L (ms⁻¹)	c_T (ms⁻¹)
Insulators				**Metallic elements**			
Glass (crown)	5660	3420		Aluminium	Al	6374	3111
Glass (heavy flint)	5260	2960		Beryllium	Be	12890	8880
Glass (pyrex)	5640	3280		Cadmium	Cd	2780	–
Quartz crystal, X-cut	5720	–		Chromium	Cr	6608	4005
Quartz fused	5970	3765		Copper	Cu	4759	2325
Concrete	4250–5250	–		Gold	Au	3240	1200
Ice (−20°C)	≈3840	–		Iron	Fe	5957	3224
				Lead	Pb	2160	7000
Plastics				Magnesium	Mg	5823	3163
Polyethylene	2000	–	3111	Manganese	Mn	4600	–
Polystyrene	2350	1120	8880	Molybdenum	Mo	6475	3505
PVC	2300	–	–	Nickel	Ni	5700	3000
Rubber	1600	–	4005	Niobium	Nb	5068	2092
				Platinum	Pt	3260	1730
				Silver	Ag	3704	1698
				Tantalum	Ta	4159	2036
				Tin	Sn	3380	1594
				Titanium	Ti	6130	3182
				Tungsten	W	5221	2887
				Uranium	U	3370	1940
				Vanadium	V	6023	2774
				Zinc	Zn	4187	2421
				Zirconium	Zr	4650	2250

(Data from Kaye & Laby, see §1.4.1)

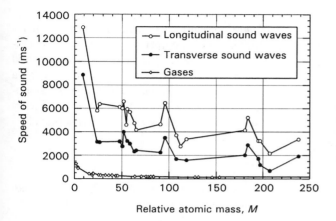

Figure 7.9 The variation of the speed of longitudinal and transverse sound waves (see Table 7.6) with relative atomic mass. For comparison, some of the data for gases given in Table 5.14 is also plotted. There appears to be a tendency towards lower speeds at larger atomic masses, similar to the tendency shown by gases in Figure 5.20. Data points are joined by straight lines to highlight trends in each data set.

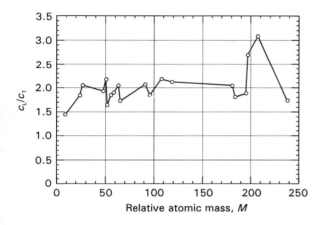

Figure 7.10 The variation in the ratio of the longitudinal to transverse sound velocities with relative atomic mass for the elements listed in Table 7.6. Typically, $c_L \approx 2c_T$, but in some cases the results differ significantly from this value.

mathematically it can be described in similar terms.

The basic idea is that if there is any distortion of the solid away from its equilibrium shape (Fig. 7.11), then the average separation of the atoms within the solid is no longer optimal: some will be slightly too close to their neighbours, and some too far apart. In either case, there will be a restoring force that will act to return the atoms to their equilibrium separations. The dynamics of the sound wave will be affected by the way the solid responds under the action of the restoring force. The two factors most critical in determining this response are:

- the restoring force per unit displacement – the natural "springiness" of the substance; and
- the density of the substance.

We will look at each of these factors in turn.

The restoring force per unit strain

The restoring force on a small region of a solid depends on the type of distortion (strain) that has taken place. The parameters that describe the restoring force per unit strain are known as the *elastic moduli* of a substance. There are three such moduli, each defined by an equation of the form

$$\text{Restoring force} = \text{Elastic modulus} \times \text{Strain} \quad (7.9)$$

Figure 7.11 (a) Two planes within a solid at their equilibrium separation and relative orientation. (b, c) The planes subject to two types of shear strain. (d) The planes subject to a compressive strain.

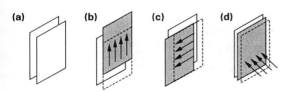

167

Example 7.7

A pulse of sound of frequency 100 MHz and lasting for 1 μs is fired into one face of an experimental copper sample of dimensions 10 mm×10 mm×10 mm. If all modes of sound propagation are excited,
(a) What is the wavelength of the sound wave?
(b) How far has the front of the pulse travelled by the time the pulse transmission is stopped?
(c) What will be the duration (width) of the pulse by the time it hits the far side of the cube and bounces back to its starting face ?

Transmitter

(a) Using $c = f\lambda$ we find

$$\lambda = \frac{c}{f}$$

and so for the longitudinal waves we have

$$\lambda_L = \frac{4759}{100 \times 10^6} = 47.6 \times 10^{-6} \, \text{m}$$

or around one-twentieth of a millimetre. Similarly, for transverse waves we have

$$\lambda_T = \frac{2325}{100 \times 10^6} = 23.3 \times 10^{-6} \, \text{m}$$

or around one fortieth of a millimetre.

(b) In 1 μs the longitudinal waves in the pulse have travelled a distance of $vt = 4759 \times 10^{-6} = 4.76$ mm. Similarly, the transverse waves in the pulse have travelled a distance of $vt = 2325 \times 10^{-6} = 2.33$ mm.

(c) The waves in the pulse will return to their original face after travelling once each way across the cube – a distance of 20 mm.

The longitudinal (*fast*) waves at the *start* of the pulse will return to the original face after a time

$$t = 20 \times 10^{-3}/4759 = 4.20 \, \mu s.$$

The transverse (slow) waves at the *end* of the pulse will return to the original face after a time

$$t = 20/v_T = 20 \times 10^{-3}/2325 = 8.60 \, \mu s.$$

The pulse width has therefore become

$$8.60 - 4.20 = 4.4 \, \mu s.$$

The three moduli are:
- *Young's modulus*, *E*, which characterizes the restoring forces appropriate to longitudinal extensions of a substance. Figure 7.12 shows two rigid planes of area *A* separated in equilibrium by a distance *a* and held together by "springs" (analogous to planes within a solid held together by atomic bonds). Young's modulus is defined by

$$\frac{F}{A} = E \frac{\Delta x}{a} \qquad (7.10)$$

- The *shear* or *rigidity modulus*, *G*, which characterizes the restoring forces appropriate to shear or transverse deformations of the substance. Figure 7.13 shows two rigid planes of area *A* held together by "springs" (analogous to planes within a solid held together by atomic bonds). The rigidity modulus is defined by

$$\frac{F}{A} = G\theta \qquad (7.11)$$

- The *bulk modulus, B*, which describes the restoring forces appropriate to volume compressions of the substance. This is also the inverse of the compressibility, *K*, of a material.

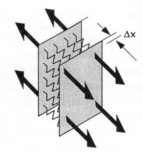

Figure 7.12 Young's modulus, *E*: two rigid planes of area *A* separated at equilibrium by a distance *a* and held together by "springs".

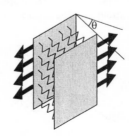

Figure 7.13 Shear or rigidity modulus, *G*: two rigid planes of area *A* held together by "springs".

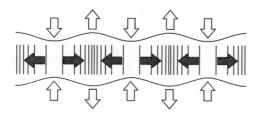

Figure 7.14 Compressive and extensive strains (grey arrows) in a long, thin rod.

If the material is easily compressed or easily sheared, then for a given strain the restoring force will be small, i.e. a high modulus indicates that the corresponding deformation of the solid is difficult and the solid has a strong tendency to "spring" back to its equilibrium position.

The density of the solid, ρ
If the density of a solid is high, i.e. if the *mass per unit volume* is high, then by Newton's second law ($F = ma$) the acceleration due to a given restoring force will be low. If the acceleration of piece of the solid is low, then it will return to its equilibrium position only slowly, and the wave disturbance will propagate only slowly.

Formulae for the speed of sound waves in solids
In experimental situations, one commonly encounters three similar expressions for the speed of sound. Given the two factors discussed above, the general form of these expressions should not be at all surprising. However, the derivation of these simple equations is rather complex and is given in Appendix 2.

The speed of a longitudinal sound wave in a long, thin rod
The speed of a longitudinal sound wave along a narrow rod is given by

$$c = \sqrt{\frac{E}{\rho}} \qquad (7.12)$$

where E is Young's modulus. The types of compressive and extensive strains that take place within the rod are illustrated by the black arrows in Figure 7.14. Note that in the regions where the substance is com-

pressed the rod tends to bulge slightly, while in the regions where the substance is extended the rod tends to "neck" slightly. It is important to realize that this is *not* the situation of a longitudinal sound wave in a bulk solid.

The speed of a transverse sound wave in a bulk solid
The speed of a transverse sound wave, c_T, in a bulk solid is given by

$$c_T = \sqrt{\frac{G}{\rho}} \qquad (7.13)$$

where G is the rigidity modulus.

The speed of a longitudinal sound wave in a bulk solid
The speed of a longitudinal sound wave in a bulk solid is given by an expression similar to Equation 7.12, but with an extra factor, γ, which takes account of the fact that the lateral accommodation (the "necking" and "bulging" shown in Figure 7.12) does not take place for longitudinal waves in a bulk solid. The speed of sound is given by

$$c_L = \sqrt{\frac{\gamma E}{\rho}} \qquad (7.14)$$

where

$$\gamma = \frac{(1-\sigma)}{(1-2\sigma)(1+\sigma)} \qquad (7.15)$$

and σ is the *Poisson ratio* of the substance. Note that the factor γ in Equation 7.14 is not related at all to the factor γ in the formula for the speed of sound in gases.

The Poisson ratio, σ
The physical significance of the Poisson ratio is illustrated in Example 7.8. The Poisson ratio quantitatively expresses the tendency of a substance to "neck" or "bulge" when strained longitudinally. Technically, it is the ratio of the *lateral strain* to the *direct strain* and has a typical values between 0 and 0.5. A substance that does not "neck" or "bulge" at all has a Poisson ratio of 0, and one that changes shape so that the volume of solid is not changed has a Poisson ratio of 0.5.

Example 7.8

Consider a rectangular bar of material of cross-sectional area x^2 and length L.

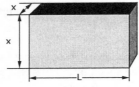

If the bar is subject to a stretching force $\mathbf{F}$ then the bar will increase its length by an amount ΔL. The extent of this stretching is determined by Young's Modulus, E, according to

$$E = \frac{\text{Tensile stress}}{\text{Tensile strain}} = \frac{F/A}{\Delta L/L}$$

However, it is perhaps not so obvious that the lateral dimensions of the bar will, in general, contract to some extent. Suppose the lateral contraction is Δx, then the ratio of the lateral strain, $\Delta x/x$, to the tensile strain, $\Delta L/L$, is known as the Poisson ratio. Evaluate this ratio for the situation depicted below if the *volume* of the bar is conserved in the extension.

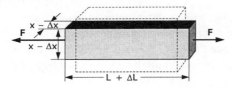

The initial volume of the bar is $x^2 L$ and the final volume of the bar is $(x - \Delta x)^2 (L + \Delta L)$. Equating these two volumes we find

$$x^2 L = (x - \Delta x)^2 (L + \Delta L)$$

Expanding the brackets

$$x^2 L = \left(x^2 - 2x\Delta x + (\Delta x)^2\right)(L + \Delta L)$$

and multiplying out the terms yields

$$x^2 L = x^2 L - 2Lx\Delta x + L(\Delta x)^2 + x^2\Delta L - 2\Delta Lx\Delta x + \Delta L(\Delta x)^2$$

Several of these terms contain the products of two small quantities ΔL and Δx.

$$x^2 L = x^2 L - 2Lx\Delta x + x^2\Delta L + \text{small quantities}$$

Neglecting these small quantities and rearranging yields

$$\frac{\Delta x}{x} = \frac{1}{2}\frac{\Delta L}{L}$$

The fraction on the left-hand side is the lateral strain and the fraction on the right is the tensile strain. By definition, therefore, the Poisson ratio in this example is just 0.5. If, as one would expect, the volume of the material expands under the tensile strain, then the lateral strain would be less than that predicted in this example and the Poisson ratio would be less than 0.5. Commonly the Poisson ratio is $\approx 1/3$.

Relationships between the different moduli

Finally, we note that for many substances the rigidity modulus and Young's modulus are related by

$$G = \frac{E}{2(1+\sigma)} \qquad (7.16)$$

Taking a typical value for the Poisson ratio, $\alpha \approx 1/3$, Eq. 7.16 predicts that, typically, $G \approx 3E/8$.

Example 7.9

The experimentally determined values of Young's modulus and the Poisson ratio for copper are E =129.8 GPa and σ = 0.343 (Kaye & Laby; see §1.4.1). What is the expected value for the speed of longitudinal and transverse sound waves in copper?

Equations 7.14 & 7.15 give us the formulae we require. Evaluating the expected value of γ we write

$$\gamma = \frac{1-\sigma}{(1-2\sigma)(1+\sigma)} = \frac{1-0.343}{(1-0.686)(1+0.343)} = 1.558$$

From Table 7.2 we find that the density of copper is 8933 kg m^{-3} and substituting into the expression for the speed of sound we find

$$c_{\text{L}} = \sqrt{\frac{\gamma E}{\rho}} = \sqrt{\frac{1.558 \times 129.8 \times 10^9}{8933}} = 4758 \text{ m s}^{-1}$$

We can estimate the speed of transverse sound waves in solids by using Equation 7.16 to determine the rigidity modulus, G, in terms of Young's modulus, E. Using the values of E and σ given, we find

$$G = \frac{E}{2(1+\sigma)} = \frac{129.8 \times 10^9}{2(1+0.343)} = 48.32 \times 10^9 \text{ Pa}$$

Substituting this value for G into Equation 7.13 we find

$$c_{\text{T}} = \sqrt{\frac{G}{\rho}} = \sqrt{\frac{48.32 \times 10^9}{8933}} = 2326 \text{ m s}^{-1}$$

These values can be compared with the experimental values given in Table 7.6 of c_{L} = 4759 ms^{-1} and c_{T} = 2325 ms^{-1} – a suspiciously good agreement!

Comparison of theory and experiment

Comparison of the speed of sound in gases and solids

Equations 5.76 & 7.12–7.14 indicate that the speed of sound in both gases and solids is determined by the ratio of the "springiness" of a substance to its density. So, if we combine the fact that the speed of sound in solids is typically one order of magnitude greater than the speed of sound in gases at around STP with the fact that the density of solids (see §7.1) is typically 10^3 to 10^4 times greater than the density of gases at STP, then together these facts imply that the "springiness" of solids is around 10^5 to 10^6 times greater than gases. Thus it requires around a million times more energy to reduce the volume of a solid by a given factor than it does to reduce the volume of a gas. It is easy to understand why this is so. Reducing the volume of gas mainly increases the number density of molecules and so increases the frequency with which they hit each other and the walls. In order to reduce the volume of a solid one needs to rearrange the valence electrons of strongly interacting atoms.

The elastic moduli of solids are not catalogued in this text, but may be found in Kaye & Laby or the *CRC handbook* (see §1.4.1). "Typical" values of E or B or G for elements or for engineering solids are of the order of 10^{10}–10^{11} Pa. Example 7.9 shows that Equations 7.13 and 7.14 predict the speed of sound with good accuracy.

The relative speed of longitudinal and transverse sound waves

Equations 7.13 and 7.14 yield the speeds of longitudinal and transverse sound waves through a bulk solid. If we take the ratio of these two equations we find that

$$\frac{c_L}{c_T} = \frac{\sqrt{\gamma E/\rho}}{\sqrt{G/\rho}} = \sqrt{\frac{\gamma E}{G}} \tag{7.17}$$

Substituting the expression for γ (Eq. 7.15) and the relationship between E and G (Eq. 7.16), we find

$$\frac{c_L}{c_T} = \sqrt{\frac{(1-\sigma)E/[(1-2\sigma)(1+\sigma)]}{E/2(1+\sigma)}} \tag{7.18}$$

Cancelling and simplifying we arrive at

$$\frac{c_L}{c_T} = \sqrt{\frac{2(1+\sigma)(1-\sigma)}{(1-2\sigma)(1+\sigma)}} = \sqrt{\frac{2(1-\sigma)}{1-2\sigma}} \tag{7.19}$$

Thus the experimental fact that, typically, $c_L \approx 2c_T$ (Fig. 7.9) may be reinterpreted as a statement that

$$4 \approx \frac{2(1-\sigma)}{1-2\sigma} \tag{7.20}$$

Rearranging

$$4 - 8\sigma \approx 2 - 2\sigma \tag{7.21}$$

and solving for α we find

$$\sigma \approx \frac{1}{3} \tag{7.22}$$

which is indeed a typical value of the Poisson ratio (see Kaye and Laby for tables; §1.4.1).

The dependence of the speed of sound in solids on atomic mass

Equations 7.12–7.14 indicate that one should expect the speed of any kind of sound wave to vary inversely as the square root of the density $c \propto \rho^{-\frac{1}{2}}$. In §7.1 (Fig. 7.2) we saw that there is a general trend amongst the elements for high atomic number elements to be relatively more dense. However, in addition to the trend to high density, there are also periodic increases and decreases in density associated with the different types, and strengths, of bonding amongst the elements. Based on the density data one would therefore expect to find a general trend towards a lower speed of sound for high atomic number elements, and fluctuations in the speed of sound associated with the density variations. This is, broadly speaking, just what is depicted in Figure 7.9.

A simple calculation

In fact we can construct a simple theory that allows us to understand the variation in the speed of sound based on the following assumptions:

- All elements have the same value of Young's modulus E of ≈ 100 GPa. This is a typical value of E (see Kaye & Laby, section 1, for details; §1.4.1) but there are considerable variations

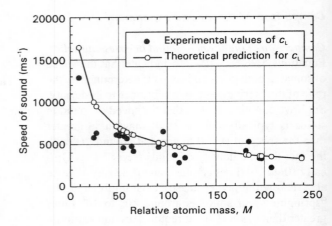

Figure 7.15 The results of a simple prediction of the speed of sound amongst the elements, making the three assumptions given in the text. This line is also drawn in Figure 7.2 and indicates the trend of the data, but not the periodic increases and decreases in density.

about this figure.

- All elements have a Poisson ratio of $\approx 1/3$
- The density of the elements is given by $\rho = 61.5A$ (Example 7.2). This line is indicated in Figure 7.2 and follows the trend of the data, but not the periodic increases and decreases in density.

Figure 7.15 shows the speed of longitudinal sound waves calculated according to Equation 7.15 and the above assumptions, and it is clear that the line captures the trend of the data rather well.

7.4 Heat capacity

Heat capacity is a measure of the rate at which the temperature of a substance rises for a given input of heat energy. It is defined in terms of the temperature rise, ΔT, resulting from an input of heat energy, ΔQ, by the ratio

$$C = \frac{\Delta Q}{\Delta T} \quad \text{joules kelvin}^{-1} \ (\text{J K}^{-1}) \qquad (7.23)$$

Equation 7.23 is the formula used to determine the heat capacity from experimental measurements of ΔQ and ΔT and is an approximation to the theoretical definition, which is the limit of Equation 7.19 as ΔT tends to zero:

$$C = \frac{\mathrm{d}Q}{\mathrm{d}T} \ \text{J K}^{-1} \qquad (7.24)$$

The heat capacity of a substance is usually quoted either for a given *mass* of material, the *specific* heat capacity (e.g. $\text{J K}^{-1}\text{kg}^{-1}$), or per *mole*, the *molar* heat capacity ($\text{J K}^{-1}\text{mol}^{-1}$). For practical calculations the specific heat capacity is usually more convenient, but from a fundamental point of view the molar heat capacity reveals far more. Remember that the molar heat capacity is the heat capacity of Avogadro's number ($N_A = 6.02 \times 10^{23}$) of atoms or molecules.

The heat capacities at constant pressure and constant volume are designated by C_P and C_V, respectively, and for solids one normally assumes that measurements are made at constant pressure unless told otherwise. The difference between C_P and C_V for solids is usually smaller than for gases because the expansivity of solids is so much smaller. Measured values of C_P are normally just a few per cent greater than C_V at around room temperature, with the difference becoming smaller at lower temperatures and increasing at higher temperatures.

7.4.1 Data on the heat capacity of the elements

The *molar* heat capacities at constant pressure C_P for the elements at 25°C (288.1 K) are given in Table 7.7 and a histogram showing the distribution of values of C_P for the solid elements is shown in Figure 7.16.

The histogram (Fig. 7.16) indicates a striking phenomenon: more than 50% of the solid elements have a heat capacity close to $25\,\text{J K}^{-1}\text{mol}^{-1}$, and nearly all elements have a heat capacity between $22\,\text{J K}^{-1}\text{mol}^{-1}$

Table 7.7 The molar heat capacities at constant pressure, C_p, for the elements at room temperature (25°C = 298.15 K)(see Fig. 7.17). The shaded elements are either liquids or gases at this temperature.

To obtain the specific heat capacity per kilogram use:

$$C_p\left(JK^{-1}kg^{-1}\right) = \frac{C_p\left(JK^{-1}mol^{-1}\right)}{A}$$

Z	Element		A	ρ (kgm^{-3})	C_p (JK^{-1}mol^{-1})	Z	Element		A	ρ (kgm^{-3})	C_p (JK^{-1}mol^{-1})
1	Hydrogen	H	1.008	89	28.824	49	Indium	In	114.8	7290	26.74
2	Helium	He	4.003	120	20.786	50	Tin	Sn	118.7	7285	26.99
3	Lithium	Li	6.941	533	24.770	51	Antimony	Sb	121.7	6692	25.23
4	Beryllium	Be	9.012	1846	16.44	52	Tellurium	Te	127.6	6247	25.73
5	Boron	B	10.81	2466	11.09	53	Iodine	I	126.9	4953	54.438
6	Carbon	C	12.01	2266	8.527	54	Xenon	Xe	131.3	3560	20.786
7	Nitrogen	N	14.01	1035	29.125	55	Caesium	Cs	132.9	1900	32.17
8	Oxygen	O	16.00	1460	29.355	56	Barium	Ba	137.3	3594	28.07
9	Fluorine	F	19.00	1140	31.300	57	Lanthanum	La	138.9	6174	27.11
10	Neon	Ne	20.18	1442	20.786	58	Cerium	Ce	140.1	6711	26.94
11	Sodium	Na	22.99	966	28.24	59	Praseodymium	Pr	140.9	6779	27.20
12	Magnesium	Mg	24.31	1738	24.89	60	Neodymium	Nd	144.2	7000	27.45
13	Aluminium	Al	26.98	2698	24.35	61	Promethium	Pm	145	7220	26.81
14	Silicon	Si	28.09	2329	20.0	62	Samarium	Sm	150.4	7536	29.54
15	Phosphorous	P	30.97	1820	23.84	63	Europium	Eu	152	5248	27.66
16	Sulphur	S	32.06	2086	22.64	64	Gadolinium	Gd	157.2	7870	37.03
17	Chlorine	Cl	35.45	2030	33.907	65	Terbium	Tb	158.9	8267	28.91
18	Argon	A	39.95	1656	20.786	66	Dysprosium	Dy	162.5	8531	28.16
19	Potassium	K	39.10	862	29.58	67	Holmium	Ho	164.9	8797	27.15
20	Calcium	Ca	40.08	1530	25.31	68	Erbium	Er	167.3	9044	28.12
21	Scandium	Sc	44.96	2992	25.52	69	Thulium	Tm	168.9	9325	27.03
22	Titanium	Ti	47.90	4508	25.02	70	Ytterbium	Yb	173	6966	26.74
23	Vanadium	V	50.94	6090	24.89	71	Lutetium	Lu	175	9842	26.86
24	Chromium	Cr	52.00	7194	23.35	72	Hafnium	Hf	178.5	13276	25.73
25	Manganese	Mn	54.94	7473	26.32	73	Tantalum	Ta	180.9	16670	25.36
26	Iron	Fe	55.85	7873	25.10	74	Tungsten	W	183.9	19254	24.27
27	Cobalt	Co	58.93	8800	24.81	75	Rhenium	Re	186.2	21023	25.48
28	Nickel	Ni	58.70	8907	26.07	76	Osmium	Os	190.2	22580	24.70
29	Copper	Cu	63.55	8933	24.44	77	Iridium	Ir	192.2	22550	25.10
30	Zinc	Zn	65.38	7135	25.40	78	Platinum	Pt	195.1	21450	25.86
31	Gallium	Ga	69.72	5905	25.86	79	Gold	Au	197.0	19281	25.42
32	Germanium	Ge	72.59	5323	23.35	80	Mercury	Hg	200.6	13546	27.98
33	Arsenic	As	74.92	5776	24.64	81	Thallium	Tl	204.4	11871	26.32
34	Selenium	Se	78.96	4808	25.36	82	Lead	Pb	207.2	11343	26.44
35	Bromine	Br	79.90	3120	75.69	83	Bismuth	Bi	209	9803	25.52
36	Krypton	Kr	83.80	3000	20.79	84	Polonium	Po	209	9400	25.75
37	Rubidium	Rb	85.47	1533	31.06	85	Astatine	At	210		
38	Strontium	Sr	87.62	2583	26.40	86	Radon	Rn	222	4400	20.786
39	Yttrium	Y	88.91	4475	26.53	87	Francium	Fr	223	2410	31.70
40	Zirconium	Zr	91.22	6507	25.36	88	Radium	Ra	226	5000	25.76
41	Niobium	Nb	92.91	8578	24.60	89	Actinium	Ac	227	10060	27.20
42	Molybdenum	Mo	95.94	10222	24.06	90	Thorium	Th	232	11725	27.32
43	Technetium	Tc	97	11496	25.88	91	Protractinium	Pa	231	15370	27.20
44	Ruthenium	Ru	101.1	12360	24.06	92	Uranium	U	238	19050	27.66
45	Rhodium	Rh	102.9	12420	24.98	93	Neptunium	Np	237	20250	29.62
46	Palladium	Pd	106.4	11995	25.98	94	Plutonium	Pu	244	19840	32.80
47	Silver	Ag	107.9	10500	25.35	95	Americium	Am	243	13670	25.86
48	Cadmium	Cd	112.4	8647	25.98	96	Curium	Cm	247	13300	27.70

(Data from Kaye & Laby; see §1.4.1)

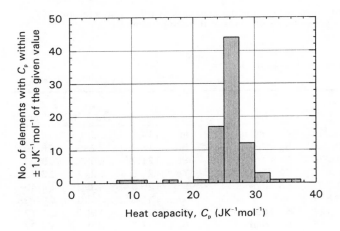

Figure 7.16 Histogram of the heat capacities at constant pressure, C_P, for the solid elements at around room temperature (25 °C = 298.15 K). More than 50% of the elements have a heat capacity close to 25 J K^{-1} mol^{-1}, and nearly all elements have C_P between 22 and 32 J K^{-1} mol^{-1}.

and $32 \, \mathrm{J \, K^{-1} \, mol^{-1}}$. That is, if one collects 1 mole of almost any element (i.e. Avogadro's number of atoms) then it takes roughly 25 J to raise its temperature by 1 K, almost independent of the type of atom, crystal structure or bonding type!

There are some exceptions. On the low side, the three light elements beryllium, boron and carbon have molar heat capacities of 16.4, 11.1 and $8.5 \, \mathrm{J \, K^{-1} \, mol^{-1}}$, respectively. On the high side, the three alkali metals rubidium, caesium and francium have molar heat capacities of 31.1, 32.2 and $31.7 \, \mathrm{J \, K^{-1} \, mol^{-1}}$ respectively, while the ferromagnetic element gadolinium has $C_P = 37.0 \, \mathrm{J \, K^{-1} \, mol^{-1}}$.

The variation of heat capacity with temperature in solids

The heat capacities of three metallic elements are depicted as a function of temperature in Figure 7.17. The data chosen are typical of the behaviour of many crystalline materials.

Figure 7.17 shows that the molar heat capacities of copper, silver and gold tend to roughly the same value at high temperatures. The behaviour below room temperature is qualitatively similar, apparently tending to zero at absolute zero. Figure 7.18 shows the heat capacities of the insulating substance sapphire (Al_2O_3). The behaviour below about 1000 K is quali-

Figure 7.17 The variation with temperature of the molar heat capacity of copper, silver and gold. All three curves tend to a value of ≈25 J K^{-1} mol^{-1} at high temperatures, which is in agreement with the data given in Table 7.7 and Figure 7.16. At low temperatures, all three curves tend to zero. (Data from Kaye & Laby; see §1.4.1)

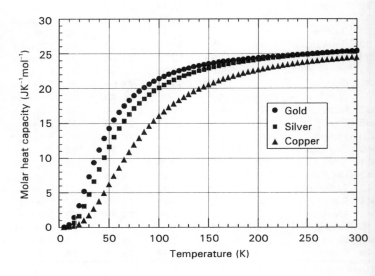

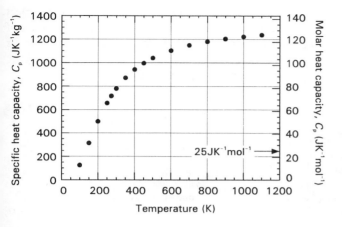

Figure 7.18 The variation with temperature of the heat capacity of sapphire (Al_2O_3). The data tends to a value of $\approx 130\,J\,K^{-1}\,mol^{-1}$ at high temperatures, which is much higher than the values given in Table 7.7 and Figure 7.17. At low temperatures the data tend to zero in a similar manner to the data for copper, silver and gold. (Data from *CRC handbook*; see §1.4.1)

tatively similar to the behaviour of the metallic elements below room temperature, apparently tending to zero at absolute zero.

become smaller below room temperature and tend to zero at low temperatures?

Summary

The main questions raised by our preliminary examination of the experimental data on C_p are:
- Why, at room temperature, are the heat capacities of the elements often close to $25\,J\,K^{-1}\,mol^{-1}$?
- Why do the heat capacities of solid elements

7.4.2 Understanding the heat capacity of solids

To answer the questions raised in the previous section we will construct a model of a solid that is simpler even than any of those discussed in Chapter 6. We will not assume any particular bonding type because the universality of the value of

---

Example 7.10

Use Figure 7.17 to work out how much heat energy is required to raise the temperature of a cube of copper of side 2.42 cm from (a) 273 K to 283 K, (b) 173 K to 183 K, (c) 73 K to 83 K, (d) 3 K to 13 K.

First of all, we work out how many moles of copper are contained in our sample. If it has side L then its volume is L^3 and its mass is ρL^3, where ρ is the density of copper (Table 7.2). The number of moles, z, in the cube is then

$$z = \frac{\rho L^3}{\text{molar mass}}$$

For the sample in question this becomes

$$z = \frac{8933 \times \left(2.42 \times 10^{-2}\right)^3}{63.55 \times 10^{-3}}$$

$$= \frac{127.1 \times 10^{-3}}{63.55 \times 10^{-3}} = 2.00\,mol$$

From Figure 7.17, we see that in the ranges indicated the heat capacity is approximately:
(a) 273 K to 283 K: $C_p \approx 24\,J\,K^{-1}\,mol^{-1}$
(b) 173 K to 183 K: $C_p \approx 21.5\,J\,K^{-1}\,mol^{-1}$
(c) 73 K to 83 K: $C_p \approx 11.8\,J\,K^{-1}\,mol^{-1}$
(d) 3 K to 13 K: too small to infer directly from the graph, but definitely $<0.1\,J\,K^{-1}\,mol^{-1}$

Thus the amount of heat energy required to heat the sample through 10 K in each range, $\Delta Q = zC_p\Delta T$, is given by $\Delta Q = 2.00 \times C_p \times 10.0\,J$.
(a) 273 K to 283 K: $\Delta Q = 2.00 \times 24 \times 10.0 = 480\,J$
(b) 173 K to 183 K: $\Delta Q = 2.00 \times 21.5 \times 10.0 = 430\,J$
(c) 73 K to 83 K: $\Delta Q = 2.00 \times 11.8 \times 10.0 = 236\,J$
(d) 3 K to 13 K: $\Delta Q < 2.00 \times 0.1 \times 10.0 < 2\,J$

Note just how small the heat capacity becomes at low temperatures!

$C_p \approx 25\,\text{J}\,\text{K}^{-1}\,\text{mol}^{-1}$ means that the value cannot be related to details of the bonding type. We will then develop the model further to try to understand the reduction of heat capacity at low temperatures.

A simple model of a solid

Our model takes into account the fact that atoms in all solids have an equilibrium position about which they vibrate. This position represents their minimum-energy position, and any displacement from this position takes them into a region where their potential energy of interaction with neighbouring atoms is higher. The atom will therefore experience a restoring force trying to return it towards its minimum-energy position. This is exactly as if the atom were held in position with microscopic springs (see Fig. 7.19). This is a fanciful idea, but it can be helpful as long as one remembers that in reality the forces involved are mainly electrostatic as described in Chapter 6.

If we consider all the "springs" in Figure 7.15 to have the same spring constant K, then we can write the potential energy of a single spring as

$$v = \frac{1}{2} K (x - x_0)^2 \qquad (7.25)$$

where $x - x_0$ is the displacement of the atom from its equilibrium position in the x direction. Similar equations apply for displacements in the y and z directions. (Displacing the atom in the x direction will

slightly stretch the y and z springs too, but we neglect this small effect in this analysis.) Taking account of the kinetic energy of motion in each of the x, y and z directions we can write down the total energy, u, associated with vibration of an atom

$$u = \frac{1}{2} K (x - x_0)^2 + \frac{1}{2} m v_x^2$$
(simple harmonic oscillator 1)

$$+ \frac{1}{2} K (y - y_0)^2 + \frac{1}{2} m v_y^2$$
(simple harmonic oscillator 2)

$$+ \frac{1}{2} K (z - z_0)^2 + \frac{1}{2} m v_z^2 \qquad (7.26)$$
(simple harmonic oscillator 3)

At any instant of time some of these terms will be large and others small, but over time we would expect the *average value* of the energy associated with each of these energy terms to be equal. Why?

There are three reasons. The first is connected to the symmetry of the situation. There is no difference in the crystal between the x, y and z directions and so there is no reason why one vibration should *on average*, have a higher energy of vibration than the other directions. If this were so at one time, then that direction of vibration would tend to lose energy to the other directions of vibration. The second reason concerns the exchange of energy between potential and kinetic energy. It is a standard exercise in analyzing simple harmonic motion to show that the average values of kinetic and potential energy are exactly equal. Finally, we remind ourselves that this is merely a special case of the equipartition of energy between degrees of freedom discussed in §2.5 and §4.3.1, complication 3.

One might suppose that if the spring constant was different in one direction then the average energy associated with vibration in that direction would be different from the other terms. However, this supposition would be mistaken! For example (Fig. 7.20), suppose the x spring was stiffer than the others, i.e. $K_x > K_y$ or K_z. Then the amplitude of x vibration would be less than in the y and z directions, but this smaller amplitude x vibration would "cost" the same amount of energy as the larger amplitude y and z vibrations because the spring constant K_x is larger.

Figure 7.19 Analogy to the situation of atoms in a crystal. The forces on an atom are such that the atom behaves *as if* it were held at its minimum-energy position by springs. (a) The atom in its equilibrium position; (b) The atom displaced from its equilibrium position. Note that the springs in (b) are all either compressed or extended compared with their equilibrium length of r_0.

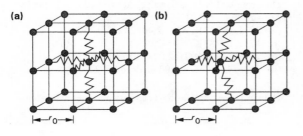

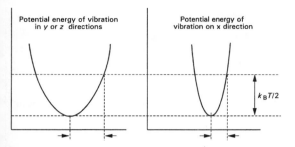

Potential energy of vibration in y or z directions

Potential energy of vibration on x direction

$k_BT/2$

Figure 7.20 An illustration of two simple harmonic potentials with different spring constants. A given amount of energy results in smaller oscillations of the stiffer-springed potential.

Thus, using the standard notation of representing the average of a quantity by a "bar" over the top of it, we expect the average of each term in Equation 7.26 to be equal, i.e.

$$\frac{1}{2}K(x-x_0)^2 = \frac{1}{2}mv_x^2 = \frac{1}{2}K(y-y_0)^2$$
$$= \frac{1}{2}mv_y^2 = \frac{1}{2}K(z-z_0)^2 = \frac{1}{2}mv_z^2 \qquad (7.27)$$

Each of the terms in Equation 7.27 constitutes a *degree of freedom* for the atom in the sense discussed in §2.5, §4.3 & §5.3.3. As you may recall from those discussions, temperature is *defined* so that the average energy per accessible degree of freedom is $k_BT/2$, where k_B is the Boltzmann constant. Thus molecules in a gas each have three degrees of freedom associated with kinetic energy of motion in the x, y, and z directions. From Equation 7.27 we can see that atoms in a solid each have six degrees of freedom associated with the kinetic *and* potential energy of motion in the x, y, and z directions. Thus, according to our simple model, if we have one mole of an element, i.e. Avogadro's number (N_A) of atoms, we expect the internal energy of that substance at temperature T to be

$$U = U_0 + N_A \times 6 \times \tfrac{1}{2}k_BT \qquad (7.28)$$

where U_0 is the cohesive energy at $T = 0\,K$. This will apply, however, only for materials with one atom per chemical formula unit. For example, 1 mole of NaCl

contains $2N_A$ atoms: N_A of Na and N_A of Cl. Generalizing Equation 7.28 to take account of solids with p atoms per chemical formula unit, we obtain a prediction for the internal energy of a solid:

$$U = U_0 + pN_A \times 6 \times \tfrac{1}{2}k_BT \qquad (7.29)$$

By remembering that $N_Ak_B = R$, the molar gas constant, we find

$$U = U_0 + 3pRT \qquad (7.30)$$

By the first law of thermodynamics (Eq. 2.65), changes in internal energy are related to the heat supplied to a substance and the work on that substance by

$$\Delta U = \Delta Q + \Delta W \qquad (7.31)$$

If we consider first the effect of a temperature change at constant volume, then $\Delta W = P\Delta V$ is zero. In this case we may differentiate Equation 7.30 to find the heat capacity at constant volume, C_V,

$$C_V = \frac{dQ}{dT} = \frac{dU}{dT} = 3pR \qquad (7.32)$$

If the heat is supplied at constant pressure, then we need to take account of the thermal expansivity and the volume dependence of the internal energy. However, although the latter effect is large, the thermal expansivity is generally rather small (Table 7.4), and at or below room temperature C_P and C_V generally differ by only a few per cent at most. In what follows we assume that $C_P \approx C_V$.

If we evaluate Equation 7.32 for elements ($p = 1$) we have a startlingly simple prediction that $C_V = 3R$, independent of the mass of atoms, type of bonding, or crystal structure! Evaluating this prediction we find $C_V = 3 \times 8.31 = 24.93\,JK^{-1}mol^{-1}$. Comparison of this result with the data in Table 7.7 shows that this estimate is accurate for C_P to within 10% for most elements at room temperature. When we consider the diverse properties of the solids involved, this is indeed a remarkable result. This result was noted as early as the nineteenth century and is called the *Law of Dulong and Petit*.

The analysis leading to Equation 7.32 allows us to understand the heat capacity of elements and compounds at around room temperature. We can even understand some of the exceptions in Figure 7.21.

177

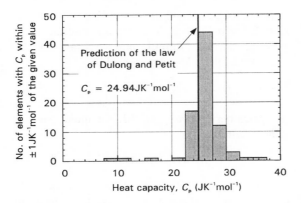

Figure 7.21 Histogram of the heat capacities at constant pressure, C_P, for the solid elements at room temperature (25 °C = 298.15 K). The bold line shows the prediction of Equation 7.32 for the heat capacities of all elements.

Example 7.12

How does the high-temperature limiting value of the heat capacity of sapphire (Al_2O_3) depicted in Figure 7.18 compare with the value predicted by the law of Dulong and Petit?

We can work out the theoretically expected value using Equation 7.32

$$C_V = 3pR$$

For Al_2O_3, there are five atoms per chemical formula unit, i.e. $p = 5$ and so we predict a heat capacity of $15R = 124.7\,JK^{-1}mol^{-1}$

There is some uncertainty in predicting the high-temperature limiting value of the heat capacity of sapphire depicted in Figure 7.18 but by my estimation the value is $\approx 127 \pm 3\,JK^{-1}mol^{-1}$, a value within 2% of the value predicted by the law of Dulong and Petit.

Three of the elements with particularly high values of C_P – the three alkali metals rubidium, caesium and francium – have molar heat capacities of 31.1, 32.2 and $31.7\,JK^{-1}mol^{-1}$ respectively. These elements melt at unusually low temperatures (39.1°C, 28.4°C and 27°C, respectively) and so at 25°C they are close to their melting points. In this region their thermal expansivity is particularly large and the difference between the experimental value of C_P and the predicted value of $\approx 3R$ is just the difference between C_P and C_V that we neglected for the other elements. The

unusual datum for the ferromagnetic element gadolinium (C_P of $37.0\,JK^{-1}mol^{-1}$) is connected with the transition to the ferromagnetic state (see §11.9)

The heat capacity of solids at low temperatures
The approach developed so far can plausibly explain the unusually high values of C_P at around room temperature, but it cannot explain the unusually low values. Neither can it explain the reduction of the heat capacity observed at low temperatures (Figs 7.17 & 7.18).

Degrees of freedom
Based on the analysis of the heat capacity of gases in §5.3, we may surmise that the reduction in heat capacity at low temperatures is because some of the six degrees of freedom available to each atom (Eq. 7.27) have become constrained, i.e. they are not *fully accessible*. It is at first difficult to see where this restriction on the accessible degrees of freedom of vibration comes from: it might appear that at low temperatures the amplitude of the vibration of the atoms of solid would be reduced, but would otherwise be similar to the situation at high temperatures. However, this is plainly not so since, if it were, the heat capacity would not be reduced at low temperatures!

The origin of the restriction on the vibration of each atom lies in the quantum mechanical nature of

Example 7.11

How does the experimental value of the heat capacity of NaCl ($0.88\,JK^{-1}g^{-1}$ at 10°C; Kaye & Laby; see §1.4.1) compare with the value predicted by the law of Dulong and Petit?

We can work out the theoretically expected value using Equation 7.32

$$C_V = 3pR$$

For NaCl, there are two atoms per chemical formula unit, i.e. $p = 2$, and so we predict a heat capacity of $6R = 49.86\,JK^{-1}mol^{-1}$.

Since 1 mol of NaCl has a mass of 22.99 + 35.45 = 58.44 g (Table 7.2) we see that 1 g is 1/58.44 = 17.1×10^{-3} mol and so the experimental value of the molar heat capacity is $58.44 \times 0.88 = 51.4\,JK^{-1}mol^{-1}$, within 3% of the predicted value of $49.86\,JK^{-1}mol^{-1}$.

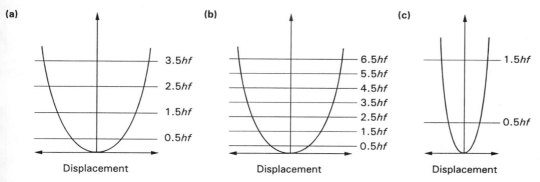

Figure 7.22 An illustration of the relationship between energy levels, particle mass and spring constant for a simple harmonic potential. If (a) represents the energy levels of a particle of mass m bonded in a particular way, then (b) would represent the energy levels of a heavier particle in the same potential, and (c) would represent the energy levels of a particle of a similar mass, but more stiffly constrained.

the vibration. If the potential energy of the atom of mass m varies with position approximately like a simple harmonic oscillator (the theory of which is summarized in §2.4), then we write

$$v = \tfrac{1}{2} K(x - x_0)^2 \tag{7.33}$$

where $x - x_0$ is the displacement of the atom from its equilibrium position. The frequency of vibration is given by

$$f_0 = \frac{1}{2\pi} \sqrt{\frac{K}{m}} \tag{7.34}$$

and the *amplitude* of vibration is quantized such that the total energy of vibration is limited to one of the values

$$u = \left(n_x + \tfrac{1}{2}\right) h f_0 \tag{7.35}$$

where h is the Planck constant and n_x is any positive integer starting from 0 (Fig. 7.22).

The transition to inaccessibility

In §5.3.3 we noted that a quantum state becomes inaccessible when the thermal energy, $k_B T$, is significantly less than the energy gap, ΔE, between quantum states. In this case ΔE is the difference in energy between successive quantum states $\Delta E = h f_0 = \hbar\sqrt{(K/m)}$. Thus we would expect to observe the reduction in accessibility of the quantum states when

$k_B T < {\sim} \hbar\sqrt{(K/m)}$ or when

$$T < \frac{\hbar}{k_B} \sqrt{\frac{K}{m}} \tag{7.36}$$

This idea, due to Einstein, captures some of the essential physics of the reduction of the heat capacity of solids at low temperatures. Equation 7.36 defines a temperature below which excited vibrational quantum states begin to become inaccessible. The Einstein theory correctly predicts that the heat capacity of solids becomes smaller below a characteristic temperature – known as the *Einstein temperature*. The Einstein temperature, θ_E, is defined as

$$\theta_E = \frac{h f_0}{k_B} = \frac{\hbar \omega_0}{k_B} = \frac{\hbar}{k_B} \sqrt{\frac{K}{m}} \tag{7.35}$$

and varies from solid to solid depending on the stiffness constant, K, of the bonds and the mass of the atoms.

The Einstein theory of the heat capacity of solids

The Einstein theory of the heat capacity is developed more fully in Appendix 5. In many ways it is similar to the classical theory developed earlier. The key difference is that the average energy associated with each (x, y or z) harmonic oscillator is no longer given by $\bar{u} = 2 \times k_B T/2$, but by a more complex expression

$$\bar{u} = \tfrac{1}{2}hf_0 + \frac{hf_0}{\exp(hf_0/k_B T)-1} \qquad (7.38)$$

Note, however, that at high temperatures, i.e. when $k_B T \gg hf_0$, the exponential in the denominator of Equation 7.38 can be expanded:

$$\bar{u} = \tfrac{1}{2}hf_0 + \frac{hf_0}{1+(hf_0/k_B T)+\ldots-1} \approx \tfrac{1}{2}hf_0 + k_B T \qquad (7.39)$$

which agrees with the classical result, except for the $hf_0/2$ term. Proceeding as we did in the classical case (Eq. 7.29), we write the molar internal energy as

$$U = U_0 + 3pN_A \bar{u} \qquad (7.40)$$

where p is the number of atoms per chemical formula unit. Recalling that for temperature changes at constant volume $dQ/dT = dU/dT$, we write

$$C_V = \frac{dU}{dT} = \frac{d}{dT}(3pN_A\bar{u}) = 3pN_A \frac{d\bar{u}}{dT} \qquad (7.41)$$

Substituting for $\bar{u}$

$$C_V = 3pN_A \frac{d}{dT}\left[\tfrac{1}{2}hf_0 + \frac{hf_0}{\exp(-hf_0/k_B T)-1}\right] \qquad (7.42)$$

and differentiating we find

$$C_V = \frac{hf_0 \exp(hf_0/k_B T)}{k_B T^2} \times \frac{3pN_A hf_0}{\left[\exp(hf_0/k_B T)-1\right]^2} \qquad (7.43)$$

Recalling that $R = N_A k_B$ this may be written as

$$C_V = 3pR\left(\frac{hf_0}{k_B T}\right)^2 \frac{(hf_0/k_B T)}{\left[\exp(hf_0/k_B T)-1\right]^2} \qquad (7.44)$$

Finally, recalling the definition of θ_E (Eq. 7.37), this may also be written as

$$C_V = 3pR\left(\frac{\theta_E}{T}\right)^2 \frac{\exp(\theta_E/T)}{\left[\exp(\theta_E/T)-1\right]^2} \qquad (7.45)$$

Figure 7.23 shows the heat capacity of copper from Figure 7.17 plotted together with the Einstein prediction for the heat capacity based on $\theta_E = 230\,K$. This value has been chosen, by trial and error, to give good agreement across most of the temperature range. It is clear that the theory does indeed capture the trend of the data. However, careful examination of Figure 7.23 shows that the agreement between theory and experiment becomes rather poor at low temperatures.

It would be nice if we could neglect the discrepancy at low temperatures between the copper data and Einstein theory in Figure 7.23. However, the dif-

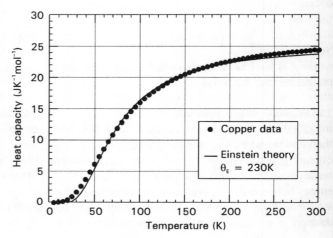

Figure 7.23 The heat capacity of copper from Figure 7.17 plotted together with the Einstein prediction for the heat capacity based on $\theta_E = 230\,K$. The predicted Einstein temperature corresponds to a spring constant of $K = 2.423\,J\,m^{-2}$, and a frequency of vibration of $f_0 = 4.79\times10^{12}\,Hz$. It is clear that the theory captures the trend of the data. However, careful examination shows that agreement between the theory and experiment becomes poor at low temperatures. Note: the Einstein theory prediction is for C_V, but the data with which it is compared is based on Cp.

Example 7.13

Assuming an Einstein temperature of 230 K, use Equation 7.37 to predict the frequency, f_0, at which the copper atoms are supposed to vibrate, and predict a value for K, the spring constant of a Cu–Cu bond.

Using Equation 7.37,

$$f_0 = \frac{k_B\theta_E}{h} = \frac{1.38\times10^{-23}\times230}{6.63\times10^{-34}} = 4.79\times10^{12}\text{ Hz}$$

Recalling that f_0 is defined (Eq. 7.34) as $f_0 = (1/2\pi)\sqrt{(K/m)}$, and rearranging to solve for K, we find

$$K = 4m\pi^2 f_0^2$$

From Table 7.2 we find that the mass of a copper atom is $63.55\times1.66\times10^{-27}$ kg and so we predict that

$$K = 4\times63.55\times1.66\times10^{-27}\times\pi^2\times\left(4.79\times10^{12}\right)^2$$

$$= 95.6\,\text{J m}^{-2}$$

This value may be compared with the value worked out from the speed of sound (§7.9; Exercise 7P30).

---

ference is important because it is a clue that there is something wrong with the way in which the Einstein theory models the vibrations within solids. In the next section, we will develop this model in order to reflect correctly the types of vibration that actually take place in solids.

Lattice waves, sound waves and phonons

The key problem with the Einstein theory as applied to ordinary solids is that it assumes that each atom is in the situation shown in Figure 7.24a and that all the atoms vibrate at the same frequency $f_0 = (1/2\pi)\times\sqrt{(K/m)}$. However, a little thought shows that the model illustrated in Figure 7.24b is likely to be more realistic. The difficulty with this new model is that the atoms no longer vibrate *independently*.

Consider the following thought experiment: imagine displacing the central atom in Figure 7.24b and then letting it go. The atom would not vibrate at frequency f_0 leaving its neighbours unaffected. What would happen is that a wave-like disturbance would spread out from the central atom. This is what is wrong with the Einstein model at low temperatures. It assumes that the vibrational energy of the lattice is held as individual and independent vibrations of atoms. In fact the energy is held in waves of displacements running through the lattice.

Now, we have already considered waves of displacement in a solid – they are nothing more than sound waves. However, the frequencies of the naturally occurring displacement waves in solids are typically of the order of f_0 (see Example 7.13), i.e. $\approx10^{13}$ Hz which is considerably greater than the frequencies considered in §7.3 (Fig. 7.25).

In seeking to describe the vibrations of the lattice at an atomic level we need to develop a language which (a) correctly describes the wave-like nature of the excitations in the lattice and (b) incorporates the quantum mechanical nature of the vibrations of the atoms. These requirements are resolved with the introduction of the concept of a *phonon*.

What is a phonon?

The concept of a phonon is analogous to the concept of a photon (see §2.3): the phonon describes the wave-like excitations of the displacements in a lattice; a photon describes the wave-like excitations of the electromagnetic field. The phonon concept allows us to describe realistically the wave-like nature of the vibrations in a lattice and yet incorporates the quantum mechanical nature of the vibrating atoms. The key features of the phonon description of lattice vibrations are outlined below and in Figure 7.26.

Figure 7.24 Analogy to the situation of atoms in a crystal. The forces on an atom are such that the atom behaves as if it were held at its minimum-energy position by springs. (a) The simple model used in our initial analysis; (b) A more realistic representation of the situation within the solid. Motion of the central atom in (b) causes all the atoms around to move as well, creating a tremendously difficult situation to analyze.

(a)   **(b)**

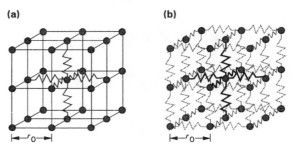

Figure 7.25 The relationship between frequency and wavelength, assuming a speed of sound of $4000\,\mathrm{m\,s^{-1}}$. For wavelengths less than an atomic spacing (typically $3\,10^{-10}\,\mathrm{m}$), the idea of a compression wave becomes meaningless, because there is nothing to be compressed! This estimates that the maximum frequency of atomic vibrations are $\approx 10^{13}\,\mathrm{Hz}$.

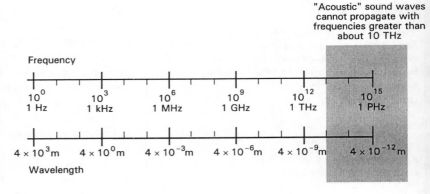

"Acoustic" sound waves cannot propagate with frequencies greater than about 10 THz

Frequency

10^0 1 Hz 10^3 1 kHz 10^6 1 MHz 10^9 1 GHz 10^{12} 1 THz 10^{15} 1 PHz

$4 \times 10^3\,\mathrm{m}$ $4 \times 10^0\,\mathrm{m}$ $4 \times 10^{-3}\,\mathrm{m}$ $4 \times 10^{-6}\,\mathrm{m}$ $4 \times 10^{-9}\,\mathrm{m}$ $4 \times 10^{-12}\,\mathrm{m}$

Wavelength

Quantized vibration amplitude

A displacement wave of frequency f in a lattice cannot have an arbitrary amplitude. Its amplitude may only increase in such a way that the energy associated with the wave is limited to one of a set of values

$$u = \left(n + \tfrac{1}{2}\right)hf \qquad (7.46)$$

where n may have the value 0, 1, 2, etc. If $n = 0$ then we say "there are no phonons in that mode of vibration". Otherwise we say "there are n phonons in that mode of vibration". Note the similarity of Equation 7.46 with Equation 7.35.

Boson nature

The terminology above allows us to treat phonons as particles with a boson nature. The "mode of vibration" represents a quantum state that is "occupied" by phonons. Importantly, this allows us to use the Bose–Einstein occupation function (Eq. 2.69) to predict the average occupancy of an individual mode of vibration as

$$f_{\mathrm{BE}}(u, T) = \bar{n}(u, T) = \frac{1}{\exp(u/k_{\mathrm{B}}T) - 1} \qquad (7.47)$$

where u is given by Equation 7.46. The chemical potential, μ, has been set equal to zero because phonons are not "conserved" – they are destroyed and created by putting energy into or out of the lattice.

No minimum phonon energy

There is no minimum phonon energy. Long wavelength waves have low frequencies, and hence hf in Equation 7.46 can take arbitrarily small values. This means there is no "energy gap" in the spectrum of possible vibrational energies of the atoms in a lattice. This is in distinct contrast with the Einstein model in which atoms were assumed to either vibrate at f_0, or not at all.

Figure 7.26 Illustration of the concept of a phonon. The filled circles represent the instantaneous positions of atoms. The empty circles represent the equilibrium positions of atoms. (a) A section of a one-dimensional crystal in which there are no phonons present. (b) There is one transverse phonon present in the lattice with a relatively long wavelength, λ_1, and a relatively low frequency, f_1. (c) There are two transverse phonons with the same wavelength, λ_1, and relatively low frequency, f_1, as in (b); note the increased amplitude of vibration as compared with (b). (d) There is one transverse phonon present in the lattice with a relatively short wavelength, λ_2, and a relatively high frequency, f_2. (e) There are two transverse phonons in the lattice: one with a relatively long wavelength, λ_1, and a relatively low frequency, f_1, and one with a relatively short wavelength, λ_2, and relatively high frequency, f_2, i.e. the sum of (b) and (d).

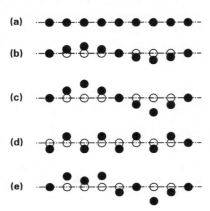

(a)

(b)

(c)

(d)

(e)

Maximum phonon energy

There is a maximum phonon energy. It makes no sense to imagine waves with a wavelength shorter than a lattice spacing i.e. $\lambda_{min} \approx 3 \times 10^{-10}$ m. Since there is a short wavelength limit to λ, there must be a consequent high frequency limit to f, and hence there is a maximum phonon energy, hf_{max}. Assuming that the speed of these very high frequency sound waves is the same as that of the more familiar sound waves, i.e. ≈ 4000 m s^{-1}, then we can use $c_{sound} = f\lambda$ to estimate the maximum phonon frequency as

$$f_{max} \approx 4000/3 \times 10^{-10} \approx 1.3 \times 10^{13} \text{ Hz.}$$

The Debye theory of the heat capacity of solids

The theory of the heat capacity of solids in which the internal energy of the lattice is held as phonons (quantized sound waves) is known as the Debye theory of the heat capacity. The theory is more fully developed in Appendix 5. In a similar manner to the Einstein model, the heat capacity is predicted in terms of a characteristic temperature, known in this case as the *Debye temperature*, θ_D. In the same way as the Einstein temperature is related to the frequency of vibration of the atoms, f_0, by Equation 7.37, so the Debye temperature is related to the maximum phonon frequency, f_D, by

$$\theta_D = \frac{hf_D}{k_B} = \frac{\hbar\omega_D}{k_B} \tag{7.48}$$

However, unlike the Einstein theory, the predictions

cannot be expressed by a closed-form formula such as Equation 7.45. Thus in Appendix 5, the predicted values for the heat capacity of solids are tabulated at $0.1\theta_D$, $0.2\theta_D$ etc.

Figure 7.27 shows the heat capacity of copper from Figure 7.17 plotted together with both the Debye and the Einstein predictions for the heat capacity. The Debye curve is based on $\theta_D = 310$ K (cf. $\theta_E = 230$ K) which has been chosen, by trial and error, to give good agreement across most of the temperature range. Both curves capture the trend of the data, but at low temperatures the Debye prediction falls off more slowly than the Einstein prediction. This is because in the Debye model (and in reality) there is a continuum of low-energy vibrational modes that can be excited at low temperatures.

Key features of the Debye theory

Debye theory predicts that at temperatures less than $\approx 0.1\theta_D$ (i.e. less than around 30 K for copper), C_V should vary as T^3 according to

$$C_V = \left(\frac{1944p}{\theta_D^3}\right) T^3 \text{ J K}^{-1} \text{ mol}^{-1} \tag{7.49}$$

where p is the number of atoms per chemical formula unit, and has the value 1 for elements. Equation 7.48 allows one to estimate θ_D by measuring the low-temperature heat capacity (see Fig. 7.31).

At $T = \theta_D$, the heat capacity is predicted to be around 95% of its classical value ($C_V = 3pR$) as predicted by the law of Dulong and Petit. Table 7.8 con-

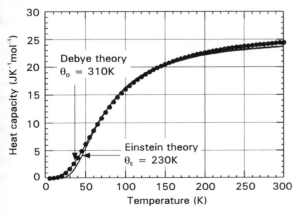

Figure 7.27 The Debye prediction for the heat capacity of copper compared with the Einstein prediction shown in Figure 7.23.

tains a selection of Debye temperatures for elements determined from their low-temperature heat capacity according to Equation 7.48. Figure 7.28 shows the predicted variation of the heat capacity for substances with Debye temperatures of 100 K, 300 K and 1000 K. We see that at room temperature, the heat capacity values will be close to $C_V = 3R$ if θ_D is of the order of room temperature or less. However, if θ_D is much greater than room temperature, then the heat capacity at room temperature will appear anomalously low. This is the origin of the anomalously low values in the histogram shown in Fig. 7.17.

Relationship between the speed of sound and θ_D

The Debye theory assumes that the thermal energy of the lattice is held in the form of high frequency sound waves. If this is so then we might expect to find a relationship between the Debye temperature of a substance and the speed of sound in that substance. Recalling the definition of θ_D (Eq. 7.48) as the temperature corresponding to maximum phonon frequency, and thus the minimum phonon wavelength, we can write

$$\theta_D = \frac{hf_D}{k_B} \approx \frac{h}{k_B \lambda_{min}} c_{sound} \qquad (7.50)$$

Since λ_{min} is of the order of a typical lattice spacing ($a \approx 3 \times 10^{-10}$ m), which is roughly the same for many elements, Equation 7.50 suggests that the Debye tem-

Table 7.8 The Debye temperatures, θ_D, of several elements, as determined by analysis of the T^3 behaviour of their low-temperature heat capacity (Eq. 7.49 and Fig 7.31).

Element	Z	θ_D (K)	Element	Z	θ_D (K)
Beryllium	4	1440	Zirconium	40	291
Magnesium	12	400	Molybdenum	42	450
Aluminium	13	428	Silver	47	225
Titanium	22	420	Cadmium	48	209
Vanadium	23	380	Tin	50	200
Chromium	24	630	Tantalum	73	240
Manganese	25	410	Tungsten	74	400
Iron	26	470	Platinum	78	240
Nickel	28	450	Gold	79	165
Copper	29	343	Lead	82	105
Zinc	30	327	Uranium	92	207

perature is roughly proportional to the speed of sound. Figure 7.29 shows a graph of the speed of transverse sound waves (Table 7.6) as a function of the Debye temperature (Table 7.8) and it is evident that there is a roughly linear relationship between the two.

I personally find Figure 7.29 fascinating. It indicates a relationship between a characteristic temperature determined from heat capacity measurements and the speed of sound determined from acoustic measurements. The fact that there is any relationship at all between the two quantities gives us confidence that our explanation of the heat capacity in terms of lattice waves (phonons) really is correct.

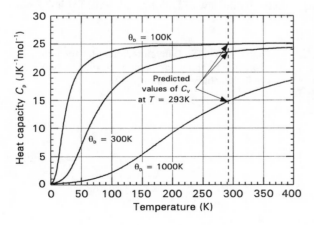

Figure 7.28 The Debye prediction for the heat capacity of solids with Debye temperatures of 100 K, 300 K and 1000 K. At room temperature the heat capacity values will be close to $C_V = 3R$ if θ_D is less than room temperature. However if θ_D is much greater than room temperature, then the heat capacity at room temperature appears anomalously low. This is the origin of the anomalously low values of C_P for beryllium, boron and carbon in Figure 7.17.

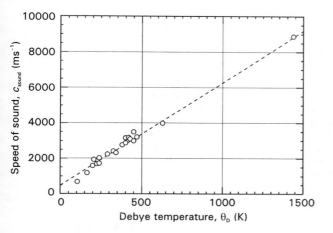

Figure 7.29 The speed of transverse sound waves in some elements (Table 7.6) plotted as a function of their Debye temperature (Table 7.8). It is clear that there is a roughly linear relationship between the two quantities. As a rule of thumb, the graph shows that, typically, the speed of transverse sound waves (in metres per second) is around a factor seven times the Debye temperature (in kelvin).

Metals

In the Debye theory of the heat capacity of solids, the thermal energy of a solid is considered to be tied up in vibrations of the atoms of the lattice in the form of lattice waves. However, as discussed in §6.5, within metals there is a high density of free electrons and this "gas" of electrons does not have a heat capacity of $3R/2$ per mole as one would expect. In fact, if we look only at high temperature data, the "electron gas" appears to have a negligibly small heat capacity. For example, C_p for copper at its Debye temperature of 343 K (70°C) is $\approx 25\,\mathrm{J\,K^{-1}\,mol^{-1}}$ which can all be plausibly understood (within the uncertainties of our study) in terms of the effect of lattice vibrations. This result is typical for metals and raises the question: what has happened to the heat capacity of the free electron "gas" that we suppose to exist within metals?

By now the answer may be becoming familiar to you: whenever the full heat capacity expected is not observed then we should look to see if some of the degrees of freedom of the system are not accessible. We saw this with diatomic, triatomic, and polyatomic molecules (§5.3.3), with atoms in solids at low temperature (§7.4) and now with electrons. For electrons we anticipate three degrees of freedom corresponding to the kinetic energy of motion in each of the x, y, and z directions. The question of what has happened to the heat capacity of the free electron "gas" may now be interpreted as the question: what has happened to these degrees of freedom?.

The answer lies in our consideration (§6.5) of the effect that quantum mechanics has on the formation of metals. We described the quantum states with the aid of a k-space graph, with the electrons occupying quantum states nearest the origin, and forming a "sphere" of occupied states (Fig. 7.30). This correctly describes the occupancy of quantum states at absolute zero. However, when the temperature is raised the picture is slightly altered because some electrons can accept energy and move into higher quantum states. However, because of the Pauli exclusion principle, most electrons are unable to accept thermal energy because there are no vacant quantum states into which they can move. Only electrons in quantum states with energies close to the Fermi energy, E_F, can accept energy, and (as we saw in §6.5.3) at room temperature $k_B T \ll E_F$.

With this model we can now understand the behaviour of the electron gas in a metal. If the gas had been a classical molecular gas, then the three degrees of freedom would have resulted in an internal energy, U, of

$$U = U_0 + N_A \times \tfrac{3}{2} k_B T = U_0 + \tfrac{3}{2} RT \; \mathrm{mol^{-1}} \quad (7.51)$$

and hence a molar heat capacity of $C_V = 3R/2$. However, the internal energy due to thermal excitations of electrons is only a small fraction of the classically expected value (Eq. 7.51). So we write

$$U = U_0 + \text{Number of electrons} \times \tfrac{3}{2} k_B T \; \mathrm{mol^{-1}}$$
$$(7.52)$$

185

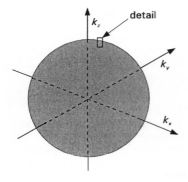

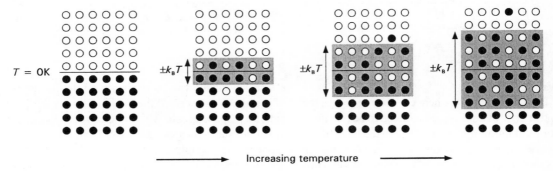

Figure 7.30 A close-up view of the occupation of quantum states in the region near the occupied quantum states with a maximum energy of E_F. Filled circles represent occupied quantum states, and empty circles represent empty quantum states. As the temperature increases, electrons in quantum states within a range $\approx \pm k_B T$ of E_F can change states. The Pauli exclusion principle prevents electrons in states much more than $\approx k_B T$ below the maximum energy from accepting energy, because there are no vacant states for these electrons to move into.

Increasing temperature

Although we do not know the precise value of the number of electrons in Equation 7.52, we can say that it should be:

- related to the number of available quantum states with energies in a small range close to E_F. This is directly proportional to the density of quantum states at the Fermi energy $g(E_F)$ (Eq. 6.73).
- proportional to $k_B T$. As shown in Figure 7.30, electrons occupying quantum states within a range $\approx k_B T$ of E_F are able to accept thermal energy.

We can thus rewrite Equation 7.51 as

$$U \approx U_0 + \left[g_m(E_F) k_B T \right] \times \tfrac{3}{2} k_B T \text{ mol}^{-1} \quad (7.53)$$

where $g_m(E_F) \Delta E$ is the number of quantum states in 1 mole of substance in the range ΔE around the Fermi energy. Hence the molar heat capacity is given by

$$C_V^{el} = \frac{dU}{dT} = 3 g_m(E_F) k_B^2 T \quad (7.54)$$

This is generally written as

$$C_V^{el} = \gamma T \quad (7.55)$$

where γ is the *coefficient of electronic heat capacity*:

$$\gamma = 3 g_m(E_F) k_B^2 \quad (7.56)$$

The rough theory outlined above predicts that the electronic heat capacity of a metal is much less than would be expected of a classical gas, and is proportional to the absolute temperature. We can test to see if this is a valid model by looking at the heat capacities of metals at low temperature.

We note that γ varies from one metal to another and depends on the number of (mainly occupied) quantum states just below the Fermi energy and the number of (mainly empty) quantum states just above the Fermi energy. Thus, determining the quantity γ by means of heat capacity measurements tells us how many quantum states there are with energies close to the Fermi energy. This quantity is just the density of electronic quantum states evaluated at the Fermi energy, i.e. $g(E_F)$. A more detailed theory of the heat capacity of metals predicts that γ is given by

Example 7.14

What is the expected value of the electronic contribution to the molar heat capacity of copper at 300 K?

We can estimate the electronic contribution to C by using Equation 7.57 and substituting for $g(E_F)$ from Equation 6.73:

$$g(E) = \frac{V\sqrt{2m^3 E}}{\pi^2 \hbar^3}$$

Equation 6.73 expresses the density of quantum states around energy E in a volume V of metal. We need first to substitute V for the molar volume V_m. Using the density data for copper (Table 7.2) (molar mass 63.55×10^{-3} kg) we find $V_m = 63.55 \times 10^{-3}/8933 = 7.11 \times 10^{-6}$ m^3. In Equation 6.87 we worked out that (according to the theory of a free electron gas) the Fermi energy in copper is 1.13×10^{-18} J (7.0 eV). Substituting these values into the equation for $g(E)$ yields

$$g_m(E_F) = \frac{7.11 \times 10^{-6}\sqrt{2 \times \left(9.1 \times 10^{-31}\right)^3 \times 1.13 \times 10^{-18}}}{\pi^2 \left(1.054 \times 10^{-34}\right)^3}$$

$$= \frac{7.11 \times 10^{-6}\sqrt{1.703 \times 10^{-108}}}{11.56 \times 10^{-102}}$$

$$= 8.026 \times 10^{41} \text{ states J}^{-1}\text{ mol}^{-1}$$

Substitution of this result into Eq. 7.57 yields

$$C_V^{el} = \frac{\pi^2}{3}\left(1.38 \times 10^{-23}\right)^2 \times 8.026 \times 10^{41} \times 300$$

$$= 0.151 \text{ J K}^{-1}\text{ mol}^{-1}$$

This is to be compared with the contribution due to the lattice of around 25 J K^{-1}mol^{-1}, i.e. the electronic contribution to the heat capacity is only around 0.6% of the total heat capacity.

$$\gamma = \frac{\pi^2}{3} k_B^2 g(E_F) \qquad (7.57)$$

which differs from the results obtained with Equation 7.56 by just a few per cent.

Low temperature heat capacity

Measurements of the heat capacity of materials at low temperatures provide a way of testing both the Debye theory of the heat capacity of solids and the theory of metallic heat capacity. In a metal at high temperatures the electronic heat capacity is much smaller than the lattice heat capacity, which makes it difficult to identify experimentally the small electronic component on top of the large lattice heat capacity. At low temperatures, i.e. less than about one-tenth of the Debye temperature, the lattice term is much smaller, and we have clear predictions for the behaviour of both the lattice and the electronic terms.

At $T \ll \theta_D$ we expect from Equation 7.48 that the

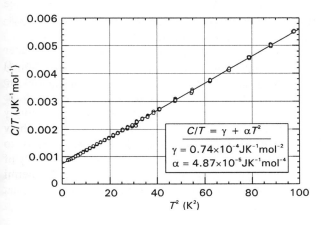

Figure 7.31 My own measurements of the low temperature heat capacity of copper. Note that, as predicted by Equation 7.58, the data conform to a straight line from the slope of which α may be determined and intercept of which γ may be determined. Note that this graph has T^2 as its x axis. The data shown therefore correspond to the temperatures between 2 K and 10 K.

$$C/T = \gamma + \alpha T^2$$
$$\gamma = 0.74 \times 10^{-4}\text{JK}^{-1}\text{mol}^{-2}$$
$$\alpha = 4.87 \times 10^{-5}\text{JK}^{-1}\text{mol}^{-4}$$

lattice heat capacity, $C_L = \alpha T^3$ and at all temperatures we expect that the electronic heat capacity $C_E = \gamma T$. Thus we predict the total heat capacity to be given by

$$C_{\text{Total}} = C_E + C_L = \gamma T + \alpha T^3 \qquad (7.58)$$

We can see clearly whether or not this is valid if we plot some data in the following form C_{Total}/T versus T^2. If the data conform to Equation 7.58, then they will form a straight line with slope α and intercept γ:

$$\left(\frac{C_{\text{Total}}}{T}\right) = \gamma + \alpha T^2 \qquad (7.59)$$

The fact that Figure 7.31 shows a straight line indicates that the heat capacity of copper does indeed conform to the theoretical expectation that $C_{\text{Total}} = C_E + C_L = \gamma T + \alpha T^3$.

7.5 Electrical properties

In this section we examine the response of solids to externally applied electric fields.

7.5.1 Introduction: conductors and insulators

When any substance is subject to an applied electric field, E, a current of electronic charge flows through the substance. The magnitude of the resultant current density, j, is characterized by the *electrical resistivity*, ρ, or the *electrical conductivity*, σ, of the substance. The two measures are the inverse of each other, $\rho = 1/\sigma$, and so for most purposes there is no advantage in using one measure or the other. The electrical resistivity and conductivity are defined by

$$j = \sigma E \qquad (7.60a)$$

and

$$E = \rho j \qquad (7.60b)$$

If the current density is measured in amperes per square metre ($A\,m^{-2}$) and the electric field in volts per metre ($V\,m^{-1}$), then the units of σ are $\Omega^{-1}m^{-1}$ or Sm^{-1}, where the symbol S stands for *seimen*, not to be confused with s for seconds. The units of resitivity are $S^{-1}m$, or more commonly, Ωm. For a sample of cross-sectional area A and length L, the resistivity is

---

Example 7.15

A piece of platinum wire is 1 m long and 1 mm in diameter. What is its electrical resistance at about room temperature?

Use Equation 7.61 ($R = \rho L/A$) and the data from Table 7.9, $\rho = 1.06 \times 10^{-7}\,\Omega m$, $L = 1\,m$, and

$$A = \pi\left(\frac{d}{2}\right)^2 = 3.142 \times \left(0.5 \times 10^{-3}\right)^2 = 7.854 \times 10^{-7}\,m^2$$

to work out that

$$R = \frac{1.06 \times 10^{-7} \times 1}{7.855 \times 10^{-7}} = 0.135\,\Omega$$

---

related to the *electrical resistance*, R, by

$$\rho = \frac{RA}{L}\,\Omega\,m \qquad (7.61)$$

The electrical conductivity and resistivity of the elements are listed in Table 7.9 and plotted as a function of atomic number in Figure 7.32.

Looking at Table 7.9, the data appear to vary almost randomly. However, in Figure 7.32 one can see that the resistivities of the elements fall into two rather distinct categories:

(a) Most elements are good conductors (i.e. metals) and have resistivities in the range $10^{-6} - 10^{-8}\,\Omega m$.

(b) The rest of the elements are considerably poorer electrical conductors. Note again the logarithmic scale of Figure 7.32.

A few elements have resistivities much greater than $1\,\Omega m$ ($\log(R) = 0$ in Figure 7.32) and are effectively *electrical insulators*. The elements Si, Ge, Se and Te have resistivities intermediate between metals and insulators and are known as *semiconductors*.

In the following sections we will examine in turn the electrical behaviour of metals and insulators, and then briefly discuss the properties of semiconductors.

7.5.2 Data on the electrical properties of metals

The most striking property of metallic conductors is their ability to conduct electricity in an arbitrarily small electric field. In other words there is no equiva-

Table 7.9 The electrical resistivity of the elements at around room temperature. Gases have been arbitrarily ascribed a resistivity of $10^{20}\,\Omega\,m$.

Z	Element		$\rho\ (\Omega m)$	$\sigma\ (Sm^{-1})$	Z	Element		$\rho\ (\Omega m)$	$\sigma\ (Sm^{-1})$
1	Hydrogen	H	10^{20}	10^{-20}	49	Indium	In	8.37×10^{-8}	1.1947×10^{7}
2	Helium	He	10^{20}	10^{-20}	50	Tin	Sn	1.1×10^{-7}	9.0909×10^{6}
3	Lithium	Li	8.55×10^{-8}	1.1696×10^{7}	51	Antimony	Sb	3.9×10^{-7}	2.5641×10^{6}
4	Beryllium	Be	4×10^{-8}	2.5×10^{7}	52	Tellurium	Te	0.00436	229.36
5	Boron	B	18000	5.5556×10^{5}	53	Iodine	I	1.37×10^{-7}	7.2993×10^{8}
6	Carbon (Diamond)	C	10^{11}	10^{-11}	54	Xenon	Xe	10^{20}	10^{-20}
7	Nitrogen	N	10^{20}	10^{-20}	55	Caesium	Cs	2×10^{-7}	5×10^{6}
8	Oxygen	O	10^{20}	10^{-20}	56	Barium	Ba	5×10^{-7}	2×10^{6}
9	Fluorine	F	10^{20}	10^{-20}	57	Lanthanum	La	5.7×10^{-7}	1.7544×10^{6}
10	Neon	Ne	10^{20}	10^{-20}	58	Cerium	Ce	7.3×10^{-7}	1.3699×10^{6}
11	Sodium	Na	4.2×10^{-8}	2.381×10^{7}	59	Praseodymium	Pr	6.8×10^{-7}	1.4706×10^{6}
12	Magnesium	Mg	4.38×10^{-8}	2.2831×10^{7}	60	Neodymium	Nd	6.4×10^{-7}	1.5625×10^{6}
13	Aluminium	Al	2.6548×10^{-8}	3.7668×10^{7}	61	Promethium	Pm	5×10^{-7}	2×10^{6}
14	Silicon	Si	0.001	1000	62	Samarium	Sm	9.4×10^{-7}	1.0638×10^{6}
15	Phosphorous	P	1×10^{-9}	1×10^{9}	63	Europium	Eu	9×10^{-7}	1.1111×10^{6}
16	Sulphur	S	2×10^{15}	5×10^{-16}	64	Gadolinium	Gd	1.34×10^{-6}	7.4627×10^{5}
17	Chlorine	Cl	10^{20}	10^{-20}	65	Terbium	Tb	1.14×10^{-6}	8.7719×10^{5}
18	Argon	A	10^{20}	10^{-20}	66	Dysprosium	Dy	5.7×10^{-7}	1.7544×10^{6}
19	Potassium	K	6.15×10^{-8}	1.626×10^{7}	67	Holmium	Ho	8.7×10^{-7}	1.1494×10^{6}
20	Calcium	Ca	3.43×10^{-8}	2.9155×10^{7}	68	Erbium	Er	8.7×10^{-7}	1.1494×10^{6}
21	Scandium	Sc	6.1×10^{-7}	1.6393×10^{6}	69	Thulium	Tm	7.9×10^{-7}	1.2658×10^{6}
22	Titanium	Ti	4.2×10^{-7}	2.381×10^{6}	70	Ytterbium	Yb	2.9×10^{-7}	3.4483×10^{6}
23	Vanadium	V	2.48×10^{-7}	4.0323×10^{6}	71	Lutetium	Lu	7.9×10^{-7}	1.2658×10^{6}
24	Chromium	Cr	1.27×10^{-7}	7.874×10^{6}	72	Hafnium	Hf	3.51×10^{-7}	2.849×10^{6}
25	Manganese	Mn	1.85×10^{-6}	5.4054×10^{5}	73	Tantalum	Ta	1.245×10^{-7}	8.0321×10^{6}
26	Iron	Fe	9.71×10^{-8}	1.0299×10^{7}	74	Tungsten	W	5.65×10^{-8}	1.7699×10^{7}
27	Cobalt	Co	6.24×10^{-8}	1.6026×10^{7}	75	Rhenium	Re	1.93×10^{-7}	5.1813×10^{6}
28	Nickel	Ni	6.84×10^{-8}	1.462×10^{7}	76	Osmium	Os	8.12×10^{-8}	1.2315×10^{7}
29	Copper	Cu	1.673×10^{-8}	5.9773×10^{7}	77	Iridium	Ir	5.3×10^{-8}	1.8868×10^{7}
30	Zinc	Zn	5.916×10^{-8}	1.6903×10^{7}	78	Platinum	Pt	1.06×10^{-7}	9.434×10^{6}
31	Gallium	Ga	2.7×10^{-7}	3.7037×10^{6}	79	Gold	Au	2.35×10^{-8}	4.2553×10^{7}
32	Germanium	Ge	0.46	2.1739	80	Mercury	Hg	9.41×10^{-7}	1.0627×10^{6}
33	Arsenic	As	2.6×10^{-7}	3.8462×10^{6}	81	Thallium	Tl	1.8×10^{-7}	5.5556×10^{6}
34	Selenium	Se	0.01	100	82	Lead	Pb	2.0648×10^{-7}	4.8431×10^{6}
35	Bromine	Br	10^{20}	10^{-20}	83	Bismuth	Bi	1.068×10^{-6}	9.3633×10^{5}
36	Krypton	Kr	10^{20}	10^{-20}	84	Polonium	Po	1.4×10^{-6}	7.1429×10^{5}
37	Rubidium	Rb	1.25×10^{-7}	8×10^{6}	85	Astatine	At	10^{20}	10^{-20}
38	Strontium	Sr	2.3×10^{-7}	4.3478×10^{6}	86	Radon	Rn	10^{20}	10^{-20}
39	Yttrium	Y	5.7×10^{-7}	1.7544×10^{6}	87	Francium	Fr	–	–
40	Zirconium	Zr	4.21×10^{-7}	2.3753×10^{6}	88	Radium	Ra	1×10^{-6}	1×10^{6}
41	Niobium	Nb	1.25×10^{-7}	8×10^{6}	89	Actinium	Ac	–	–
42	Molybdenum	Mo	5.2×10^{-8}	1.9231×10^{7}	90	Thorium	Th	1.3×10^{-7}	7.6923×10^{6}
43	Technetium	Tc	2.26×10^{-7}	4.4248×10^{6}	91	Protractinium	Pa	1.77×10^{-7}	5.6497×10^{6}
44	Ruthenium	Ru	7.6×10^{-8}	1.3158×10^{7}	92	Uranium	U	3.08×10^{-7}	3.2468×10^{6}
45	Rhodium	Rh	4.51×10^{-8}	2.2173×10^{7}	93	Neptunium	Np	1.22×10^{-6}	8.1967×10^{5}
46	Palladium	Pd	1.08×10^{-7}	9.2593×10^{6}	94	Plutonium	Pu	1.46×10^{-6}	6.8493×10^{5}
47	Silver	Ag	1.59×10^{-8}	6.2893×10^{7}	95	Americium	Am	6.8×10^{-7}	1.4706×10^{6}
48	Cadmium	Cd	6.83×10^{-8}	1.4641×10^{7}					

(Data from Kaye & Laby; see §1.4.1)

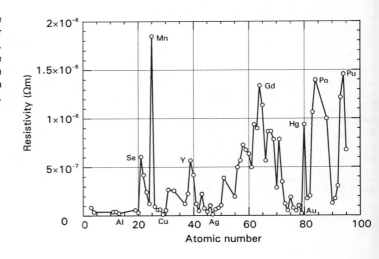

Figure 7.32 The logarithm of the resistivities of the elements plotted as a function of atomic number. The elemental gases are plotted as having a resistivity of $10^{20}\,\Omega$m. The data for metals is shown on an enlarged linear scale in Figure 7.33.

lent of the "breakdown" effect observed in electrical insulators (see §7.5.6).

Data on the resistivity of the elements

From Table 7.9, it is clear that most elements are metals, with resistivities at around room temperature of between $10^{-6}\,\Omega$m and $10^{-8}\,\Omega$m. Figure 7.33 shows detail from the lower part of Figure 7.32 on a linear scale. The data show significant non-random variations, but there are too many details to attempt an explanation of the conductivity of each element. The structure in the data is reminiscent of that in the density data for the elements shown in Figure 7.1, but the peaks and troughs in the data are not so regular.

If we take the best conductors from Table 7.9, we find that they are silver, copper and gold (the noble metals), closely followed by aluminium. These four elements are marked on Figure 7.33, and the temperature dependence of their resistivity is shown in Figure 7.34 along with that of platinum. The variation is broadly linear, with the resistivity tending towards small values at low temperatures.

Data on the resistivity of alloys

Alloys are mixtures of elemental metals but, as illustrated in Figure 7.35, the mixtures are at the atomic level. They are usually made by melting metallic elements together and mixing them while they are mol-

Figure 7.33 The resistivities of the metallic elements plotted on a linear scale as a function of atomic number. Elements with points off the scale of the graph have not been plotted. The data for all the elements are shown on a logarithmic scale in Figure 7.32.

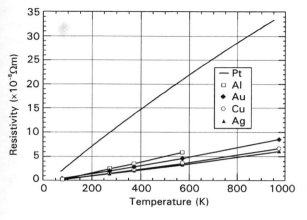

Figure 7.34 The resistivities of five metallic elements plotted on a linear scale as a function of temperature. The data for platinum are the result of many measurements at closely spaced temperatures and are plotted as a continuous curve. The data for aluminium, copper, silver and gold consist of just a few points and the lines drawn merely connect the data points.

ten. The electrical resistivity of three alloys is compared with the electrical resistivity of their constituent elements in Table 7.10. Note that in all cases the resistivity of the alloy is considerably greater than the resistivity of either component element. This is particularly striking for the Pt(10% Ir), Pt(10% Rh) alloys; replacing 1 in 10 platinum atoms with an iridium atom causes the resistivity to more than double. This is despite the fact that iridium has a lower resistivity than platinum.

Example 7.16

Consider two wires of equal length and circular cross-section, one made from aluminium and one from copper. What will be the ratio of diameters of the wires if the resistance of the wires is equal?

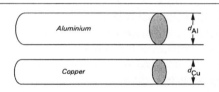

For a wire of length L and cross-sectional area $A = \pi d^2/4$ the electrical resistance is given by

$$R = \frac{\rho L}{\pi d^2/4}$$

If the two wires have equal resistance ($R_{Cu} = R_{Al}$) and length then

$$\frac{\rho_{Cu} L}{\pi d_{Cu}^2/4} = \frac{\rho_{Al} L}{\pi d_{Al}^2/4}$$

which simplifies to

$$\left(\frac{d_{Al}}{d_{Cu}}\right)^2 = \frac{\rho_{Al}}{\rho_{Cu}}$$

or

$$\frac{d_{Al}}{d_{Cu}} = \sqrt{\frac{\rho_{Al}}{\rho_{Cu}}}$$

Substituting $\rho_{Al} = 2.655\times10^{-8}$ and $\rho_{Cu} = 1.673\times10^{-8}$ from Table 7.19 we find

$$\frac{d_{Al}}{d_{Cu}} = \sqrt{\frac{2.655\times10^{-8}}{1.673\times10^{-8}}} = 1.26$$

and so the diameter of aluminium wire would have to be about 26% greater than the copper cable.

The volume of aluminium used would thus be greater in the ratio of = 1.59. However, the densities of the aluminium (2698 kg m^{-3}) and copper (8933 kg m^{-3}) are such that the aluminium cable would be lighter in the ratio

$$\frac{\text{density}\,(Al)}{\text{density}\,(Cu)} \times \left(\frac{d_{Al}}{d_{Cu}}\right)^2 = \frac{2698}{8933} \times 1.59 = 0.49$$

In other words, an aluminium cable of similar electrical resistance to a copper cable would be slightly larger, but would weigh only around half as much. In some circumstances this proves to be an important engineering advantage.

191

(a)   **(b)**

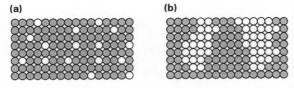

Figure 7.35 Illustration of the difference between (a) a random binary alloy and (b) a mixture of two elements.

Table 7.10 The resistivities (Ωm) of three alloys (centre) and their component elements (left and right) at around room temperature.

Component 1	Alloy	Component 2
Cu	Cu(Zn)	Zn
1.55×10^{-8}	6.3×10^{-8}	5.5×10^{-8}
Pt	Pt(10% Ir)	Ir
9.81×10^{-8}	24.8×10^{-8}	4.7×10^{-8}
Pt	Pt(10% Rh)	Rh
9.81×10^{-8}	18.7×10^{-8}	4.3×10^{-8}

(Data from Kaye & Laby; see §1.4.1)

Superconductivity

Many metals and alloys when cooled to temperatures within a few degrees of absolute zero display an extraordinary set of properties, including the ability to conduct electricity with no detectable resistivity. This phenomenon – known as *superconductivity* – occurs only below a certain critical temperature, T_C. The critical temperatures of a variety of elemental superconductors and alloys are shown in Table 7.11.

Table 7.11 Examples of substances that display superconducting behaviour below the transition temperature T_C, shown.

Substance	T_C (K)
Low-temperature superconductors: elements	
Aluminium	1.75
Lead	7.2
Niobium	9.25
Tin	30.5
Vanadium	5.4
Low-temperature superconductors: alloys	
V_3Si	17.1
Nb_3Sn	18.3
High-temperature superconductors	
$YBa_2Cu_3O_{7-\delta}$	92

(Data from Kaye & Laby; see §1.4.1)

Interestingly, the three best elemental conductors (copper, silver and gold) do not become superconductors (or, if they do, their transition temperatures are below a few microkelvin). Furthermore, it is interesting to note that the conductivity of a substance in the superconducting state, while not known to be strictly infinite, is better than that of pure copper at 4 K by a factor in excess of 10^{15}. In other words, the conductivity in the superconducting state is better than the conductivity in the metallic state by a factor similar to that by which the conductivity in the metallic state is better than the conductivity of some insulators (see Table 7.13).

Summary

So the main questions raised by our preliminary examination of the experimental data on the electrical resistivities of metals are:

- why is there no "breakdown field" for metals?
- why are most elements metals?
- why do the resistivities of elements vary roughly linearly with temperature?
- why is the resistivity of an alloy larger than the resistivities of its component metals?
- why do some metals become superconducting at low temperatures?

7.5.3 Understanding the electrical properties of metals

To answer these questions we need to develop further the quantum mechanical description of electrons that we began in §6.5.

Quantum theory of conduction in metals

The basis of the quantum mechanical approach to the problem is illustrated in Figure 7.36. We consider that the primary effect of the electric field is to alter the balance of occupancy between travelling-wave states with velocities parallel, or anti-parallel, to the applied electric field, **E**.

When no electric field is applied (Fig. 7.36a), electrons occupy the travelling-wave quantum states inside the Fermi sphere on a k-space graph. Outside the sphere, the travelling-wave states are unoccupied. Note that since electrons do not, on average, leave the

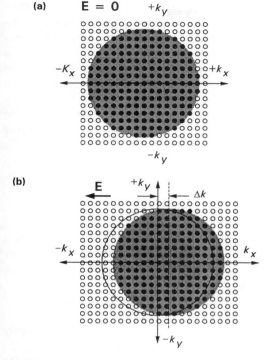

Figure 7.36 Illustration of the effect of an electric field on the occupation of quantum states in a simple model of a metal. Each small circle represents an allowed travelling-wave solution to the Shrödinger equation. Filled circles represent occupied states and empty circles represent unoccupied states. (a) The case when no electric field is applied; and (b) the case when an electric field is applied in the negative-x direction. Note that in (b) the distribution is not symmetric but is shifted by an amount Δk and so has a net excess of electrons with positive k vectors.

metal, the average electron velocity must be zero.

When an electric field is applied, electrons occupying quantum states with energies near the Fermi energy are able to make transitions into unoccupied quantum states just above the Fermi energy. The net effect of this is to change the balance of occupied and unoccupied states, as shown in Figure 7.36b.

Scattering

Inside a metal, electrons undergo several processes that are grouped together under the term "scattering". Scattering in this context is a process whereby an electron in one quantum state, say k_1, makes a transition to a quantum state k_2 where the factors that deter-

mine the relationship between k_1 and k_2 have a large random element. Scattering occurs when the quantum state occupied by an electron is no longer the "appropriate" quantum state to occupy. For example:

(a) *Electron–phonon scattering.* Suppose an electron is in a quantum state k_1 appropriate to a region of the crystal with electron density n. Suppose also that a lattice distortion (a phonon) temporarily increases or decreases the lattice spacing and hence the equilibrium number density of electrons. Such a situation might cause an electron to make a transition from k_1 to another quantum state k_2 which is "more appropriate" to the new situation. However, since the lattice vibrations are highly random, the new quantum state will, in general, only have a weak correlation with k_1. This process is described as electron–phonon scattering.

(b) *Electron–impurity scattering.* Suppose an electron is in a quantum state k_1 appropriate to a region of the crystal with number density n when it encounters a region of the crystal with an impurity. In general, the impurity will have a complicated distribution of electronic charge around it that may cause an electron to make a transition from k_1 to another quantum state k_2 which is "more appropriate" to the new situation. This process is described as electron–impurity scattering.

It is important to note that in the absence of either of the above processes electrons would stay in their quantum state indefinitely. Thus in a perfect crystal, with no phonons (i.e. at $T = 0\,\mathrm{K}$), and no impurities (and no surfaces, as in this context surfaces count as an electron scatterer), the conductivity would be infinite.

Working out the conductivity

According to the de Broglie hypothesis, an electron in a travelling-wave quantum state with wave vector k has momentum

$$|\mathbf{p}| = \hbar k = \frac{h}{2\pi} \times \frac{2\pi}{\lambda} = \frac{h}{\lambda} \qquad (7.62)$$

and the effect of the applied electric field is to change this momentum in accord with Newton's second law

$$\mathbf{F} = \frac{d\mathbf{p}}{dt} \approx \frac{\Delta\mathbf{p}}{\Delta t} \qquad (7.63)$$

193

We consider that each electron can be accelerated under the action of the applied electric field for some time average, Δt, after which it is "scattered". We consider that each scattering event destroys the momentum that the electron has gained in the time Δt. (This is in fact not quite correct, but any other assumption leads to a considerably more complex theory. We note, therefore, that the value of Δt which we deduce from our theory may be slightly longer than the actual time between scattering events. This is because it may take several real scattering events to completely destroy the momentum acquired by an electron.) In between the scattering events, the electron changes its momentum by

$$\Delta \mathbf{p} = \mathbf{F}\Delta t \qquad (7.64)$$

and since the force on each electron is $q\mathbf{E}$, where q is the charge on the electron, we have

$$\Delta \mathbf{p} = q\mathbf{E}\Delta t \qquad (7.65)$$

Now a change in momentum $\Delta \mathbf{p}$ corresponds to a change in wave vector $\hbar \Delta \mathbf{k}$, so every electron in the metal will, on average, shift its wave vector by an amount $\Delta \mathbf{k}$ (Fig. 7.36) given by

$$\Delta \mathbf{k} = \frac{q\mathbf{E}\Delta t}{\hbar} \qquad (7.66)$$

Notice that Δk is in the opposite direction to $\mathbf{E}$ because the electron charge q is negative. The momentum shift $\Delta \mathbf{p}$ corresponds to a velocity shift $\Delta \mathbf{v}$ given by $\Delta \mathbf{v} = \Delta \mathbf{p}/m$, i.e.

$$\Delta \mathbf{v} = \frac{q\mathbf{E}\Delta t}{m} \qquad (7.67)$$

Notice that $\Delta \mathbf{v}$ is, on average, the same for *every* electron and is known as the *drift velocity*.

The drift velocity
In the absence of an applied electric field, we expect that the average velocity of electrons is zero, i.e

$$(\mathbf{E} = 0) \qquad \bar{\mathbf{v}} = \frac{\sum\limits_{i=1}^{N} v_i}{N} = 0 \qquad (7.68)$$

since in the absence of an applied electric field, on

average, as many electrons are travelling in one direction as another. Now when an electric field has been applied the average velocity is no longer zero, but just $\Delta \mathbf{v}$ from Equation 7.67, i.e.

$(\mathbf{E} \neq 0)$

$$\bar{\mathbf{v}} = \frac{\sum\limits_{i=1}^{N}(\mathbf{v}_i + \Delta \mathbf{v})}{N} = \frac{\sum\limits_{i=1}^{N}\mathbf{v}_i + \sum\limits_{i=1}^{N}\Delta \mathbf{v}}{N} \qquad (7.69)$$

$(\mathbf{E} \neq 0)$

$$\bar{\mathbf{v}} = 0 + \frac{\sum\limits_{i=1}^{N}\Delta \mathbf{v}}{N} = \frac{N\Delta \mathbf{v}}{N} = \Delta \mathbf{v} \qquad (7.70)$$

Current flow
If we consider a metal sample through which a current i is flowing (Fig. 7.37), we can calculate the relationship between the current density and the average speed of the charged particles carrying the current. If we consider a particular cross-section of area A then the total current that flows past the cross-section per second is, by the definition of current, just i coulomb. However, we can also see that, on average, all the charge within a volume $A\bar{v}$ will flow past the cross-section per second. If the number density of current carriers is n and each carrier has charge q, then the total charge in volume $A\bar{v}$ is just $nqA\bar{v}$.

We thus find $i = nqA\bar{v}$, and after substituting for $\bar{v}$

Figure 7.37 Electrons flowing through a piece of metal. In one second all the charge within a distance $\bar{v}$ "upstream" of a cross-section perpendicular to the current flow will pass the cross-section. In this case the volume of charge that will flow past the cross-sectional area is $A\bar{v}$. The amount of charge in this volume is $qnAv_{avg}$.

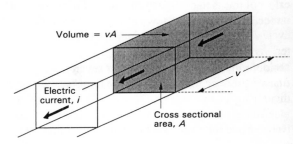

Volume = vA

Electric current, i

v

Cross sectional area, A

Example 7.17

Estimate the scattering time Δt for electrons in copper at room temperature.

From Table 7.2 we estimate a number density of electrons as $n = 8.417 \times 10^{28} \, \text{m}^{-3}$. Using Equation 7.73 we estimate Δt according to

$$\Delta t = \frac{m\sigma}{nq^2} = \frac{9.1 \times 10^{-31} \times 5.98 \times 10^7}{8.417 \times 10^{28} \times \left(1.60 \times 10^{-19}\right)^2}$$

$$\approx 2.5 \times 10^{-14} \, \text{s}$$

Where we have taken the value $\sigma = 5.98 \times 10^7$ siemens for copper at 20°C from Table 7.10.

from Equation 7.67 we have

$$i = nqA \times \frac{q\mathbf{E}\Delta t}{m} \qquad (7.71)$$

Rearranging this to give an expression for the current density, j, we find

$$j = \frac{i}{A} = \left(\frac{nq^2\Delta t}{m}\right)\mathbf{E} \qquad (7.72)$$

Comparing with Equation 7.60a we see that the quantity in brackets is the conductivity σ shown in Table 7.9. Hence

$$\sigma = \frac{nq^2\Delta t}{m} \qquad (7.73)$$

The scattering time Δt

We will now compare Equation 7.73 for σ with the experimental results. The quantities n, q and m are well known or can be fairly easily estimated. However, the quantity Δt, the average time taken to "scatter", is not easy to estimate. What we can do is to use measured values of conductivity to predict values for Δt, and then see whether we can understand the predicted values.

Thus according to this theory, at around room temperature electrons remain in their travelling-wave

quantum states for approximately 10^{-14} s to 10^{-13} s before being scattered. This might seem like a very short time, but in fact it is quite reasonable if one considers how we expect the electron to be scattered. In §7.2.2 we saw that the naturally occurring displacement waves in a lattice had frequencies of the order 10^{13} Hz. Thus 10^{-13} s is a typical time for a region of solid to be compressed and then expand. If it is indeed these atomic vibrations that scatter electrons then we would expect scattering to take place after the positions of the atoms have changed substantially, i.e. after some fraction of an oscillation time.

The questions raised by the data

Absence of a "breakdown field" for metals

Within this model we can understand the first question raised by the data: Why there is no "breakdown field" for metals? The reason for this lies in the extreme closeness in energy of the quantum states (Fig. 7.36): no matter in what direction an electric field is applied, electrons in occupied states can easily access vacant states and so change their momentum.

Commonness of metallic behaviour among elements

The question raised in §7.5.2 of why most elements are metals is what is known by lecturers as a "good question": it is very difficult to answer directly. As discussed in §6.5, the origin of metallic behaviour lies in the tendency for the outer (valence) electrons on an atom to lower their energy by spreading their wave function, thus allowing motion from one atom to another. This tendency is quite general and will occur unless electrons can lower their energy more effectively in some alternative way. As you will be aware, it is actually rather rare to come across elements naturally, and those one does find are rarely metallic. Normally, elements form chemical compounds in which the outer electron of one element finds a low energy quantum state to occupy on another atom. The outer electrons are then bound to one atom or the other – or tied up in covalent or ionic bonds.

In elements, however, there is only one type of atom. Hence electron transfer does not occur because the electron will merely find itself in an identical situation on the neighbouring atom. Thus in elements, the formation of delocalized states is one of a limited number of strategies for reducing electron energies. If we chemically combine metallic elements with the common insulating elements, e.g. oxygen or nitrogen, then in general the metallicity is destroyed as the valence electrons find low-energy, localized quantum states.

The temperature dependence of the resistivity

When an electron scatters from a lattice distortion as described above, the lattice gives up momentum $\Delta \mathbf{p}$ and energy ΔE to the electron. However, as we saw in §7.4.2, the lattice can only lose energy in quantized amounts. In order to scatter an electron, the amplitude of vibration of a displacement wave with frequency f such that $\Delta E = hf$ must be reduced such that the energy tied up in the wave falls from $(n_{ph} + \frac{1}{2})hf$ to $(n_{ph} - 1 + \frac{1}{2})hf$. Clearly this process cannot take place if $n_{ph} < 1$. Furthermore, it seems plausible that the process becomes more likely as the number of phonons, n_{ph}, associated with that mode of vibration increases. Thus at high temperatures, where there are many phonons associated with each mode of vibration of the lattice, electrons are more likely to be scattered, and the "scattering time", Δt in Equation 7.73, is reduced.

The average number of phonons associated with a displacement wave of frequency f is given by the Bose–Einstein occupation factor as

$$n_{ph} = \frac{1}{\exp(hf/k_B T) - 1} \qquad (7.47)^*$$

If the temperature is such that $k_B T \gg hf$ then we may expand the denominator of Eq. 7.47 to yield

$$n_{ph} \approx \frac{1}{\left[1 + (hf/k_B T) + \ldots\right] - 1} \approx \frac{k_B T}{hf} \qquad (7.74)$$

This indicates that at high temperatures, the number of phonons in any mode of vibration is proportional to the absolute temperature. Following on from the arguments above, we would therefore expect the "scattering time", Δt, to decrease as $1/T$, and hence the resistivity to increase as T. This is indeed broadly what happens (Fig. 7.34).

The argument above is certainly correct at temperatures such that $k_B T$ is greater than the maximum

phonon energy, i.e. when $T > \theta_D$. However, Figure 7.34 indicates that the linear behaviour actually continues well below θ_D. This indicates that the approximation in Equation 7.74 remains valid even below θ_D. This implies that the phonons that are actually giving rise to most of the resistivity of these metals have frequencies well below the maximum possible phonon frequency.

Alloys

The resistivity of an alloy is larger than the resistivities of its component elements. For example, data for an alloy of platinum and its neighbour in the periodic table, iridium, extracted from Table 7.10 is shown below:

These data are difficult to understand at first.

Pt	Pt(10% Ir)	Ir
$9.81 \times 10^{-8}\,\Omega\,\mathrm{m}$	$24.8 \times 10^{-8}\,\Omega\,\mathrm{m}$	$4.7 \times 10^{-8}\,\Omega\,\mathrm{m}$

Surely adding 10% of "low resistivity iridium" to "higher resistivity platinum" will lower the overall resistivity? The answer to this question is "No", but in order to understand it we have to think again about the processes by which electron waves are scattered.

We recall that a low resistivity is achieved by having a perfect lattice, and it does not matter by what process this perfect lattice is disrupted – lattice waves or impurities of any kind. Any disruption will cause an increase in resistivity. The appearance of a plane of atoms in an alloy of two elements A (e.g. platinum) and B (e.g. iridium) is illustrated in Figure 7.38.

The resistivity of pure A or pure B arises entirely from lattice waves (phonons) within the material, as discussed previously. In addition to this, the resistivity of alloys has scattering contributions from each of the impurities. For dilute impurities it is fairly easy to identify which is the "pure" substance and which is the "impurity". However, for non-dilute alloys estimates of the resistivity are extremely difficult. It is worth noting that the resistivity data for Pt(Ir) is evaluated at 20°C and thus consists of contributions both from impurities and from lattice vibrations.

In Example 7.18 we consider the data given in Table 7.10 in more detail to see if we can understand roughly the magnitudes of the increase in resistivity.

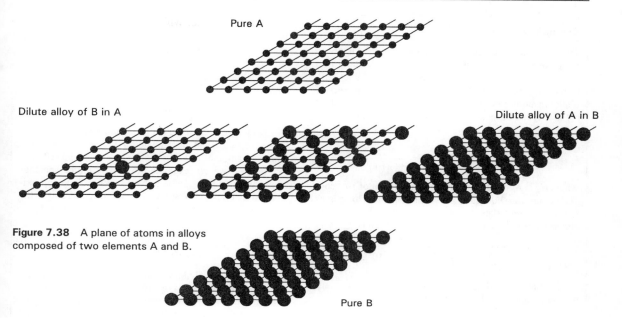

Pure A

Dilute alloy of B in A

Dilute alloy of A in B

Figure 7.38 A plane of atoms in alloys composed of two elements A and B.

Pure B

Superconductors

The phenomenon of superconductivity is too complex for us to attempt even the most simple analysis of the data given in Table 7.11. However, the phenomenon is so fascinating that a few words must be written on the subject. The key feature of all theories of superconductivity is that the electron gas described in §6.5 undergoes a kind of "condensation" at low temperatures. The electron "fluid" that "condenses out" from the electron gas has the unusual properties that are characteristic of the superconducting state.

In the transition of a normal *molecular* gas to a *molecular* liquid, the cause of the condensation is a weak attractive interaction between molecules. Similarly, a weak attraction between electrons is the origin of condensation into the superconducting state. However, for electrons in a metal it is not obvious what the origin of this weak attractive interaction between electrons might be. In fact, so elusive was the source of this interaction, that it took scientists around 40 years to work out what it was! In low-temperature superconductors some electrons within a metal can interact with other electrons by distorting the lattice of ions – the lattice *mediates* an attractive interaction between electrons. In technical terms the electron– electron attraction is actually an *electron–phonon– electron* interaction. Thus metals with a strong electron–

phonon interaction and hence a high resistivity are more likely to become superconducting. This is why the best conductors (Cu, Ag, Au) – whose electrons are least cattered by phonons – do not become superconducting.

7.5.4 Data on insulators in weak electric fields

Conduction and polarization

The simplest electrical property of electrical insulators (or *dielectrics*) is that for small electric fields they are poor electrical conductors! Neglecting the tiny electric current that flows in response to the applied electric field, the main effect of the electric field is to *polarize* the material. This weakens applied electric fields within the material to an extent measured by the *relative dielectric permitivity* or *dielectric constant*, ε, of the material (§2.3).

From Table 7.12 it seems that typical values of ε are in the range 2–10 for a variety of common solids. This is around 1000 times greater than typical values for gases (see §5.6.1), but of a similar order to that found for liquids (see §9.8.3). However, materials such as strontium titanate and strontium zirconate show dramatically enhanced dielectric constants, i.e.

197

Example 7.18

The discussion of the resistivity of alloys of Pt(Ir) has been rather qualitative. Let us see if we can understand in a more quantitative way the results given in Table 7.10. Consider pure platinum: we can estimate the number of platinum atoms per cubic meter from the atomic weight ($M = 195.1\,u$) and the density ($\rho = 21450\,\mathrm{kg\,m^{-3}}$; Table 7.2) by using

$$\text{Atomic number density} = \frac{N_A\rho}{M \times 10^{-3}}$$

where N_A is Avogado's number and the factor of 10^{-3} converts grams into kilograms. Assuming platinum to be in its normal divalent state, the number density of conduction electrons, n, will be just twice this amount. Thus

$$n = \frac{2 \times 6.022 \times 10^{23} \times 21450}{195.1 \times 10^{-3}} = 1.32 \times 10^{29}\,\mathrm{m^{-3}}$$

We can now estimate the scattering time in pure platinum at around 20°C by using Example 7.17:

$$\Delta t = \frac{9.1 \times 10^{-31}}{\rho \times 1.324 \times 10^{29} \times \left(1.6 \times 10^{-19}\right)^2}$$

and hence

$$\Delta t = \frac{9.1 \times 10^{-31}}{9.81 \times 10^{-8} \times 1.324 \times 10^{29} \times \left(1.6 \times 10^{-19}\right)^2} = 2.74 \times 10^{-15}\,\mathrm{s}$$

Let us now evaluate the distance that an electron at the Fermi surface of a metal will travel in a time Δt. This will be the equivalent of the *mean free path*, λ_{mfp}, that we mentioned when discussing the behaviour of gases (§4.4). We estimate the Fermi velocity, v_F, by first estimating the Fermi wave-vector k_F from Equation 6.80:

$$k_F = \left(3n\pi^2\right)^{\frac{1}{3}} = \left(3\pi^2 \times 1.324 \times 10^{29}\right)^{\frac{1}{3}}$$

$$= 1.577 \times 10^{10}\,\mathrm{m^{-1}}$$

and hence the v_F from Equation 6.86

$$v_F = \frac{\hbar k_F}{m} = \frac{1.054 \times 10^{-34} \times 1.577 \times 10^{10}}{9.1 \times 10^{-31}}$$

$$= 1.83 \times 10^6\,\mathrm{m\,s^{-1}}$$

Now we can combine our estimate for v_F with our estimate for Δt to yield an estimate for $\lambda_{\mathrm{mfp}} = v_F\Delta t$, which in this case amounts to $1.83 \times 10^6 \times 2.73 \times 10^{-15} = 4.99 \times 10^{-9}\,\mathrm{m}$

Finally, we can estimate how many lattice spacings this corresponds to. To do this we make a crude approximation that the crystal structure is simple cubic and so the atomic number density is just $1/a^3$. Hence a is given by

$$a = \left(\frac{1}{6.6 \times 10^{28}}\right)^{\frac{1}{3}} = 2.47 \times 10^{-10}\,\mathrm{m}$$

Thus for pure platinmum at 20°C an electron travels on average a distance of $4.99 \times 10^{-9}\,\mathrm{m}$ which corresponds to roughly $4.99 \times 10^{-9}/2.47 \times 10^{-10} = 20.2$ lattice spacings. Using similar techniques one can show that for Pt(10% Ir) the mean free path corresponds to approximately eight lattice spacings of the platinum lattice.

Is this a reasonable result? If we consider that electrons are scattered whenever they encounter an iridium atom, and that 1 in 10 platinum atoms has been randomly replaced by iridium, then we can estimate the number of lattice spacings an electron will travel before, on average, it will "hit" an impurity. Imagining that one starts at a platinum atom the chance that a neighbouring atom will be platinum is 0.9; similarly, the chance that the neighbouring two atoms will be platinum atoms is $0.9 \times 0.9 = 0.81$, and the chance of N atoms in a row being platinum is 0.9^{N-1}. If we estimate the value of n that gives a 50% chance of striking an iridium atom we find that N lies between 7 and 8, which is in rough agreement with the experimental value.

Example 7.19

A capacitor is made from two plates of metal of area 1 cm² separated by 0.1 mm. Calculate the capacitance if the intervening material is (a) air, (b) PTFE or (c) strontium titanate.

The capacitance of a parallel plate capacitor is given by

$$C = \frac{\varepsilon\varepsilon_0 A}{L}$$

which for the capacitor in question is

$$C = \varepsilon\frac{8.854 \times 10^{-12} \times 10^{-4}}{10^{-4}} = 8.854 \times 10^{-12}\,\mathrm{F}$$

$$C = 8.854\varepsilon\ \mathrm{pF}$$

From Table 5.16 we find $\varepsilon(\mathrm{air}) \approx 1.000$ and from Table 7.13 we find $\varepsilon(\mathrm{PTFE}) \approx 2.1$ and $\varepsilon(\mathrm{SrTiO_3}) \approx 200$. Substituting for ε yields
(a) $C = 8.86\,\mathrm{pF}$ \qquad Air
(b) $C = 18.6\,\mathrm{pF}$ \qquad PTFE
(c) $C = 1771\,\mathrm{pF}$ \qquad Strontium titanate

Table 7.12 The relative dielectric permittivity, ε, of various insulators (and semiconductors). The relative permittivity of vacuum is exactly 1. All measurements refer to 20°C, but are insensitive to small changes (≈ ±10°C) around this temperature.

Material		ε
Elements		
Silicon	Si	11.9
Germanium	Ge	16.0
Ceramics		
Alumina	Al_2O_3	8.5
Strontium titanate	$SrTiO_3$	200
Strontium zirconate	$SrZrO_3$	38
Glass		
Quartz	SiO_2	4.5
Borosilicate glass	SiO_2 with BO	4–5
Lead glass	SiO_2 with PbO	7
Plastics		
Polyethylene		2.3
Polystyrene		2.6
Polytetrafluoroethylene	PTFE	2.1
Polyamide	Nylon	3–4

(Data from Kaye & Laby; see §1.4.1)

an applied electric field has an enormous effect on the distribution of electric charge within the material.

The main questions raised by our preliminary examination of the experimental data on the electrical properties of insulators are:

- Why is the difference of the dielectric constant from unity (ε−1) in solids typically 1000 times greater than for gases?
- Why is the dielectric constant of some solids is anomalously large?

7.5.5 Understanding the behaviour of insulators in weak electric fields

In this section we consider the data on the properties of insulators under weak *static* electric fields. The response to rapidly varying electric fields is covered in §7.7 on the optical properties of insulators.

Comparison with gases

In the discussion of the electrical properties of gases (§5.6) we saw that in order to understand the dielectric constants of gases we had to divide gas

molecules into two classes: those with a permanent "built-in" electric dipole moment, called *polar* molecules, and those without any "built-in" electric dipole moment, called *non-polar* molecules. In an applied electric field both types of molecule acquire *induced* electric dipole moments. In addition to this, polar molecules also experience a torque that allows them to rotate in order to line up with the applied electric field. This additional mechanism usually gives polar molecules a significantly stronger response to applied fields than non-polar molecules.

In a solid, the mechanism of molecular rotation is, in general, not possible: atoms and molecules within a solid are usually unable to rotate because of the constraints of the bonds to neighbouring atoms. Bearing this in mind we will make our first attempt to understand the behaviour of the dielectric constant of solids by considering a solid to be the limit of a very dense gas, in which the molecules cannot rotate. Clearly this is not a sophisticated model of a solid, but as we shall see it is able to explain the first question raised by the data: Why is the difference of the dielectric constant from unity (ε−1) in solids typically 1000 times greater than for gases?

Equation 5.97 predicts that for a non-polar gas

$$\varepsilon - 1 = \frac{n\alpha}{\varepsilon_0} \qquad (7.75)$$

where n is the number density and α is the polarizability of a single molecule. For gases we found that values of $10^4(\varepsilon-1)$ were between 1 and 10 for non-polar molecules (Table 5.16) at STP. The number density of gas molecules at STP is given by (Eq. 4.49)

$$n = \frac{P}{k_B T} \qquad (7.76)$$

which evaluates to

$$n = \frac{1.013 \times 10^5}{1.38 \times 10^{-23} \times 273.15} = 2.68 \times 10^{25} \text{ molecules m}^{-3} \qquad (7.77)$$

For solids the number densities are closer to 10^{29} m^{-3} and so, considering a solid as dense gas, we expect to find the dielctric constants enhanced by a factor of approximately $10^{29}/2.68\times10^{25}\approx3700$. Since the values of $\varepsilon-1$ for gases are between 10^{-3} and 10^{-4}, our predicted range of values of $\varepsilon-1$ is between 3.7 and 0.37. Comparison with the data in Table 7.12 shows that this is indeed the correct order of magnitude for the dielectric constants of most insulators.

Having understood the order of magnitude of $\varepsilon-1$ as being due to the number density change between a solid and a gas, we can now interpret differences in ε between different solids as being due to differences in either the number density of the atoms in the solid or α the "molecular polarizability" of a solid. Note that for solids other than molecular solids one can no longer identify individual molecules within the solid and so the molecular polarizability refers to the polarizability of a chemical formula unit of the solid. However, the molecular polarizability of a solid still depends on the same factors that determine the polarizability of molecules.

Anomalously large values of ε

The remaining question to be answered from §7.3.4 is why the dielectric constant of some solids is anomalously large: for example strontium titanate ($SrTiO_3$) has a value of $\varepsilon-1$ of ≈200. Following Equation 7.75 we see that, since n cannot increase by a factor of the order of 20 from one material to another, the origin of the enhanced value of $\varepsilon-1$ must lie in an enhanced "molecular" polarizability, α.

The origin of the enhancement of the polarizability is but one of the extraordinary properties of $SrTiO_3$. The atoms in $SrTiO_3$ are both covalently and ionically bonded and can arrange themselves in two different crystal structures that have nearly the same cohesive energy. One arrangement is favoured at high temperatures and the other at low temperatures, with the crossover between the two structures being at around 125 K. Recall that in the presence of an applied electric field all the positive ions move one way and all the negative ions move in the opposite direction. Because of the delicate balance between the two crystal structures, an applied electric field can cause the ions within the crystal structure to move anomalously large distances and hence develop an anomalously large electric dipole moment.

7.5.6 Data on insulators in strong electric fields

Resistivity

As mentioned in the previous section, insulators are poor conductors of electricity, but their conductivity, although small, is still finite (Table 7.13).

Table 7.13 Typical orders of magnitude for the resistivity of some insulating substances at around room temperature. The data correspond to values of ρ determined 1 min after the electric field is applied.

Insulator	ρ (Ωm)	Insulator	ρ (Ωm)
Alumina , Al_2O_3	10^9–10^{12}	Paper	$\approx10^{10}$
Quartz, SiO_2	$\approx10^{16}$	PTFE	10^{15}–10^{19}
Diamond, C	10^{10}–10^{11}	Polystyrene	10^{15}–10^{19}
Boron, B	10^{10}–10^{11}	Varnish	10^7
Iodine, I_2	10^{13}	Soil	10^2–10^4
Glass	10^9–10^{12}	Distilled water	10^2–10^5

(Data from Kaye & Laby; see §1.4.1)

The measured resistivity of an insulator depends strongly on its purity and the temperature at which the measurement is made. In general, increasing the temperature lowers the resistivity dramatically. The measured resistivity also depends on how long after applying the electric field one waits before making the measurement, and whether the surface of the experimental sample is rough or smooth. Given these qualifications, the data in Table 7.13 should be considered only as an indication of the typical order of magnitude of the result to be expected.

Dielectric strength

If the strength of the electric field used to determine the resistivity of an insulator is increased, then eventually the current rises dramatically – often catastrophically destroying the substance being measured. This phenomenon is known as *electrical breakdown*. The value of electric field at which breakdown occurs – known as the *dielectric strength* of a substance – is of the order $10\,MV\,m^{-1}$, i.e. a voltage of $\approx10\,kV$ across a 1 mm piece of material could cause a spark to travel through the material. For comparison, the dielectric strength of air at 25°C and normal atmospheric pressure (0.1013 MPa) is $3.13\times10^6\,V\,m^{-1}$ and so solid insulators have dielectric strengths that are of a similar

Example 7.20

An electric field of 1 kV is applied across the two faces of a sheet of glass that is 1 mm thick and of cross-sectional area 1 cm×1 cm.

(a) Estimate the current that flows across the sheet 1 minute after the electric field is applied.
(b) How many electrons per second does this correspond to?
(c) Estimate the number density of electrons able to conduct electricity.

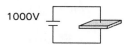

1000V

(a) From Table 7.13, the expected resistivity is 10^9–$10^{12}\,\Omega\,\text{m}$. To estimate the resistance of the glass we use

$$R = \frac{\rho L}{A}$$

with $L = 10^{-3}\,\text{m}$ and $A = 10^{-4}\,\text{m}^2$. Substituting $\rho = 10^9$ we find

$$R = \frac{10^9 \times 10^{-3}}{10^{-4}} = 10^{10}$$

and so we expect a resistance of between 10^{10} and $10^{13}\,\Omega$. For a voltage of 1 kV Ohm's Law tells us that

$$I = \frac{V}{R} = \frac{10^3}{10^{10}} = 10^{-7}\,\text{A}$$

and so we expect a current of between 0.1 nA and 0.1 μA.

(b) A current of I ampere corresponds to $I/(1.6\times10^{-19})$ electrons per second that in this case evaluates to between $\approx 10^9$ and 10^{12} electrons per second.

(c) We use the general formula (Eq. 7.73)

$$\sigma = \frac{ne^2\Delta t}{m}$$

to evaluate n, the charge carrier density. In this formula, m and e are the mass and charge of the electron, and Δt is the time over which a carrier is free to be accelerated by the electric field. Estimating Δt to have a similar value to that found in metals, and substituting in orders of magnitude only with $\sigma = 1/\rho \approx 10^{-9}\,\text{S}$ we have

$$n = \frac{\sigma m}{e^2 \Delta t} \approx \frac{10^{-9} \times 10^{-30}}{10^{-19} \times 10^{-19} \times 10^{-14}} \approx 10^{13}\,\text{m}^3$$

This result may be compared with the number density of atoms of $\approx 10^{29}\,\text{m}^3$.

order of magnitude – although actually slightly greater – than those found for gases at around room temperature and pressure (Table 7.14).

So the main questions raised by our preliminary examination of the experimental data on the electrical properties of insulators are:
- What processes cause conduction in insulators?
- Why can the dielectric strengths of solid insulators exceed those of gases at around room temperature and pressure by a factor as great as 10?

Table 7.14 Typical values of the dielectric strength in some insulating substances.

Insulator		$\times 10^6\,\text{V}\,\text{m}^{-1}$
Alumina	Al_2O_3	10–35
Sapphire	Al_2O_3	17
Quartz	SiO_2	25–40
Beryllia	BeO	10–14

(Data from Halliday, Resnick & Walker, *Fundamentals of physics* (Chichester: John Wiley.))

Example 7.21

The voltage supplied to our domestic appliances has a root-mean-square (RMS) value of typically 230 V AC. Some wires in a heater element are encased in alumina. Using the data in Table 7.14, what is the thinnest film of alumina that can be used to encase the wires and yet keep the electric field at a factor of 10 below the breakdown field of alumina?

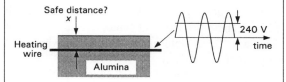

Assuming that the voltage on the heating wire oscillates sinusoidally (a good approximation), the peak value of the voltage on the heating wire is $\sqrt{2}\times230 = 325\,\text{V}$. Thus the electric field between the wire and the surface of the heater block is of the order of $(325/x)\,\text{V}\,\text{m}^{-1}$. From Table 7.14, we see that the dielectric strength of alumina varies between 10 and 35 MV m^{-1}. Applying a safety factor of 10 to the lower value, we require that the actual electric field be less than 1 MV m^{-1}, i.e.

$$\frac{325}{x} \leq 10^6$$

$$x \approx \frac{325}{10^6} = 325 \times 10^{-6}\,\text{m} = 0.325\,\text{mm}$$

7.5.7 Understanding the data on insulators in strong electric fields

The conduction process in insulators

In insulators valence electrons are bound to their "parent" atoms and cannot, in general, move in response to an applied electric field. In order to free the electrons to move through the crystal, one must supply some energy, ΔE, to move an electron from its quantum state around one atom into a vacant quantum state around a neighbouring atom. In metals, ΔE is infinitesimally small, but in an insulator this is usually a relatively large energy – typically several electron-volts. For comparison, at room temperature the typical vibrational energy of an atom is $k_B T$, which is around 1/40 of an electron-volt. The probability of an electron spontaneously moving from its orbital to an unoccupied orbital on a neighbouring atom therefore contains a Boltzmann-factor term (see §2.5)

$$\text{Probability} \approx \text{factor} \times \exp\left(\frac{-\Delta E}{k_B T}\right) \quad (7.78)$$

For a typical value of $\Delta E = 3\,\text{eV}$, the exponential Boltzmann factor evaluates to $\approx \exp(-120) \approx 10^{-52}$. This makes the process of excitation to a quantum state in which electrical conduction is possible extremely unlikely. However, unlikely as it is, it still occurs. There is a low, but finite, rate at which electrons are "freed" into quantum states in which they can conduct electricity.

However a "free electron" in an insulator will leave an uncompensated positive charge behind, and will relatively quickly return to a localized quantum state similar to that from which it derived. In other words, carriers are not only created relatively rarely but they are also destroyed relatively quickly after their creation. As Example 7.20 shows, for a substance with a resistivity of $10^9\,\Omega\text{m}$, on average there are only $\approx 10^{13}$ free electrons per cubic metre. This may be compared with a typical value of $\approx 10^{29}$ atoms per cubic metre, which corresponds, on average, to only one carrier per 10^{16} atoms. The chance that an electron on a particular atom will be excited in such a substance is about a billion times less than the chance of winning the UK national lottery!

This description of the conduction process allows us to understand directly why insulators are poor conductors. Furthermore, the mechanism described above also allows us to understand why the conductivity depends strongly on temperature. Assuming $\Delta E = 3\,\text{eV}$, we may evaluate the Boltzmann factor

$$\exp\left(\frac{-\Delta E}{k_B T}\right) \quad (7.79)$$

for $T = 293\,\text{K}$ (20°C) and $T = 303\,\text{K}$ (30°C). At 20 °C the factor evaluates to 2.5×10^{-52} and at 30°C it evaluates to 1.25×10^{-50}. Thus heating the substance by 10°C would increase the rate of carrier formation by a factor of ≈ 100.

Dielectric strength

The detailed mechanism of dielectric breakdown in solids differs considerably from that in gases (see §5.6.4); however, the two processes do have some features in common. In both processes there is a relatively low density of charge carriers in equilibrium and breakdown occurs when each carrier can, on average, create at least one more carrier before it is removed from the conduction process, i.e. both processes are *avalanche* effects. In gases, the electrons and ions create new carriers by ionization of neutral molecules with which they collide. In solids, the electrons create a region that is slightly hotter than the equilibrium temperature, making it considerably more probable that a new carrier will be formed locally.

Consider the case of an insulator with an excitation energy of $\Delta E = 3\,\text{eV}$. Suppose that a carrier is created and travels one lattice spacing, a, before being lost to the conduction process. In an electric field, $E\,(\text{V m}^{-1})$, it will have acquired an extra Ea electron-volts of energy from the electric field. For $E \approx 10^7\,\text{V m}^{-1}$, which is typical of the values in Table 7.15, and $a \approx 0.3\,\text{nm}$, this amounts to $3 \times 10^{-3}\,\text{eV}$ of energy. If this energy is given up to vibrations of the atom on which it now resides, this is equivalent to local heating. At $T \approx 293\,\text{K}$, $k_B T \approx 1/40\,\text{eV} = 25 \times 10^{-3}\,\text{eV}$. Thus locally the effective temperature is increased such that $k_B T \approx 25 \times 10^{-3} + 3 \times 10^{-3}\,\text{eV}$ i.e. $T \approx 325\,\text{K}$. The Boltzmann factor gov-

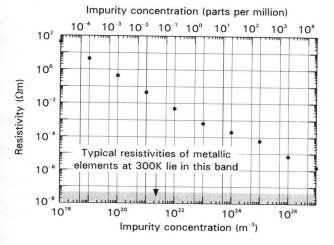

Figure 7.39 The resistivity at 300 K of silicon alloyed with small amounts of phosphorus, both plotted on logarithmic axes. Alloying at this low level (the highest impurity concentration shown corresponds to just over one phosphorus atom for every 100 silicon atoms) is generally referred to as *doping*. Note that the doping phosphorus dramatically reduces the resistivity of the silicon even at low levels. However, even at its lowest value the resistivity of the silicon is 100 times higher than that of typical metallic elements (see Fig. 7.33). (Data from Sze; see §1.4.1)

erning the rate of carrier formation is thus increased to $\approx 3 \times 10^{-47}$, i.e. it is now around 10^5 more likely than previously that an electron carrier will be created in this region.

This amplification of probability is exactly the kind of process that gives rise to avalanche, and will eventually create a hot, high carrier density region (a plasma) within the solid. Typically this region will grow rapidly and breakdown will take place, potentially vaporizing the insulator in the process.

The physics of conduction and breakdown in insulators is considerably more complex than has been indicated above. However, even in this simple analysis we have been able to understand the fundamental processes underlying electrical conduction and breakdown. We will consider these conduction processes again when we discuss conduction in semiconductors (see §7.5.9)

7.5.8 Data on semiconductors

Semiconductors are indicated in Figure 7.32 as having a resistivity at room temperature of around $1\,\Omega m$. This is around a million times higher than the resisitivity of metals, but more than a million times lower than the values for the insulators discussed in §7.5.6. As with insulators, the resistivity of semiconductors depends strongly on the purity and temperature of the semiconductor.

Figure 7.39 illustrates the sensitivity of the resis-

tivity of silicon to phosphorus impurities. The main feature to note is that low levels of the impurity – even at the level of a few parts per million – significantly *reduce* the resistivity. This is in direct contrast to metals, where we saw that resistivity is *increased* by impurities. The highest concentration shown in Figure 7.39 corresponds to a replacement of around 1% of silicon atoms with phosphorous atoms and the resistivity is about 100 times higher than the resistivity of typical metals.

Figure 7.40 illustrates the sensitivity of the resistivity of silicon to temperature. The main feature to note is that the resistivity falls dramatically as the temperature is increased. This is in direct contrast to metals, where we saw that the resistivity *increases* as the temperature increases.

So the questions raised by our preliminary examination of the experimental data on the electrical resistivities of semiconductors are:

- Why does the resistivity of silicon decrease when small amounts of phosphorus are added as impurities?
- Why do the resistivities of semiconductors increase strongly with decreasing temperature.

7.5.9 Understanding the electrical properties of semiconductors

We approach the behaviour of semiconductors by considering them to be essentially insulators. As dis-

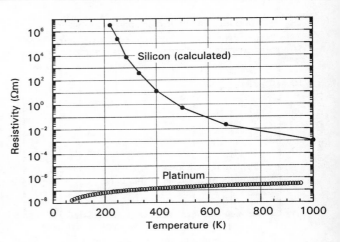

Figure 7.40 Calculation of the resistivity of relatively pure silicon plotted on a logarithmic scale versus temperature. Also shown, for contrast, is the resistivity of platinum (see also Fig. 7.34). Note: the silicon data have been calculated from data tabulated in various sources; they should be taken as indicative of the trend in the data only. (Calculated from data in S. M. Sze 1985, *Semiconductor devices: physics and technology*, New York, John Wiley.)

cussed in §7.5.7, the conduction process in insulators is dramatically different from that appropriate to metals. When considering insulators we saw that the main effect of an applied electric field was to polarize the electronic structure around the atoms of the substance. Only occasionally did local fluctuations conspire to excite an electron from its lowest energy bound state to a state in which conduction was possible. The difference between the terms "insulator" and "semiconductor" lies only in the magnitude of ΔE, the energy required to excite an electron into a quantum state in which electrical conduction is possible.

Background theory

The creation of charge carriers

The arrangement of quantum states on an atom of semiconducting substance is shown in Figure 7.41. A low energy quantum state, generally corresponding to an electron taking part in a covalent bond, is occupied by an electron. There also exists a higher energy quantum state that is, in general, empty.

The relative probability of occupancy of the two quantum states is given (see §2.5) by a Boltzmann factor. Thus the relative probability, P, that the higher energy of the two quantum states is occupied is given by

$$P = \exp\left(\frac{-\Delta E}{k_{\mathrm{B}}T}\right) \qquad (7.80)$$

For semiconductors, the *energy gap*, ΔE is rather less than for insulators, being typically $\approx 1\,\mathrm{eV}$. However, the Boltzmann factor in Equation 7.80 is still extremely small – at room temperature it is typically $\exp(-40) \approx 4 \times 10^{-18}$. This is much greater than for insulators, but still leads to a slow rate of excitation of electrons to the upper quantum state.

Note that when an electron is in the upper state it is able to conduct electricity. In an applied electric field the electron may move into an excited quantum state on a neighbouring atom which is almost certainly empty, and so move through the crystal (Fig. 7.42). This contrasts with the situation when there are no excited electrons. In this case electrons cannot move easily onto neighbouring atoms because there is no low energy quantum state available for them. In order

Figure 7.41 An illustration of the situation on an atom of semiconducting substance at $T = 0\,\mathrm{K}$. A low energy quantum state is occupied, and a higher energy quantum is empty. The energy difference ΔE between the two states is very much greater than $k_{\mathrm{B}}T$. The figure is drawn approximately to scale, so that if ΔE is 1 eV then $k_{\mathrm{B}}T \approx 1/40\,\mathrm{eV}$, i.e the value of $k_{\mathrm{B}}T$ at around room temperature.

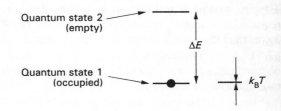

(a)

(b)

(c)

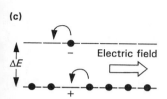

Electric field

(d)

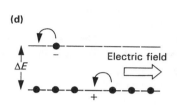

Electric field

Figure 7.42 An illustration of the situation within a semiconducting substance: (a) at T = 0 K; (b) above T = 0 K. (c, d) An applied electric field can now cause electrons to move to neighbouring quantum states. Note that the effect of the field is to make the electron in the upper state move one way, and the uncompensated positive charge in the lower quantum state move in the opposite direction.

to move into an upper quantum state an electron requires ΔE of energy, which it has to acquire at the expense of the electric field. If the spacing between atoms is a, then the electric field would have to exceed $\approx \Delta E/a \, \mathrm{V \, m^{-1}}$, which amounts to $\approx 10^9 \, \mathrm{V \, m^{-1}}$ – well in excess of the breakdown field.

So thermal excitation of electrons into the upper electron state makes them available for the electric field to draw them into a conduction process. Note also that the atom that the electron has left now has a vacant lower energy quantum state. This allows neighbouring electrons to move into this quantum state as shown in Figure 7.42c,d. In this process the vacant lower quantum state is transferred in the same direction as the applied electric field. Since this vacant state has an uncompensated positive electric charge, the excitation of single electron allows conduction in the lower quantum states in a way which is distinctly different from conduction in the upper quantum states.

Carrier recombination

Note that if no external electric field acts on the semiconductor then the excited electron will still be tied to its "original" atom by the uncompensated positive charge. After a short time the electron will, in general, return to its low energy quantum state, giving up its excess energy in the form of a photon and/or a phonon. This process is called "carrier recombination" and "annihilates" two charge carriers.

Terminology

The lower energy quantum states are generally referred to as *valence states* and the upper quantum states as *conduction states*. Electrons in the conduction states are called *electron carriers* and the vacant valence states left behind by excited electron carriers are called *hole carriers*.

In equilibrium

At temperature T in a semiconductor, electron and hole carriers are created at a characteristic rate that is proportional to $\exp(-\Delta E/k_{\mathrm{B}}T)$. After a short while these carriers annihilate. Thus at any instant there is an equilibrium concentration of electron carriers and hole carriers that are able to conduct electricity. These carriers form two dilute interpenetrating charged "gases" within the semiconductor, as shown in Figure 7.43. Note that because of the way in which they are created, the number densities of hole and electron carriers are equal; and because of the exponential dependence of the creation rate, the number density of carriers increases with temperature.

Figure 7.43 A region of a semiconductor with the lattice of atoms represented by a grey box. The number densities of hole and electron carriers are equal, and increase strongly with temperature.

(a) Low temperature

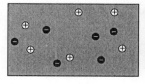

(b) High temperature

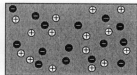

Conductivity

Consider a single charge carrier with charge q within the semiconductor. Under the action of an applied electric field, the carrier makes transitions from quantum state to quantum state, and so moves through the crystal (see Fig. 7.42c,d). The rate at which these transitions are made depends on the details of the quantum states involved and, in general, is not equal for the hole and electron carriers. Since the rate at which the electron is accelerated by an electric field would normally be determined by its mass, we characterize the rate at which these transitions are made by means of an *effective mass*, m^*. In a manner similar to the case for electrons in metals, we write the drift velocity of the charge carrier as

$$F = m^* \frac{\Delta v}{\Delta t} \qquad (7.81)$$

and hence

$$\Delta v = \frac{qE}{m^*} \Delta t \qquad (7.82)$$

Using a similar argument to that following Figure 7.37 we may write the current density due to a carrier density, n, of such carriers as

$$j = nq \left(\frac{q \Delta t}{m^*} \right) E \qquad (7.83)$$

If we have two such "gases" of carriers, one of which is composed of hole carriers, then the current density will be given by

$$j = n_h q \left(\frac{q \Delta t_h}{m_h^*} \right) E + n_e q \left(\frac{q \Delta t_e}{m_e^*} \right) E \qquad (7.84)$$

The quantity $q\Delta t/m^*$ is known as the *carrier mobility*, μ and so in terms of μ we may write an expression for the conductivity as

$$\sigma = n_h q_h \mu_h + n_e q_e \mu_e \qquad (7.85)$$

where the subscripts h and e refer to hole and electron carriers respectively.

Comparison with experiment

We can understand the temperature dependence of the resistivity of semiconductors in terms of a temperature dependence of the terms in Equation 7.85. Since the charge, q, on either type of carrier is unlikely to change with temperature, we concentrate our attention on the temperature dependence of the mobility or the carrier density.

The temperature dependence of the mobility

The mobility of a carrier is defined by

$$\mu = \frac{q\Delta t}{m^*} \qquad (7.86)$$

From our study of metals, we would expect the carrier lifetime Δt to change with temperature. However, we would expect, in general, that the carrier lifetime would get shorter at high temperatures, which would cause an increase in resistivity at higher temperatures – the opposite of what is observed. The details of the quantum states that determine m^* do not change strongly with temperature and we cannot explain the changes in resistivity as being due to changes in m^*.

The temperature dependence of the carrier density

The source of the temperature dependence of the resistivity lies in the temperature dependence of the number density of charge carriers. As we saw above, the rate at which electrons make transitions to conduction states and hence the creation rate of hole and an electron carriers, is proportional to $\exp(-\Delta E/k_B T)$. However, the recombination rate does not increase so dramatically, which leads to an exponential temperature dependence of the equilibrium number density of charge carriers.

The estimate in Example 7.22 is extremely instructive. There are roughly 10^{29} atoms per cubic metre in silicon, and so a carrier density of 2.4×10^{17} per cubic metre corresponds to around one carrier for 10^{12} atoms. Even given the uncertainties and assumptions involved in this calculation, the key result remains: the electron and hole carriers that occur spontaneously within silicon exist at an extremely low density.

Example 7.22

Estimate the number density of electron and hole carriers in pure silicon at 300 K.

Let us first estimate the mobility of the carriers according to Equation 7.86. Since we do not have direct information on these quantities, we must make some plausible guesses. We estimate that the effective masses of both types of carrier are of the order of the mass of a free electron, m_e, and that the scattering lifetime of carriers is similar to that which we worked out for copper in Example 7.17 ($\approx 2.5 \times 10^{-14}$ s). We thus write the mobility of either hole or electron carriers as

$$\mu_e = \mu_h \approx \mu \approx \frac{1.6 \times 10^{-19} \times 2.5 \times 10^{-14}}{9.1 \times 10^{-31}} = 4.4 \times 10^{-3}\,\mathrm{C\,s\,kg^{-1}}$$

(Note: this is in fact a significant underestimate of the mobility.) Recalling that we expect $n_h = n_e$, we therefore write Equation 7.85 as

$$\sigma = 2nq\mu$$

where n is the number density of either holes or electrons. Rearranging to solve for n we find

$$n \approx \frac{\sigma}{2q\mu} \approx \frac{1}{2\rho q\mu}$$

$$= \frac{1}{2 \times 3 \times 10^3 \times 1.6 \times 10^{-19} \times 4.4 \times 10^{-3}} \approx 2.4 \times 10^{17}\,\mathrm{m^{-3}}$$

where we have used $\rho \approx 3 \times 10^3$ at around 300 K read from Figure 7.40.

The effect of impurities on the resistivity of silicon

Example 7.22 indicates that the number density of either hole or electron carriers in silicon at room temperature is extremely small, being around 1 carrier for every 10^{12} silicon atoms. This is the number of carriers that exist due to processes *intrinsic* to the silicon lattice itself. When an impurity is added to silicon in such a way that it replaces the silicon atom within the lattice, then quantum states in the region of the impurity are altered. For some impurities, this alteration is such as to make it much easier to create either a hole (a p-type impurity) or an electron carrier (an n-type impurity). If each impurity atom creates one carrier, impurities present in silicon at a level of 1 impurity atom per 10^9 silicon atoms will create around 1000

times more carriers than occur intrinsically within the silicon. This is the origin of the extreme sensitivity of the resistivity of semiconductors to impurities. The excess carriers that are created as a result of impurities are known as *extrinsic* carriers.

The effect of n-type (donor) impurities

Let us consider the way in which the quantum states within a lattice are affected when an atom of phosphorous is substituted for an atom of silicon (Fig. 7.44). Phosphorous has five valence electrons (one more than silicon) and one extra nuclear charge than silicon. The extra electron will occupy the next available quantum state, which will be at roughly the same energy as the empty conduction states in silicon. However, in general, the quantum state is at a slightly lower energy than the conduction states in silicon because of the attractive effect of the extra nuclear charge.

Note now that the energy required to remove this extra electron and place it into a conduction

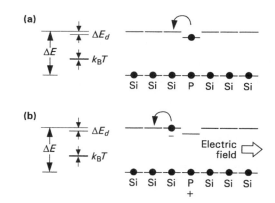

Figure 7.44 The mechanism by which a donor impurity causes an increase in the number of carriers available for conduction. (a) At $T = 0\,\mathrm{K}$, the extra electron on the phosphorous impurity is in a quantum state with an energy just below the conduction quantum states on neighbouring ions. The energy, ΔE_d, required to "ionize" the impurity and place the donated electron into the conduction states is much less than the band gap, ΔE. Thus at relatively low temperatures the electron leaves its localized quantum state and an electron carrier is created in the conduction band. (b) Unlike the case of a carrier arising from processes intrinsic to the silicon, the electron carrier in the conduction states does not have a partner hole carrier in the valence states. There is, however, an immobile net positive charge left on the phosphorous impurity.

state on the silicon atom, ΔE_d, is much less than the energy, ΔE, required to remove an electron from a valence state and place it into a conduction state. The reduction is in fact quite dramatic. For example, for silicon the energy gap ΔE is around 1.1 eV, but ΔE_d for a phosphorus impurity is only 0.044 eV. At room temperature, where $k_B T$ is around 0.025 eV, it is highly likely that the phosphorous impurity electron will be able to move easily into the conduction states. Thus each phosphorous impurity effectively *donates* one electron to the number density of electron carriers. But note that no equivalent hole carrier has been created.

So we see that phosphorous impurities in silicon at a level of 1 part per billion can easily create sufficient electron carriers to dominate the conduction process completely.

The effect of p-type (acceptor) impurities

Let us consider the way in which the quantum states within a lattice are affected when an atom of aluminium is substituted for an atom of silicon (Fig.

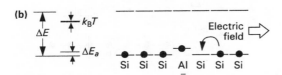

(a) ΔE $k_B T$ ΔE_a Si Si Si Al Si Si Si

(b) ΔE $k_B T$ ΔE_a Electric field ⟹   Si Si Si Al Si Si Si

Figure 7.45 The mechanism by which an acceptor impurity causes an increase in the number of carriers available for conduction. (a) At $T = 0$ K, the quantum state on an aluminium atom equivalent to the valence state in silicon is unoccupied. The energy of this state is just above the valence quantum states on neighbouring ions. The energy, ΔE_a, required by an electron to occupy this quantum state is much less than the band gap, ΔE. At relatively low temperatures, this capture can occur with high probability, resulting in the creation of a hole carrier in the valence states of the silicon. (b) Unlike the case of a carrier arising from processes intrinsic to the silicon, the hole carrier in the valence states does not have a partner electron carrier in the conduction states. There is, however, an immobile net negative charge left on the aluminium impurity.

7.45). Aluminium has three valence electrons (one less than silicon) and one less extra nuclear charge. Thus the valence quantum state that is occupied in silicon is empty in aluminium. The empty aluminium valence state will be at roughly the same energy as the occupied valence states in silicon. However, in general, the quantum state is at a slightly higher energy than the valence states in silicon because the attractive effect of the nuclear charge in aluminium is less than in silicon.

Note now that the energy required to remove an electron in a valence state in silicon and place it into the empty state on the aluminium atom, ΔE_a, is much less than the energy, ΔE, required to remove an electron from a valence state and place it into a conduction state. The reduction is again quite dramatic. For example, for silicon the energy gap ΔE is around 1.1 eV, but ΔE_a is only 0.057 eV. At room temperature where $k_B T$ is around 0.025 eV, it is highly likely that the electrons will be able to move easily into this *acceptor* state. Note that each aluminium impurity effectively *accepts* an electron and hence creates a hole carrier in the valence states of silicon, but no equivalent electron carrier has been created.

7.6 Thermal conductivity

Thermal conductivity, κ (pronounced "kappa"), is defined by the equation

$$\frac{dQ}{dt} = -\kappa A \frac{dT}{dx} \qquad (7.87)$$

where:
dQ/dt is the rate of heat flow (W),
   κ is the thermal conductivity ($W m^{-1} K^{-1}$),
   A is the cross-sectional area across which heat is flowing (m^2), and
dT/dt is the temperature gradient ($K m^{-1}$).

The minus sign indicates that heat flows against the temperature gradient, from high temperatures to low temperatures.

7.6.1 Data on the thermal conductivity of solids

The thermal conductivity data of some solid elements is given in Table 7.15, and summarized in Figure 7.46. The most striking feature of the data is the astonishing thermal conductivity of diamond, which is between 2 and 5 times greater than that for copper at room temperature.

Perhaps, surprisingly, the metallic elements do not show up as being dramatically better conductors of heat than the electrically insulating elements. However, there is a distinct difference between the temperature dependence of thermal conductivity. In particular, the thermal conductivity of insulators falls more rapidly than that of metals as the temperature is increased.

We should note, however, that the elemental insulators listed in Table 7.15 refer to specialized materials that are rather different from the day-to-day substances that we refer to as "insulators". The thermal conductivities of some miscellaneous substances (including substances we would more commonly refer to as thermal insulators such as wool) are given in Table 7.16. Note that in the alloy section of Table 7.16, the thermal conductivity of bronze (made from 90% copper and 10% tin) is closer to that of tin than copper.

So the main questions raised by our preliminary examination of the experimental data on the thermal conductivity of solids are:

• Why do the thermal conductivities of elements fall into two classes, electrical conductors and electrical insulators, distinguished by the temperature dependence of their thermal conductivity?

• Why, in general, does the thermal conductivity of a solid increase on cooling? For some materi-

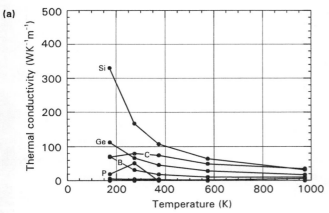

(a)

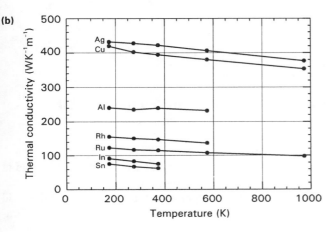

(b)

Figure 7.46 Selected data from Table 7.15. (a) The data for all the insulators and semiconductors listed in Table 7.15, with the exception of diamond. The thermal conductivity of diamond is so much greater than any other material that it distorts the scaling of the graph. The two unlabelled curves close to the temperature axis are for tellurium and sulphur. (b) The data for a selection of metals. The vertical scale is the same as in (a) showing that a typical value of the metallic conductivity is slightly greater for insulators, but more striking is the different temperature dependences shown by metals and insulators. The metals show a less pronounced temperature dependence. In general, the thermal conductivity of metals appears to improve with decreasing temperature. However, two of the insulators show peaks in thermal conductivity with the thermal conductivity, becoming worse as the temperature is lowered.

Table 7.15 The thermal conductivity, κ, of solid elements as a function of absolute temperature. The shaded entries refer to data above the melting temperature of the element, i.e. to the liquid.

Element	Material type*	κ (WK⁻¹m⁻¹) 173.2K	273.2K	373.2K	573.2K	973.2K
Lithium, Li	M	94	86	82	47	59
Beryllium, Be	M	367	218	168	129	93
Boron, B	I	72	32	19	11	10
Carbon (graphite), C	I	70–220	80–230	75–195	50–130	35–70
Carbon (diamond), C	I	1700–4900	1000–2600	700–1700	–	–
Sodium, Na	M	141	142	88	78	60
Magnesium, Mg	M	160	157	154	150	–
Aluminium, Al	M	241	236	240	233	92
Silicon, Si	SC	330	168	108	65	32
Phosphorous, P	I	20	13/0.25	0.18	0.16	–
Sulphur, S	I	0.39	0.29	0.15	0.17	–
Potassium, K	M	105	104	53	45	32
Scandium, Sc	M	15	16			
Titanium, Ti	M	2622	21	19	21	
Vanadium, V	M	32	31	31	33	38
Chromium, Cr	M	120	96.5	92	82	66
Manganese, Mn	M	7	8	–	–	–
Iron, Fe	M	99	83.5	72	56	34
Cobalt, Co	M	130	105	89	69	53
Nickel, Ni	M	113	94	83	67	71
Copper, Cu	M	420	403	395	381	354
Zinc, Zn	M	117	117	112	104	66
Gallium, Ga	M	43	41	33	45	–
Germanium, Ge	SC	113	67	46.5	29	17.5
Selenium (c-axis), Se	I	6.8	4.8	4.8		
Rubidium, Rb	M	59	58	32	29	22
Yttrium, Y	M	16.5	17	–	–	–
Zirconium, Zr	M	26	23	22	21	23
Niobium, Nb	M	53	53	55	58	64
Molybdenum, Mo	M	145	139	135	127	113
Technetium, Tc	M	–	51	50	50	
Ruthenium, Ru	M	123	117	115	108	98
Rhodium, Rh	M	156	151	147	137	–
Palladium, Pd	M	72	72	73	79	93
Silver, Ag	M	432	428	422	407	377
Cadmium, Cd	M	100	97	95	89	445
Indium, In	M	92	84	76	42	–
Tin, Sn	M	76	68	63	32	40
Antimony, Sb	M	33	25.5	22	19	27
Tellurium(c-axis), Te	I	5.1	3.6	2.9	2.4	6.3

* I, insulator; M, metal; SC, semiconductor. (Data from Kaye & Laby; see §1.4.1)

als there appears to be a limit to this increase and the thermal conductivity reaches a peak and then gets smaller as the temperature is lowered.

- Why are the thermal conductivities of the elemental solids large compared with those of gases (Table 5.4)? However, many common materials such as wool or paper have values that are similar to those of gases.

7.6.2 Understanding the thermal conductivity of solids

Introduction

Thermal energy can be transferred through a solid in two quite distinct processes.

In the first process, heating of a region of the solid causes an increase in the amplitude of atomic vibra-

Table 7.16 Thermal conductivity, κ, of miscellaneous materials: metallic alloys, refractory materials (i.e. those suitable for use in high temperatures without degradation), and a selection of everyday materials.

	κ (W K⁻¹ m⁻¹)						
	173.2 K	273.2 K	373.2 K	573.2 K	873.2 K	973.2 K	1473.2 K
Brass (Cu 70%, Zn 30%)	89	106	128	146	–	–	–
Bronze (Cu 90%, Sn 10%)	–	53	60	80	–	–	–
Carbon steel	48	50	48.5	54.5	–	30.5	–
Silicon steel	–	25	28.5	31	–	28	–
Stainless steel	–	24.5	25	25.5	–	24.8	–
Alumina (Al₂O₃)	–	40	28	–	9.2	–	5.7
Beryllia (BeO)	–	300	213	–	61	–	22
Fire brick	–	–	–	–	1.1	–	1.3
Silica (SiO₂) fused quartz	–	1.33	1.48	–	2.4	–	–
Zirconia (ZrO₂)	–	–	1.8	–	2.0	–	2.2

Thermal conductivities at room temperature.

Material	κ (W K⁻¹ m⁻¹)	Material	κ (W K⁻¹ m⁻¹)	Material	κ (W K⁻¹ m⁻¹)
Brick wall	≈1	Porcelain	1.5	Glass wool	0.037
Plaster	≈0.13	Rubber	≈0.2	Cotton wool	0.03
Timber	≈0.15	Polystyrene	≈0.1	Sheep's wool	0.05
Balsa wood	≈0.06	Glass (crown)	1.1	Nylon	0.25
Paper	0.06	Glass (flint)	0.85	Epoxy resins	≈0.2
Cardboard	0.21	Glass (pyrex)	1.1	Cellular polystyrene	≈0.04

(Data from Kaye & Laby; see §1.4.1)

Example 7.23

Performing experiments well below room temperature is one of the author's favourite pastimes. In such experiments it is important that not too much heat flows down any electrical connections to the cold experimental region. To avoid this, wires are "heat sunk" at several convenient temperatures. This involves connecting the wires thermally – but not electrically – to points in the cryostat where there is sufficient cooling to cope with the heat flowing down the wire without warming up. One way of doing this is to use contact pads made of beryllium oxide. The pads consist of a thin layer of beryllium oxide coated with metal on each side to which electrical connections can be soldered.

If the heat sink is kept at 77.3 K (the temperature of boiling nitrogen liquid at atmospheric pressure) and 0.8 W of heat flows down the wire from room temperature, estimate the temperature at point A in the diagram below if the contact pads are 5 mm×5 mm×0.5 mm in size.

If we consider first heat flowing through the beryllia alone, we use the thermal conductivity equation

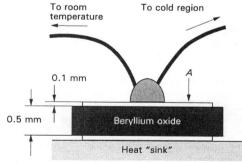

$$\frac{dQ}{dt} = -\kappa A \frac{dT}{dx}$$

with $dQ/dt = 0.8$ W and $A = 5$ mm×5 mm $= 25\times10^{-6}$ m², then it only remains to estimate κ in order to evaluate the temperature gradient across the beryllia contact pad. Comparing the data for beryllia with the data for the elements given in Figure 7.46, we make an estimate of $\kappa = 300 \pm 100$ W K⁻¹ m⁻¹. We thus find that the temperature gradient across the chip is

$$\frac{dT}{dx} = -\frac{\kappa A}{(dQ/dt)} = \frac{(300 \pm 100) \times 25 \times 10^{-6}}{0.8} = 9.4 \pm 3.1 \times 10^{-3} \, \text{K m}^{-1}$$

Since the beryllia pad is 0.5 mm thick this represents a temperature difference across the two surfaces of the pad of

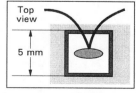

$$\Delta T \approx \frac{dT}{dx} \Delta x = 9.4 \pm 3.1 \times 10^{-3} \times 0.5 \times 10^{-3} \approx 5 \, \mu\text{K}$$

Clearly, even despite the large uncertainty in the thermal conductivity of beryllia, we may rest assured that the upper surface of the pad will be at a temperature extremely close to 77.3 K even when there is a substantial heat flow down the wires leading to room temperature.

tions in that region. The mechanism of heat transfer is simply that an atom A with a large vibrational amplitude interacting with a neighbouring atom B with a smaller vibrational amplitude tends to lose energy until the average energies per degree of freedom of A and B are equal. In terms of phonons (see §7.2), the increased amplitude of atomic vibration is described as the creation of phonons – technically an increased number density of phonons, $n_{ph}(T)$. These phonons then travel at the speed of sound to the colder regions of the lattice taking with them the excess energy of vibration from the hotter regions.

In the second process, which takes place only in metals, the free electrons within the metal can accept a small amount of thermal energy and then carry it to colder regions of the metal. Notice that in metals this second process takes place *in addition* to the first process and so, other things being equal, we would expect metals to have a higher thermal conductivity than insulators.

Background theory

Both of the above processes may be analyzed in a manner analogous to that used to examine the thermal conductivity of gases (see §5.4 and Fig. 7.47). In these analogies the first process described above is analyzed as a *gas of phonons* and the second process as a *gas of electrons*.

As in §5.4.2, we consider a gas in which there is a

Figure 7.47 Simplified illustration of the calculation of the thermal conductivity of a gas. The figure shows three planes in the gas perpendicular to a temperature gradient. The separation of the planes is λ_{mfp}, the mean free path of the particles of gas. Thus particles that are travelling in the appropriate direction in either the top or the bottom plane will (probably) cross the central plane before colliding. Energy is carried across A in both directions, but on average more energy flows across A from the plane in the hotter region of the gas. The analysis of the thermal conductivity centres on the problem of evaluating the net energy flow across an area A.

temperature gradient, and consider particles striking an area A perpendicular to the temperature gradient. On average, particles striking A will have a travelled a mean free path, λ_{mfp}, without any "collisions". They thus bring with them the thermal energy characteristic of the region about a mean free path distant. In our analogy with the molecular gas we assume that:

(a) The average speed of the particles, $\bar{v}$, is relatively weakly affected by temperature (true in both the following cases).

(b) The number of particles crossing a unit area per second is given by Equation 4.51 as $n\bar{v}/4$, where n is the number density of the particles. We will consider that, in general, the number density of particles may change with temperature, i.e. $n = n(T)$.

(c) Each particle brings with it, on average, an amount of energy $\bar{E}_{th}$ which may be passed on to other particles in a collision.

Combining these three assumptions, we estimate that the rate of energy flow due to particles crossing area A from the hotter side is

$$\frac{dQ}{dt} = \frac{1}{4} A\bar{v}n(T+\Delta T)\bar{E}_{th}(T+\Delta T) \qquad (7.88)$$

and

$$\frac{dQ}{dt} = \frac{1}{4} A\bar{v}n(T-\Delta T)\bar{E}_{th}(T-\Delta T) \qquad (7.89)$$

from the colder side. The net rate of energy flow is therefore just the difference between Equations 7.88 and 7.89:

$$\frac{dQ}{dt} = \frac{1}{4} A\bar{v}\left[n(T+\Delta T)\bar{E}_{th}(T+\Delta T) \right. $$
$$\left. -n(T-\Delta T)\bar{E}_{th}(T-\Delta T)\right] \qquad (7.90)$$

Writing the product $n(T) \times \bar{E}_{th}(T)$ as a single quantity $n\bar{E}_{th}(T)$

$$\frac{dQ}{dt} = \frac{1}{4} A\bar{v}\left[n\bar{E}_{th}(T+\Delta T) - n\bar{E}_{th}(T-\Delta T)\right] \qquad (7.91)$$

and expanding $n\bar{E}_{th}(T)$ to first order, we find

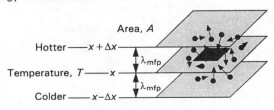

Area, A

Hotter —— $x + \Delta x$

Temperature, T —— x λ_{mfp}

   λ_{mfp}

Colder —— $x - \Delta x$

$$\frac{dQ}{dt} = \frac{1}{4} A \bar{v} \left(\left\{ n\bar{E}_{th}(T) + \frac{d[n\bar{E}_{th}(T)]}{dT} \Delta T \right\} \right.$$

$$\left. - \left\{ n\bar{E}_{th}(T) - \frac{d[n\bar{E}_{th}(T)]}{dT} \Delta T \right\} \right) \quad (7.92)$$

Subtracting yields

$$\frac{dQ}{dt} = \frac{1}{4} A \bar{v} \left\{ 2 \frac{d[n\bar{E}_{th}(T)]}{dT} \Delta T \right\} \quad (7.93)$$

which may be rewritten in terms of the temperature gradient as

$$\frac{dQ}{dt} = \frac{1}{2} A \bar{v} \frac{d[n\bar{E}_{ph}(T)]}{dT} \frac{dT}{dx} \lambda_{mfp} \quad (7.94)$$

Comparing this with the standard form of Equation 7.87 yields an estimate for the thermal conductivity of

$$\kappa = \frac{1}{2} \bar{v} \lambda_{mfp} \frac{d[n\bar{E}_{ph}(T)]}{dT} \quad (7.95)$$

So we conclude that the thermal conductivity of a gas depends on:

(a) the mean speed, $\bar{v}$, of the particles of the gas;
(b) the mean free path λ_{mfp} of the particles of the gas; and
(c) the *rate* at which the product $n\bar{E}_{ph}(T)$ changes with temperature.

In this text we restrict ourselves to trying to understand the temperature dependence of the thermal conductivity rather than its absolute magnitude.

Insulators

For insulators we model the vibrations of the lattice as a gas of phonons, and in this section we will work out the temperature dependence of the thermal conductivity of a phonon gas. The data given in Table 7.15 refer mainly to temperatures that are greater than (or at least of the order of) the Debye temperature, θ_D (Table 7.10) and so we treat the three factors in Equation 7.95 as follows.

(a) We approximate the speed of all phonons to c_{sound} – an average of the speeds of sound for different modes. The speed of sound in a solid is roughly temperature independent.

(b) We consider that at high temperatures the main source of scattering of phonons is other phonons. (This is in fact correct, but it is rather difficult to establish that it should be so. At low temperatures, when the amplitude of atomic vibrations is small, the main sources of phonon scattering are impurities, crystalline defects and surface irregularities.) Considering the phonon gas in the same light as a molecular gas (see §4.4.4) we expect that the mean free path for phonons will be inversely proportional to the number density of phonons, i.e.

$$\lambda_{mfp}^{ph} \propto \frac{1}{n_{ph}(T)} \quad (7.96)$$

(c) The number of phonons in each mode of energy, E_{ph}, is given by the Bose–Einstein factor (Eq. 7.47), which we rewrite here in the form

$$n_{ph}(E_{ph}, T) = \frac{1}{\exp(E_{ph}/k_B T) - 1} \quad (7.97)$$

When T is high enough that $k_B T$ is greater than the maximum phonon energy, $\approx k_B \theta_D$, then we can expand the exponential to first order to give

$$n_{ph}(E_{ph}, T) \propto \frac{1}{\left(1 + \frac{E_{ph}}{k_B T} + \dots \right) - 1} \propto \frac{k_B T}{E_{ph}} \quad (7.98)$$

So for all phonon modes, $n_{ph}(E_{ph}, T)E_{ph}$ is proportional to temperature. Since this happens for each individual phonon mode, the product $n_{ph}(T)E_{ph}$ involving the total phonon density must also increase linearly as the absolute temperature increases.

Combining only the temperature dependences of these factors in Equation 7.95 for κ

$$\kappa = \frac{1}{2} c_{sound} \lambda_{mfp}^{ph} \frac{d[n\bar{E}_{ph}(T)]}{dT} \quad (7.99)$$

we arrive at

213

$$\kappa \propto \text{Constants} \times \frac{1}{\text{Constants} \times T} \times \frac{\text{d}(\text{Constants} \times T)}{\text{d}T} \tag{7.100}$$

which simplifies to

$$\boxed{\kappa \propto \frac{\text{Constants}}{T}} \tag{7.101}$$

for insulators at $T > \theta_D$.

Hence at high temperatures we expect to find $\kappa \propto T^{-1}$ This agrees, at least qualitatively, with the high temperature data on insulators shown in Figure 7.46.

Metals

For metals we need to work out the thermal conductivity of an electron gas such as that described in §6.5. Once again we note that the data given in Table 7.15 refer mainly to temperatures that are greater than (or at least of the order of) θ_D and so bearing this in mind and we treat the three factors in Equation 7.95 as follows.

(a) We approximate the speed of all electrons to the Fermi speed v_F. As in our discussion of the electrical conductivity in a metal, only those electrons in quantum states with energies close to the Fermi energy, E_F, are able to accept thermal energy.

(b) We consider that at high temperatures the main source of scattering of electrons is phonons. From the discussion of electrical conductivity, we concluded that the scattering time for electrons, Δt, varied inversely with absolute temperature (Eq. 7.74). Since the speed of the electrons is temperature independent, we conclude that the mean free path of electrons will vary inversely with temperature, i.e.

$$\lambda_{\text{mfp}}^e = v_F \Delta t \propto \frac{1}{T} \tag{7.102}$$

(c) As discussed in §7.2 on the electronic heat capacity of solids, only a small fraction of electrons is able to accept thermal energy from the lattice. This fraction is proportional to $k_B T$ and the typical energy of an electron in this fraction is $E_F + k_B T$. Note that only the latter component

of this energy may be given up in thermal interactions, and so the thermal energy, $\bar{E}_{\text{th}}$, of electrons in the fraction is $k_B T$. Hence the product $n\bar{E}_{\text{th}}$ is

$$n\bar{E}_{\text{th}} \propto \text{constants} \times k_B T \times k_B T \tag{7.103}$$

Combining only the temperature dependences of these factors in Equation 7.95 for κ

$$\kappa = \frac{1}{2} v_F \lambda_{\text{mfp}}^e \frac{\text{d}\left[n\bar{E}_{\text{th}}(T) \right]}{\text{d}T} \tag{7.104}$$

we arrive at

$$\kappa \propto \text{Constants} \times \frac{1}{T} \times \frac{\text{d}\left(\text{Constants} \times T^2 \right)}{\text{d}T} \tag{7.105}$$

which after differentiation yields

$$\boxed{\kappa \propto \text{Constants}} \tag{7.106}$$

for metals at $T > \theta_D$

Hence we conclude that at high temperatures we should expect to find that κ is temperature independent. This agrees, at least qualitatively, with the high temperature data on metals shown in Figure 7.46. We see, however, that the metallic data do show a small decrease with increasing temperature. We can understand this as being due to the fact that in metals thermal conduction also occurs through the phonon gas described in the previous section. The thermal conductivity of the phonon gas is proportional to $1/T$, which explains the slight fall of κ with T. A rough estimate from Figure 7.46 indicates that at around room temperature most (around 95% or so) of the thermal conductivity of elemental metals is due to conduction by electrons.

Low temperatures

Several of the assumptions that allowed us to predict the temperature dependence of κ in metals and insulators at high temperatures become invalid at temperatures much less than λ_{mfp}. For example, at low temperatures the primary source of scattering of both electrons and phonons are impurities, not phonons as we assumed above.

So, for example, in metals the mean free path for

electrons becomes fixed at low temperatures, and thus Equation 7.105 becomes

$$\kappa \propto \text{Constants} \times \text{Constants} \times \frac{d\left(\text{Constants} \times T^2\right)}{dT}$$

$$(7.107)$$

which predicts that $\kappa \propto T$ at low temperatures, i.e. the thermal conductivity of a metal will become poorer at low temperatures.

Similarly, in insulators, the mean free path for phonons also becomes fixed at low temperatures, and so we expect

$$\kappa \propto \text{Constants} \times \frac{d\left[n\overline{E}_{ph}(T)\right]}{dT} \qquad (7.108)$$

It may be shown that at low temperatures $n_{ph} \propto T^3$ and $\overline{E}_{ph} \propto T$. We therefore expect

$$\kappa \propto \text{Constants} \times \frac{d\left(\text{Constants} \times T^3 \times T\right)}{dT}$$

$$\propto \text{Constants} \times T^3 \qquad (7.109)$$

and again the thermal conductivity falls at low temperatures.

The fact that the thermal conductivity falls at low temperatures for both insulators and metals (coupled with the high-temperature dependences worked out in the previous sections) leads to the phenomenon of a peak in the thermal conductivity at temperatures well below λ_{mfp}. This peak may be seen for some elements in Figure 7.46, but more normally occurs at temperatures of a few tens of kelvin.

Inhomogeneous materials

The final questions raised by the data concern the order of magnitude of the data given in Tables 7.15 and 7.16 when compared with those of gases (Table 5.10). The reason why the typical thermal conductivities of solids ($\approx 10\,\text{WK}^{-1}\text{m}^{-1}$) are so much greater than those of gases ($\approx 10^{-2}\,\text{WK}^{-1}\text{m}^{-1}$) is simply that the density of atoms present in a solid is about 1000 times greater than in a molecular gas.

However, in inhomogeneous materials such as paper or wool, gases (usually air) may be trapped within the solid. Such materials may have a thermal conductivity that differs only slightly from that of a gas. This low thermal conductivity arises because, although the conductivity, κ, of the solid itself (the individual wool or paper fibres) may be relatively good, the material is constituted such that heat must take an extremely long path through the substance, with an extremely small effective cross-sectional area. For example, in wool, on heating one end of a fibre the heat may quickly transfer along the whole length of the fibre. However, each fibre only touches its neighbouring fibres at a few points, which reduces its effective cross-sectional area dramatically. If the reduction in thermal conduction is sufficiently large, then we may neglect the heat flowing through the fibres of the inhomogeneous substance. In this case the weak thermal conduction through the trapped gas in the material may prove the dominant thermal conduction process.

7.7 Optical properties

The optical properties of solids are extremely varied: from strongly absorbent, to transparent, to strongly reflecting. This indicates considerable variety in the way solids respond to oscillating electromagnetic fields with frequencies between $400 \times 10^{12}\,\text{Hz}$ (red light) and $750 \times 10^{12}\,\text{Hz}$ (blue light).

The optical properties of solids are rather difficult to characterize succinctly, since typically this requires a specification of their absorption and reflection across the optical frequency range. However, for many purposes, substances may be characterized by two parameters: the *refractive index* of the material and the *absorption coefficient*.

7.7.1 Data on the optical properties of metals

Metal samples thicker than $\approx 10\,\text{nm}$ are opaque to visible light, and are strongly reflective. With the exceptions of copper and gold, all the metallic elements have the same "silvery" colour, that is they are strongly reflective at all optical wavelengths, (Table 7.17).

Example 7.24

A laser beam with an intensity of 4 MW mm⁻² and a beam radius of 7.3 μm is bounced off a silver-coated glass mirror. Will the mirror surface be damaged?

The beam intensity is 4×10^6 W mm⁻², i.e. 4×10^{12} W m⁻², and its cross-section is $\pi r^2 = \pi (7.3 \times 10^{-6})^2 = 5.33 \times 10^{-11}$ m². Thus the total power in the beam is

$$(4 \times 10^{12}\,\text{W m}^{-2}) \times (5.33 \times 10^{-11}\,\text{m}^2) = 213\,\text{W}.$$

The reflectivity of the mirror is 98% and so the power dissipated in the mirror is $0.02 \times 213 \approx 4.3$ W.

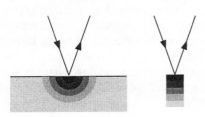

Now this is certainly not a great deal of power, but the beam is highly concentrated, and so the temperature might rise significantly at the mirror surface. The geometry of the situation makes the exact problem difficult to solve, but if we imagine the situation shown, where the beam delivers its power to the end of a rod shaped sample then we can calculate the temperature gradient along this rod fairly easily using Equation 7.87:

$$\frac{dQ}{dt} = -\kappa A \frac{dT}{dx}$$

Using:
(a) an approximate value for the area of the rod of $A = \pi r^2 = 5.33 \times 10^{-11}$ m²,
(b) a value of κ for glass from Table 7.16 of ≈ 1 W K⁻¹ m⁻¹,
(c) the beam power 4.3W,

we find
$$4.3 = 5.33 \times 10^{-11} \frac{dT}{dx}$$

Hence
$$\frac{dT}{dx} = \frac{4.3}{5.33 \times 10^{-11}} \approx 10^{11}\,\text{K m}^{-1}$$

If we assume that 1mm or so below the surface the temperature is around room temperature, then the temperature at the mirror surface would be $\approx 10^{11} \times 10^{-3} = 10^8$ K. Even allowing for much stronger cooling it is clear that such a beam could not be bounced off any solid surface without vaporizing the solid.

Table 7.17 The reflectivity of polished surfaces of several metals (the figures represent averages across the optical spectrum).

	Reflectivity (%)
Steel	58
Aluminium	92
Silver	98
Copper	67
Gold	81

7.7.2 Data on optical properties of insulators

The refractive index, n_{light}, of a transparent substance is related to the speed of light, v, in the substance by

$$n_{\text{light}} = \frac{c}{v} \tag{7.110}$$

where c is the speed of light in a vacuum. The symbol n_{light} is used instead of the more usual n to prevent confusion with n, the number density of atoms. The measured refractive index, n_{light}, of various transparent substances is plotted as a function of wavelength in Figures 7.48 and 7.49. The figures show that n_{light} increases (i.e. the light slows down) at shorter wavelengths. The glasses in Figure 7.48 are composed mainly of SiO_2, with a variety of oxides, typically BaO or PbO, added at various concentrations. The glasses shown represent the range of refractive indices available to optical designers. The high-refractive-index glasses have a tendency to tarnish (oxidize) quickly, which means they are not used unless they are essential to an optical design. It is clear from both Figures 7.49 and 7.50 that for transparent substances in general the refractive index is greater for shorter wavelengths than longer wavelengths. Interestingly, the *rate* at which the refractive index increases with wavelength also increases as the wavelength is reduced.

So the questions raised by our preliminary examination of the experimental data on the optical properties of solids are:

- Why are some insulators transparent to visible light? However, light travels through these materials with a speed between ≈50% and ≈70% of its speed in free space.
- Why does the speed of light depend on the wave-

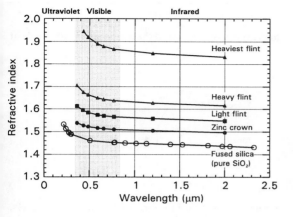

Figure 7.48 The refractive index of a variety of optical glasses as a function of wavelength. The shaded region corresponds to the visible region of the spectrum. The longer wavelengths are therefore in the infrared region of the spectrum, and the shorter wavelengths are in the ultraviolet region. (Data from Kaye & Laby; see §1.4.1)

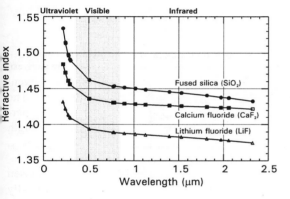

Figure 7.49 The refractive index of three transparent insulators as a function of wavelength. The SiO_2 curve is also plotted in Figure 7.48 for comparison. The shaded region corresponds to the visible region of the spectrum. The longer wavelengths are therefore in the infrared region of the spectrum, and the shorter wavelengths are in the ultraviolet region. (Data from Kaye & Laby; see §1.4.1)

length of light and become slower as the wavelength decreases?

• Why are metals highly reflective?

7.7.3 Understanding the data on the optical properties of solids

Before we can address the questions raised by the data, we need to develop a theory to describe the response of the electric charges within both metals and insulators to the oscillating electric field of a light wave.

We have considered previously what happens to metals and insulators when they are subject to *static* electric fields (§7.5). We saw that:

• in metals, a static electric field causes a flow of electrons that are essentially free to move through the metal;

• in insulators (aside from a tiny fraction of mobile electrons), most electrons remain attached to their "parent" atoms or molecules, but the charge within each atom or molecule becomes displaced or *polarized*.

At optical frequencies the distinction between metallic and insulating response remains, but now the currents and polarizations that are caused by the electric field of a light wave are *oscillating* currents and *oscillating* polarizations. In attempting to understand the optical properties in terms of these oscillating currents and polarizations we are assuming (correctly) that the electric component of the electromagnetic field produces much larger effects in atoms than does the magnetic component.

Insulators

In considering the optical properties of insulators, we

217

Example 7.25

Snell's law relates the deviation of a light ray at an interface to the refractive indices of the materials on either side of the interface.

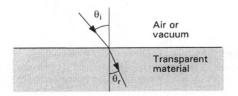
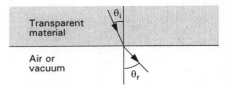

If the refractive indices of the materials on the incident and refracted sides of the interface are n_i and n_r, respectively, then Snell's law predicts that the incident angle, θ_i and the refracted angle θ_r are related by

$$\frac{\sin\theta_r}{\sin\theta_i} = \frac{n_i}{n_r}$$

For blue light ($\lambda \approx 0.4\ \mu m$) incident on a glass surface inclined at 45°, work out the angle of refraction and angle of deviation ($\theta_i - \theta_r$) of the light if the glass surface is made from (a) crown glass, (b) heavy flint glass and (c) the heaviest flint glass. Repeat the exercise for red light ($\lambda \approx 0.7\ \mu m$).

Reading from Figure 7.48 we can assemble the following table of approximate refractive indices:

	Crown glass	Heavy flint glass	Heaviest flint glass
Red light	1.51	1.63	1.88
Blue light	1.53	1.68	1.94

So for crown glass, substituting the values into Snell's law we get

$$\frac{\sin\theta_r}{\sin(45°)} = \frac{1}{1.51} = 0.6623$$

where we have assumed that the refractive index of air is ≈ 1 (Table 5.17). Solving for θ_r we find

$$\sin\theta_r = 0.6623 \times \sin(45°)$$
$$\theta_r = \sin^{-1}(0.6623 \times 0.7071) = \sin^{-1}(0.4683) = 27.93°$$

This situation is illustrated below, where it is clear that the *angle of deviation* is $45 - 27.93 = 17.07°$.

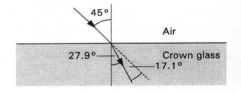

Repeating this calculation for the other glasses yields the table of angles of deviation:

	Crown glass	Heavy flint glass	Heaviest flint glass
Red light	17.07°	19.29°	22.91°
Blue light	17.47°	20.11°	23.62°
Difference	0.40°	0.82°	0.71°

The difference between the angles through which red and blue light are refracted is calculated in the last row of the table. Note that the glass with the greatest angle of deviation does not produce the greatest *dispersion* between the red and blue light.

neglect the tiny fraction of mobile electrons within the insulator. We consider that the main effect of light is to cause oscillations of the electric charges trapped around each atom or ion. Consider a "shell" of electrons with mass M and electric charge Q dis-tributed around the atom (Fig. 7.50). We assume that the "shell" moves essentially rigidly with respect to the rest of the atom, which amounts to assuming that the electric forces within the shell are much stronger than the external field of the light wave. We note also

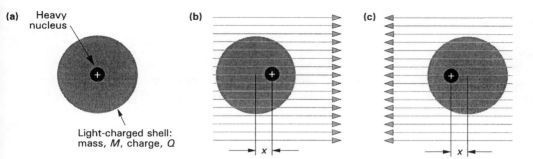

(a) Heavy nucleus

Light-charged shell: mass, M, charge, Q

(b)

(c)

x

x

Figure 7.50 The polarization of an atom in an applied electric field. (a) The electron shells around an atom are much lower in mass than the nucleus. (b, c) When the direction of the applied electric field is reversed, the heavy nucleus moves much less than the charged electron shells.

that the "shell" will have a much smaller mass than will the atom as a whole.

The electric field due to the light is $E = E_0\cos(\omega t)$ and so the shell experiences a force $F = QE_0\cos(\omega t)$. In addition to this force there is a restoring force due to the internal electric fields within the atom. We do not know the precise form of this force, but we do know that the force acts to restore the charge around the atom to its initial position. We assume that the force has the form $-Kx$, where K is the "spring constant" acting to restore the shell of charge to its optimum position, and x is the displacement of the charge shell. Assuming Newton's second law

$$F = ma \qquad (7.111)$$

we write

$$-Kx + QE_0 \cos(\omega t) = M\frac{d^2x}{dt^2} \qquad (7.112)$$

Since this is a differential equation we must assume a solution, and then show that our solution is appropriate. So assuming that the displacement of the charge x is given by

$$x = x_0 \cos(\omega t + \varphi) \qquad (7.113)$$

where x_0 is the amplitude of the charge displacement, we deduce that

$$\frac{dx}{dt} = -\omega x_0 \sin(\omega t + \varphi) \qquad (7.114)$$

and so

$$\frac{d^2x}{dt^2} = -\omega^2 x_0 \cos(\omega t + \varphi) \qquad (7.115)$$

Substituting Equations 7.113 and 7.115 into Equation 7.112, we arrive at

$$-Kx_0 \cos(\omega t + \varphi) + QE_0 \cos(\omega t) = \\ -M\omega^2 x_0 \cos(\omega t + \varphi) \qquad (7.116)$$

Rearranging this in two stages

$$QE_0 \cos(\omega t) = Kx_0 \cos(\omega t + \varphi) \\ -M\omega^2 x_0 \cos(\omega t + \varphi) \qquad (7.117)$$

$$QE_0 \cos(\omega t) = x_0 \cos(\omega t + \varphi)(K - M\omega^2) \qquad (7.118)$$

we arrive at an expression for x_0

$$x_0 = \frac{QE_0}{(K - M\omega^2)}\left[\frac{\cos(\omega t)}{\cos(\omega t + \varphi)}\right] \qquad (7.119)$$

This equation describes a situation in which the amplitude of the charge displacement grows resonantly large when the frequency of the light is such that $K = M\omega^2$. The theory as we have developed it predicts (incorrectly) that the charge displacement will be infinite when $K = M\omega^2$. This is because we have neglected the *damping term* in our differential equation (Eq. 7.112) (for further details see §2.4.2 on the simple harmonic oscillator). Because of this difficulty we must only apply Equation 7.119 at frequencies that are much less than the resonant frequency. We note, however, that the amplitude of oscillation will still grow large when $K = M\omega^2$, but that the actual amplitude of the oscillations will depend on the extent of damping of the charge oscillations.

219

Low frequency approximation

As long as we work well below the resonant frequency, the phase difference, ϕ, between the forcing field and the charge displacement stays small (see Fig. 2.7) and the time-dependent factor (in brackets in Equation 7.119) is ≈ 1. With this proviso, we write the amplitude of the charge oscillations as

$$x_0 = \frac{QE_0}{\left(K - M\omega^2\right)} \quad (7.120)$$

The resonant frequency is reached when ω satisfies the condition $K = M\omega^2$, i.e. when $\omega = \omega_0$ as given by

$$\omega_0 = \sqrt{\frac{K}{M}} \quad (7.121)$$

Substituting Equation 7.121 into Equation 7.120 we arrive at

$$x_0 = \frac{QE_0}{M\left(\omega_0^2 - \omega^2\right)} \quad (7.122)$$

If this is the amplitude of the oscillations of charge displacement, then the amplitude of the electric dipole moment caused by this displacement is

$$p_0 = Qx_0 \approx \frac{Q^2 E_0}{M\left(\omega_0^2 - \omega^2\right)} \quad (7.123)$$

If we recall the definition of the polarizability of an atom, α,

$$p_0 = \alpha E_0 \quad (7.124)$$

then by comparing Equations 7.123 & 7.124 we see that

$$\alpha \approx \frac{Q^2}{M\left(\omega_0^2 - \omega^2\right)} \quad (7.125)$$

Recalling that the approximations we have made restrict the theory to frequencies such that $\omega^2 \ll \omega_0^2$, we can write the molecular polarizability at such low frequencies as

$$\boxed{\alpha \approx \frac{Q^2}{M\omega_0^2}} \quad (7.126)$$

when $\omega \ll \omega_0$. Note that at low frequencies, the molecular polarizability, α, still depends on the value of the resonant frequency, ω_0.

The dielectric constant and the refractive index

Recalling from §5.6.2 (Equation 5.97) that the dielectric constant of a substance consisting of n non-interacting molecules per unit volume may be expressed as

$$\varepsilon - 1 = \frac{n\alpha}{\varepsilon_0} \quad (7.127)$$

then substituting for α this becomes

$$\varepsilon - 1 = \frac{nQ^2}{\varepsilon_0 M\omega_0^2} \quad (7.128)$$

Now the speed of light in free space may be expressed as

$$c = \frac{1}{\sqrt{\varepsilon_0 \mu_0}} \quad (7.129)$$

and similarly, the speed of light in a medium may be expressed as

$$v = \frac{1}{\sqrt{\varepsilon \varepsilon_0 \mu \mu_0}} \quad (7.130)$$

where ε is the dielectric constant and μ is the magnetic permeability of the medium. Taking the ratio of Equations 7.129 and 7.130 yields the refractive index, n_{light},

$$n_{light} = \frac{c}{v} = \frac{\sqrt{\varepsilon \varepsilon_0 \mu \mu_0}}{\sqrt{\varepsilon_0 \mu_0}} = \sqrt{\varepsilon \mu} \quad (7.131)$$

If the magnetic permeability is ≈ 1 then we expect to find

$$n_{light} \approx \sqrt{\varepsilon} \quad (7.132)$$

Substituting for ε from Equation 7.128 we find

$$\boxed{n_{light}^2 = \varepsilon \approx 1 + \frac{nQ^2}{\varepsilon_0 M\omega_0^2}} \quad (7.133)$$

when $\omega \ll \omega_0$.

This expression predicts the "low frequency" value of the refractive index in terms of some simple quantities and ω_0, the frequency of the resonant vibration of the electron shell within the atom.

Finding the resonant frequency

We can now solve for ω_0 in terms of the experimental values of the "low frequency" value of the refractive index. Rearranging Equation 7.133 as an expression for ω_0

$$\omega_0 = \sqrt{\frac{nQ^2}{\varepsilon_0 M \left(n_{\text{light}}^2 - 1 \right)}} \qquad (7.134)$$

and then for f_0, the frequency rather than the angular frequency,

$$f_0 = \frac{1}{2\pi} \sqrt{\frac{Q^2}{\varepsilon_0 M}} \sqrt{\frac{n}{\left(n_{\text{light}}^2 - 1 \right)}} \qquad (7.135)$$

and finally for the wavelength, $\lambda_0 = c/f_0$ of the radiation involved, we predict

$$\lambda_0 = \frac{c}{f_0} = \left(\frac{2\pi c \sqrt{\varepsilon_0 M}}{Q} \right) \sqrt{\frac{n_{\text{light}}^2 - 1}{n}} \qquad (7.136)$$

Evaluating this numerically for an electron shell with just a single electron

$$\lambda_0 = \left(\frac{2\pi \times 3.0 \times 10^8 \times \sqrt{8.85 \times 10^{-12} \times 9.1 \times 10^{-31}}}{1.6 \times 10^{-19}} \right)$$
$$\times \sqrt{\frac{n_{\text{light}}^2 - 1}{n}} \qquad (7.137)$$

$$\boxed{\lambda_0 = 3.34 \times 10^7 \sqrt{\frac{n_{\text{light}}^2 - 1}{n}}} \qquad (7.138)$$

Single electron in shell

where n_{light} is the refractive index and n is the number density of atoms.

Example 7.26 shows that we can understand the range of experimental values of the refractive index

by assuming that atoms contain charged "shells" that have resonances in the ultraviolet region of the spectrum ($\lambda_0 = 0.1\,\mu\text{m}$ to $0.4\,\mu\text{m}$). Recall that visible light has wavelengths ranging from $\approx 0.4\,\mu\text{m}$ (violet) to $\approx 0.7\,\mu\text{m}$ (red).

The questions raised by the data

In the light of the theory outlined above we will look in turn at the questions raised by our examination of the data.

The value of the refractive index

The reason why insulators are transparent to visible light is that there are no mechanisms for absorbing the light within the insulator. This is because the resonant frequencies of electron shells typically lie in the ultraviolet region of the spectrum. Thus the amplitude of the charge oscillations induced by the (relatively) low frequency oscillations of the electric field of the light wave are (relatively) small.

In a pure glass (mainly SiO_2) the resonant frequencies of electron shells lie in the far ultraviolet. However, impurities with many electron shells (such as lead) have lower resonant frequencies, much closer to the visible region of the spectrum. As Equation 7.133 suggests, lowering ω_0 has the effect of increasing the refractive index, and so adding such impurities to glass increases the overall refractive index of the mixture. However, the refractive index cannot be increased indefinitely.

Example 7.26

A typical number density of atoms in a solid is $n_0 \approx 5 \times 10^{28}\,\text{m}^{-3}$. For solids with refractive indices $n = 1.2, 1.5$ and 2, evaluate the expected value of the wavelength of radiation corresponding to the resonant frequency of the charges within the solid.

Using Equation 7.138 for $n \approx 5 \times 10^{28}\,\text{m}^{-3}$ and $n_{\text{light}} = 1.2$, we write

$$\lambda_0 = 3.34 \times 10^7 \sqrt{\frac{1.2^2 - 1}{5 \times 10^{28}}}$$

which evaluates to $\lambda_0 = 0.099 \times 10^{-6}\,\text{m}$. Similarly for $n = 1.5$ we find $\lambda_0 = 0.167 \times 10^{-6}\,\text{m}$ and for $n = 2$ we find $\lambda_0 = 0.259 \times 10^{-6}\,\text{m}$.

If the resonant frequency is lowered too much, then the substance will start to absorb light at frequencies around ω_0. The theory outlined above has ignored the role of absorption, but we can see fairly directly how this absorption occurs. If the oscillations of a charge shell grow large, then nearby atoms will be set into motion by their electrical interactions with the oscillating charge shell. This dissipates the energy of oscillation in the form of lattice waves (phonons). In addition, an oscillating charge radiates electromagnetic energy at the frequency at which it is oscillating in a process known as Rayleigh scattering (see §5.7.3). Both processes limit the usable refractive index of a transparent substance to $n < \approx 2$.

Quantum mechanics

In discussing the dynamics of a "charged shell" inside an atom under the action of an electromagnetic field, we need to be aware that the laws of quantum mechanics will limit the realm in which the classical approach we have taken is applicable. In particular, it is important to understand that the process that we have called *resonance* in our classical vocabulary describes a process that is described in a quantum mechanical vocabulary as a *transition between quantum states* (Fig. 7.51). Thus the resonances that occur in the ultraviolet region are properly described as transitions between quantum states by electrons in the outer parts of the atom.

Infrared response

The shells of charge *within* each atom of a substance

Figure 7.51 Classical and quantum mechanical descriptions of resonance. (a) Classically, the amplitude of an oscillation grows dramatically when the oscillator is subject to a force at frequency $\omega = \omega_0 = \sqrt{(K/M)}$. (b) Quantum mechanically, when $\omega = \omega_0$ photons with energy $\hbar\omega$ are absorbed and the oscillator makes transitions between its quantum states.

(a) Classical resonance   **(b) Quantum resonance**

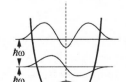

are not the only charged objects that can move in response to the oscillating electric field in a light wave: the ions themselves can move as a whole. However, the process in which whole ions move in response to an oscillating electric field occurs at much lower frequencies. Equation 7.136 gives the wavelength of the radiation that will cause resonance in a bound particle of mass M and charge Q. Example 7.26 shows that λ_0 for an electron lies in the ultraviolet region of the spectrum. However, although an ion has a similar magnitude of charge to an electron, an ionic mass may be 10^4 times greater than an electron mass. Substituting $M \approx 10^4 m_e$ in Equation 7.136, we find

$$\lambda_0 \approx 3.34 \times 10^9 \sqrt{\frac{n_{\text{light}}^2 - 1}{n}} \qquad (7.139)$$

Assuming that the refractive index and the number density of ions are similar to those in Example 7.26, the resonant wavelength is 100 times longer (of the order of $10\,\mu m$) and lies in the infrared part of the spectrum. Thus at optical frequencies (of the order of 10 to 100 times greater than the resonant frequencies corresponding to ionic vibration) the electric field oscillates so quickly that the relatively heavy ions have no time to respond before the field reverses and a new oscillation begins. Thus at optical frequencies only the electrons are light enough to oscillate in response to the field and so we may ignore the polarizability of the lattice in our considerations.

The visible range of the electromagnetic spectrum lies between the region in which ionic vibrations absorb radiation (infrared), and that in which electronic vibrations within atoms absorb radiation (ultraviolet). This is not a coincidence. The considerations outlined above apply to ions in molecules as well as ions in solids. Our eyes have evolved so as to be useful to us, and they would be of precious little use if the gas in which we lived, and the fluid in which we evolved, were opaque. We have evolved sensitivity in the range of the electromagnetic spectrum in which gases are transparent.

Variation of n_{light} with wavelength

The dependence of the speed of light on wavelength arises quite naturally from the equation for $\alpha(\omega)$. Rearranging Equation 7.125

$$\alpha(\omega) \approx \frac{Q^2}{M\omega_0^2\left[1-(\omega/\omega_0)^2\right]} \qquad (7.140)$$

and expanding the bracket to first order in ω/ω_0, we obtain

$$\alpha(\omega) \approx \frac{Q^2}{M\omega_0^2}\left[1+\left(\frac{\omega}{\omega_0}\right)^2+\ldots\right] \qquad (7.141)$$

which should be valid as long as $\omega/\omega_0 \ll 1$. Comparing this with Equation 7.126 for the zero-frequency polarizability

$$\alpha(\omega) \approx \alpha(\omega = 0)\left[1+\left(\frac{\omega}{\omega_0}\right)^2+\ldots\right] \qquad (7.142)$$

and rearranging as an expression for λ yields

$$\alpha(\lambda) \approx \alpha(\lambda = \infty)\left[1+\left(\frac{\lambda_0}{\lambda}\right)^2+\ldots\right] \qquad (7.143)$$

This revised wavelength-dependent polarizability $\alpha(\lambda)$ feeds through to an expression for the refractive index in the same way that the infinite wavelength polarizability $\alpha(\lambda = \infty)$ fed through to the infinite wavelength refractive index (Eqs 7.127–7.133). We thus find that the wavelength-dependent refractive

index is given by

$$\boxed{n_{\text{light}}(\lambda) \approx n_{\text{light}}(\lambda = \infty)\left[1+\left(\frac{\lambda_0}{\lambda}\right)^2\right]} \qquad (7.144)$$

This predicts that if we plot $n_{\text{light}}(\lambda)$ as a function of $1/\lambda^2$ then we should expect a straight-line variation with an intercept of $n_{\text{light}}(\lambda = \infty)$ and a slope $\lambda_0^2 n_{\text{light}}(\lambda = \infty)$.

Figure 7.52 shows the refractive index data from Figure 7.48 plotted versus $1/\lambda^2$. We see that the data conform rather well to straight lines and that analysis of the slopes predicts similar values of λ_0 to those inferred from an analysis of the $n_{\text{light}}(\lambda = \infty)$ theory. Of the two predictions for λ_0, the slope analysis is to be preferred, because in order to evaluate Equation 7.138 we needed to assume a certain number of electrons per shell – a number that cannot be estimated accurately without some detailed knowledge of the charge distribution of the substance under examination.

Before proceeding, it is worth reflecting on the astonishing results predicted here. Based on little more than the theory of a harmonic oscillator and Coulomb's law, we have plausibly explained the optical properties of glasses! The tiny masses involved in atomic-size oscillators have shifted the resonant frequencies to the ultra-violet regime, but the basic results still apply.

Figure 7.52 The data in Figure 7.48 replotted to show the refractive index of various glasses plotted as a function $1/\lambda^2$. The straight lines are least-squares fits to the data. The equations of the lines and the value of λ_0 inferred from the slope of the lines is given in the inset. In Example 7.26 we found that $n(\infty) \approx 1.5$ implied $\lambda_0 \approx 0.167\,\mu m$, whereas from the present data we deduce a value for λ_0 of $0.059\,\mu m$. This difference arises because in Example 7.26 we assumed just one electron per atom and used only an approximation for the number density.

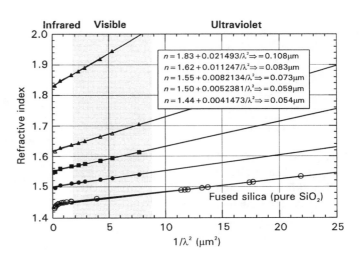

223

Metals

The main question raised by our examination of the optical properties of metals is why are metals highly optically reflective? We consider this question in a largely qualitative manner.

Insulators can be optically transparent if the charges within the insulators are bound sufficiently tightly that their resonant frequencies lie well within the ultraviolet region of the spectrum. If the resonant frequency is too close to the optical region of the spectrum, then visible light will be able to cause relatively large charge oscillations, that can dissipate the energy of the wave. In contrast, it is easy for the electric charges within metals to move in response to the electric field of the light wave.

The quantitative analysis of electromagnetic wave propagation at vacuum–metal interfaces is technically rather complex, and so in this section we present a qualitative description of the processes involved. We envisage reflection as a three-stage process.

Stage 1: The skin depth

In the first stage (Fig. 7.53a) the incoming electromagnetic wave is heavily damped as it enters the metal. This damping occurs as the electric field does work in accelerating the many free electrons in the surface of the metal. The more free electrons, the greater the damping of the wave. The electric field decays over a distance known as the *skin depth*, δ, which is related to the conductivity, σ, of the metal by (Bleaney & Bleaney, see §1.4.1)

$$\delta = \sqrt{\frac{2}{\mu_0 \sigma \omega}} \qquad (7.145)$$

Substituting for μ_0 yields

$$\delta = \sqrt{\frac{2}{4\pi \times 10^{-7} \times 2\pi}} \sqrt{\frac{1}{f\sigma}} = 503\sqrt{\frac{1}{f\sigma}} \quad (7.146)$$

Note that the damping of the wave is *not* due to dissipation into heat of the electron's energy. If that were so then skin depth would be shorter in more resistive substances, but Equation 7.146 shows that δ behaves in exactly the opposite manner.

For aluminium, $\sigma \approx 4 \times 10^7 \, \mathrm{Sm^{-1}}$ at around room

temperature (Table 7.9) and for light $f \approx 5 \times 10^{14}$ Hz so Eq. 7.146 predicts a skin depth, δ, of ≈ 4 nm. Since the wavelength of light with this frequency is ≈ 600 nm, we see that light penetrates the metal for only a small fraction of its wavelength.

Stage 2: current flow

In the second stage (Fig. 7.53b) oscillating currents flow in the small region of depth δ in which the electric field is not zero.

(a) Incoming wave

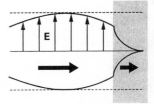

(b) Charge oscillation due to incoming wave (a)

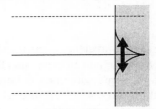

(c) Radiated (reflected) wave due to charge oscillations (b)

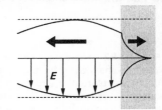

Figure 7.53 Illustration of the processes underlying the reflection of light from a metallic surface. (a) The incoming electromagnetic wave is strongly damped as it enters the metal. The length over which the field decays is known as the *skin depth*, δ, and is related to the conductivity, σ, of the metal as shown in Equation 7.145. (b) In the thin region of depth, δ, in which the electric field is not zero, oscillating currents will flow. The currents are driven by the electric field of the light wave. (c) The oscillating currents radiate electromagnetic waves. Waves moving into the solid are strongly absorbed, but waves moving away from the surface can propagate easily.

Stage 3: re-radiation

In the third stage (Fig. 7.53c) the oscillating currents will radiate electromagnetic waves. The wave radiated into the metal decays strongly, but the wave radiated back out of the metal can propagate. Since the currents in Stage 2 have not been strongly dissipated because the conductivity of the metal is so high, most of the energy is re-radiated.

The colour of copper

Figure 7.53 allows us to understand where the burnished red colour of copper arises. In a thin layer at the metal surface, the copper ions will be subjected to an oscillating electric field. As we saw above, electron shells within ions typically have resonances in the ultraviolet region of the spectrum. However, in copper, a resonance occurs in the blue region of the visible spectrum. At these resonances there is strong absorption, and so when white light illuminates the surface of copper the blue end of the spectrum is absorbed and the red and orange part of the spectrum is reflected.

Quantum mechanically, the resonance in the blue region of the spectrum corresponds to an electronic transition within the copper ions in which an electron moves from a quantum state that is part of a covalent bond to a state on the surface of the "Fermi sphere". This transition occurs for light with wavelengths $\lambda < 6.2 \times 10^{-7}$ m (blue) which corresponds to frequencies of more than 4.8×10^{14} Hz. Photons with these frequencies have an energy, $E = hf$, given by

$$E = hf = 6.6 \times 10^{-34} \times 4.8 \times 10^{14}$$
$$= 3.2 \times 10^{-19} \text{ J} = 2.0 \text{ eV} \qquad (7.147)$$

By examining the frequencies at which light (ultraviolet, optical and infrared) is transmitted through solids, one can determine experimentally the key features of the energy differences between the quantum states in solids.

Semiconductors

In discussing electrical conduction (§7.5.9) we categorized semiconductors as electrical insulators in which electrical carriers could be created relatively easily – either thermally or by the addition of impurities. However, even in the most highly doped semi-conductors, the carrier density was only ≈1% of that in typical metals. So in discussing the optical properties of semiconductors we expect to find properties characteristic of insulators, with additional characteristics appropriate to a metal with a low carrier density.

Both these expectations are borne out, but there are additional processes that take place in semiconductors that make them especially complicated to describe, but potentially especially useful. These extra complications arise because the "resonances" which occur in the ultraviolet region of the spectrum in transparent insulators occur in the optical or infra-red regions of the spectrum in semiconductors. We recall (§7.7.5) that these "resonances" correspond to processes in which electrons make transitions between quantum states within the solid. In semiconductors, the key transitions are made by electrons in quantum states in covalent bonds (valence states) to other states in which they can move through the semiconductor (conduction states). The energy above which such electronic transitions become possible is called the *energy gap* of a semiconductor and has a typical value of ≈1 eV.

Optical absorption

Since the conductivity of semiconductors is much less than that of metals, the skin depth (Eq. 7.145) is much longer than for metals, and the electric field of a light wave can penetrate a semiconductor for a distance of typically a few wavelengths. In this distance, it is able to interact with atoms and excite electrons from valence states into conduction states. This creates electron and hole carrier pairs, which in the presence of an electric field will be drawn apart and contribute to the current flow through the semiconductor. Thus a component of the current flow through a semiconductor may be made light dependent.

Optical emission

Consider a situation within a semiconductor in which conduction states are populated with electron carriers to a level above their equilibrium density. After a short while, the electron carriers will find hole carriers and annihilate. The excess energy of the electron carrier may be emitted in the form of phonons and photons. The photon energy will correspond to the energy gap of the semiconductor, typically ≈1 eV.

> **Example 7.27**
>
> **Silicon has a band gap of around 1.13 eV. Estimate the minimum frequency of light that could be detected or emitted by a silicon-based electrical device.**
>
> The frequency, f, is related to the band gap energy by $E = hf$, and so we write
>
> $$f = \frac{E}{h} = \frac{1.13 \times 1.6 \times 10^{-19}}{6.62 \times 10^{-34}} = 2.7 \times 10^{14}\,\text{Hz}$$
>
> This corresponds to a frequency in the infrared region of the spectrum.

The situation in which an excess of carriers is created may occur at the junction between two pieces of semiconductor, one which has mainly n-type impurities, and the other with mainly p-type impurities. Such a device is called a *pn junction* or a *diode* (a general name for any two-terminal electrical device). Among the many uses of pn junctions, perhaps the most technologically significant is their ability to emit light.

7.8 Magnetic properties

7.8.1 Introduction

All substances respond to an applied magnetic field by becoming *magnetized*, i.e. they behave as if they were a bar magnet themselves. The quantitative measure of the extent to which they behave like a bar magnet is called the *magnetization* of the substance. The responses of different substances to applied magnetic fields normally belong to one of three broad categories.

(a) The "bar magnet" created in the sample lines up so that the magnetic field inside the material is increased. The energy of the atoms in the material is lowered by applying a field and so the sample tries to move into regions of high magnetic field. Such materials are called *paramagnets*.

(b) The "bar magnet" created in the sample lines up so that the field inside the material is decreased. The energy of the atoms in the material is increased by applying a field and so the sample tries to move away from regions of high mag-

netic field. Such materials are called *diamagnets*.

(c) The sample can possess a non-zero magnetization *in the absence of any applied magnetic field*. Such materials are called *ferromagnets*.

By far the majority of substances fall into categories (a) and (b). In either category, when the sample is removed from the magnetic field, the magnetization returns to zero. Other more unusual forms of magnetic response do exist, but their phenomenology is rather complex and it is not appropriate to discuss them here. The response of diamagnetic and paramagnetic substances to an applied magnetic field is shown schematically in Figure 7.54. The key feature of ferromagnetic behaviour is illustrated in Figure 7.55.

7.8.2 Quantitative magnetic measurements

Magnetic field

The magnetic field in a region of space, also referred to as the *magnetic induction,* has the symbol **B** and is measured in tesla (T).

One may also define a second field, **H**, known as the *applied magnetic field*, which is measured in ampere per metre (A m^{-1}). The existence of two ways of describing magnetic phenomena has led to enormous confusion. From the point of view of the solid-state physicist, there are two points to bear in mind.

(a) Microscopically, the magnetic field to which electrons, protons and neutrons respond is the *local* value of **B**, the *magnetic induction* in their vicinity.

(b) Inside a sample exposed to a magnetic field, the value of **B** differs from the value outside the sample because of the magnetic response of the substance. This effect is illustrated for diamagnets and paramagnets in Figures 7.54a and 7.54b respectively, although the effect is largest in ferromagnets (Fig. 7.55) which can experience large values of internal magnetic field in the absence of *any* applied field. Thus the value of **B** inside the sample is not the same as the value of **B** which was applied to the sample. In the context of magnetic measurements, the field **H** may be regarded as a fictional field which is not affected by the magnetization of the material.

(a) A diamagnetic sample in a magnetic field

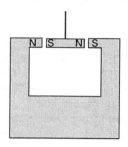

(a) A paramagnetic sample in a magnetic field

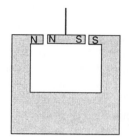

(b) The lines of flux of the field due to the "bar magnet" tend to *decrease* the field inside the material and *increase* the field on either side of the material

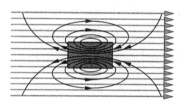

(b) The lines of flux of the field due to the "bar magnet" tend to *increase* the field inside the material and *decrease* the field on either side of the material

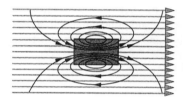

(c) The flux density inside the material is *less* than the flux density of the applied field, i.e. the sample has "repelled" flux lines

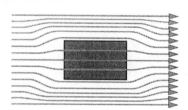

(c) The flux density inside the material is *greater* than the flux density of the applied field, i.e. the sample has "concentrated" flux lines.

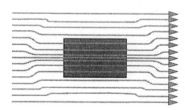

Figure 7.54 Illustration of the response of diamagnetic and paramagnetic substances to an applied magnetic field. Note that both types of substance have no magnetic moment in the absence of an applied field. (a) Samples suspended in a magnetic field. (b) The applied magnetic field is indicated by the shaded arrowed lines. The magnetic field induced in the sample (full black lines) has the form of the field due to a bar magnet. (c) The net magnetic field.

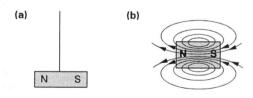

Figure 7.55 Illustration of the key properties of a ferromagnetic material. (a) A sample suspended in free space showing a magnetic moment in the absence of an applied field. (b) The lines of magnetic flux around a ferromagnetic material; note that there is no applied magnetic field.

The magnetic field **B** inside a material is given by

$$\mathbf{B} = \mu_0(\mathbf{H} + \mathbf{M}) \qquad (7.148)$$

where μ_0 is a constant with the value $4\pi \times 10^{-7}$ henry per metre ($\mathrm{Hm^{-1}}$), and **M** is the magnetization of the material, defined below.

There are several ways of specifying the extent of the magnetic effect induced in a sample of material. These are described below.

Magnetic moment

The extent of the magnetic response of a sample can be determined by measuring the distortion of the magnetic field around the edge of a magnetized sample. The "magnetic strength" exhibited in a particular sample is specified in terms of the equivalent distortion that would be achieved by an electric current flowing in a small loop at the position of the sample. The strength of the magnetic field created by a current loop is specified in terms of its *magnetic moment*, **m**, (sometimes called the *magnetic dipole moment*), which is given by

$$\mathbf{m} = i \times \mathbf{A} \qquad (7.149)$$

Thus the magnetic moment is measured in ampere square metres (Am^2) and points in the direction indicated by a right-hand screw rotating in the same sense as the current.

Magnetization

In a fixed applied field, the magnetic moment (defined in Eq. 7.149) of a large sample will (obviously) be larger than the magnetic moment of a small sample of the same substance. Thus one defines the *magnetic moment per unit volume*, the *magnetic moment per kilogram* and the *magnetic moment per mole*. The most common of these is the specification of the magnetic moment per unit volume, which is known as the *magnetization*, **M**:

$$\mathbf{M} = \frac{\mathbf{m}}{V} \qquad \text{(units: } \frac{Am^2}{m^3} = Am^{-1}) \quad (7.150)$$

Susceptibility

Both the magnetic moment (Eq. 7.149) and the magnetization (Eq. 7.150) of a sample depend on the magnetic field that the sample experiences: the magnetization in a large magnetic field will be larger than in a small magnetic field. If we divide the magnetization by the applied field **H**, then we obtain a measure of the strength of magnetic response, which does not depend on applied field or sample size. If this is done, one obtains the *magnetic susceptibility per unit volume*, or the *volume susceptibility* for short. However, frequently one encounters:

(a) the *mass susceptibility*, which is the magnetic moment per kilogram, divided by the applied field; or

(b) the *molar susceptibility*, which is the magnetic moment per mole, divided by the applied field.

All these susceptibilities are given the symbol, χ, pronounced "khi", subscripted with an appropriate letter. We thus have

(a) χ_v the *volume susceptibility*, which is the *magnetic moment per unit volume*, divided by the applied field **H**, or

$$\chi_v = \frac{m}{\text{Sample volume} \times H} \qquad (7.151)$$

$$\left(\text{Units: } \frac{Am^2}{m^3 Am^{-1}} = \text{dimensionless} \right)$$

or $\chi_v = \dfrac{M}{H}$ $\left(\text{Units: } \dfrac{Am^{-1}}{Am^{-1}} = \text{dimensionless} \right)$

$$(7.152)$$

(b) χ_m, the *molar susceptibility*, which is the *magnetic moment per mole*, divided by the applied field **H**, or

$$\chi_m = \frac{m}{\text{Number of moles in sample} \times H}$$

$$\left(\text{Units: } \frac{Am^2}{mol \times Am^{-1}} = m^3 mol^{-1} \right)$$

$$(7.153)$$

(c) χ_{kg}, the *mass susceptibility*, which is the *magnetic moment per unit mass*, divided by the applied field **H**, or

$$\chi_{kg} = \frac{m}{\text{Sample mass} \times H}$$

$$\left(\text{Units: } \frac{Am^2}{kg \times Am^{-1}} = m^3 kg^{-1} \right)$$

$$(7.154)$$

Example 7.28

The molar susceptibilities of copper and of cerium are $-6.87 \times 10^{-11} \, m^3 \, mol^{-1}$ and $3.04 \times 10^{-8} \, m^3 \, mol^{-1}$ respectively. Samples of each substance are subjected to a magnetic induction of $B_0 = 0.1 \, T$. What is the magnetic field inside each sample?

In order to calculate the field, **B**, inside the sample we need to use Equation 7.148

$$\mathbf{B} = \mu_0 (\mathbf{H} + \mathbf{M})$$

and so we need first to calculate the applied field, **H**, and the magnetization, **M**. The applied field is unaffected by the presence of a sample and so it is the same as the field which would be there in the absence of sample, i.e. if $\mathbf{M} = 0$. Thus

$$\mathbf{H} = \frac{\mathbf{B}_0}{\mu_0}$$

The magnetization is not quite so straightforward to calculate. We are given the molar susceptibility, χ_M, i.e. the magnetic moment per mole (per unit field). In order to calculate the magnetic moment per unit volume – the magnetization, **M** – we need to know the density, ρ, and molar mass, A, of the substance. The volume susceptibility χ_V is then given in terms of χ_M by

$$\chi_V = \frac{\rho \chi_M}{A} = \frac{\mathbf{M}}{\mathbf{H}} \text{ and so } \mathbf{M} = \frac{\mathbf{H} \rho \chi_M}{A}$$

and hence the magnetic induction field inside the sample is given by

$$\mathbf{B} = \mu_0 \left(\frac{\mathbf{B}_0}{\mu_0} + \frac{\mathbf{B}_0 \rho \chi_M}{\mu_0 A} \right) = \mathbf{B}_0 + \frac{\mathbf{B}_0 \rho \chi_M}{A}$$

From Table 7.18 we find that 1 mol of copper has a mass of $A = 63.55 \times 10^{-3} \, kg$ and a density of $\rho = 8933 \, kg \, m^{-3}$ and hence the difference in magnetic field between the inside and outside of a sample of copper $(\mathbf{B} - \mathbf{B}_0)$ is

$$\Delta \mathbf{B} = \frac{\mathbf{B}_0 \rho \chi_M}{A} = \frac{0.1 \times 8933 \times (-6.87) \times 10^{-11}}{63.55 \times 10^{-3}} = -9.66 \times 10^{-7} \, T$$

Similarly, from Table 7.19, 1 mol of cerium has a mass of $A = 140.1 \times 10^{-3} \, kg$ and a density of $\rho = 6711 \, kg \, m^{-3}$ and hence the difference in magnetic field between the inside and outside of a sample of cerium is

$$\Delta \mathbf{B} = \frac{\mathbf{B}_0 \rho \chi_M}{A} = \frac{0.1 \times 6711 \times 3.04 \times 10^{-8}}{140.1 \times 10^{-3}} = 1.46 \times 10^{-4} \, T$$

Thus for a sample of copper in an applied field of 0.1 T, the field inside the sample is slightly less than the field outside the sample by around $10^{-6} \, T$, or roughly 1 part in 10^5 of the applied field. For the cerium sample, the field inside the sample is slightly more than the field outside the sample by around $10^{-4} \, T$, or roughly 1 part in 10^3 of the applied field.

7.8.3 Data on the elements

For paramagnetic and diamagnetic substances the magnetic moment induced by an applied field is, in general, proportional to the applied magnetic field, at least for weak applied fields. From the definitions of magnetic susceptibility (Eq. 7.151–7.154), it can be seen that the susceptibility will then be independent of the strength of the applied magnetic field.

The *molar magnetic susceptibility*, χ_M (Eq. 7.153), of the elements at around room temperature is shown in Table 7.18 and summarized in Figure 7.56. The shading in Table 7.18 corresponds to the shaded regions in Figure 7.56. The shaded data indicated by B–D in Figure 7.56 mark the first and second row of transition elements, and the lanthanide series, respectively. The unshaded band, A, marks the ferromagnetic elements iron, cobalt and nickel.

The data indicate that some elements exhibit weak diamagnetism and others a rather stronger paramagnetism. Ferromagnetism appears to be a relatively rare magnetic phenomenon – at least at room temperature. However, the clearest feature of Figure 7.56 is that elements with atomic numbers, Z, from 57 to 70 – known as the lanthanide series of elements – have by far the strongest magnetic response – and this magnetic response is paramagnetic. Furthermore, it is clear that first row ($Z = 21$ to 30) and second row ($Z = 39$ to 48) transition elements also have a relatively strong paramagnetic response. The third row transition series ($Z = 71$ to 80) also has a relatively strong response, but this does not show up so clearly in comparison with the nearby lanthanide series.

In large magnetic fields, the proportionality between the induced magnetic moment and applied field is *sometimes* broken. However, the characteristic deviations from direct proportionality with field vary with temperature in a complex manner.

The main questions raised by our preliminary examination of the experimental data on the magnetic properties of solids are:

- Why is paramagnetism strongest in elements belonging to the transition series and especially strong in the lanthanide series of elements?
- Why may elements outside the transition series be either slightly diamagnetic or slightly paramagnetic?
- Why do most elements acquire only small magnetic moments in an applied magnetic field but a

Table 7.18 The molar magnetic susceptibility of the elements at around room temperature. The data are summarized in Figure 7.56. The shading in the table corresponds to the shaded regions in Figure 7.56, and highlights sets of elements in which there is a stronger than normal magnetic response.

Z	Element	A	ρ (kg m^{-3})	χ_M (m^3 mol^{-1})	Z	Element	A	ρ (kg m^{-3})	χ_M (m^3 mol^{-1})
1	Hydrogen, H	1.008	89	–	51	Antimony, Sb	121.7	6692	-1.22×10^{-9}
2	Helium, He	4.003	120	–	52	Tellurium, Te	127.6	6247	-4.98×10^{-10}
3	Lithium, Li	6.941	533	1.78×10^{-10}	53	Iodine, I	126.9	4953	-5.58×10^{-10}
4	Beryllium, Be	9.012	1846	-1.17×10^{-10}	54	Xenon, Xe	131.3	3560	-5.51×10^{-10}
5	Boron, B	10.81	2466	-8.43×10^{-11}	55	Caesium, Cs	132.9	1900	3.72×10^{-10}
6	Carbon, C	12.01	2266	-7.57×10^{-11}	56	Barium, Ba	137.3	3594	2.61×10^{-10}
7	Nitrogen, N	14.01	1035	–	57	Lanthanum, La	138.9	6174	1.53×10^{-9}
8	Oxygen, O	16	1460	–	58	Cerium, Ce	140.1	6711	3.04×10^{-8}
9	Fluorine, F	19	1140	–	59	Praseodymium, Pr	140.9	6779	6.30×10^{-8}
10	Neon, Ne	20.18	1442	-8.48×10^{-11}	60	Neodymium, Nd	144.2	7000	7.07×10^{-8}
11	Sodium, Na	22.99	966	2.02×10^{-10}	61	Promethium, Pm	145	7220	–
12	Magnesium, Mg	24.31	1738	1.65×10^{-10}	62	Samarium, Sm	150.4	7536	2.29×10^{-8}
13	Aluminium, Al	26.98	2698	2.08×10^{-10}	63	Europium, Eu	152	5248	4.27×10^{-7}
14	Silicon, Si	28.09	2329	-5.06×10^{-11}	64	Gadolinium, Gd	157.2	7870	Ferro
15	Phosphorus, P	30.97	1820	-3.41×10^{-10}	65	Terbium, Tb	158.9	8267	1.83×10^{-6}
16	Sulphur, S	32.06	2086	-1.95×10^{-10}	66	Dysprosium, Dy	162.5	8531	1.30×10^{-6}
17	Chlorine, Cl	35.45	2030	–	67	Holmium, Ho	164.9	8797	9.05×10^{-7}
18	Argon, A	39.95	1656	–	68	Erbium, Er	167.3	9044	5.57×10^{-7}
19	Potassium, K	39.1	862	2.62×10^{-10}	69	Thulium, Tm	168.9	9325	3.21×10^{-7}
20	Calcium, Ca	40.08	1530	5.61×10^{-10}	70	Ytterbium, Yb	173	6966	3.13×10^{-9}
21	Scandium, Sc	44.96	2992	3.96×10^{-9}	71	Lutetium, Lu	175	9842	2.28×10^{-10}
22	Titanium, Ti	47.9	4508	1.92×10^{-9}	72	Hafnium, Hf	178.5	13276	9.46×10^{-10}
23	Vanadium, V	50.94	6090	3.20×10^{-9}	73	Tantalum, Ta	180.9	16670	1.94×10^{-9}
24	Chromium, Cr	52	7194	2.31×10^{-9}	74	Tungsten, W	183.9	19254	7.36×10^{-10}
25	Manganese, Mn	54.94	7473	6.59×10^{-9}	75	Rhenium, Re	186.2	21023	8.49×10^{-10}
26	Iron, Fe	55.85	7873	Ferro	76	Osmium, Os	190.2	22580	1.24×10^{-10}
27	Cobalt, Co	58.93	8800	Ferro	77	Iridium, Ir	192.2	22550	3.21×10^{-10}
28	Nickel, Ni	58.7	8907	Ferro	78	Platinum, Pt	195.1	21450	2.54×10^{-9}
29	Copper, Cu	63.55	8933	-6.87×10^{-11}	79	Gold, Au	197	19281	-3.51×10^{-10}
30	Zinc, Zn	65.38	7135	-1.44×10^{-10}	80	Mercury, Hg	200.6	13546	–
31	Gallium, Ga	69.72	5905	-2.72×10^{-10}	81	Thallium, Tl	204.4	11871	-6.40×10^{-10}
32	Germanium, Ge	72.59	5323	-9.64×10^{-11}	82	Lead, Pb	207.2	11343	-2.88×10^{-10}
33	Arsenic, As	74.92	5776	-6.87×10^{-11}	83	Bismuth, Bi	209	9803	-3.52×10^{-9}
34	Selenium, Se	78.96	4808	-3.16×10^{-10}	84	Polonium, Po	209	9400	–
35	Bromine, Br	79.9	3120	–	85	Astatine, At	210	–	–
36	Krypton, Kr	83.8	3000	–	86	Radon, Rn	222	4400	–
37	Rubidium, Rb	85.47	1533	2.13×10^{-10}	87	Francium, Fr	223	–	–
38	Strontium, Sr	87.62	2583	1.16×10^{-9}	88	Radium, Ra	226	5000	–
39	Yttrium, Y	88.91	4475	2.40×10^{-9}	89	Actinium, Ac	227	10060	–
40	Zirconium, Zr	91.22	6507	1.53×10^{-9}	90	Thorium, Th	232	11725	1.67×10^{-9}
41	Niobium, Nb	92.91	8578	2.56×10^{-9}	91	Protactinium, Pa	231	15370	–
42	Molybdenum, Mo	95.94	10222	1.15×10^{-9}	92	Uranium, U	238	19050	5.14×10^{-9}
43	Technetium, Tc	97	11496	3.01×10^{-9}	93	Neptunium, Np	237	20250	–
44	Ruthenium, Ru	101.1	12360	5.43×10^{-10}	94	Plutonium, Pu	244	19840	7.73×10^{-9}
45	Rhodium, Rh	102.9	12420	1.40×10^{-9}	95	Americium, Am	243	13670	1.22×10^{-8}
46	Palladium, Pd	106.4	11995	7.13×10^{-9}					
47	Silver, Ag	107.9	10500	-2.45×10^{-10}					
48	Cadmium, Cd	112.4	8647	-2.48×10^{-10}					
49	Indium, In	114.8	7290	-8.04×10^{-10}					
50	Tin, Sn	118.7	7285	-4.75×10^{-10}					

(Data from Emsley; see §1.4.1)

(a)

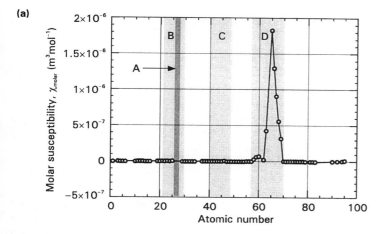

(b)

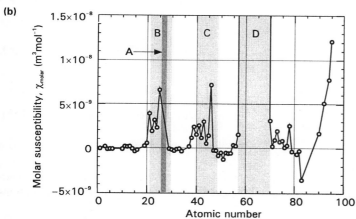

Figure 7.56 Summary of the molar magnetic susceptibility data for the solid elements shown: (a) at a large scale (b) on a detailed scale. The bands B–D mark the first and second row of transition elements, and the lanthanide series. Band A marks the ferromagnetic elements iron, cobalt and nickel. Points greater than zero correspond to a paramagnetic response; points less than zero correspond to a diamagnetic response.

few elements, such as iron, exhibit permanent magnetic moments?

7.8.4 Understanding the magnetic properties of the elements

We will analyze the magnetic response of solids as the sum of several relatively independent terms, some of which are diamagnetic, some paramagnetic and some ferromagnetic. In any particular case the balance between the different terms determines the overall response. The different magnetic responses arise from electrons in different situations within the solid. In general we can categorize electrons as being:

- core electrons, in filled electron shells;
- core electrons, in partly filled electron shells; or

- conduction electrons.

However, for both core electrons and conduction electrons, we have to consider:

- the alteration of the motion of charged particles in an applied magnetic field, and
- the reorientation of the intrinsic magnetic moment of electrons in an applied magnetic field.

These two contributions, generally referred to as the *orbital* and *spin* contributions, respectively, must be considered for both core electrons and conduction electrons. For example, conduction electrons generally have a diamagnetic orbital response, but a paramagnetic spin response.

The diversity of magnetic properties is so great, that it will be as well to map out where our discussion will lead before we actually begin. Table 7.19 summarizes the magnetic responses of non-interacting elec-

Table 7.19 Summary of the response of non-interacting electrons to applied magnetic fields (see Table 7.20 for more details). The type of response is listed with the name of the scientist most closely associated with it.

	Spin response	Orbital response
Core electrons (filled shells)	No response	Diamagnetism (Larmor)
Core electrons (partly filled shells)	Paramagnetism (Curie)	Paramagnetism (Curie)
Conduction electrons	Paramagnetism (Pauli)	Diamagnetism (Landau)

trons in solids. In any particular material the balance between paramagnetic and diamagnetic responses is slightly different.

The origin of the ferromagnetic response, which is seen in just a few elements, lies in the *electrical* interaction between electrons – often, but not exclusively, *core electrons*. This is discussed at the end of this section (see Ferromagnets, p. 240). Our discussion of the magnetic properties of solids will therefore cover each of the three "subsystems" first, assuming no electrical interaction between electrons on neighbouring atoms.

Filled electron shells

The *spin response* of a filled electron shell is easy to calculate: it is zero, i.e. filled electron shells do not acquire any net magnetic moment in the presence of an applied field due to electron spin. This can be seen by noting that, in any filled shell, electrons occupy quantum states in pairs with opposite spin. Thus in zero applied field the net magnetic moment due to spin is zero. Furthermore, the Pauli exclusion principle prevents the spins from reorienting and thus the spins are not able to acquire a net magnetic moment in an applied field.

A semi-classical calculation

The orbital response of a filled electron shell is not so straightforward to calculate but we can approach the problem semi-classically as follows. We consider electrons in atomic orbitals to move in circles of radius r. The magnetic moment due to a current i flowing in a circle of area A is

$$\text{Magnetic moment} = iA \qquad (7.155)$$

If the "current" is due to the rapid orbit with angular frequency ω of an electron with charge q moving around a circular path of radius r (Fig. 7.57a), then Equation 7.155 becomes

$$\text{Magnetic moment} = \frac{q\omega}{2\pi} \times \pi r^2 \qquad (7.156)$$

which simplifies to

$$\text{Magnetic moment} = \frac{1}{2} q\omega r^2 \qquad (7.157)$$

Now the electron moves in a circular path due to a centripetal force, F_0, which in our case will be the electrical attraction between the electron and its ion core. We can relate F_0 and ω by the standard relationship

$$F_0 = mr\omega^2 \qquad (7.158)$$

where m is the electron mass.

Now consider the situation when a magnetic field, B, is applied perpendicular to the orbit (Fig. 7.57b). In the presence of an applied magnetic field there is an extra force on the particle which may act either with F_0 or against it. In either case its effect is mainly to alter the orbital frequency – the area of the orbit remains unaffected to first order. Thus we write

$$F_0 - qvB = mr(\omega + \Delta\omega)^2 \qquad (7.159)$$

Expanding to first order in $\Delta\omega$,

$$F_0 - qr\omega B = mr\omega^2 + 2mr\omega\Delta\omega + \dots \qquad (7.160)$$

and recalling that $F_0 = mr\omega^2$, we write

$$-qr\omega B = 2mr\omega\Delta\omega + \dots \qquad (7.161)$$

and solving for $\Delta\omega$ we arrive at

$$\Delta\omega = \frac{-qB}{2m} \qquad (7.162)$$

$\Delta\omega$ is also known as the *cyclotron* or *Larmor frequency*. Now the increased (or decreased) orbital frequency causes a change in the magnetic moment of the orbital, since the magnetic moment $= q\omega r^2/2$ (Eq. 7.157). Substituting Equation 7.162 in Equation 7.157 we write

$$\boxed{\Delta(\text{Magnetic moment}) = -\frac{1}{4}\frac{q^2B}{m}r^2} \qquad (7.163)$$

which yields the change in magnetic moment of an orbital due to an applied magnetic field B. Note that in the case shown in Figure 7.57:

- The magnetic moment of the orbit initially points in the opposite direction to the magnetic field.
- The effect of the magnetic field (Figs 7.5a, c) is to *increase* the orbital speed of the electron, and hence to increase the current, and hence to increase the size of the magnetic moment that opposes the applied magnetic field. The field effect on the orbit is, therefore, diamagnetic.

However, in fully filled electron shells, for every electron orbiting in one sense there is an equivalent electron orbiting in the opposite sense. For an electron orbiting in the opposite sense to that in Figure 7.57:

(a) The magnetic moment of the orbit initially points in the same direction as the magnetic field.

Figure 7.57 The force F_0 (a) keeps a charge particle in a circular orbit of radius r at speed v. (b) In the presence of additional magnetic field **B**, the electron experiences an additional force qvB, as shown. Note that for electrons q is negative and so the actual force of magnitude $|qvB|$ actually acts in the opposite direction to that shown.

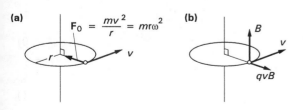

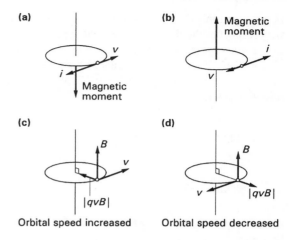

Orbital speed increased   Orbital speed decreased

Figure 7.58 (a&b) The directions of the magnetic moment resulting from electron orbits in opposite directions. In the presence of a magnetic field **B**, the additional force (Fig. 7.57b) acts (c) to increase the orbital speed of the electron whose moment opposes **B**, and (d) to decrease the orbital speed of the electron whose moment is parallel to **B**. The sign of electron charge has been taken into account in calculating the directions shown.

(b) The effect of the magnetic field (Figs 7.58b, d) is to *decrease* the orbital speed of the electron, and hence to decrease the current and decrease the size of the magnetic moment in the same sense as the applied magnetic field. The field effect on the orbit is, therefore, also diamagnetic.

Thus no matter what the orientation of the magnetic field, its net effect on the orbital moment is diamagnetic. Note that the magnetic moment of the orbital is generally much larger than the diamagnetic alteration to its value. However, in filled electron shells, each orbital moment is paired with an equal and opposite one, and so only the diamagnetic effect is observed. Diamagnetism arising from this origin is known as *Larmor* diamagnetism.

More detailed predictions

Electron orbits in an atom are not simple circles, but three-dimensional charge distributions. To take account of this, in place of r^2 we write $\overline{r^2}$, which is the average radius of an orbit perpendicular to the magnetic field:

$$\Delta(\text{Magnetic moment}) = -\frac{1}{4}\frac{q^2B}{m}\overline{r^2} \qquad (7.164)$$

233

If the electron orbital has no particular orientation with respect to the direction of a field applied along, say, **z**, then we expect to find $\overline{x^2} = \overline{y^2} = \overline{z^2}$. Since $\overline{r^2} = \overline{x^2} + \overline{y^2}$, if the average radius of the orbital is $\overline{\rho}$ then $\overline{r^2} = \overline{x^2} + \overline{y^2} + \overline{z^2}$ and so $\overline{r^2} = 2\overline{\rho^2}/3$. We can thus rewrite Equation 7.164 as

$$\Delta(\text{Magnetic moment}) = -\frac{1}{4}\frac{q^2 B}{m} \times \frac{2}{3}\overline{\rho^2} \quad (7.165)$$

$$\Delta(\text{Magnetic moment}) = -\frac{q^2 B}{6m}\overline{\rho^2} \quad (7.166)$$

If we express this in terms of the applied field, **H**, this becomes

$$\Delta(\text{Magnetic moment}) = -\frac{q^2 \mu_0 H}{6m}\overline{\rho^2} \quad (7.167)$$

Now, each atom has $\approx Z$ filled electron orbitals, and so we expect each orbital to aquire a moment

$$\Delta(\text{Magnetic moment}) = -\frac{Z q^2 \mu_0 H}{6m}\overline{\rho^2} \quad (7.168)$$

where $\overline{\rho^2}$ is now an average value for each type of atom. In one mole of substance we have N_A such magnetic moments, and so dividing by H we obtain

$$\chi_{\text{molar}} = -\frac{N_A Z q^2 \mu_0}{6m}\overline{\rho^2} \quad (7.169)$$

Approximating $\rho \approx 0.1\,\text{nm}$ for all orbitals (which is likely to be correct within a factor of 2 or 3 either way), Equation 7.169 evaluates to

$$\chi_{\text{molar}} \approx -Z\frac{6.02 \times 10^{23} \times \left(1.6 \times 10^{-19}\right)^2 \times 4\pi \times 10^{-7}}{6 \times 9.1 \times 10^{-31}}$$
$$\times \left(10^{-10}\right)^2$$

$$\qquad (7.170)$$

$$\chi_{\text{molar}} \approx 3.6Z \times 10^{-11}\,\text{m}^3\text{mol}^{-1} \quad (7.171)$$

This rough prediction for the Larmor diamagnetic response of all atoms is compared with the experimental data in Figure 7.59. We see that the prediction appears to be slightly greater than the diamagnetic response of any particular element, but does form a conceivable "baseline" – or background – diamagnetic suspectibility. In order to make a more realistic calculation of this background diamagnetic response found in all matter, one can consult tabulations of the calculated values of $\overline{\rho^2}$ for atomic orbitals. These values would show more realistic variations across the periodic table.

Partly filled electron shells

As we mentioned in the text following Figure 7.58, the magnetic moment due to an electron in an orbital is generally much larger than the diamagnetic change in the magnetic moment when a magnetic field is applied. If the magnetic moment is not compensated by an equivalent orbital, then the main effect of the applied field is to apply a *torque* to the magnetic moment, which tends to align it with the applied field. This paramagnetic response can only occur in

Figure 7.59 Estimate of the Larmor diamagnetic contribution to the susceptibility of the elements according to Equation 7.171.

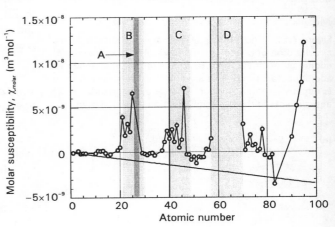

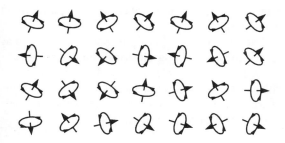

Figure 7.60 An illustration of the random orientations of permanent magnetic moments on atoms. The magnetic moment is represent schematically by an arrow and a loop indicating the sense in which an equivalent current would flow.

partly filled electron shells, but when it does occur it is generally considerably stronger than the background diamagnetic response of the filled orbitals on the atom or ion.

Thus if we have partly filled electron shells, in the absence of an applied field the magnetic moments on each atom or ion will point randomly in all directions (Fig. 7.60). An applied magnetic field will tend to align the magnetic moments to some extent, in competition with the thermal disorder that tends to randomize the atomic orientations.

Curie law

If all the magnetic moments, m, in one mole of substance were parallel to one another, then the total magnetic moment would be $N_A m$. The effect of the thermal disorder reduces the total magnetic moment below this figure, and so we can write

$$\text{Molar magnetic moment} = \text{Fraction} \times N_A m \quad (7.172)$$

where the fraction is zero in the absence of a magnetic field, and has a maximum value of 1 when all the atomic magnetic moments are aligned. The value of the fraction will depend on the ratio of the magnetic energy of a magnetic dipole in a field to the thermal energy of the magnetic dipole. As an approximation we therefore write

$$\text{Molar magnetic moment} = \frac{\text{Magnetic energy}}{\text{Thermal energy}} \times N_A m$$

$$(7.173)$$

Recalling that the potential energy of a magnetic dipole in a field is $-\mathbf{m} \cdot \mathbf{B}$ which is $\approx mB$, (§2.3.4), and the thermal energy associated with atomic rotation is $\approx k_B T$, we write

$$\text{Molar magnetic moment} \approx \left(\frac{mB}{k_B T} \right) \times N_A m \quad (7.174)$$

We expect this expression to be valid only when $mB \ll k_B T$. In this case we can write $B \approx \mu_0 H$:

$$\text{Molar magnetic moment} \approx \left(\frac{\mu_0 mH}{k_B T} \right) \times N_A m$$

$$(7.175)$$

so the molar susceptibility (Eq. 7.153) is given by

$$\chi_{\text{molar}} \approx \frac{\mu_0 N_A m^2}{k_B T} \quad (7.176)$$

A more detailed theory (Bleaney & Bleaney, see §1.4.1) predicts an expression which differs from Equation 7.176 by a factor of $1/3$:

$$\chi_{\text{molar}} = \frac{1}{3} \frac{\mu_0 N_A m^2}{k_B T} \, \text{m}^3 \text{mol}^{-1} \quad (7.177)$$

(a)

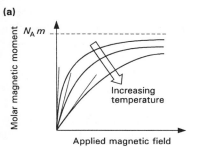

(b)

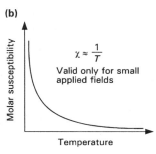

$$\chi \approx \frac{1}{T}$$

Valid only for small applied fields

Figure 7.61 Illustration of the (Pierre) Curie law behaviour. (a) Three magnetization curves at three different temperatures. Note that at the highest fields the magnetic moment eventually saturates. (b) The variation of the initial slopes in (a) with temperature.

235

Example 7.29

Vanadium is an element that has unfilled electronic orbitals. What is the value of the magnetic moment on each atom ?

If we rearrange Equation 7.177 to solve for m we can write

$$m = \sqrt{\frac{3k_B}{\mu_0 N_A}} \times \sqrt{\chi_{molar} T}$$

Enumerating this yields

$$m = \sqrt{\frac{3 \times 1.38 \times 10^{-23}}{4\pi \times 10^{-7} \times 6.02 \times 10^{23}}} \times \sqrt{\chi_{molar} T}$$

which evaluates to

$$m = 7.40 \times 10^{-21} \sqrt{\chi_{molar} T} \ \text{J T}^{-1}$$

Substituting the value of the molar susceptibility of vanadium (from Table 7.19) = $3.20 \times 10^{-9} \text{m}^3 \text{mol}^{-1}$ at 293 K) we arrive at

$$m = 7.40 \times 10^{-21} \sqrt{3.20 \times 10^{-9} \times 293} = 7.17 \times 10^{-24} \ \text{J T}^{-1}$$

Magnetic moments on atoms and ions are normally expressed in smaller units than joules per tesla (J T^{-1}) known as Bohr magnetons, μ_B, defined by

$$\mu_B = \frac{e\hbar}{2m_e} = 9.274 \times 10^{-24} \ \text{J T}^{-1}$$

In these units the magnetic moments of the atoms of a substance may be written as

$$m = 798 \sqrt{\chi_{molar} T} \quad \text{Bohr magneton}$$

which evaluates to

$$m = 0.77 \ \mu_B$$

Note that this answer has neglected to take account of the weak diamagnetic contribution to the molar susceptibility that arises from filled shells (see the previous section on Filled electron shells).

Equation 7.177 expresses the *Curie law* of magnetic susceptibility (Fig. 7.61) – named after Pierre Curie, the husband of Marie (see also §5.6.2).

The magnitudes of atomic magnetic moments

As shown in Example 7.29, typical values of atomic magnetic moments deduced from a Curie law analysis of the transition series elements are of the order of $1 \mu_B$. The magnetic moments arise because of the or-bital and spin orientations of electrons in partly filled electron shells.

Each electron within a shell chooses its orbit so as to minimize the strong electrical repulsion between itself and the other electrons in the shell. In partly filled shells with several electrons, the minimum-energy configuration of electrons can have a consid-erable resultant magnetic moment.

Conduction electrons

We have described in the previous section the mag-netic response of electrons in orbitals around atoms or ions. This response occurs in all matter. However, in metals and semiconductors there are additional contributions to the magnetic response of a substance due to the conduction electrons. It is found that this response is generally rather weak, but that the orbital motion of conduction electrons respond diamagneti-cally and the spins of the conduction electrons respond paramagnetically.

Conduction electron diamagnetism

In §6.5 and §7.5.3 we described the way in which conduction electrons occupied plane wave quantum states. Each of these quantum states corresponded to a electron travelling in a straight line. We envisaged that electrons would stay in a quantum state until a *scattering event* occurs, after an average time τ, which will transfer the electron to a new quantum state. However, when a magnetic field **B** is applied, the Lorentz force $q\text{v} \times \text{B}$ causes the electron paths to change from straight lines to arcs of circles as shown in Figure 7.62.

Figure 7.62c shows clearly that at high fields the electron trajectories become significantly curved and one can imagine that at much larger fields electrons will travel in complete circles before scattering. The magnetic force that acts perpendicular to the electron trajectories causes electrons to move in circular orbits with an angular frequency ω_c given by Equation 7.162 (where the symbol $\Delta\omega$ was used instead of ω_c)

$$\omega_c = \frac{eB}{2m_e} \tag{7.178}$$

Thus the number of radians that an electron orbits before scattering is, on average, $\omega_c \tau$. If the magnetic

(a)

B = 0

(b)

B ⊙

(c)

B ⊙

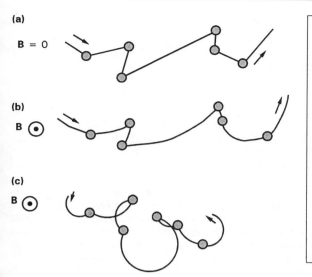

Figure 7.62 Illustration of the effect of a magnetic field on typical conduction electron trajectory. (a) The trajectory in zero field; the spheres indicate a scattering event. (b) When a weak field is applied there is a slight curvature of the electron paths. (c) When a much stronger field is applied, the electron trajectories are significantly affected.

field is such that $\omega_c\tau \gg 1$ then a variety of interesting phenomena occur, but space constraints do not permit their inclusion here.

The magnitude of the diamagnetic effect is extremely difficult to calculate. Here we merely note that it is (a) relatively small, and (b) typically $\approx 1/3$ of the paramagnetic response of the electrons described in the following section.

Conduction electron (Pauli) paramagnetism

Now we consider the effect of a magnetic field on the spin of conduction electrons. The origin of this interaction with the field is that electrons have a magnetic moment which is described in relation to an internal "spin" degree of freedom. The magnetic moment, μ_e, points in the opposite direction to the spin and is given by

$$\mu_e = \frac{g_s\mu_B}{\hbar}\mathbf{S} \qquad (7.179)$$

where:
- g_s is the spin g factor, which has a value of 2.002. Its value is usually taken as 2 for work in solids.

Example 7.30

Atoms that possess intrinsic magnetic moments typically have moments of the order of 1 Bohr magneton. Estimate the current, i, that must flow around an area, A, of the same order as an atomic cross-sectional area in order to produce a magnetic moment of 1 μ_B.

We remind ourselves of Equation 7.149, $m = iA$. We can estimate A as being $\approx \pi r^2$, which for $r \approx 10^{-10}$ m is $\approx 10^{-20}$ m^2. Since $1\,\mu_B = 9.27 \times 10^{-24}$ A m^2 we have

$$i = \frac{m}{A} \approx \frac{9.27 \times 10^{-24}}{10^{-20}} \approx 10^{-3}\,\text{A}$$

Thus to produce a magnetic moment typical of the values found in atoms requires currents of the order of 1 mA to flow around a loop of atomic dimensions.

Example 7.31

How large must a magnetic field be in order to ensure the condition $\omega_c\tau \gg 1$ is satisfied?

We must have (from Eq. 7.178).

$$\frac{eB\tau}{2m} \gg 1$$

which requires

$$B \gg \frac{2m}{e\tau}$$

Recalling from our studies of the resistivities of metals that at room temperature $\tau \approx 10^{-14}$ s we write

$$B \gg \frac{2 \times 9.1 \times 10^{-31}}{1.6 \times 10^{-19} \times 10^{-14}} \approx 1100\,\text{tesla}$$

It is currently technologically impossible to achieve a field of this magnitude. Steady field values are limited to less than 20 T or 30 T, and pulsed values to less than ≈ 100 T. Thus, in order to study matter when $\omega_c\tau \gg 1$ the scattering time τ must be reduced by using pure samples of substances at low temperatures (a few kelvin or less).

The significance of the spin g factor taking the value 2 is that magnetic moments arising from the intrinsic spin are twice as large as would be expected from the equivalent amount of non-spin (i.e. orbital) angular momentum.
- μ_B is a Bohr magneton ($= 9.274 \times 10^{-24}$ A m^2 or J T^{-1}).

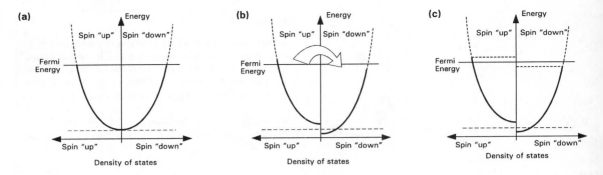

Figure 7.63 Illustration of the origin of conduction electron (Pauli) paramagnetism. (a) When no magnetic field is applied, as many electrons have spin up as spin down because the energy of these two states is the same. (b) Electrons with spins parallel to the magnetic field, **B**, have a lower energy than electrons with spins antiparallel to **B**. The figure shows a hypothetical situation that might occur if the field were applied suddenly before the electrons had a chance to change their spin state. In the situation shown, the metal would still have no net magnetic moment. However, the "spin up" electrons can clearly lower their energy by reversing their spin and occupying spin-down states that have a higher kinetic energy. (c) The process described in (b) has been completed. There is now a net excess of electrons having "spin down", i.e. spin antiparallel to **B**. It is this imbalance that we can detect as the paramagnetic response of conduction electrons. The magnitude of the response will clearly be larger if we have a large density of states at the Fermi energy.

- **S** is the spin of the electron, which has the values $\pm\hbar/2$.

The electron spin can orient itself in one of two ways with respect to an applied field, and the energy of interaction with the magnetic field, **B**, is therefore

$$u = \mu_e \cdot \mathbf{B} \tag{7.180}$$

$$u = +\frac{g_s\mu_B}{\hbar}\mathbf{S}\cdot\mathbf{B} \tag{7.181}$$

and an electron can lower its energy if it orients itself so that its magnetic moment is *parallel* to **B**, i.e. so that its spin is *antiparallel* to **B**. Since **S** takes the values $\pm\hbar/2$, the energy of the electron will be

$$u = +\frac{2\mu_B}{\hbar}\times\left(\pm\frac{1}{2}\hbar B\right) = \pm\mu_B B \tag{7.182}$$

Now, conduction electrons in a metal are not able to just change their spin orientation at will because, as outlined in §6.5, the electrons are "packed" two at a time into k-states, one spin up and one spin down. In general, the Pauli exclusion principle prevents an electron from changing its spin state because the quantum state with the same k-state but opposite spin is already occupied. As shown in Figure 7.63, only those electrons in k-states close to the Fermi energy

can change their spin state. They do this by occupying k-states with higher *kinetic energy*, but opposite spin.

When the electrons change spin, the balance between "spin up" and "spin down" is altered and the metal acquires a net magnetic moment. Since more electron magnetic moments are parallel to the magnetic field than antiparallel to it, the metal has a paramagnetic response, normally called *Pauli paramagnetism*. We can calculate the magnetization if we look at the imbalance between "spin-up" and "spin-down" electrons.

Net magnetic moment per unit volume =

$$\left(\text{Extra "spin-down" electrons}\right)\times\left(+\mu_B\right)$$
$$-\left(\text{Lost "spin-up" electrons}\right)\times\left(-\mu_B\right)$$
$$\tag{7.183}$$

$$M = \left[\frac{g(E_F)}{2}\times\mu_B B\right]\times\mu_B + \left[\frac{g(E_F)}{2}\times\mu_B B\right]\times\mu_B \tag{7.184}$$

$$M = \left[\frac{g(E_F)}{2}\times\mu_B B + \frac{g(E_F)}{2}\times\mu_B B\right]\times\mu_B \tag{7.185}$$

$$M = g(E_F)\mu_B^2 B \qquad (7.186)$$

where $g(E_F)$ is the density of states at the Fermi energy, and so $g(E_F)/2$ is the density of either "spin-up" or "spin-down" states at the Fermi energy.

The Pauli susceptibility for ciopper is calculated in Example 7.32. It is of a similar order of magnitude to, but rather smaller than, the diamagnetic contribution (Eq. 7.171). Note that the experimental value of c_M ($-6.87 \times 10^{-11}\,m^3\,mol^{-1}$) is the sum of Pauli susceptibility ($+8.67 \times 10^{-11}\,m^3\,mol^{-1}$) and the diamagnetic response of the orbitals and conduction electrons.

Summary

We now turn to the main questions raised by our preliminary examination of the experimental data on the magnetic properties of solids.

The strength of paramagnetism in the transition and lanthanide series

We saw that in all matter there is a weak background diamagnetic response due to filled electron shells, and that in addition to this there is a paramagnetic response from electrons in partly filled shells. In the transition series of elements the $3d$, $4d$ and $5d$ electron shells are partly filled, and so we can understand that the paramagnetism of the elements in these series should be rather strong. Understanding the detailed behaviour requires a theory of the origin of atomic magnetic moments which is beyond the scope of this text.

We saw that the lanthanide elements display an especially strong paramagnetic response. This response is due to the partial filling of the $4f$ electron shell. We recall in our discussion of the density anomalies in the lanthanide series (§7.1.2) that the $4f$ orbital lies *inside the atom* and is not involved in bonding with neighbouring atoms. This is in contrast with the transition elements whose d shells are involved in bonding to neighbouring atoms. Electrons involved in bonding are not able to reorient their orbital motion in order to align with the field – the electric forces of bonding are much stronger than the magnetic forces trying to orient the orbits. Thus for such electrons only their spin magnetic moment aligns with the applied field. The inability of the orbital angular momentum of bonded

---

Example 7.32

Estimate the molar susceptibility due to the spins of the conduction electrons in copper according to Equation 7.186.

We first note that the susceptibility will eventually be shown to be rather small, and so we may equate the magnetic induction field, $\mathbf{B}$, in Equation 7.186, with $\mu_0 H$. Hence the molar susceptibility $\chi_m = M/H$ may be written as

$$\chi_{molar}^{Pauli} = \frac{M}{H} = \mu_0 \mu_B^2 g_m(E_F)$$

The density of electronic states at the Fermi energy, $g(E_F)$, is given in the free electron approximation by

$$g(E_F) = \frac{V\sqrt{2m^3 E_F}}{\pi^2 \hbar^3}$$

This has been evaluated for copper in Example 7.14 where we estimated the electronic contribution to the heat capacity of copper. Using that estimate of

$$g(E_F) = 8.026 \times 10^{41} \text{ states J}^{-1}\,\text{mol}^{-1}$$

and recalling that a Bohr magneton, μ_B, has a value $9.27 \times 10^{-24}\,J\,T^{-1}$, we estimate the Pauli contribution to the magnetic susceptibility of copper as

$$\chi_{molar}^{Pauli} = 4\pi \times 10^{-7} \times \left(9.274 \times 10^{-24}\right)^2 \times 8.026 \times 10^{41}$$
$$= 8.67 \times 10^{-11}\,m^3\,mol^{-1}$$

---

electrons to reorient themselves is known as *quenching* of the orbital magnetic moment.

However, in the lanthanide series, there is no quenching of the orbital angular momentum. This, coupled with the fact that the orbital angular momentum of electrons in the 4-f shells is rather large, gives rise to the extraordinarily large magnetic susceptibilities of the elements in the lanthanide series.

In conclusion, we note that the measured values of susceptibility and the inferred values of the magnetic moments give many clues about the detailed occupancy of electronic orbitals in solids.

The occurrence of paramagnetism and diamagnetism outside the transition series

Outside the transition series, there are no contributions to the magnetic response from partly filled orbitals. The susceptibility is therefore determined by the balance between the remaining modes of response. In insulators there is only the Larmor diamagnetic response, but in metals there is an additional (generally paramagnetic) response due to the conduction electrons.

The smallness of most susceptibilities

The magnetic responses of substances that we have discussed so far are, in general, rather weak. The basic reason for this is that the magnetic energy is generally rather small in comparison with the other energies involved. In Larmor diamagnetism, the magnetic force on the electron competes with the electronic forces on the electron (F_0 in Eq. 7.158). In Curie paramagnetism, there is competition between thermal disorder and the magnetic torques on atoms. Thus at low temperatures the Curie susceptibility can become much larger than Figure 7.56 would indicate. Similarly, in Pauli paramagnetism the comparison is between the magnitude of the magnetic energy and the Fermi energy. Finally, in Landau diamagnetism the competition is between the magnitude of electronic speeds of the order of the Fermi velocity, and the Lorentz force on the electrons. In all these responses, the magnetic energy is generally small compared with the other (mainly electronic) energies involved.

Ferromagnets

For the reasons mentioned above, we can be even more surprised at the gigantic magnitude of the magnetic response of ferromagnets. We do not have space to discuss the experimental data at length, but it is important to note that the origin of this behaviour is in the Coulomb interaction between electrons in the solid, not the magnetic interaction of these electrons with an external magnetic field.

The Curie theory of the paramagnetic properties of the atoms with partly filled orbitals implicitly assumes that there are no interactions of any kind between neighbouring magnetic moments. Inevitably there will of course be some kind of interaction between the atomic magnetic moments. In ferromagnets these interactions cause neighbouring atomic magnetic moments to align parallel to one another (Fig. 7.64).

However – and this is frequently not appreciated – the interactions between neighbouring magnetic magnetic moments are *not* magnetic. They arise from consideration both of quantum mechanics and the Coulomb electrical interactions. That magnetic forces cannot give rise to ferromagnetism may be appreciated by anyone who has played with a pair of bar magnets. Such magnets will spontaneously try to align such that the north-seeking and south-seeking poles of the magnet are together (Fig. 7.64b) which in atomic terms corresponds to neighbouring atomic magnetic moments pointing in *opposite* directions (Fig. 7.64c). Clearly such forces cannot give rise to the ferromagnetic structure illustrated in Figure 7.64a.

The interactions which give rise to ferromagnetism are of the same nature as those which cause magnetic moments *within* atoms. Consider two electrons in a partly filled orbital on an atom. Since the electrons repel one another, they correlate their motion so as to minimize this repulsion. In classical terms, one way of doing this is to orbit the atom in the same sense, but 180° out of phase so that the electrons are always on opposite sides of the atom. Quantum mechanically this corresponds to occupying quantum states with parallel angular momentum. It is considerations of this type which give rise to the occurrence of permanent magnetic moments on atoms with partly filled orbitals.

In a ferromagnetic substance, these correlations persist from each atom to its immediate neighbour.

Table 7.20 Summary of the response of non-interacting electrons to applied magnetic fields. The type of response is listed with the name of the scientist most closely associated with it.

	Core electrons (Filled shells)	Core electrons (Partly-filled shells)	Conduction electrons	Nuclei
Spin response	No response	Paramagnetism (Curie) $\chi \approx 1/T$	Paramagnetism (Pauli)	Weak paramagnetism (Curie) $\chi \approx 1/T$
Orbital response	Diamagnetism	Paramagnetism (Curie) $\chi \approx 1/T$	Diamagnetism (Landau)	Weak paramagnetism (Curie) $\chi \approx 1/T$

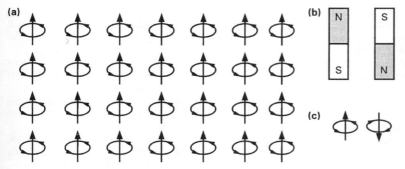

For example, electrons on one atom may be able to minimize their coulomb repulsion of electrons on a neighbouring atom by arranging that their orbits are in the same sense as those of electrons on the neighbouring atom. If this is so for all neighbours within the solid, then the magnetic dipoles on each atom will be aligned *despite* the fact that, magnetically, they would rather be oriented anti-parallel to their neighbours. This essentially Coulomb-driven alignment of neighbouring orbitals is known as *ferromagnetism*. This is the mechanism whereby a substance possesses a permanent magnetic dipole moment in the absence of an applied field. The ferromagnetically ordered arrangement of atomic dipole moments may be destroyed if the temperature is raised sufficiently. The temperature required to destroy the spontaneous magnetization of a substance is known as the *Curie temperature*, and is large for substances in which there is a large Coulomb interaction between electrons on neighbouring atoms (known technically as the *exchange interaction*). Many substances have weak exchange interactions and become magnetically ordered at low temperatures. Only rarely do substances have sufficiently strong exchange interactions to allow ferromagnetism to survive at room temperature and above.

Finally, we note that sometimes electrons on one atom may minimize their Coulomb repulsion of electrons on the neighbouring atoms, by arranging that their orbits are in the opposite sense to those of electrons on neighbouring atoms i.e. they have a negative exchange energy. This phenomenon gives rise to a different (but related) type of magnetic order known as *antiferromagnetism*.

7.9 Exercises

Exercises with a P prefix are "normal" problems. Those with a C prefix are best solved numerically using a computer program or spreadsheet.

Density

P1. The graph of density versus atomic number has "shoulders" on the edges of peaks e, f and g. To what do these shoulders correspond on the periodic table (Fig. 2.2)?

C2. Replot the graph of density versus atomic number (Fig. 7.1) up to element 20 to see the detail (Table 7.1). Is there still evidence of periodic behaviour?

P3. What are the densest and second densest elements? Are these also the elements with the greatest *number density* of atoms? Would you expect to find that a compound or alloy of the densest element was denser than the element itself (Table 7.2)?

C4. Which element has the greatest *number* density of atoms? (Table 7.2 & Example 7.1)

P5. Work out approximately (a) the number density, (b) the molar density, (c) the molar volume, (d) the atomic volume and (e) the typical separation between atoms in (i) tungsten and (ii) aluminium. (Table 7.2 & Example 7.1).

P6. What types of wood will sink in water (Table 7.1)? What types of wood would sink in ethanol (Tables 7.1 & 9.2)?

P7. The density of the element with atomic number 50 is the same as that predicted by the simple theory of Example 7.2. Does this imply that element 50 has a simple cubic crystal structure with a lattice spacing of 0.3 nm? If not, what does it imply (Fig. 7.2 & Example 7.1)?

Thermal expansivity

P8. Of the elements listed in Table 7.4A, which has (a) the largest and (b) the second largest thermal expansivity ?

P9. Estimate a typical figure for the linear thermal

expansivity of brick and cement. (Table 7.4). A house has walls made from brick and cement and the mean temperature of the walls can change by around 10°C between summer and winter. Estimate (a) the change in mean height of a house 10 m high and (b) the change in height of a door frame 2.2 m high (Example 7.5).

P10. Show that the area expansivity of a substance is given by 2a (Example 7.4).

P11. Example 7.4 shows that the volume expansivity may be given by $b = 3a$. Show that this same relationship holds for a cuboid of initial dimension $x \times y \times z$. Show that if the linear expansivity in each direction is different, the volume expansivity is given by the sum of the linear expansivities in each of the three directions.

P12. Most elemental solids melt when the amplitude of atomic vibration is about 5% of the separation between atoms (Lindemann theory of melting, §11.3.2). Use order-of-magnitude estimates from Table 11.1 & Table 7.4 to estimate how much an elemental solid expands before it melts.

P13. Two cubes of copper and nylon are each machined to have sides of length 34.4 mm at 20°C. The cubes are then cooled to the boiling temperature of liquid nitrogen (Table 11.2). For each cube calculate (a) the percentage change in volume, (b) the percentage change in surface area and (c) the percentage change in length of side (Table 7.4 & Examples 7.4–7.6).

P14. Estimate the percentage change in the average separation of two atoms in a piece of copper when the temperature is changed from (a) 20°C to 100°C, (b) 20°C to 1000°C, and (c) 20 °C to 10 K. (Table 7.4 & Examples 7.5 & 7.6).

P15. Write an explanation (about half a page) for a non-scientific friend outlining what an *alloy* is, and explaining why the thermal expansion of alloys is not always given by the average of those of its component metals. (Table 7.5).

P16. Write a briefing paper for a fellow student explaining what *invar* is (Table 7.4B) and how its thermal expansivity is related to that of iron and nickel. By considering the brief outline of the origin of ferromagnetic interactions at the end of Section 7.8, speculate as to how the anomalous pair potential shown in Figure 7.8 may be produced.

P17. Write an explanation (about half a page) for a non-scientist friend explaining why plastics expand more than crystalline solids? Show the explanation to the friend and ask them to ask the first question which comes into their mind on reading it. Answer the question. (§7.2)

Sound

P18. What is the speed of the longitudinal sound waves in (a) domestic glassware, (b) ice, (c) polyethylene, (d) aluminium, (e) copper and (f) lead (Table 7.6).

P19. Typically, what is the *ratio* of the speed of longitudinal

sound waves to the speed of transverse sound waves in elements (Fig. 7.10)? Name (a) one element in which the ratio lies close to this typical value, (b) one element in which the ratio lies well below this value, and (c) one element in which the ratio lies well above this value. Evaluate the Poisson ratio for each element (a) to (c) (Eq. 7.15).

P20. Based on the data given in Table 7.6, estimate very roughly the speed of sound in wood. Musical instruments can be made by striking blocks of wood that are free to "ring". If the fundamental resonance of a block of length L occurs when $L = \lambda/2$, estimate the length of a block that will resonate at 440 Hz (the note A above middle C).

P21. Write an explanation for a colleague summarizing the definitions of the shear modulus, G, and Young's modulus, E. Explain briefly why one might reasonably expect the shear modulus of a solid to be less than Young's modulus (Fig. 7.12 & 7.13).

P22. Based on Equations 7.13–7.15, estimate the shear modulus, G, Young's modulus, E, and the Poisson ratio, σ, of copper, silver and gold (Tables 7.2 & 7.6).

P23. Describe an experiment to demonstrate directly to a non-scientist friend that sound travels faster through solids than through gases (Tables 5.14 & 7.6).

Thermal properties

P24. What is the molar heat capacity of (a) gold and (b) neodymium at around room temperature (Table 7.7)?

P25. Which elements have the highest and lowest molar heat capacity ($J K^{-1} mol^{-1}$) at 298 K (Table 7.7)?

C26. Which elements have the highest and lowest *specific* heat capacity ($J K^{-1} mol^{-1}$) at 298 K (Tables 7.2 & 7.7)?

P27. Estimate C_P for copper, silver and gold at the boiling temperature of liquid nitrogen (Fig. 7.17 & Table 11.2).

P28. Liquid helium costs around £3.00 per litre and is used routinely to cool apparatus for use in low-temperature experiments. The latent heat of liquid helium is around 2000 J per liquid litre. Estimate roughly how much it would cost to cool 1 kg of copper from room temperature to the boiling temperature of liquid helium (Fig. 7.17 & Table 11.2). In fact, a simple calculation overestimates the amount of helium required: can you suggest why?

P29. Following Example 7.13 & Figures 7.19 & 7.23, estimate an Einstein temperature for gold. Based on that estimate, calculate the Einstein frequency and the spring constant between gold atoms.

P30. By considering the spring constant, K, estimated in Example 7.13, and the model of a solid sketched in Figure 7.12, show that Young's modulus for a substance may be estimated as $E \approx K/a$, where a is the lattice spacing (≈ 0.3 nm). How well does the estimate for E derived from analysis of the heat capacity data (Example 7.13) tie up with E estimated from the speed of sound data (Tables 7.1 & 7.6)?

P31. Figure 7.29 indicates a roughly linear relationship between the Debye temperature of a substance and the speed of sound waves in a substance. Estimate the constant of proportionality using Equation 7.50 and discuss how well it agrees with the experimental value of approximately 7. Use this correlation to estimate the Debye temperature for niobium and seek confirmation of your estimate in the scientific literature (Tables 7.6 & 7.8).

P32. Compare the heat capacity predicted by Equation 7.49 with the tabulated values of the Debye function (Appendix 5). Up to what fraction of θ_D is the equation accurate to within (a) 1% and (b) 10%?

P33. Explain to a friend what is meant by the terms *phonon* and *photon* and outline the correct usage of each term.

P34. What is the ratio $k_B T/E_F$ for a typical metal at 10K, 100K and 1000K (Eqs 6.80–6.83).

P35. Estimate the ratio of $C_{el}/C_{lattice}$ for silver at (a) room temperature and (b) 1K (Example 7.14, Table 7.8 & Eq. 7.49).

P36. A single photon of energy 3×10^4 eV is absorbed in a block of silicon ($\theta_D \approx 630$ K) of volume 1 cm^3. Estimate roughly the temperature rise if the silicon is held at initial temperatures of (a) 1mK, (b) 10mK and (c) 100mK (Eq. 7.49 & Table 7.2). Could such a device be used as a photon detector at any of these temperatures?

Electrical properties

P37. Which four elements are the best electrical conductors at room temperature (Table 7.9 & Fig. 7.33)? Are these still the best conductors at a temperature of 1 K (Table 7.11)? Among the lanthanide elements (Fig. 7.1) which is the worst and which the best electrical conductor (Table 7.9)?

P38. What is the resistivity of (a) copper, (b) brass and (c) zinc at around room temperature (Table 7.10)?

P39. A copper wire is 50km in length, has a diameter of 10cm, and has a potential difference of 3×10^5 V across it. Estimate the current through the wire and the power dissipated per metre (Table 7.9 & Example 7.16).

P40. The resistivity, ρ_z, of an element with atomic number Z is required in a calculation, but your tables record only the value of ρ_{z+1} for the element with atomic number $Z+1$. How good a guide is this to the likely value of ρ_z (Figs 7.32 & 7.33)? If you had to choose a single figure as a ballpark estimate of the resistivity of *all* elemental metals, what value would you choose?

P41. The resistivity, ρ, of rhodium is 4.51×10^{-8} Ωm at room temperature. Estimate ρ at 77K. (Hint: look at Figure 7.34 and make some assumptions.)

P42. Work out the scattering time for electrons in Au, Cu, Zn, Cu(Zn) and Nd (Example 7.16). Discuss briefly the origin of the differences in Δt.

P43. Ask your tutor why metallic behaviour is common amongst the elements. Write down the answer, think about it, and then send it to me.

P44. Element A has resistivity ρ_A and element B has resistivity ρ_B. Sketch how you would expect the resistivity of a random alloy $A_x B_{1-x}$ to vary for $0 < x < 1$ (Table 7.10).

P45. Work out the thickness of a parallel-plate capacitor made with a quartz dielectric with an area 10mm^2 and a capacitance of 1nF (Example 7.19). If the capacitor has a voltage of 100 V across its plates, roughly what is the current (known as the *leakage current*) which flows through the capacitor (Table 7.13 and Example 7.20)?

If an AC voltage at a frequency of 1kHz is now applied to the capacitor, the reactive current through the capacitor has a magnitude $2\pi f V/C$. Compare the magnitudes of the capacitative and leakage currents through the capacitor. A good dielectric substance for a capacitor has a low leakage current and a high dielectric constant. In these terms is quartz or polystyrene the better material with which to make a capacitor (Table 7.13)?

P46. A parallel-plate capacitor is made with a quartz dielectric of area 10mm^2 and capacitance 1nF (Example 7.19). What is the maximum voltage that may be applied to the capacitor (Table 7.14)? How would you expect this maximum voltage to change with temperature (§7.5.7)?

Semiconductors

P47. What is the resistivity of silicon with 1 part per million phosphorus impurity at around 300K? How many phosphorus atoms per cubic metre does this correspond to (Figure 7.40)?

P48. What is a typical value of the energy gap for a semiconductor (Fig. 7.41 and Eq. 7.80)?

P49. Estimate the number density of electron and hole carriers in pure silicon at 1000K (Example 7.22). How does this compare with the number density of carriers in copper at 1000K?

P50. Based on the discussion around Figures 7.44 and 7.45, suggest elements that would make good donor and acceptor dopants for (a) silicon and (b) germanium. Speculate which dopants would have the smallest activation energies, i.e. have quantum states with an energy most similar to the host material.

Thermal conductivity

P51. Roughly 1kW of heat flows through the base of a cast iron frying pan (diameter 30cm, thickness 5mm) and heats the oil beneath some sausages to around 200°C. Estimate the temperature of the underneath of the frying pan. The handle is 15cm long, 2cm in diameter, and made from epoxy resin. Make some simple assumptions and estimate the temperature half way down the handle. (Tables 7.15 & 7.16).

P52. The room in which I am sitting has one outside wall which has an area of approximately $3\,m \times 4\,m$. Roughly half of this area is taken up with glass $5\,mm$ thick. Neglecting leaks around the window frame and assuming that the wall has two layers of bricks (approximately $20\,cm$ thickness) but no cavity, estimate the rate at which energy must be dissipated inside the room in order to keep it at $23°C$ when it is $-2°C$ outside.

Based on this calculation, would you recommend that I invest in double-glazing? Describe with the aid of a sketch how even a thin cavity of trapped air would help improve the thermal insulation of both the brick wall and the window (Table 5.11).

Optical properties

P53. What is the average optical reflectivity of aluminium (Table 7.17)?

C54. The reflectivity of glass to light normally incident upon it is given by the formula

$$R = \left(\frac{n_{light}^{vacuum} - n_{light}^{glass}}{n_{light}^{vacuum} + n_{light}^{glass}} \right)^2$$

so that, for example, when $n_{light} = 1.5$, $R = 0.04$ and so 4% of the incident light intensity is reflected. Using a spreadsheet or otherwise, plot R as a function of n_{light} for n_{light} in the range 1–3. Since glass with $R > 0.1$ is not useful for many applications, what is the maximum available refractive index?

P55. If the refractive index of glass were the same for all wavelengths across the spectrum, would a high refractive index prism split white light into different colours (Example 7.25)?

P56. Diamonds are commonly used in jewellery. Suggest which optical properties of diamonds (if any) are responsible for this popularity. Justify your suggestion.

P57. What conclusions can be drawn from the level of agreement between the theory of the refractive index and the experimental data in Figure 7.52?

P58. Estimate the wavelength of UV resonance (or UV electronic transition) for glass with a refractive index of 1.3 (Example 7.26). By considering the legend to Figure 7.52, revise your estimate to produce a more realistic value for λ_0. From your estimate of λ_0 produce an estimate of f_0, the resonant frequency. By considering Equation 2.31, and supposing only a single electron to be excited, estimate the "spring constant", K, with

which the electron is bound within the glass. Compare your answer with the result of Exercise P6 in Chapter 2.

P59. Estimate the skin depth of copper at a frequency of (a) $50\,Hz$ (mains frequency), (b) $50\,MHz$ (the clock frequency of a computer), (c) $10\,GHz$ (microwave), (d) $10\,THz$ (infra red) and (e) $1000\,THz$ (optical) (Eq. 7.146). How thick should the aluminium screening around a computer be in order to reduce the strength of the radiated electric field by a factor of 100?

Magnetic properties

P60. List three diamagnetic and three paramagnetic elements (Table 7.18).

P61. What is the maximum susceptibility of an element in the second transition row of the periodic table (Y to Cd) (Fig. 2.2 & 7.56 & Table 7.18).

P62. Estimate the Larmor diamagnetic susceptibility (Eq. 7.169) for (a) scandium (b) neodymium and (c) germanium.

P63. What is the magnetic susceptibility of titanium? What is the expected magnetic moment of a titanium sample of volume $1\,mm^3$ held in a magnetic field of $1\,T$? What is the magnetization, M, of the titanium? What is the applied field, H, inside the sample? What is the magnetic field, B, inside the sample? What is the magnetic moment per atom?

P64. Superconductors can display perfect diamagnetism, i.e. the internal magnetic field B is always zero. Use Equations 7.148 & 7.152 to show that the volume susceptibility $\chi_v = -1$.

P65. Explain why the torque on a sample of copper in a large magnetic field is essentially zero (Eq. 2.20).

P66. Review the method of operation of a vibrating sample magnetometer (VSM) (Fig. 3.11). At room temperature a sample of ferromagnetic nickel has a magnetic moment of $3 \times 10^{-3}\,Am^{-2}$ and induces a root-mean-square (rms) voltage of $1.3\,V$ in a VSM pickup coil. What voltage would you expect from samples of (a) silicon and (b) uranium, each with a volume of $10\,mm^3$, in a magnetic field of $1\,T$. (Table 7.18)?

P67. Explain to a colleague who is not studying physics that all substances, even apparently non-magnetic ones (such as plastics), become magnetized in the presence of an applied magnetic field. How would you demonstrate this using apparatus costing less than £100 (Eq. 2.16–2.21)? (Hint: think of the principle of "null methods" or "balance methods" used for detecting small changes described in Chapter 3.)

CHAPTER 8

Liquids: background theory

8.1 Introduction

If we raise the temperature of a solid it will, commonly, become liquid, and then evaporate to become a gas. In this sequences of states (solid–liquid–gas) the liquid state is intermediate between the solid and gaseous states. This intermediate position is a reflection of the fact that the arrangement of atoms or molecules in a liquid is, in general, intermediate between the order of the solid state, and the random molecular motions of the gaseous state. In what follows we will discuss liquids as being *structurally* intermediate between solids and gases.

In general, we will find that most properties of liquids can be understood by explanations that begin by assuming that *either* liquids are similar to solids but more disordered and slightly less dense, *or* by saying that liquids are similar to gases but more ordered and much more dense. Figure 8.1 shows our imagined picture of the situation of atoms or molecules in the liquid state. It might represent the result of taking a "snapshot" of a microscopic amount of liquid. The atoms are vibrating about their average positions rather like the equivalent picture for a solid (Figure 6.1). The atoms are highly constrained by the closeness of neighbouring atoms and so their vibrational frequencies are similar to those in solids: around 10^{13} Hz. The equivalent picture taken, say, one vibrational period ($\approx 10^{-13}$ second) later would look *in detail* almost exactly the same. However, the equivalent picture taken, say, 10^{-10} s later (after around a thousand or so atomic oscillations) would typically look *qualitatively* similar, but in detail the picture would look completely different: the average positions of most the atoms would have shifted. This changing structure gives rise to difficulties in adopting a universally

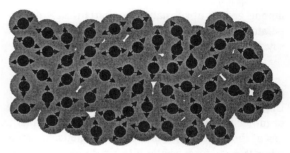

Figure 8.1 A schematic illustration of the motion of atoms in a liquid. Note the small separation between the atoms, and the random orientation of the vibrations of the molecules. The atoms themselves are illustrated schematically as a central darkly shaded region, where the electron charge density is high, and a peripheral lightly shaded region. The electric field in this peripheral region significantly affects the motion of neighbouring atoms, and disturbs the electronic charge density of neighbouring atoms.

appro-priate model of liquids. The changing structure can be seen clearly by using the computer simulation listed in Appendix 4.

8.1.1 A perfect liquid?

When we discussed the properties of gases we were able to arrive at the theory of a "perfect gas" which for many purposes was a good approximation to the properties of real gases. However, there was no single model of a "perfect solid" which could explain the massively diverse properties of solids.

Liquids fall into an intermediate category, and we will discuss their properties in terms of two simplified models. One model describes the *structure* of a liquid and will be used to understand properties such as the density. The other model describes the *dynamics* of the liquid molecules and will be used to understand properties such as the viscosity. We will find that we are able to understand many of the properties of real liquids in terms of these models. However, the models are so simplified that we will not really be able to "believe" them in the way that we believe the model of a perfect gas. The models capture just one or two of the key features of liquid behaviour and ignore many properties of molecules that make up the liquid. The predictions of the models tend to be rather qualitative – allowing us to examine trends among groups of substances, or variations with temperature – rather than predicting that the viscosity of, say, water at temperature T will be X.

8.1.2 Bonding in liquids

Liquids consist of a "condensed" collection of atoms or molecules with less average kinetic energy than a gas of the molecules, but too much kinetic energy to allow them to form a solid. We can divide the electrostatic bonding mechanisms into the same four categories as for solids (molecular, ionic, covalent and metallic) plus one extra category, hydrogen bonding, which is discussed below (Fig. 8.2).

8.1.3 Hydrogen bonding

Hydrogen bonding is of special importance in discussing organic liquids, and also our most precious liquid – water. It occurs in substances the molecules of which contain the chemical group OH: a combination of a hydrogen atom and an oxygen atom. This combination occurs in water (H_2O or HOH), in alcohols such as methanol (CH_3OH) or ethanol (C_2H_5OH), and in numerous other organic molecules. Hydrogen bonding also occurs to a lesser extent in bonds between hydrogen and nitrogen, and hydrogen and a halogen florine, chlorine or bromide.

In liquids made from organic molecules, the *atoms within the molecules* are held together by primarily covalent bonding. However, the molecules are attracted to each other by the relatively weak *Van der Waals* force. The hydrogen bond is an attractive mechanism that acts between different molecules in addition to the Van der Waals force. It is much stronger and more directional than the Van der Waals force, similar in effect to the covalent bonding that occurs within molecules.

The OH covalent bond within a water molecule is highly asymmetric, with the centre of charge symmetry being much closer to the oxygen rather than the hydrogen atom. This leaves the hydrogen atom

(a)   **(b)**   **(c)**   **(d)**

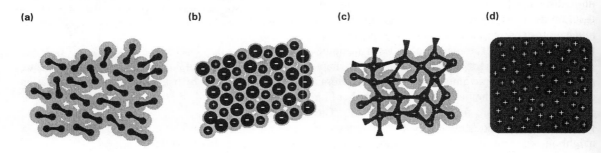

Figure 8.2 (a) In molecular liquids the entities that make up the liquid (atoms or molecules) are essentially the same as the entities that made up the gas. Internally the atoms that comprise the molecules are bound within each molecule by primarily covalent bonding. Externally the molecules are bound together by the Van der Waals force (see Section 6.2). (b) In ionic liquids the entities that make up the liquid are ions rather than atoms or molecules. Locally, most ions experience a situation similar to that experienced within a solid, but the regular periodicity of the solid lattice is absent. (c) In covalent liquids, the entities that make up the liquid (atoms or molecules) are greatly altered from their state in the gas. In particular, there is a high electronic charge density on some regions in between the mean positions of the atoms. Locally, most ions experience a situation similar to that experienced within a solid, but the regular periodicity of the solid lattice is absent. (d) In metallic liquids, the electrons from the outer parts of the atoms can move anywhere within the liquid and are not attached to any individual atom. Note that the electrons are still free to move from ion to ion in any direction, even though the regular periodicty of the lattice has been destroyed.

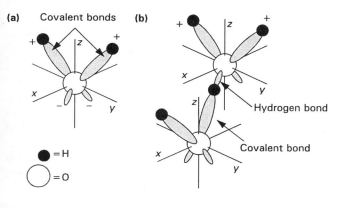

(a) Covalent bonds **(b)**

$\bullet$ = H
$\bigcirc$ = O

Hydrogen bond

Covalent bond

Figure 8.3 A qualitative indication of the structure of a hydrogen bond between water molecules. (a) The distribution of electric charge within an isolated water molecule: the hydrogen atoms are slightly positively charged and the oxygen atoms slightly negatively charge. The excess charge on the oxygen atom resides in two orbitals oriented so as to minimize their Coulomb repulsion from the covalent bonds. (b) A second molecule may orient itself so as to place its hydrogen atoms close to the negatively charged electron orbitals around the oxygen atom. The two oxygen atoms are linked by a hydrogen atom and two bonds— one is covalent and the other is known as a *hydrogen bond*.

slightly positively charged, and the oxygen atom negatively charged. The extra electron charge density around the oxygen atom distributes itself so as to minimize its Coulomb repulsion from the other electrons on the oxygen atom. The electron density in the region of an oxygen atom in a water molecule is shown in Figure 8.3(a). Note that the charge density forms two lobes on the other side of the oxygen atom from the covalent bond regions.

A second water molecule may orient itself with respect to the first so that one of its hydrogen atoms is close to electron orbitals around the oxygen atom. This structure is known as a hydrogen bond and has many of the features of a covalent bond. For example, although it has only around one-tenth the strength of the OH covalent bond, it is rigid like a covalent bond. By "rigid" we mean that the O–H–O link (Fig. 8.3b) has a minimum energy when the three atoms are in line. Hydrogen bonding will be discussed further in §9.2.2.

8.1.4 Organic liquids

The term "organic" (in this context) originally referred to substances that originated from a once-living organism. In modern parlance the term has a technical sense in which it refers to substances the molecules of which contain *both* carbon and hydrogen atoms. Thus carbon tetrachloride (CCl_4) and ammonia NH_3 are *inorganic*, but methane (CH_4) is organic. The distinction between organic and inorganic molecules is important in this context as data books frequently separate substances along these lines, and we shall follow suit in discussions of the

experimental data in Chapter 9. Understanding the bonding in organic substances is particularly important, since much of the experimental data on liquids refers to organic liquids. The key features of this bonding may be fairly easily stated.

- *Intra*molecular bonding
  The atoms within each molecule are held together by primarily covalent bonds that are extremely strong, and highly directional. This gives the molecules of organic substances characteristic shapes.
- *Inter*molecular bonding
  The molecules of the substance are (in general) attracted to each other by much weaker forces, primarily the non-directional Van der Waals force, which is sometimes augmented by directional hydrogen bonds.

The combination of strong intramolecular bonding and relatively weak intermolecular bonding results, as we shall see, in a complex array of properties.

8.2 The structure of liquids

Having reviewed what actually holds liquids together, we will now develop a simple and more general model that treats the origin of the bonding as mere detail and concentrates on the positioning of a single average molecule with respect to its neighbours within the liquid.

I have said previously that the structure of a liquid is similar to that of a disordered solid, but that the structure of a liquid changes on a time-scale of a fraction of a nanosecond. Given this, it might at first seem that it would be impossible to say anything *quantitative* about the structure of liquids. In fact one can make rather precise statements about liquid structure, as long as one is content with descriptions of the *average* structure. Obviously this will not describe the many individual and unique situations in which each molecule is placed. But we have seen in previous chapters that many properties of materials depend on average properties of a substance (e.g. average speed, average separation and average energy). So if we can calculate these averages, we may well be able to make progress in understanding the relationship between liquid structure and the properties of liquids. With the aim of quantitatively describing the average arrangement of molecules in a liquid, we use two related mathematical functions called the *radial distribution function*, $N(r)$, and the *radial density function*, $\rho(r)$.

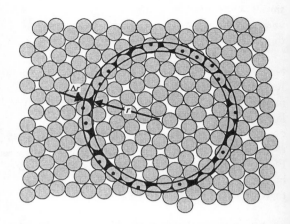

Figure 8.4 The radial distribution function describes the *average* distribution of molecules around any particular molecule. My count of the number of molecules reveals that there are 15 molecules in the ring shown.

8.2.1 The radial distribution function for spherical molecules

The radial distribution function is a mathematical function that describes the *average* distribution of molecules around a particular molecule. It specifies the average number of molecules that are found at a certain distance from a particular molecule. For spherical molecules the function answers the question: "*on average*, how many molecules have centres which lie within a spherical shell of inner radius r and thickness dr centred on a particular molecule?"

The idea is complicated to depict in three dimensions, but is illustrated in Figure 8.4 for a two-dimensional liquid-like arrangement of molecules. In two dimensions we ask the question: "*on average*, how many molecules have centres which lie within an annular area of inner radius r and thickness dr centred on a particular molecule?"

The number of molecules with centres in the annular area of inner radius r and thickness dr increases as r increases, because annular areas of larger radius have a larger area and so tend to have the centres of more molecules within them. In two dimensions, an annulus has an area $2\pi r\,dr$, so the number of molecules within it increases roughly linearly with r. In three dimensions, a spherical shell has an area $4\pi r^2\,dr$, so the number of molecules within it increases as r^2. However, in addition to the

(a)

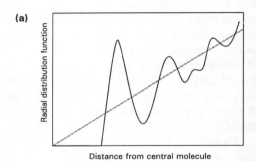

Distance from central molecule

(b)

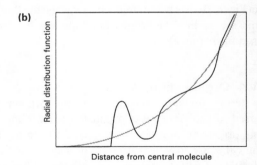

Distance from central molecule

Figure 8.5 A qualitative illustration of the radial distribution function for simple two-dimensional (a) and three-dimensional (b) liquid-like structures. Note the linear trend of the two-dimensional function, and the quadratic trend of the three-dimensional function. The construction of a radial distribution function for a two-dimensional liquid-like structure is shown in Example 8.1.

smooth increases, one can discern shorter range variations in the functions in either two or three dimensions.

The significance of the radial distribution may become clearer if we work out the radial distribution function for the two-dimensional liquid-like arrangement of circles shown in Figure 8.4. Within the annular area shown, lie the centres of 15 circles. If we count the numbers within similar rings of different radii then we obtain results similar to those shown in Figure 8.5. The process of constructing a radial distribution function for a two-dimensional liquid similar to that above is outlined in Example 8.1.

8.2.2 The radial density function

Of more fundamental significance in the study of liquids is a function closely related to the radial distribution function known as the *radial density function* (RDF), $\rho(r)$. This function charts local variations in the average density of the substance as a function of distance from the centre of a molecule. At large distances from a particular molecule this function tends to the value of the bulk density of the substance. However, on an atomic scale, we see peaks and troughs corresponding to the average separations of nearest and next-nearest neighbour molecules. In Example 8.1 and Table 8.1 we work out both the

Table 8.1 Data from Example 8.1 on the two-dimensional radial density function for a solid, liquid and gas: this is an approximation to the radial density function for a substance. (see also Figure 8.6).

Ring	Radius* r	Gas† N	$N/(2\pi r\Delta r)$	Liquid† N	$N/(2\pi r\Delta r)$	Solid† N	$N/(2\pi r\Delta r)$
1	1.5	0	0	0	0	0	0
2	2.5	0	0	0	0	0	0
3	3.5	0	0	4	0.182	6	0.273
4	4.5	0	0	2	0.071	0	0
5	5.5	0	0	1	0.029	0	0
6	6.5	2	0.049	4	0.098	12	0.294
7	7.5	1	0.021	6	0.127	0	0
8	8.5	0	0	3	0.056	0	0
9	9.5	3	0.050	6	0.101	12	0.201
10	10.5	1	0.015	6	0.091	6	0.091
11	11.5	1	0.014	7	0.097	0	0

*The mean radius of each ring in Example 8.1. The width of each ring, Δr, is 1.

†N, the number of circles (atoms) in each ring. $2\pi r \Delta r$, the approximate area of each ring.

Example 8.1

The radial distribution function for a two-dimensional solid, liquid, and gas.

Superimposed on the drawings is a ring structure that allows us to determine how many circles (atoms) have centres that lie within any particular annular area. The sums from each annular area are shown totalled in each figure and are given in Table 8.1.

Solid

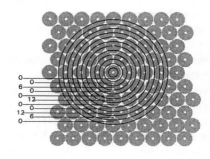

Liquid

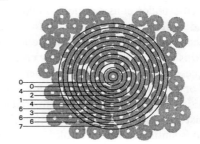

Gas

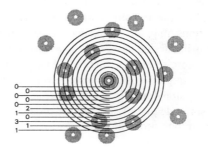

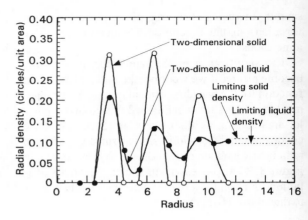

Figure 8.6 The radial density function of a two-dimensional "liquid" and "solid" (see Example 8.1 and Table 8.1). The peaks in the solid data correspond to nearest neighbours, next-nearest neighbours, etc. Note that these peaks are maintained in the liquid state, but are smoothed by the increased disorder in the liquid state. The figure also shows the limiting values of the macroscopic liquid and solid density.

radial density and radial distribution functions for a two-dimensional solid, liquid and gas.

Having worked out the RDF, the question arises of how to interpret it. The first point apparent from Figure 8.6 is that the RDF for a liquid is rather similar to the RDF for a solid. This indicates that around each molecule there remains some degree of order that was typical of the solid state. However, that observable periodicity is reduced as one moves further from the molecule under consideration, unlike a crystalline solid where the positions of the molecules remain correlated over long distances. Positional order of this type is called *short-range order*.

The question remains of precisely over what range the RDF of a liquid loses its correlations. The rate at which this occurs depends on the temperature of the liquid. Just above the melting temperature, the RDF of the liquid is similar to that of a solid, but as the temperature is increased the RDF becomes increasingly smoothed out, with the correlations between the positions of the molecules becoming less pronounced. Just below the transition to the gaseous state the correlations may be weakened considerably. This matter is illustrated more fully in Figure 9.22.

The exercise of constructing the two-dimensional RDF in Example 8.1 is clearly somewhat artificial, but the conclusion drawn is in fact rather general: liquids retain some degree of short-range order in the positioning of their molecules.

8.2.3 Non-spherical molecules

Many substances that form liquids around room temperature are organic in nature and have molecules that are not in the least spherical in shape. Such molecules may have rings of atoms that make them essentially planar in shape, or they may have chains of atoms that make them essentially linear in shape. There are also many complicated combinations of rings and chains that are not well described by the idea of the simple spherical molecule that we used in the previous section.

Thus for organic molecules, the RDF can describe the positional correlations of the centres of molecules, but does not include information about the relative *orientations* of neighbouring molecules. Consider a molecule such as that illustrated schematically in Figure 8.7. The picture is intended to illustrate a molecule which is not symmetric, and which has covalent bonds within the molecule that give it a relatively fixed shape.

Figure 8.8 illustrates two such molecules in several different relative orientations but separated by the same distance between centres. It is not difficult to imagine that the interaction energy of one molecule with the other depends not only on the separation of the two molecules, but also on their relative orientation. This orientation dependence of the interaction energy is particularly significant if the molecules are *polar* (see §5.6.2), i.e. if they have some

Figure 8.7 An asymmetrical molecule containing covalent bonds that impart to it a relatively fixed shape.

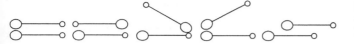

Figure 8.8 Pairs of asymmetrical molecules in different relative orientations but separated by the same distance between centres.

regions that are electrically positive and others that are electrically negative. The hydrogen bond illustrated in Figure 8.3 is a special example of interactions between polar molecules.

The importance of the molecular shape will be seen when we compare data on organic liquids with data derived from molten metals which are better described by the spherical molecules model that we considered in §8.2.2.

Figure 8.9 For the sake of discussion we suppose that the molecules such as the ones illustrated in Figure 8.7 have their lowest energy when they are oriented as shown here.

8.2.4 Liquid crystals

Many molecules that have some of the properties of the molecule shown in Figure 8.7 exhibit a strong tendency to align themselves with their neighbours. For example, they may prefer to orient themselves so that their long axis is parallel with that of their neighbours. If we arbitrarily define "the centre" of a molecule as the position of its centre of mass, then we find that the RDF is highly anisotropic. In some directions it may show essentially crystalline order, while in other directions it indicates order more typical of a liquid. States with this mixed order are known as *liquid crystals* or *mesophase states*. The prefix *meso* is from the Greek for "middle" or "intermediate" and indicates that such states are structurally mid way between the solid and liquid state.

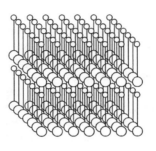

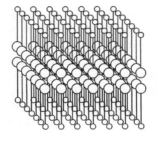

Figure 8.10 Two plausible solid structures that might be formed from molecules such as those illustrated in Figure 8.9.

Solid and liquid structures

We can look at the generic types of structure that arise in liquid crystal mesophases by considering the interactions between molecules such as the one illustrated in Figure 8.7. The molecules will have some relative orientation in which they can minimize their energy of interaction. For the purposes of this example, let us suppose that they have a low energy when they are aligned parallel, as indicated below in Figure 8.9. The *solid* formed from such molecules might look something like either of the options in Figure 8.10. In the liquid state (Fig. 8.11) the positions of the molecules are disordered in a similar way to a "normal liquid" and the orientations of the molecules are also highly randomized.

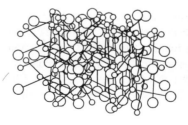

Figure 8.11 A plausible liquid structure that might be formed from molecules such as those illustrated in Figure 8.9.

Liquid crystal structures

There are several possible states – known as *mesostates* or more commonly *liquid crystal states* – intermediate between the order of the solid and the disorder of the liquid. Perhaps the simplest state to

251

(a) **(b)**

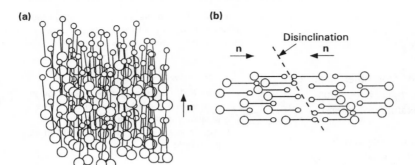

Figure 8.12 A conceivable nematic liquid crystal structure of molecules that have a minimum-energy configuration like that illustrated in Figure 8.9: (a) the vector **n** is the *director* which describes the direction of the texture of the structure; (b) a region in which two directors meet at a disinclination.

describe is the *nematic state*. The word derives from the Greek *nema*, meaning "thread", and the state describes thread-like molecules that remain roughly parallel to one another, but the positions of their centres are disordered. The orientation of the molecules within the liquid is indicated by a vector **n** known as the *director*. In general, the director will change from one region to another, producing the equivalent to crystalline defects in a solid that are known as *disinclinations*. Figure 8.12 illustrates a nematic liquid crystal state of molecules which have a minimum-energy configuration like that illustrated in Figure 8.9.

One special type of variation of **n** has important applications in the use of liquid crystals in digital displays, and is known as a *cholesteric* liquid crystal. The term cholesteric derives from *chole* the Greek prefix for the "bile" or "gall duct" from which these chemicals were derived. *Stereo*, in this sense, refers to the three-dimensional waxy nature of the substance. Within any layer of the structure, the substance looks like a nematic liquid crystal structure, but the director *rotates* from one region of the crystal to another. As illustrated in Figure 8.13, the rotation of the director is perpendicular to the plane containing the nematic director. Molecules such as those illustrated in Figure 8.9 would be unlikely to form a cholesteric phase liquid crystal.

The final type of liquid crystal state is known as a

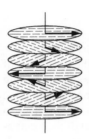

Figure 8.13 A cholesteric liquid crystal. Within any layer of the structure the substance appears to be nematic. To appreciate the structure consider, for example, the top layer on the diagram that has a director pointed to the right. If we consider a neighbouring layer, its nematic director is rotated with respect to the top layer. Similarly, the layer below is rotated with respect to the second layer. In this way the substance has a director that rotates in a plane the normal of which is perpendicular to all the directors.

smectic state in which molecules form layers similar to the layers indicated in the solid state (Figure 8.14). The word smectic derives from the Greek *smektikos* meaning "to wash", because of the soap-like consistency of substances in this state. Different varieties of smectic liquid crystal are distinguished by the different relative orientations of the director **n** with respect to the layers of the structure, and by the degree of order within each layer. Molecules such as those illustrated in Figure 8.9 could quite feasibly form smectic structures.

With a little imagination, readers may be able to envisage other combinations of positional and orientational disorder that constitute what are known to be a multitude of different liquid crystal states.

Table 8.2 The transition temperatures of some substances which form liquid crystal mesophases.

Ethyl-anisal-*p*-aminocinnamate:

Crystal $\xleftrightarrow{\text{83°C}}$ Smectic B $\xleftrightarrow{\text{91°C}}$ Smectic A $\xleftrightarrow{\text{118°C}}$ Nematic phase $\xleftrightarrow{\text{139°C}}$ Liquid

Cholestrol benzoate ($C_{34}H_{50}O_2$; relative molecular mass 491):

Crystal $\xleftrightarrow{\text{146°C}}$ Cholesteric $\xleftrightarrow{\text{178.5°C}}$ Liquid

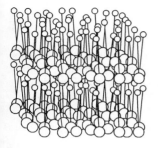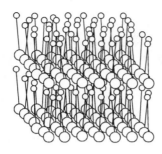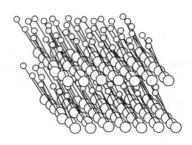

Figure 8.14 Some smectic liquid crystal structures. Smectic B is the most solid-like of the liquid crystal states, having positional order within each layer of the smectic structure. Smectic A is a structure in which this positional order *within* the layer is lost, but the orientational order within the layer is retained along with the layer structure itself. In Smectic C the layer structure remains, but the molecules orient themselves at an angle with respect to the layers.

The liquid crystal phases discussed above generally exist for only limited temperature ranges (often only a few degrees Celsius) in between the solid and liquid states.

Summary
At the level of this book it is impossible to survey all the types of liquid crystal state or their properties, but it is impossible to pass by without pointing out that they are a natural – and common – feature of liquids whose molecules have anisotropic properties. Many living organisms are constructed from large anisotropic molecules in a near-fluid state, and hence liquid crystal structures have a profound, but still poorly understood, effect on the functioning of biological systems, including our own bodies.

8.3 The dynamics of a liquid: the cell model

In the previous section we considered descriptions of the structure of the liquids. We introduced the radial density function as a method of describing the structure of a liquid of spherical molecules. We then considered some of the factors relevant to non-spherical molecules, and saw that ultimately these factors could give rise to liquid crystal structures.

However, in order to describe phenomena that are characteristic of the liquid state such as *viscosity*, which describes the ease with which a liquid changes shape, we need to develop an understanding of the way in which the structure of a liquid *changes*. We will do this by first considering how the structure of a liquid of spherical molecules might change, and then see how our conclusions would be affected if the molecules of the liquid were non-spherical.

8.3.1 Spherical molecules

The following model of liquids, known generally as the *cell model*, is an attempt to develop a near-universal model of liquids. It assumes that each molecule is constrained by its neighbours into a "cell" of roughly atomic dimensions: this is certainly the case in real liquids. Each molecule vibrates within a "cage" created by its neighbours. However, as consideration of Figure 8.15b shows, molecules such as B take part in the formation of several cells. In the figure, B is part of the "cell wall" constraining molecules A and C amongst others. In addition, *B* is itself constrained within a cell (not shown in Figure 8.15) in which A and C form part of the "cell wall". In order to break through this kind of complex analysis we need to make a dramatic simplification.

The simplification of the cell model is that it considers only the properties of an *average molecule* in an *average cell*. The cell model has four parameters (Fig. 8.16):

- the average size of the cell, a in Figure 8.15 b;
- the size of the energy barrier, ΔE_h, which must be overcome in order to move a molecule from

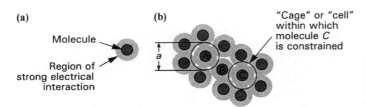

(a)

Molecule

Region of
strong electrical
interaction

(b)

a

"Cage" or "cell"
within which
molecule C
is constrained

Figure 8.15 The cell model of a liquid: (a) a simple representation of a molecule; (b) the way in which closely packed molecules form a cell, or cage, around other molecules.

one cell to another (the subscript "h" is intended to remind the reader of the "hopping" process by which molecules move through the liquid);

- the energy cost, ΔE_e, of entirely removing a molecule from the liquid (the subscript "e" on the ΔE_e is intended to remind the reader of "escape" or "evaporation" through which molecules leave the liquid); and

- the energy cost, ΔE_s, of moving a molecule from the body of the liquid to the surface of the liquid (the subscript "s" stands for "surface").

It is important to realize that, as illustrated in Figure 8.16, the actual process by a which a molecule moves through a liquid is not a single-particle process. It will, in general, involve the chance motions of several molecules conspiring to allow a molecule to move. However, the simplification of the cell model is that it considers only the motion of a single *average* particle moving through an *average* potential that is empty of other molecules. The cell model thus represents a considerable simplification of the real situation. However, it does capture the essence of the situation, in that molecules tend to vibrate about a position for a certain time, but are still able to move away from their neighbours, although with a relatively low probability. The chance of escape from a cell is determined by the strength of the interactions between molecules, by the shape of the molecules

and by the temperature.

If the molecules were completely trapped in their cells then the model would describe a solid – a set of molecules vibrating about fixed positions. The solid would lack the crystalline order we normally assumed in Chapter 7 and would be described as an *amorphous* solid.

8.3.2 The potential energy in the liquid state

Let us consider how the (electrical) potential energy of an average molecule in the liquid state will vary with position throughout the liquid. As illustrated in Figure 8.17, we expect a potential that repeats throughout the liquid, since the *average* experience of each molecule will be similar, independent of where it is in the liquid.

Figure 8.17 indicates two different potentials, one with a barrier to hopping, ΔE_h, from cell to cell much greater than the other. Since a substance in which atoms remain in their original potential cells is a solid, we expect the liquid with the larger ΔE_h will be more solid-like i.e. less able to change its shape easily, than the one with the smaller ΔE_h. The ease of changing shape is related to the *viscosity* of the substance (see §9.5.1), and so we expect that a small ΔE_h will give rise to a low viscosity.

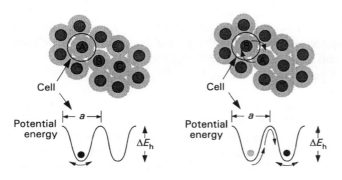

Cell

Potential
energy

a

ΔE_h

Cell

Potential
energy

a

ΔE_h

Figure 8.16 Illustration of the significance of the potential energy in the cell model of a liquid. (a) A typical molecule, A, is shown trapped by its neighbours. This is represented on the potential energy diagram by a single *average* molecule vibrating inside a fixed *average* potential well. (b) Occasionally, the vibrations of neighbouring molecules conspire to allow a molecule to change its cell. Note that, although this process of moving from cell to cell will normally involve two or more molecules, it is represented on the potential energy diagram as the motion of the single representative particle.

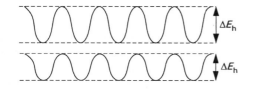

Figure 8.17 The variation of the potential energy of an average molecule with position according to the cell model of a liquid. The upper curve represents a liquid in which the "hopping" process described in Figure 8.16 is more difficult than for the liquid represented by the lower curve.

As the temperature increases, the probability per unit time that a molecule will escape from its cell also increases, since the average kinetic energy of the molecule is $\approx k_B T$. We thus expect that the viscosity of a liquid will be reduced at higher temperatures, a prediction that is compared with experimental data in §9.5.2

The energy required to activate "hopping" from cell to cell within a liquid may be compared to the energy required to remove a molecule from the liquid altogether, ΔE_e (Fig. 8.18). In general, a molecule in the liquid will be interacting with a number of other adjacent molecules, typically between 8 and 11. The bonds with all these molecules must be broken if the molecule is to escape from the body of the liquid. The energy ΔE_e required to do this will thus, in general, be rather greater than ΔE_h. The ease with which a molecule can leave the body of the liquid is related to the vapour pressure that a liquid holds above its surface. This matter is discussed more fully in §11.4.2. However, we can say already that we expect that substances with a high ΔE_h (i.e. a high viscosity) will

tend to have a high ΔE_e and hence a low vapour pressure at a given temperature.

Finally, we consider the situation of a molecule at the surface of a liquid. This molecule will only be bonded to perhaps 5–6 others as opposed to between 8 and 11 others for a molecule in the body of the liquid. Its binding energy might therefore only be $\approx 50\%$ of that of a molecule in the body of the liquid, and thus we might expect that it will cost ΔE_s (roughly 50% of ΔE_e) to take a molecule from the body of the liquid and place it at the surface. This energy cost is the origin of the surface energy – or *surface tension* – of a liquid and is discussed more fully in the comparison with experimental surface energy data in §9.5.3.

Figure 8.19 encapsulates the essence of the cell model of liquid dynamics. The three activation energies will have very different values for, say, molten sodium and molten aluminium, but we might expect that for each substance the *ratio* of the activation energies $\Delta E_h : \Delta E_s : \Delta E_e$ should be similar. Furthermore, we might hope to find some similarities between the behaviour of molten metals and very different liquids such as water or organic liquids.

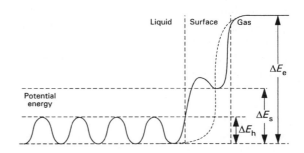

Figure 8.19 The variation of the potential energy of an average molecule with position according to the cell model of a liquid, showing the relative values of ΔE_h, ΔE_s and ΔE_e. The curve is likely to be most appropriate for spherical models which can "hop" past one another relatively easily. The curve should be contrasted with the one shown in Figure 8.20, which is likely to be appropriate for non-spherical molecules.

8.3.3 Non-spherical molecules

The dynamics of liquids of non-spherical molecules is likely to be more complex than the dynamics of the simple liquids discussed above. However, if we stay

Figure 8.18 The variation of the potential energy of an average molecule with position according to the cell model of a liquid showing the relative values of ΔE_h and ΔE_e.

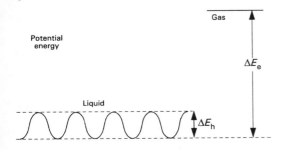

within the framework of the cell model then we expect to find just two significant differences between non-spherical and spherical molecules.

First, we expect that the "hopping" or "swapping places" process discussed for spherical molecules will be considerably more difficult. Long molecules will become "tangled" and "hooked" in a way that has no analogy in the simple liquids. We thus expect that for a given binding energy, ΔE_e, substances with non-spherical molecules will have relatively large values of ΔE_h.

Second, we expect that it will be considerably easier to place a molecule on the surface because it will be possible to arrange the orientation of the molecule so that most of it stays "under" the surface, i.e. still within the body of the liquid. Since this *can* happen relatively easily, we can expect that it *will* happen, and so we expect that, for a given binding energy ΔE_e, substances with non-spherical molecules will have relatively small values of ΔE_s.

Thus a cell model picture for non-spherical molecules might now look something like Figure 8.20. Note the changes in the relative magnitudes of ΔE_h and ΔE_s compared with those for spherical molecules in Figure 8.19.

Figure 8.20 describes a liquid that forms surfaces relatively easily, but which is more viscous than one would expect on the basis of its binding energy. The extent to which this change in the relative magnitudes of ΔE_h and ΔE_s actually occurs in real liquids is discussed in §9.6.

Finally, before leaving this description of the cell model of liquids, we stress again that the parameters of the cell model are not single-particle parameters; they merely parameterize complex many-molecule correlations in rather simple terms.

8.3.4 Non-Newtonian liquids

The two options outlined in Figures 8.19 and 8.20 for spherical and non-spherical molecules show some of the flexibility of the cell model. It can describe liquids that result from a wide variety of different molecular constituents. However, it is common to find liquids that are not well described by any variant of the cell model. These liquids are frequently:

(a) formed from very long molecules, often polymers, which may be hundreds of times longer than they are wide. Clearly such a molecule will not "hop" from cell to cell. Its motion is dominated by the nature of its entanglements with its neighbours;

(b) dissolved in a solvent – a more "normal" liquid made of smaller molecules.

It is not possible to discuss the dynamics of these complex liquids here, but we can point out some interesting properties of such liquids and invite the reader to experiment with the following.

(a) Soup: when stirring a smooth thick soup in a bowl, if one stops stirring one finds that the soup may "spring back" and briefly rotate in the opposite direction. This is quite contrary to the expected behaviour for a simple fluid which would continue to rotate. This phenomenon is known as *visco elasticity*.

(b) Non-drip paint: paints are composed of pigments in specially formulated "carrier liquids". Some carrier liquids are manufactured such that in the low-stress situation of a can, or on a paintbrush, they behave as a solid and retain their shape, i.e. they do not drip. Indeed, they may wobble like a jelly, indicating elastic behaviour and some sense of "shape". However, under the stress of brush stroke they behave as a liquid and

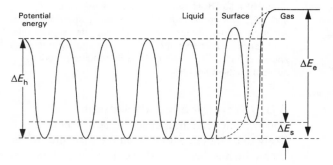

Figure 8.20 The variation of the potential enery of an average molecule with position according to the cell model of a liquid, showing the relative values of ΔE_h, ΔE_s and ΔE_e. For non-spherical molecules, which cannot "hop" past one another easily, the curve shown is likely to be more realistic than the one shown in Figure 8.19.

leave the brush. Substances behaving in this way are known as *thixotropic*.

(c) Cornflour and water: pour a small amount of cornflour (a natural medium/long chain polymer) onto a flat surface. Add water a drop at a time and stir thoroughly. Eventually a state is reached in which the substance may be "torn" in two or "broken" if pulled sharply, but which heals and flows like a liquid over the time-scale of a few seconds.

Controlling the properties of substances with many molecules in solution is a problem of great complexity, and one that still presents many challenges for scientists in both academic and industrial concerns.

8.4 Exercises

Exercises with a P prefix are "normal" problems. Those with a C prefix are best solved numerically using a computer program or spreadsheet.

C1. Use the computer program listed in Appendix 4 to note the key features of the dynamics of molecules in a simple two-dimensional liquid. Contrast these with the dynamics of molecules in (a) a simple two-dimensional solid and (b) a simple two-dimensional solid gas.

C2. Pause the computer program listed in Appendix 4 and examine the instantaneous positions of molecules in a simple two-dimensional liquid. Describe to what extent the figures in Example 8.1 and Figure 8.4 reflect the patterns seen in the simulation.

C3. Using the computer program listed in Appendix 4, look for the occurrence of processes in which molecules swap places in a manner similar to that described in Figure 8.16.

P4. Construct (a) the radial density function and (b) the radial distribution function for the data listed in Table 8.1.

CHAPTER 9

Liquids: comparison with experiment

Most elements are in the solid state at room temperature and only enter the liquid state at elevated temperatures (see Table 11.2). For this reason, data on elemental liquids is somewhat rarer than data on the properties of substances that are in their liquid state at around room temperature. For this reason we shall commonly refer to the organic liquids mentioned in §8.1.4.

9.1 Mechanical properties

The mechanical properties of liquids are, in general, intermediate between those of gases and solids. Recall that the most striking property of matter in its solid state is that it displays both a well-defined volume *and* shape, which is in contrast with matter in its gaseous state which expands to fill a container of any shape or volume. In line with its intermediate status,

matter in its liquid state has a relatively well-defined volume, but no well-defined shape.

9.2 Density

Liquids have roughly similar densities to solids, with typical values between 0.5 and 20 times that of water. Historically, the gram was defined as the mass of $1\,cm^3$ of water at 4°C and so we find that the density of water lies close to $1\,g\,cm^{-3}$, or, in the rather incongruous SI units, $1000\,kg\,m^{-3}$. We thus find liquid densities in the range $500\,kg\,m^{-3}$ to $20\,000\,kg\,m^{-3}$.

9.2.1 Data on the density of liquids

Table 9.1 contains data on the measured density of about half the elements measured in the liquid state at their melting temperature. The table also shows the

Figure 9.1 Histogram of the ratio of the density in the liquid phase at the melting temperature to the density in the solid phase of elements at 25 °C. The data is for the 45 elements, of which 41 expand and 4 contract as they enter the liquid phase.

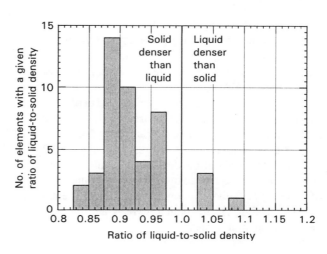

Table 9.1 The density of elements at their melting temperatures in the liquid state and the ratio of the liquid density to the density of the solid (at 25 °C; see Table 7.2). Note that, while most elements expand on melting, silicon, gallium, germanium and bismuth (shaded) contract on melting.

Z	Element	A	Liquid density ($\times 10^3$ kg m^{-3})	Liquid density/ solid density	Z	Element	A	Liquid density ($\times 10^3$ kg m^{-3})	Liquid density/ solid density
3	Lithium	6.941	0.516	0.968	44	Ruthenium	101.1	10.9	0.889
5	Boron	10.81	2.08	0.843	45	Rhodium	102.9	10.85	0.874
11	Sodium	22.99	0.930	0.962	46	Palladium	106.4	10.70	0.892
12	Magnesium	24.31	1.58	0.909	47	Silver	107.9	9.30	0.886
13	Aluminium	26.98	2.40	0.889	48	Cadmium	112.4	8.02	0.927
14	Silicon	28.09	2.525	1.080	50	Tin	118.7	6.98	0.958
16	Sulphur	32.06	1.819	0.872	51	Antimony	121.7	6.49	0.970
19	Potassium	39.10	0.824	0.955	52	Tellurium	127.6	5.77	0.924
20	Calcium	40.08	1.365	0.892	55	Caesium	132.9	1.845	0.971
22	Titanium	47.90	4.13	0.916	56	Barium	137.3	3.323	0.925
23	Vanadium	50.94	5.55	0.878	72	Hafnium	178.5	12.0	0.904
25	Manganese	54.94	6.43	0.860	73	Tantalum	180.9	15.0	0.900
26	Iron	55.85	7.1	0.901	74	Tungsten	183.9	17.6	0.914
28	Nickel	58.70	7.8	0.875	75	Rhenium	186.2	18.8	0.894
29	Copper	63.55	8.00	0.895	76	Osmium	190.2	20.1	0.890
30	Zinc	65.38	6.60	0.925	77	Iridium	192.2	20.0	0.887
31	Gallium	69.72	6.1136	1.035	78	Platinum	195.1	19.7	0.918
32	Germanium	72.59	5.53	1.038	79	Gold	197.0	17.32	0.898
34	Selenium	78.96	4.00	0.832	81	Thallium	204.4	11.29	0.951
37	Rubidium	85.47	1.47	0.959	82	Lead	207.2	10.69	0.942
40	Zirconium	91.22	5.80	0.891	83	Bismuth	209.0	10.05	1.025
41	Niobium	92.91	7.83	0.913	92	Uranium	238.0	17.907	0.940
42	Molybdenum	95.94	9.35	0.915					

(Data from *CRC handbook*; see §1.4.1)

ratio of the stated liquid density to the density in the solid state at 25°C reported in Table 7.2. The thermal expansion of the solid between 25°C and the melting temperature is generally less than 1% and does not alter the general conclusion drawn from Table 9.1 and illustrated in Figure 9.1: liquids are generally around 10% less dense than the corresponding solids. Note, however, that four elements (silicon, germanium, gallium and bismuth) are a few per cent more dense in the liquid state.

Table 9.2 contains data on the measured density of substances, mainly organic, that are liquids at around room temperature. The densities are all within a factor 2 of the density of water. Interestingly, Figure 9.2 illustrates the fact that the densities of similarly constituted organic liquids do not increase with the mass of a molecule. This is in contrast with the elemental solids (see Fig 7.1) for which there is a strong dependence on atomic mass.

Water is a more complicated liquid than the sim-

Table 9.2 The density of substances that are liquid at room temperature (only the last three entries in the table are inorganic).

Substance	Chemical formula	MW	Density (kg m^{-3})
Methanol	CH_3OH	32	791 (20°C)
Ethanol	C_2H_5OH	46	789 (20°C)
Propan-1-ol	C_3H_7OH	60	804 (20°C)
Propan-2-ol	C_3H_7OH	60	786 (20°C)
2-Methylpropan-1-ol	C_4H_9OH	74	817 (20°C)
2-Methylpropan-2-ol	C_4H_9OH	74	789 (20°C)
Butan-1-ol	C_4H_9OH	74	810 (20°C)
Butan-2-ol	C_4H_9OH	74	808 (20°C)
2-Methylbutan-1-ol	$C_5H_{11}OH$	88	816 (20°C)
2-Methylbutan-2-ol	$C_5H_{11}OH$	88	809 (20°C)
Pentanol	$C_5H_{11}OH$	88	813 (20°C)
Octanol	$C_8H_{17}OH$	130	827 (20°C)
Aniline	C_6H_7N	86	1026 (15°C)
Acetone	C_3H_6O	58	787 (25°C)
Benzene	C_6H_6	78	879 (20°C)
Carbon disulphide	CS_2	76	1293 (0°C)
Carbon tetrachloride	CCl_4	153.6	1632 (0°C)
Water (see Table 9.3)	H_2O	18	1000 (0°C)

MW, relative molecular mass.
(Data from Kaye & Laby; see §1.4.1)

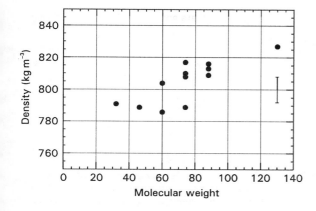

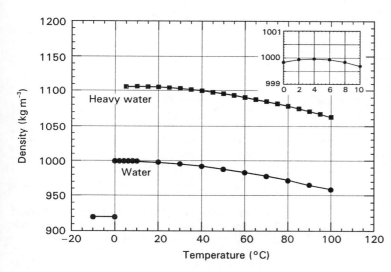

Figure 9.2 A graph of selected data from Table 9.2 showing the density of liquids with an OH group (known as *alcohols*) as a function of the relative molecular mass of the molecules of the liquid.

Figure 9.3 The density of water (H₂O) and heavy water (D₂O) as a function of temperature at atmospheric pressure. Data points plotted below 0°C refer to the density of ice. The inset shows the density of water between 0°C and 10°C, showing the weak maximum in density at 3.98°C.

ple entry in Table 9.2 might imply. Table 9.3 and Figure 9.3 show the density of ice and water as function of temperature from −10°C to 100°C. Note that water – like silicon, germanium, gallium and bismuth (Table 9.1) – is more dense in its liquid phase than its solid phase. The density in the liquid phase is roughly constant, but shows a non-linear 4% fall in density over the range to 100°C. The maximum density occurs at 3.98°C. The non-linear thermal expansion is in contrast with the strikingly linear behaviour of liquid mercury (Fig. 9.4).

Using the data from Tables 9.1 and 9.2 we can work out both the number of molecules per unit volume and the number of atoms per unit volume in the liquid state.

So the main questions raised by our preliminary

Table 9.3 The density of water (H₂O) and heavy water (D₂O) as a function of temperature at atmospheric pressure.

T(°C)	H₂O	D₂O	T (°C)	H₂O	D₂O
0	999.84	–	40	992.22	1100.0
2	999.94	–	45	–	1097.9
4	999.97	–	50	988.04	1095.7
5	–	1105.6	55	–	1093.3
6	999.94	–	60	983.20	1090.6
8	999.85	–	65	–	1087.8
10	999.70	1106.0	70	977.77	1084.8
15	–	1105.9	75	–	1081.6
20	998.20	1105.3	80	971.79	1078.2
25	–	1104.4	85	–	1074.7
30	995.65	1103.2	90	965.31	1071.1
35	–	1101.7	95	–	1067.4
			100	958.36	1063.5

(Data from Kaye & Laby; see §1.4.1)

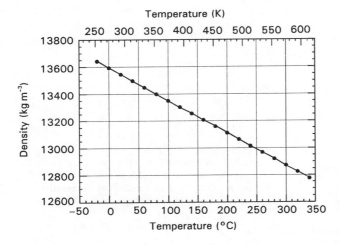

Figure 9.4 The density of mercury in its liquid state plotted as a function of temperature. Note the linearity and large magnitude of the thermal expansion: the density changes by about 7%. (Data from Kaye & Laby; see §1.4.1)

Example 9.1

Using the data from Table 9.2, work out the number of molecules per unit volume and the number of atoms per unit volume in liquid ethanol and liquid pentanol.

The molecular mass of ethanol is 46, i.e. there are Avogadro's number of ethanol molecules in 46×10^{-3} kg of ethanol. The volume of 46×10^{-3} kg of ethanol is

$$V = \frac{46 \times 10^{-3}\,\mathrm{kg}}{789\,\mathrm{kg\,m^{-3}}} = 5.83 \times 10^{-5}\,\mathrm{m^3}$$

which is around $58\,\mathrm{cm^3}$. The number density of ethanol molecules is therefore

$$n = \frac{N_A}{V} = \frac{6.02 \times 10^{23}}{5.83 \times 10^{-5}} = 1.033 \times 10^{28}\,\text{molecules m}^{-3}$$

Each ethanol molecule consists of two carbon atoms, six hydrogen atoms and one oxygen atom, i.e. nine atoms in total. Thus the density of atoms is approximately $9n = 9.29 \times 10^{28}$.

The molecular mass of pentanol is 88, i.e. there are Avogadro's number of pentanol molecules in 88×10^{-3} kg of pentanol . The volume of 88×10^{-3} kg of pentanol is

$$V = \frac{88 \times 10^{-3}\,\mathrm{kg}}{813\,\mathrm{kg\,m^{-3}}} = 1.082 \times 10^{-4}\,\mathrm{m^3}$$

which is around $108\,\mathrm{cm^3}$. The number density of pentanol molecules is therefore

$$n = \frac{N_A}{V} = \frac{6.02 \times 10^{23}}{1.082 \times 10^{-4}} = 5.56 \times 10^{27}\,\text{molecules m}^{-3}$$

Each pentanol molecule consists of five carbon atoms, 12 hydrogen atoms and one oxygen atom, i.e. 18 atoms in total. Thus the density of atoms is approximately $18n = 1.00 \times 10^{29}$.

The atomic number density of both types of liquid is close to the typical atomic number density of a solid. The molecular number density for pentanol is around half the value for ethanol because the pentanol molecules are around twice as large and of twice the mass, leading to a similar mass per unit volume for the two liquids.

This implies that the separation between atoms *within* molecules is not greatly different from the separation between atoms in neighbouring molecules.

---

examination of the experimental data on the density of liquids are:

- Why the densities of most liquids of a similar order of magnitude to those of solids?
- Why is the density of most liquid elements about 10% lower in the liquid state than in the solid state?

- Why is the density of a few liquids around 10% higher than in the solid state?
- Why, within a group of similar liquids such as the alcohols, is there only a weak dependence of density on molecular mass?

9.2.2 Understanding the density data

Normal liquids

Our discussion of the density of liquids begins with the basic assumptions about the structure of liquids outlined in §8.2. Based on this picture of a liquid as a disordered solid with a detailed structure that changes every few nanoseconds or so (Fig. 8.1) we can see immediately why liquids might, in general, be less dense than solids: the packing of molecules is less "efficient", leaving "holes" in the structure. We can take this supposition one stage further by following the lessons of §8.2, and directly evaluate the "area" density of two-dimensional liquids and solids. We can then extend the two-dimensional example to three dimensions.

Figures 9.5 & 9.6 represent a hypothetical two-dimensional substance in its solid and liquid states, respectively. The area density of spheres in this substance may be evaluated by considering the number of spheres within a given area. The odd-shaped area shown in Figures 9.5 & 9.6 is chosen in order to eliminate the edge effects (whole rows of spheres falling just in, or just out of the counting area) that occur if one uses a regularly shaped counting area. The solid density evaluates to 136 spheres per area and the liquid density evaluates to 126 spheres per area. Taking the ratio indicates that the liquid density is around $126/136 \approx 93\%$ of the solid density –an area density decrease of 7%. This is for two dimensions: if we apply the consideration of Example 7.4 we would find the three-dimensional density decrease to be about 1.5 times this amount, (around 11%).

This is in good agreement with the general level of density changes observed as solids change into liquids. The simplicity of the explanation makes it rather compelling, but raises a more difficult question: How can we understand those few substances in which the density increases on melting?

Substances that contract on melting

The answer lies again in the efficiency with which molecules pack together. We have seen that substances that bond covalently form crystal structures of relatively low density (see §6.4). This is because, in terms of energy, it is better for neighbouring molecules to adopt the optimum orientation and separation rather than merely to get as close as possible (see Fig. 6.10). Similar considerations also apply to ionically bonded substances. This is in contrast to the situation in substances that are held together by Van der Waals or metallic bonding, which seek as many near neighbours as possible.

In the liquid phase of covalently bonded substances, the random and constantly changing structure leads to a difficulty in maintaining the correct orientation with respect to neighbouring molecules.

Figure 9.5 This figure may be used to calculate the "area density" of close-packed two-dimensional circles, which are analogous to a two-dimensional solid. The curved area is the same in this figure as in Figure 9.6 and contains the centres of 136 circles. The curved area is chosen to avoid counting bias near the edge of the area.

Figure 9.6 This figure may be used to calculate the "area density" of loosely-packed two-dimensional circles, which are analogous to a two-dimensional liquid. The curved area is the same in this figure as in Figure 9.5 and contains the centres of 126 circles.

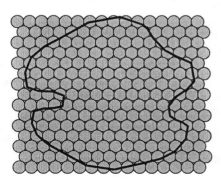

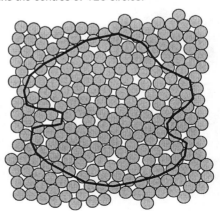

As a consequence of this, the open structure of the solid collapses as the substance enters the liquid state. If this hypothesis is correct then we ought to find that the substances that contract on melting have primarily covalent bonding: is this the case? Yes, and no.

The first two examples (silicon and germanium) do fall clearly into this category. In §6.4 we mentioned that they are covalently bonded solids and form an open crystal structure based around the tetrahedral symmetry of the electron orbitals. Water is a more complicated case, but the hydrogen bonding present in water is certainly highly directional, as discussed below. Our last example (gallium) is more puzzling. Gallium is a metal and so ought to have at least some component of non-directional bonding. However, the general run of the argument concerning the density of liquids gives us some faith that the fact that gallium contracts as it enters the liquid state must indicate that it too must have a directional (i.e. covalent) bonding mechanism in addition to its metallic bonding.

Water

In water, the directional bonds that make the solid state (ice) lower in density than the "collapsed" liquid state are the hydrogen bonds between the molecules.

In the "collapsed" liquid state a great deal of the structure present in ice remains, and advanced studies indicate that only a few per cent of the hydrogen bonds are bent (not broken) in the liquid state. The process that leads to the melting of ice is the progressive weakening of the ice structure (Fig. 9.7) by thermal vibrations that cause the hydrogen-bonded part of the O=O linkages to bend. It is as if in the skeletal steel frame of a skyscraper a few of the steel girders were replaced with rubber ones: large parts of the structure would remain intact, but the structure as a whole would collapse. The melting of ice is a similar, though less spectacular, phenomenon.

The density of organic liquids

Finally, we come to the question of why, within a group of similar liquids such as the alcohols, there is only a weak dependence of density on molecular mass (Table 9.2 & Figure 9.2). Molecules that differ in mass by a factor of 4 form liquids whose densities change by less than 5%. The conclusion to be drawn from this striking fact is that the separation between atoms in different molecules is similar to the separation between atoms within a molecule. This is at first rather surprising.

Figure 9.7 A qualitative indication of the structure of ice. (a) A central water molecule linked to two others. Concentrating on the oxygen atoms, one can see that each oxygen atom can link to four other oxygen atoms arranged roughly tetrahedrally around it. Two of these links are of the form O—covalent bond—H—hydrogen bond— O and two are of the form O—hydrogen bond—H— covalent bond—O. One example of each form an O—O linkage (thick straight line). (c) The structure of ice indicated by circles representing only the oxygen atoms; the lines represent either of the two types of O—O linkage discussed. Different bonds have been drawn either as thin, bold or shaded lines to emphasize the structure. In fact all the bonds are equivalent.

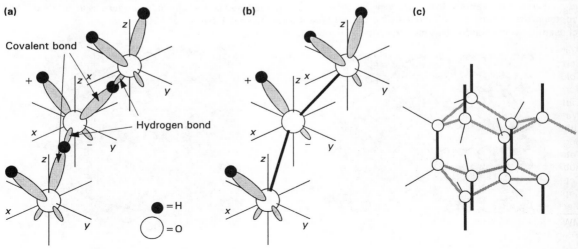

As mentioned in §8.1.4, the forces acting between organic molecules are relatively weak. The alcohols listed in Table 9.2 do not have a strong contribution from hydrogen bonding, and so a good deal of their bonding energy is the weak Van der Waals force due to the fluctuations of electronic charge on the molecules. There is also a term which results from the slightly polar nature of the carbon–hydrogen C–H bonds which constitute most of the bonds in the molecule. The forces acting between atoms within a molecule are the strong covalent bonds due to either C–H or C–C bonds.

However the larger molecules have more "surface area" and so more molecules can be attracted through the non-directional Van der Waals mechanism. This factor will compensate somewhat for the weakness of the interaction. We should also realize that, although the atomic separation between atoms within a molecule and atoms in neighbouring molecules is similar, the strength of their interactions differs dramatically. The intermolecular forces of the atoms are overcome at the boiling temperature ($\approx 300\,K$), whereas the intramolecular forces are not overcome until the molecule decomposes (at about several thousand kelvin).

9.3 Thermal expansivity

The increase in volume on heating a liquid held at a constant pressure, P_0, with initial volume V_0 can be expressed as

$$V = V_0(1 + \beta \Delta T) \qquad (9.1)$$

where ΔT is the change in temperature from the initial temperature, and β is the coefficient of volume expansivity.

9.3.1 Data on the thermal expansivity of liquids

The data given in Table 9.4 show that the volume thermal expansivity of liquids is, in general, rather greater than that of solids (see Table 7.4), but still considerably less than for gases (see Table 5.4). This is true for organic, inorganic and elemental liquids. Notice that some liquids display anomalous thermal expansivities close to their melting temperature; for example, water (Fig. 9.3) displays a negative thermal expansion (it contracts) between 0 C and 4°C. Other liquids, such as mercury (Fig. 9.4) display highly linear behaviour from their melting temperature up to their boiling temperature.

Table 9.4 The coefficient of volume expansivity β for various liquids at temperatures around room temperature. The double lines separate organic, inorganic and elemental liquids. The shaded column shows the value of the volume expansivity of the corresponding solid substance.

Substance	Chemical formula	MW	T (°C)	β (°C⁻¹) liquid	β (°C⁻¹) solid
Acetic acid	CH_3COOH	60	20	107×10^{-5}	–
Acetone	CH_3COCH_3	58	20	143×10^{-5}	–
Methanol	CH_3OH	46	20	119×10^{-5}	–
Ethanol	C_2H_5OH	32	20	108×10^{-5}	–
Aniline	C_6H_7N	86	20	85×10^{-5}	–
Benzene	C_6H_6	78	20	121×10^{-5}	–
Toluene	$C_6H_5CH_3$	92	20	107×10^{-5}	–
Carbon disulphide	CS_2	76	20	119×10^{-5}	–
Carbon tetrachloride	$CCl4$	154	20	122×10^{-5}	–
Water	H_2O	18	20	21×10^{-5}	–
Lithium	Li	23	400–1125	19×10^{-5}	168×10^{-6} (20°C)
Sodium	Na	39	96.5	25×10^{-5}	212×10^{-6} (20°C)
Potassium	K	85.5	64–1400	29×10^{-5}	249×10^{-6} (20°C)
Rubidium	Rb	133	39	30×10^{-5}	270×10^{-6} (20 °C)
Copper	Cu	63.6	1084	$10 \times 10{-5}$	49.5×10^{-6} (20°C)
Copper	Cu	63.6	1084	$10 \times 10{-5}$	60.9×10^{-6} (527°C)
Mercury	Hg	200.6	0–100	18.1×10^{-5}	–

MW, relative molecular weight.
(Data from Kaye & Laby; see §1.4.1)

In summary, these results indicate that for elements and organic liquids, a typical volume expansivity varies between $10^{-3}\,^\circ C^{-1}$ and $10^{-4}\,^\circ C^{-1}$. Note, however, that the volume expansivity of molecular liquids (with the exception of water) is considerably greater than that seen for the molten metals.

So the main questions raised by our preliminary examination of the experimental data on the expansivity of solids are:

- Why is the expansivity of liquids (in general) larger than for solids, but smaller than that shown by gases?
- Why do particular values of α for each material arise?
- What is the origin of the thermal expansivity anomaly in water?

9.3.2 Understanding the thermal expansivity of liquids

In §7.2 we explained the expansivity of solids as being due to the asymmetry of the potential energy variation between atoms. At low temperatures, the atoms vibrate with a low amplitude around the minimum of the interatomic potential. These low amplitude vibrations are essentially symmetric about the position of mean separation. As the temperature increases, the atomic vibrations become systematically larger in amplitude and increasingly asymmetric in nature. This asymmetry increases the mean separation of the atoms of the substances and leads to the phenomenon of thermal expansion. Thus the key element in understanding the thermal expansivity of solids lay in understanding the detailed manner in which the potential energy of each pair of atoms varies as a function of their average separation.

The atoms in a liquid are held together by exactly the same forces that hold the atoms of a solid together, and so we may reasonably expect the thermal expansivities of liquids to be similar to those of solids. Experimentally we find that the thermal expansivities of liquids are similar to those of the equivalent solids (see Table 9.4) but with a tendency to be slightly larger. This may be understood as being due to the fact that the thermal expansivity of solids tends to increase with temperature, and the liquid data are necessarily obtained at a higher temperature than the equivalent solid data.

Example 9.2

A liquid-in-glass thermometer consists of a reservoir of liquid connected to a thin tube called a *capilliary tube*. The volume of the liquid reservoir changes with temperature and this change in *volume* is reflected in a variation in the *length* of the liquid column in the capilliary tube.

A given liquid-in-glass thermometer has a reservoir of volume $1\,cm^3$ and is connected to a capilliary tube with an internal diameter of $0.1\,mm$. For mercury and ethanol, the two most commonly used working liquids, work out the sensitivity of such a thermometer i.e. the number of millimetres per degree Celsius that the liquid surface moves along the tube.

The basic equation is (Eq. 9.1)

$$V = V_0 (1 + \beta \Delta T)$$

and we need first to evaluate the volume change $V - V_0 = \Delta V$ of the liquid for a change in temperature of $1\,^\circ C$. Rearranging Equation 9.1 gives

$$V - V_0 = \Delta V = V_0 \beta \Delta T$$

This volume change is accommodated by the expansion of the fluid up the capilliary tube. The extra volume is therefore given by $\pi r^2 \Delta L$, where ΔL is the distance the fluid expands up the tube and r is the radius of the capilliary tube. We thus have

$$\Delta V = V_0 \beta \Delta T = \pi r^2 \Delta L$$

and so rearranging, with $\Delta T = 1\,^\circ C$ this gives

$$\Delta L = \frac{V_0 \beta}{\pi r^2}$$

Substituting $r = 0.05 \times 10^{-3}$ m and $V_0 = 1\,cm^3 = 10^{-6}\,m^3$,

$$\Delta L = \frac{10^{-6}\beta}{\pi\left(0.05 \times 10^{-3}\right)^2} = 127.3\beta$$

If we substitute the β values for mercury ($18.1 \times 10^{-5}\,^\circ C$) and ethanol ($108 \times 10^{-5}\,^\circ C$) we find

$$\Delta L(\text{Mercury}) = 0.023\,m = 2.3\,cm$$

$$\Delta L(\text{Ethanol}) = 0.137\,m = 13.7\,cm$$

Both displacements are easily appreciable and both liquids are commonly used in liquid-in-glass thermometers. A fuller calculation would take into account the expansion of the tube and the container of the liquid reservoir which tend to reduce slightly the sensitivity of the thermometers. Note that water could not be used as the working liquid in such a thermometer because its thermal expansivity (which corresponds to the *slope* of the data in Figure 9.3) varies considerably across the small temperature range in which it is a liquid. This is in contrast to the data for mercury shown in Figure 9.4. Furthermore, cooling below $0\,^\circ C$ would cause the liquid to freeze and the expansion of water on freezing would almost certainly break the reservoir container.

However, there is one further way in which a liquid can expand which is unavailable to a solid: it can change its structure in a slow undramatic manner proceeding from a solid-like structure near the melting temperature to a more gas-like structure near the boiling temperature. The data in Table 9.4 show that in elements for which data are available for both for liquids and solids, this method of expansion is a relatively small component of the total thermal expansivity.

However, the thermal expansivity of the organic liquids is nearly three times larger than elemental liquids that melt at a similar temperature. This is because the molecules of organic liquids are attracted to one another by the weak Van der Waals forces, this making it relatively easy for molecules to move apart and adopt structures that optimize their energy. Since the covalent bonds within the molecules are considerably stronger than the Van der Waals bonds between molecules, molecules tend to retain their optimal shape on heating but to change their relative orientation and separation as a function of temperature.

The final question concerns the thermal expansivity anomaly of water which causes water to become slightly denser as it warms from 0°C to 3.98°C. The origin of this anomaly is already clear from Figure 9.3 which shows the anomaly on the same scale as the 10% volume collapse that ice undergoes on melting.

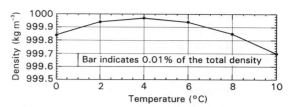

Figure 9.8 Detailed view of the thermal expansivity anomaly in water. Note the tiny scale of the anomaly. (See also Fig. 9.3)

The process which leads to the melting of ice is the progressive weakening of the ice structure (Fig. 9.7) by thermal vibrations that cause the hydrogen-bonded part of the oxygen–oxygen linkages to bend. This process of bending and collapse continues in the liquid state, but is in competition with the normal processes of thermal expansion discussed above. Thus the maximum in the density arises because at around 4°C the process of collapse of the liquid structure is exactly balanced by the tendency to expand thermally .

9.4 Speed of sound

Sound travels considerably faster through liquids than gases: the maximum speed in Table 9.5 is

Table 9.5 The velocity of sound in solids showing c_L the velocity of longitudinal waves. For the elements, where possible, the data for the solid state (taken from Table 7.6) is also included in the shaded column for comparison.

Material	T(°C)	c_L (m s^{-1})	Material		T (°C)	c_L (m s^{-1}) liquid	c_L (m s^{-1}) solid
Organic liquids			*Elements*				
Acetic acid	20	1173	Cadmium	Cd	360	2150	2780
Acetone	20	1190	Copper	Cu	1350	3350	4759
			Gallium	Ga	50	2740	
Methanol	20	1121	Mercury	Hg	20	1454	
Ethanol	20	1162	Silver	Ag	1150	2630	3704
Propanol	20	1223	Tin	Sn	240	2470	3380
Butanol	20	1258	Zinc	Zn	450	2700	4187
iso-Pentanol	20	1255					
Hexanol	20	1331	Hydrogen	H2	−258	1242	
Heptanol	20	1343	Helium	He	−269	211	
			Nitrogen	N2	−189	745	
Water	0	1402	Oxygen	O2	−186	950	
Ice	−20	3840	Sodium	Na	110	2520	
			Potassium	K	80	1869	
			Rubidium	Rb	50	1427	
			Caesium	Cs	40	980	

(Data from Kaye & Laby; see §1.4.1)

around $3350\,\mathrm{m\,s^{-1}}$ for molten copper, which may be compared with a maximum of around $1000\,\mathrm{m\,s^{-1}}$ for hydrogen gas (see Table 5.14).

9.4.1 Data on the speed of sound

Examination of the data given in Table 9.5 shows that the speed of sound in a liquid is around 30% less than the speed of sound in the corresponding solid. Furthermore, the type of wave which propagates has more in common with the single type of longitudinal wave which propagates in a gas. Shear waves (Figs 7.8b,c) are heavily damped in liquids and we do not consider such waves here.

The main question raised by our preliminary examination of the experimental data on the velocity of sound in liquids is: why is the speed of sound in liquids around 30% slower than in the equivalent solid below its melting temperature?

9.4.2 Understanding the data on the speed of sound

Having examined the speed of sound in both gases and solids we are now fairly confident that we understand well the physics of sound propagation. We understand that the speed of sound in a medium depends on two quantities: the compressibility of the medium and its density. The agreement with data for gases and solids leads us to rephrase our original question as follows: is the lower speed of propagation in a liquid as compared with a solid due to a change in compressibility or a change in density?

However, in liquids a complicating factor arises concerning the character of the waves in the liquid. In this section we will first deal with this complication, and then in turn address the effects of density and bulk modulus.

The effect of the character of the sound waves

In solids, the speed of a longitudinal compressive wave, c_L, is given (Eqs 7.14 & 7.15) as:

$$c_L = \sqrt{\frac{\gamma E}{\rho}} \tag{9.2}$$

where ρ is the density of the substance, E is its Young's modulus, and γ is a factor related to the

Poisson ratio, σ, of the substance given by

$$\gamma = \frac{(1-\sigma)}{(1-2\sigma)(1+\sigma)} \tag{9.3}$$

The reader is reminded that γ in this context has no connection with the ratio of heat capacities that occurs in formulae for the speed of sound in gases.

Our first step is to write Young's modulus in terms of the bulk modulus, B, using a standard equation (Eq A2.33 in Appendix 2):

$$B = \frac{E}{3(1-2\sigma)} \tag{9.4}$$

Substituting Equation 9.4 into Equation 9.2 yields and expression for the speed of longitudinal sound waves in a bulk solid:

$$c_L = \sqrt{\frac{3B(1-\sigma)}{\rho(1+\sigma)}} \tag{9.5}$$

Now for solids, the Poisson ratio has a typical value of $\sigma \approx 1/3$, which implies that

$$c_L^{\mathrm{solid}} \approx \sqrt{\frac{3B(\frac{2}{3})}{\rho(\frac{4}{3})}} = \sqrt{\frac{3}{2}}\sqrt{\frac{B}{\rho}} \approx 1.22\sqrt{\frac{B}{\rho}} \tag{9.6}$$

However, for liquids we expect the Poisson ratio to be very close to 0.5. This is the value of the Poisson ratio relevant to compressions and rarefactions that are volume conserving. We expect this because liquids have the ability to change shape relatively easily (see Example 7.8). Substituting $\sigma = \frac{1}{2}$ we arrive at

$$c_L^{\mathrm{liquid}} \approx \sqrt{\frac{3B(\frac{1}{2})}{\rho(\frac{3}{2})}} = \sqrt{\frac{B}{\rho}} \tag{9.7}$$

which is similar to the equivalent result for a gas (see §5.5.2).

Let us first make the (incorrect) assumption that the densities and bulk moduli of a substance in its solid and liquid states are equal. If this were so, then the expected difference in the speed of sound would arise entirely from the changed character of the sound waves in the liquid. Taking the ratio of Equations 9.6 and 9.7 predicts that the ratio $c_L^{\mathrm{liquid}}/c_L^{\mathrm{solid}}$ would be approximately $1/1.22 \approx 0.82$, i.e. we would expect the speed of sound in liquids to be about 18% lower than in the equivalent solid.

The effect of density

In §9.2 we saw that the densities of liquids are generally of the order of 10% less than those of solids due to the less efficient "packing" of atoms in the liquid state. According to Equation 9.7 the speed of sound in liquids should therefore be increased as compared to the speed of sound in an otherwise equivalent solid by a factor of ≈ 1.05. So, taking into account both the change in the character of the sound waves in the liquid and the change in density, we would expect the speed of sound in liquids to be $1.05 \times 0.82 \approx 0.86$ i.e. about 14% lower than in the equivalent solids.

The effect of the bulk modulus

Consideration of both the character of waves in a liquid and of the changes in the density of the liquid, has not explained the experimental fact of a 30% reduction in the speed of sound in liquids. So we must assume that the extra lowering in the speed of sound of a liquid is due to a lowering of the bulk modulus of the liquid.

Taking the ratio of Equations 9.7 and 9.6 yields an expression for the ratio of the speed of sound in the liquid and the solid:

$$\frac{c_L^{\text{liquid}}}{c_L^{\text{solid}}} = \sqrt{\frac{B^{\text{liquid}}}{\rho_{\text{liquid}}}} \bigg/ \sqrt{\frac{3}{2}\sqrt{\frac{B^{\text{solid}}}{\rho_{\text{solid}}}}}$$

$$= \sqrt{\frac{2}{3}}\sqrt{\frac{B^{\text{liquid}}}{B^{\text{solid}}} \times \frac{\rho_{\text{solid}}}{\rho_{\text{liquid}}}} \qquad (9.8)$$

Experimentally, we have

$$c_L^{\text{liquid}}/c_L^{\text{solid}} \approx 0.7 \quad \text{and} \quad \rho_{\text{solid}}/\rho_{\text{liquid}} \approx 1.1,$$

and so Equation 9.8 simplifies to

$$0.7 = \sqrt{\frac{2}{3} \times 1.1}\sqrt{\frac{B^{\text{liquid}}}{B^{\text{solid}}}} \qquad (9.9)$$

Solving for $B^{\text{liquid}}/B^{\text{solid}}$ yields

$$\frac{B^{\text{liquid}}}{B^{\text{solid}}} \approx 0.7^2 \times \frac{3}{2.2} \approx 0.67 \qquad (9.10)$$

Equation 9.10 states that liquids are around one-third more compressible (less springy) than the equivalent solids. In retrospect, this does not seem implausible. The inefficient packing of atoms in a liquid leads to an extra 10% of empty space within it, compared to the corresponding solid. Thus under pressure we would expect the liquid to relinquish some of this "empty space" (to compress) relatively easily. Thus the same factors that enable us to understand the lower density of the liquid state also enable us to understand the higher compressibility. What we lack, however, is a way of estimating the difference in compressibility between the liquid and solid on a similar simple model to our calculation of the density difference between liquids and solids. Unfortunately, such a calculation is rather difficult and we shall not attempt it here.

A simple calculation

What we can do is to attempt to take account of the empty space within a liquid in a more simple-minded approach. We consider a molecule trapped within its cell in the liquid. Energy may be transmitted through the cell in two modes, one characteristic of a gas and the other more typical of a solid:

- In the first mode the molecule may move across its cell from one side to the other. Energy gained on one side of the cell is thus transmitted, as in a gas, through molecular motion, albeit with an extremely short mean free path.
- In the second mode, vibrations of one molecule may be transmitted to its neighbours, as in a solid.

Consider the time it takes for energy to move across a cell of width $a + \Delta a$, where a is the typical atomic separation in the solid. We must add two terms: the time for a short gas-like trajectory through a distance Δa, plus the time for solid-like transmission through the molecule itself with a dimension of the order a (Fig. 9.9).

Figure 9.9 A molecule has more space available to it in the liquid state than in the solid state. We imagine that the molecules move freely as in a gas across the extra space Δa within the liquid state.

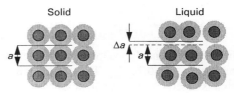

The time, Δt taken for transmission in the liquid across a cell of approximate width $\approx a + \Delta a$ is thus

$$\Delta t = \frac{\text{Distance}}{\text{Speed}} = \frac{a}{c_{\text{solid}}} + \frac{\Delta a}{c_{\text{gas}}} \qquad (9.11)$$

and thus the speed of sound in a liquid should be

$$c_{\text{liquid}} = \frac{\text{Distance}}{\Delta t} = \frac{a + \Delta a}{\left(\dfrac{a}{c_{\text{solid}}} + \dfrac{\Delta a}{c_{\text{gas}}} \right)} \qquad (9.12)$$

which simplifies in two stages:

$$c_{\text{liquid}} = \frac{a\left(1 + \dfrac{\Delta a}{a}\right)}{\dfrac{a}{c_{\text{solid}}}\left(1 + \dfrac{\Delta a c_{\text{solid}}}{a c_{\text{gas}}}\right)} \qquad (9.13)$$

and

$$c_{\text{liquid}} = c_{\text{solid}} \times \frac{\left(1 + \dfrac{\Delta a}{a}\right)}{\left(1 + \dfrac{\Delta a c_{\text{solid}}}{a c_{\text{gas}}}\right)} \qquad (9.14)$$

In order to estimate c_{liquid} we need to estimate $\Delta a/a$ and $c_{\text{solid}}/c_{\text{gas}}$. We can estimate the first factor from the density change on entering the liquid state, which is typically 10% (see §9.2). Thus we estimate that the volume of the "cell" is of the order of $(a + \Delta a)^3$ which is around 10% larger than a^3, the volume occupied by a molecule in the solid state. Noting that $(a + \Delta a)^3 \approx a^3 + 3a^2 \Delta a = a^3(1 + 3\Delta a/a)$, we estimate $\Delta a/a \approx 0.1/3 = 0.033$.

The ratio of $c_{\text{solid}}/c_{\text{gas}}$ is difficult to estimate directly here because there are no common entries in the tables for the velocity of sound in the gaseous and solid states. However, the solid values are all about $3000\,\text{m s}^{-1}$, whereas the values for gases are more typically about $300\,\text{m s}^{-1}$. We thus estimate that $c_{\text{solid}}/c_{\text{gas}} \approx 10$.

Using these estimates in Equation 9.14 we find

$$c_{\text{liquid}} \approx c_{\text{solid}} \frac{(1 + 0.033)}{(1 + 0.033 \times 10)} \approx 0.78 c_{\text{solid}} \qquad (9.15)$$

which agrees reasonably well with our experimental finding that $c_{\text{liquid}} \approx 0.7 c_{\text{liquid}}$. The calculation, even though it is based on crude estimates, has produced a reasonable estimate of the speed of sound in a liquid in terms of the speed of sound in gases and solids. This reflects the intermediate status of liquids, and the difficulty in approaching phenomena in liquids that have mixed solid-like and gas-like characters. Imperfect though it is, we will use this model again when we consider the transport of heat energy through a liquid (see §9.7.4).

9.5 Special properties of liquids

For the sake of consistency and ease of comparison I have tried, where possible, to list the properties of solids, liquids and gases in the same format in Chapters 5, 7 and 9, respectively. However, there are some properties of the liquid state that are particularly characteristic of that state, and it would be an omission to neglect them.

Perhaps the two most characteristic liquid properties are the ability of a liquid to flow and its tendency to form smooth, rounded surfaces (Fig. 9.10). The property of flow is characterized by the *viscosity* of a liquid, and this gives a numerical measure of the distinction between a liquid and a solid. The property of smooth surface formation is characterized by a surface energy or *surface tension*. This gives a numerical measure of the distinction between a liquid and a gas that requires essentially no energy to form a new "surface".

Figure 9.10 Characteristic features of a liquid.

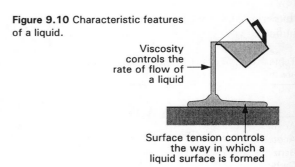

Viscosity controls the rate of flow of a liquid

Surface tension controls the way in which a liquid surface is formed

9.5.1 Data on the viscosity of liquids

I mentioned at the start of this chapter that liquids possess a relatively well-defined volume, but no well-defined shape. However, not all liquids can change their shape with similar ease, and the quantitative measure of the ease with which they can flow from one shape to another is the viscosity of the liquid.

The viscosity of liquids may be measured in a variety of ways. The methods depend on, for example: the rate of damping of oscillations of objects in the liquid; the speed with which objects fall through the liquid; or the rate of flow of the liquid through constrictions.

Formally, one defines viscosity in terms of the rate of transfer of momentum (or the *force*) exerted on a plane of liquid when a second plane of liquid is moved with velocity v. If the second plane is far away from the first, we expect the transfer of momentum to be smaller than if the plane is nearby. Also, the quicker the first plane moves, the more momentum will be transferred to adjacent planes. For many liquids, the force between the planes is:

(a) proportional to the velocity v, and
(b) inversely proportional to the separation x of the two plates.

We thus write

$$F = \eta \frac{v}{x} \qquad (9.16)$$

where the constant of proportionality η is defined as the viscosity. Liquids for which this definition holds are known as *Newtonian liquids*. Figure 9.11 illustrates this definition.

Figure 9.11a is a schematic of the definition of viscosity. Moving a plate through a liquid will drag a few atomic layers adsorbed to its surface with it and thus create a moving liquid layer. A plate moving as indicated will tend to drag a second plate with it, cre-

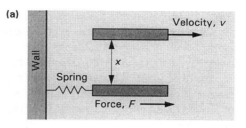

(a)

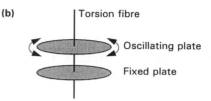

(b)

Figure 9.11 (a) Cross-section of two solid planes within a liquid. The lower plane experiences a force due to the motion of the upper plane which may be calculated according to Equation 9.16. (b) Measurement of the damping of torsional oscillations of discs presents a more practical method for determining the viscosity.

ating a force F on the second plate, which could be measured using a spring balance apparatus. Figure 9.11b illustrates an apparatus that might be used in practice. One plate oscillating above another would be damped by (i.e. lose momentum to) the liquid trapped between it and a fixed plate. Observations of the rate of damping allow a direct inference of the viscosity according to the definition given by Equation 9.10.

The viscosity of various substances is recorded in Table 9.6 and selected data from the table are plotted in Figure 9.12. Figure 9.13 shows the detailed variation of the viscosity of water with temperature. It is

Table 9.6 The viscosity η of various substances in their liquid state as a function of the temperature. The units of η are mPa s, so, for example, the viscosity of mercury at 50°C is 1.401×10^{-3} Pa s.

| | Temperature (°C) | | | | | | | | | | | | |
	−100	−50	0	25	30	50	75	100	400	600	700	800	1100
Acetic acid	–	–	–	1.116	1.037	0.792	0.591	0.457	–	–	–	–	–
Acetone	–	–	0.402	0.310	0.295	0.247	0.200	0.165	–	–	–	–	–
Benzene	–	–	–	0.603	0.562	0.436	0.332	0.263	–	–	–	–	–
Carbon disulphide	2.132	0.796	0.445	0.357	0.343	–	–	–	–	–	–	–	–
Methanol	–	2.258	0.797	0.543	0.507	0.392	0.294	0.227	–	–	–	–	–
Ethanol	98.96	8.318	1.873	1.084	0.983	0.684	0.459	0.323	–	–	–	–	–
Sodium	–	–	–	–	–	–	–	0.680	0.286	0.215	0.192	0.174	–
Potassium	–	–	–	–	–	–	–	0.458	0.224	0.172	0.155	0.141	–
Mercury	–	–	1.616	1.528	1.497	1.401	1.322	1.255	–	–	–	–	–
Tin	–	–	–	–	–	–	–	–	1.33	1.04	0.950	0.890	0.780

(Data from Kaye & Laby; see §1.4.1)

clear from all these data that the viscosity of liquids depends strongly on temperature, becoming more viscous as a liquid is cooled. Figure 9.12 shows viscosities varying strongly with temperature – the data for ethanol varies by at least two orders of magnitude. Similarly, the data for water in Figure 9.13 show a fall of around one order of magnitude between 0°C and 100°C. We note from the data that substances with higher melting temperatures tend to have higher viscosities. So, for example, sodium is more viscous than potassium, and ethanol is more viscous than methanol. However, the correlation could also be connected to molecular mass.

So the main question raised by our preliminary examination of the experimental data on the viscosity of liquids is:

- Why does the viscosity decrease strongly with increasing temperature?

9.5.2 Understanding the viscosity of liquids

Our approach here will be to use the cell model of a liquid to discuss the viscosity data. The fundamental process that takes place when a liquid changes shape is such that one layer of molecules has to "slide past" another layer. In terms of the cell model this process requires molecules to "hop" from one cell to another. Since the energy required to activate this hopping is ΔE_h, the probability that any particular molecule will

Figure 9.12 The viscosity of various substances (from Table 9.6) plotted as a function of the temperature. Note that the vertical scale is logarithmic, covering three orders of magnitude. (Data from Kaye & Laby; see §1.4.1)

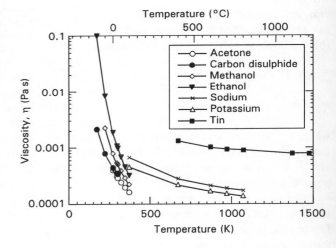

Figure 9.13 The viscosity of water plotted as a function of the temperature. Note that the vertical scale is linear.

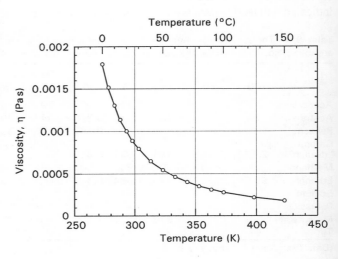

be able to hop into an appropriate cell is (see §2.5)

$$P(E_h) = \text{Constant} \times \exp(-\Delta E_h / k_B T) \quad (9.17)$$

Thus if we consider that in order to achieve a certain change of shape within the liquid a certain (large) number of individual molecular hops is necessary, then the more molecules that have energy greater than ΔE_h the easier the change will be. If molecules have energy less than ΔE_h then they will be unable to move from their local cell and the liquid will behave in a solid-like manner.

The ease of flow of the liquid – its *fluidity* – is defined as $1/\eta$. From the considerations above we expect the fluidity to vary as

$$\frac{1}{\eta} \propto \exp(-\Delta E_h / k_B T) \quad (9.18)$$

or that the viscosity will vary as

$$\eta \propto \exp(+\Delta E_h / k_B T) \quad (9.19)$$

This is an easily testable proposition. If the viscosity varies as Equation 9.13 suggests, then a plot of $\ln(\eta)$ versus $1/T$ should yield a straight line. Figure 9.14 shows the data from Table 9.6 replotted in this form. According to Figures 9.14 & 9.15, the data collated in Table 9.6 represent good evidence for the exponential dependence of viscosity on inverse temperature, and hence indirectly for the cell model of a liquid.

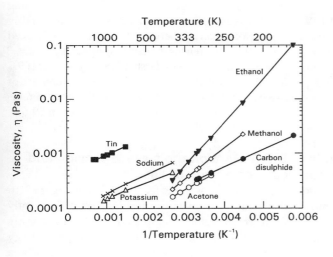

Figure 9.14 The viscosity of various substances from Table 9.6 plotted on a logarithmic axis as a function of inverse temperature. Note that the vertical scale covers three orders of magnitude in viscosity. The lines are fits to the data points, with slopes as indicated in Table 9.7.

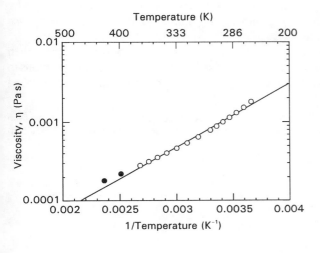

Figure 9.15 The viscosity of water plotted on a logarithmic axis as a function of the inverse temperature. The two filled circle (•) data points are taken under pressure above 100°C. The line is a fit to the data taken between 0°C and 100°C and has a slope of 1862 K. The data conform fairly closely to the line, but clear signs of curvature are evident.

Table 9.7 Analysis of the slopes found in Figures 9.15 & 9.16 in terms of Equation 9.13.

Substance	Slope (K)	ΔE_h (J)	ΔE_h (meV)
Acetone	907	12.52×10^{-21}	78.1
Carbon disulphide	736	10.16×10^{-21}	63.4
Methanol	1271	17.55×10^{-21}	109.5
Ethanol	1845	25.47×10^{-21}	159.0
Sodium	769	10.62×10^{-21}	66.3
Potassium	661	9.126×10^{-21}	57.0
Tin	688	9.50×10^{-21}	59.3
Water	1862	25.7×10^{-21}	160

We can evaluate the parameter ΔE_h from the slopes of the data in Figures 9.14 and 9.15. Since we have plotted $\ln(\eta)$ versus $1/T$, we expect the slope to be just $\Delta E_h / k_B$. The results in units of millielectron volts are shown in Table 9.7.

The values in Table 9.7 are of the order $0.1\,eV$ with values ranging from 1/6 to 1/20 of an electron volt. Now if the cell model really does make sense, then we should be able to broadly understand the magnitude of the predicted activation energies, ΔE_h. The values of ΔE_h are considered in the light of estimates for ΔE_s and ΔE_e in §9.6.

Before discussing the values deduced in Table 9.7, it is important to keep in mind that ΔE_h is not the activation energy for a single molecule to "hop" its way through the liquid. Rather it is an energy that reflects the probability of complex multi molecule correlations occurring that conspire to allow a molecule to "jiggle" through the liquid.

Looking at the data in Table 9.7 alone we see that, broadly, the values of ΔE_h seem reasonable. For example:

(a) ΔE_h for sodium is greater than ΔE_h for potassium, in line with the fact that the interatomic bonding in sodium is slightly stronger than in potassium, as evidenced by the boiling temperatures of the two metals (potassium 1047K, sodium 1156K).

(b) ΔE_h for methanol (CH_3OH) is less than for ethanol (CH_3CH_2OH), which is a larger, heavier and more awkwardly shaped molecule.

(c) ΔE_h for ethanol (CH_3CH_2OH) is similar to that of water. Here we see that the ease of motion of a molecule through the liquid arises both from bonding considerations and from considerations of molecular shape. The water molecule is much

smaller than the ethanol molecule, and so should find it easier to move through its liquid. However, water experiences strong hydrogen bonding which slows down its motion in comparison with the weaker bonding in ethanol.

There are, however, data in Table 9.7 that at first sight seem surprising. For example, ΔE_h for ethanol (boiling temperature 345K) is nearly three times greater than for tin (boiling temperature 2543K). To understand this we need once again to consider that the relative ease with which a molecule moves is a product of:

(a) geometrical considerations: it is easier for the small essentially spherical tin atom to move through liquid tin than it is for the oddly shaped ethanol molecule to move through liquid ethanol;

(b) bonding strength considerations: the bonding between ethanol molecules is considerably weaker than it is between tin atoms, as evidenced by the differences in their boiling temperatures.

However, in order to understand the data in Table 9.7 in full we need to consider the data in the context of experimental values of the other cell model parameters, ΔE_s and ΔE_e. This task is undertaken in §9.6.

9.5.3 Data on the surface tension and surface energy of liquids

We are familiar with the fact illustrated in Figure 9.16 that liquids on an impermeable surface frequently form droplets rather than spreading out in a uniformly thin film. This tendency arises because it takes energy to form a surface on a liquid. The energy per unit area required to form new surface on a liquid is denoted by γ (pronounced "gamma") and known as the *surface energy* or more commonly, the *surface tension*.

If no energy were required to form a surface, then liquids would seek to minimize other contributions to their energy – regardless of the amount of surface

Figure 9.16 (a) A (hypothetical) liquid with no surface tension; (b) a real liquid showing the effect of surface tension.

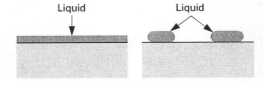

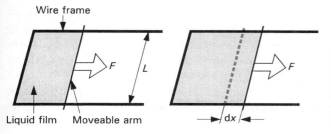

Figure 9.17 A conceptually simple, but impractical, apparatus for investigating the surface tension of liquids. Moving the arm by dx increases the area of a liquid film by 2L dx. The factor 2 arises because new surface area is created on both the upper and lower surfaces.

area created in this process. So, for example, a liquid poured onto an impermeable surface would form a layer approximately one molecule thick which minimizes the gravitational potential energy of the liquid. As pointed out in Figure 9.16, this does not happen, and small amounts of water form *droplets* with a restricted surface area. An experiment – primarily a thought experiment – which makes clear the meaning of surface tension is depicted in Figure 9.17.

The apparatus consists of a liquid film formed across a wire frame. If the arm, which is initially balanced, is moved slowly to a new position such that the area of the film increases by an amount $L\Delta x$, then extra liquid surface area has been created both on the top and the bottom of the film. The total extra area $\Delta A = 2L\Delta x$ requires an energy $F\Delta x = 2\gamma L\Delta x$. Thus the work, ΔW, done by the force in creating this surface is

$$\Delta W = 2\gamma L\Delta x = \gamma\Delta A \qquad (9.20)$$

When balanced, the force F applied to the moveable arm counteracts the tendency of the liquid film to contract and thus minimize its surface area. The magnitude of the force is given by dW/dx which in this case is $\Delta W/\Delta x$. From Equation 9.20,

$$F = -\frac{\Delta W}{\Delta x} = -\frac{2\gamma L\Delta x}{\Delta x} = -2\gamma L \qquad (9.21)$$

Comparing Equations 9.20 & 9.21 we see that the surface tension – the force per unit length of exposed surface – and the surface energy per unit area are numerically equal, and have the units of force/length or energy/area = Nm^{-1}.

Table 9.8 lists the surface tensions of various substances and Figure 9.18 shows the variation of the surface tension with temperature for selected substances from Table 9.8. Figure 9.19 shows the variation in the surface tension of water with temperature

Table 9.8 The surface energy, (or surface tension), γ of various substance in their liquid state as a function of the temperature. The units of γ are mNm^{-1}, so for example, the surface energy of water at 20°C is $722.75\times10^{-3}Nm^{-1}$.

Substance	Temperature (°C)	γ ($\times10^{-3}mNm^{-1}$)
Acetic acid	20	27.59
Acetone	20	23.46
Benzene	20	28.88
Carbon disulphide	20	32.32
Methanol	20	22.50
Ethanol	20	22.39
Water	20	72.75
Sodium	100	209.9
Potassium	65	110.9
Mercury	25	485.5
Lead	350	444.5
Aluminium	700	900
Gold	1100	1120

on a more detailed scale.

It is clear from the data that there is a considerable range of values of surface tension. There appears to be some relationship between the melting temperature and the surface tension that the substance exhib-

Figure 9.18 The surface energy, or surface tension, of various substances from Table 9.8 as a function of temperature.

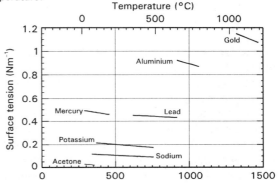

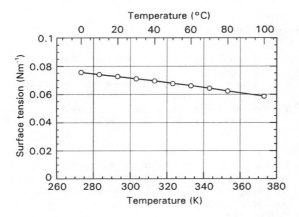

Figure 9.19 The surface energy or surface tension of water as a function of temperature. (Data from Kaye & Laby; see §1.4.1)

its, i.e. high melting temperature substances appear at the top of Figure 9.18. Mercury does not exactly fit in with this trend, but the other molten metals fit the trend quite well.

The second striking feature of Figures 9.18 & 9.19 is that all the data show a negative slope, indicating that the surface energy becomes less at higher temperatures. The slopes also appear to become steeper as the melting temperature of the substance increases. The variation with temperature appears to be approximately linear for the data shown, but the data for water show a slight curvature.

Thus the main questions raised by our preliminary examination of the experimental data on the surface energy of liquids are:

(a) Why do substances with high melting temperatures tend to have high surface energies?

(b) Why does the surface tension of liquids decline approximately linearly with temperature?

9.5.4 Understanding the surface tension of liquids

Our approach here, as with our approach to understanding the viscosity data, will be to use the cell model of a liquid to discuss the data. In Figure 9.20 we consider the situation of molecule A near the surface of the liquid in comparison to molecule B which is deeper within the liquid. On average, molecules such as A have around half the number of nearest

neighbours as do molecules such as B. Thus molecules such as A may be considered as around "halfway to being free of the liquid". In the cell model of a liquid we assume that this will take an energy, ΔE_s, of the order of 50% of ΔE_e.

So if we consider that it costs ΔE_s to have a single molecule at the surface of a liquid, then in order to evaluate the surface energy of a liquid we need only to count the number of molecules such as A present per unit surface area (Fig. 9.21). If we consider that each molecule is constrained within a cell of dimensions of the order a, then the number of molecules per unit surface area will be $\approx 1/a^2$ and so we estimate that the surface energy g is given simply by

$$\gamma = \frac{\Delta E_s}{a^2} \qquad (9.22)$$

We can estimate a for a liquid by considering that a molecule of mass m has a "volume" of the order of a^3. Following this we can estimate the density ρ of the substance to be m/a^3. The mass of molecule is Mu

Figure 9.20 A two-dimensional illustration of the situation of molecules near the surface of a liquid. In three dimensions, molecules such as A still interact with roughly half as many molecules as a molecule within the liquid such as B.

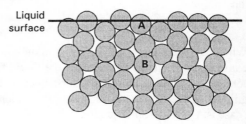

Figure 9.21 A view looking down on the surface of a liquid. In order to estimate the surface energy we need to count how many molecules such as A there are per unit area. Roughly speaking, the number will be $\approx 1/a^2$.

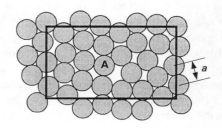

where M is its relative molecular mass and u is an atomic mass unit. Substituting for a^2 in Equation 9.22 produces an estimate for γ:

$$\gamma = \Delta E_s \left(\frac{\rho}{Mu} \right)^{\frac{2}{3}} \qquad (9.23)$$

Equation 9.23 allows us to estimate the surface energy γ in terms of a few well known properties of a substance, and a theoretical property ΔE_s which we have no independent method of estimating. In any case, due to the "handwaving" nature of the arguments that preceded Equation 9.22 we would not be surprised if Equation 9.23 were out by a factor 50% or so. However, we can still make progress in assessing to what extent our theory of surface tension makes sense.

The relative value of ΔE_s

Although we cannot predict γ from experimental values of ΔE_s, we can use experimental values of γ to predict ΔE_s according to

$$\Delta E_s = \gamma \left(\frac{Mu}{\rho} \right)^{\frac{2}{3}} \qquad (9.24)$$

Based on the discussion in §8.3 we expect to find that for any particular liquid ΔE_s should be:
- around 50% of ΔE_e, as estimated from vapour pressure data (see §11.4.2), and

- a few times greater than ΔE_h, as estimated from viscosity data (see §9.5.2).

Values of ΔE_s deduced from surface tension experiments are collated for several liquids in Table 9.9. The conclusions of Table 9.9 are broadly in line with our expectations: the surface energy of liquid metals is considerably higher than the surface energy of organic liquids. There are, however, one or two interesting comparisons to be made. For example, in §9.5.2 we saw that the activation energy for "hopping" was broadly similar for water and ethanol. However, the data for the surface energy indicate that it takes only 2/3 of the energy to place an ethanol molecule on the surface compared to that required to place a water molecule on the surface.

However, in order to appreciate the subtleties of interpretations of these derived quantities, it is best to look at the relationships between ΔE_e, ΔE_s and ΔE_h for different liquids. These data are collated in Table 9.10 and are discussed in §9.6 on the cell model.

The temperature variation of surface energy

The second point to note is that Equation 9.24 predicts that the surface energy is temperature independent. This is not in fact quite true, but the temperature dependence is much weaker than the exponential temperature dependence seen in the viscosity (see §9.5.1) or the vapour pressure (see §11.4.2). Is it possible to understand the roughly linear decline in surface

Table 9.9 The value of ΔE_s (calculated using Equation 9.24) deduced from surface tension data on various substances in their liquid state (see Table 9.8).

Substance	Chemical formula	MW	ρ (kg m^{-3})	T (°C)	γ (mN m^{-1})	ΔE_s (meV)
Acetic acid	CH_3COOH	60	1 049	20	27.59	35.9
Acetone	CH_3COCH_3	58	790	20	23.46	36.0
Benzene	C_6H_6	78	877	20	28.88	50.4
Carbon disulphide	CS_2	76	1 293	20	32.32	42.8
Methanol	CH_3OH	32	791	20	22.50	23.2
Ethanol	C_2H_5OH	46	789	20	22.39	29.5
Water	H_2O	18	1 000	20	72.75	43.7
Sodium	Na	23	930	100	209.9	156
Potassium	K	39	824	65	110.9	127
Mercury	Hg	201	13 600	25	485.5	256
Lead	Pb	207	10 690	350	444.5	281
Aluminium	Al	27	2 400	700	900	396
Gold	Au	197	17 320	1100	1120	496

MW, relative molecular weight.

energy with temperature? If we consider the data for water (Fig. 9.19) we see that γ falls by around 25% as the temperature increases from 273 K to 373 K. This decrease is much too large to be explained in terms of the 4% density variation (Table 9.2) and so we need to try to understand this reduction in the cost of forming a surface as being due to a reduction in ΔE_s itself. Understanding this reduction is rather complex, but essentially results from the changing structure of the liquid.

In §8.2 we described the average structure of a liquid in terms of the radial density function. Figure 9.22 illustrates the changes in the radial density function that take place as a liquid is warmed. The correlations between the average positions of the molecules are weakened, leading to a depression of the peaks and a rise in the troughs in Figure 9.22. It can be seen that, while still maintaining the same average density, the average separation in molecules is increased. This results in a reduction in the overall binding energy because there is a reduction in the average number of molecules at the optimum separation, and an increase in the number of molecules with slightly greater than the optimum separation.

These changes result in a reduction in all the activation energies in the cell model of a liquid. As the temperature rises it becomes easier for a molecule to:

- hop from cell to cell;
- leave the liquid completely; or
- move to the surface.

Figure 9.22 Schematic illustration of the radial density function at different temperatures in the liquid phase. For comparison, the figure also shows the radial density function of the solid and of a (hypothetical) gas-like state with the same average density as the liquid.

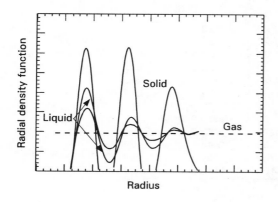

However, the activation energies change only relatively slowly with temperature, and so are difficult to discern in the temperature dependence of the viscosity or the vapour pressure because they are overwhelmed by a strong exponential temperature dependence. Thus it is only in the surface energy data that the slow decrease in activation is observed.

A second factor affecting ΔE_s is the effect of the vapour above the surface of the liquid. The density of the vapour rises approximately exponentially with temperature, but for water between 0°C and 100 °C is still only of the order of one–hundredth of the density of the liquid. Even so, the surface molecules do interact with the molecules in the gas phase, albeit more weakly than they do with the molecules in the liquid phase. We thus expect the surface energy to be reduced according to

$$\Delta E_s = \Delta E_s^0 - \text{Gas factor} \qquad (9.25)$$

where ΔE_s^0 is the surface energy in the absence of any vapour above the liquid and the gas factor is related to:

- the average distance from a surface molecule to a gas molecule: we might expect the interaction energy to vary as approximately $1/r^6$ if the interaction has a Van der Waals origin (see §6.2.2). This distance will be related to the density of molecules in the gas and we would expect that as the number density of gas molecules increases the strength of the interaction with surface molecules will increase;

- the number density of gas molecules: in addition to the distance factor discussed above, the greater the number density of molecules in the gas the greater the number of interactions that a surface molecule may have with molecules in the gas.

The interaction between surface molecules in the liquid and molecules in the vapour only becomes significant as we approach the temperature – known as the *critical temperature* – at which the density of molecules in the gas becomes equal to the density of molecules in the liquid. At this temperature the surface energy of the liquid–gas interface falls to zero. The surface tension data are considered again in §11.4 where we discuss the data on the critical temperature and pressure of liquids.

9.6 The cell model: experimental results collated

Using the cell model of a liquid we have made predictions for the temperature dependence of the viscosity and the surface tension and seen broad agreement between the model and the data. Furthermore, in §11.4 we use the model to understand the dependence on temperature of the vapour pressure above a liquid surface. By analyzing these experiments we can deduce estimates for all three activation energies in a simple cell model of the liquid: ΔE_e, ΔE_s and ΔE_h.

In order to show more clearly the relationships between ΔE_e, ΔE_s and ΔE_h for the different liquids, the data for ΔE_e given in Table 9.9 are presented in summary form in Table 9.10 along with estimates of ΔE_h from Table 9.7 and ΔE_s from Table 11.3. When normalized to put $\Delta E_e = 1$, we see a surprising degree of consistency emerge concerning the relative values of ΔE_e, ΔE_s and ΔE_h. We note immediately that, as might have been anticipated, the relative values are completely different for liquids composed of molecules (e.g. organic liquids) and liquids composed of atoms (e.g. elemental metallic liquids).

The metallic data fit the simple picture envisaged in our simplest version of the cell model (Fig. 8.19) rather well. We see that across the metals, which have a wide range of melting temperatures, there is a broad degree of consistency in the relative magnitudes of the activation parameters. The energy required to place an atom on the surface of liquid is rather less than we had anticipated, being $\approx 1/6$ rather than $\approx 1/2$ but the activation energy required for molecules to "hop" or "swap" cells, ΔE_h, is always around three times smaller than ΔE_s.

The data for organic liquids also shows consistent relative magnitudes of the activation energies, but notably ΔE_h is considerably greater than ΔE_s. This implies that the situation for organic molecules is akin to that envisaged in our second version of the cell model (Figure 9.18). In short, for these liquids it is relatively easy to place molecules at the surface, but more difficult for molecules to move through the liquid than in a metallic liquid.

In discussing the viscosity data we have already mentioned that the shape of the molecules will significantly affect their ability to move through the

Table 9.10 Collated values of ΔE_e, ΔE_s and ΔE_h from Tables 9.7, 9.9 & 11.3, and the values of ΔE_s and ΔE_h normalized for each liquid.

Substance	Chemical formula	ΔE_e (meV)	ΔE_s (meV)	ΔE_h (meV)	$\Delta E_e/\Delta E_e$	$\Delta E_s/\Delta E_e$	$\Delta E_h/\Delta E_e$
Acetic acid	CH_3COOH	391	36	114	1	0.092	0.292
Acetone	CH_3COCH_3	319	36	78	1	0.113	0.245
Benzene	C_6H_6	373	50	106	1	0.134	0.284
Carbon disulphide	CS_2	–	43	63	1	–	–
Methanol	CH_3OH	379	23	110	1	0.061	0.290
Ethanol	C_2H_5OH	423	30	159	1	0.071	0.376
Water	H_2O	405	44	160	1	0.109	0.395
Sodium	Na	954	156	66	1	0.164	0.0692
Potassium	K	786	127	57	1	0.162	0.0725
Mercury	Hg	579	256	23	1	0.442	0.0397
Tin	Sn	2720	–	59	1	–	0.0217
Lead	Pb	1742	281	98	1	0.161	0.0563
Aluminium	Al	2790	396	96	1	0.142	0.0344
Gold	Au	3220	496	175	1	0.154	0.0543
Copper	Cu	3030	–	–	1	–	–
Silver	Ag	2520	–	–	1	–	–
Helium		0.82	–		1		
Neon		20	–		1		
Argon		67	–		1		
Krypton		107	–		1		
Xenon		152	–		1		

Summary of the data for organic and metallic liquids

Liquid type	$\Delta E_e/\Delta E_e$	$\Delta E_s/\Delta E_e$	$\Delta E_h/\Delta E_e$
Organic	1	$\approx 1/10$	$\approx 1/3$
Metallic	1	$\approx 1/6$	$\approx 1/20$

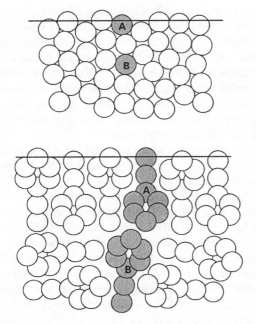

Figure 9.23 Qualitative illustration of the formation of surfaces in: (a) liquids made of atoms, such as liquid metals; (b) liquids consisting of molecules, such as organic molecules. The shapes of the molecules in (b) are entirely illustrative and the gaps between the molecules are left to allow easy identification of the molecules. In (b) the cost of forming a surface may be relatively small if the molecules orient themselves appropriately. However, for organic liquids "hopping" is considerably more difficult owing to the larger size of the molecules and their often complicated shape.

liquid. However, the shape of molecules will also have a significant effect on the ease with which surfaces are formed (Fig. 9.23). For non-spherical molecules we may reasonably expect that at the surface the molecules will orient themselves so as to minimize their potential energy. Thus it is likely that most of the molecule will arrange to stay "beneath the surface" of the liquid, "exposing" only a small part of the molecule and so incurring only a fraction of the full cost of placing a molecule completely at the surface.

The cell model: the verdict.

The cell model provides a flexible framework for discussing processes in liquids. However, its generality and applicability to a wide range of liquids make the attempt to tune the parameters of the model to explain detailed features of the behaviour of particular liquids relatively futile. In order to understand the details of the dynamics of molecules in liquids, and to predict quantitative values for the properties of liquids, we need to concern ourselves much more with the details of molecular shapes and interactions. This is something that extends well beyond the scope of this book, and is still an active area of research.

Perhaps the greatest "danger" in applying the cell model to real liquids is that people will believe that liquids really do behave like that. I feel obliged to stress yet again that the "hopping" processes envisaged in the cell model are complicated many-molecule correlated motions, and the parameters deduced characterize this complex motion, for which the simple one-molecule picture is merely an analogy.

In recent years the most direct way of tackling this problem has been by direct computer simulations of liquids. The forces between pairs of molecules are first estimated using quantum mechanical calculations of electron wave functions and charge densities. Then the motions of molecules are charted by using nothing more complicated than Newton's laws, albeit applied many millions of times, to calculate the motions of the molecules.

A QBasic computer program simulating a liquid of just a few molecules is listed in Appendix 4.

9.7 Thermal properties

9.7.1 Data on the heat capacity of liquids

Heat capacity is a measure of the rate at which the temperature rises for a given input of heat energy. It is defined in terms of the temperature rise, ΔT, resulting from an input of heat energy, ΔQ, by the ratio

$$C = \frac{\Delta Q}{\Delta T} \text{ J K}^{-1} \qquad (9.26)$$

Equation 9.26 is the formula used to determine experimentally the heat capacity from experimental measurements of ΔQ and ΔT. The equation is an approximation to the theoretical definition which is the limit of Equation 9.26 as ΔT tends to zero

$$C = \frac{dQ}{dT} \ \mathrm{J\,K^{-1}} \qquad (9.27)$$

The heat capacity of a material is usually quoted either for a given mass of material, as the specific heat capacity (e.g. $\mathrm{J\,K^{-1}\,kg^{-1}}$), or per mole , as the molar heat capacity ($\mathrm{J\,K^{-1}\,mol^{-1}}$). For practical calculations the specific heat capacity is usually more convenient, but from a fundamental point of view, the molar heat capacity reveals far more. Remember that the molar heat capacity is the heat capacity of 6.02×10^{23} atoms or molecules.

The conditions under which the measurements are made is also important, although not as important as for gases. Because the expansivity of liquids is small compared with gases, measurements made at constant pressure do not normally differ greatly from those made at constant volume. A typical level of agreement would be a few per cent at room temperature, becoming less at lower temperatures, and increasing at higher temperatures. The heat capacities at constant pressure and constant volume are designated C_P and C_V, respectively, and one normally assumes that measurements are made at constant pressure, unless told otherwise.

The data in Table 9.11 do not show any immediately striking trends, but it is still interesting to search for trends within the data. If we plot the data as a function of the relative molecular mass of the molecules of the liquid we arrive at Figure 9.24. No par-

ticularly striking trends are evident, except perhaps a broad trend,with clear exceptions, towards increasing heat capacity for high-mass molecules. But does this broad trend arise because the molecules are increasing in mass? Or is this trend related to the number of atoms in a molecule? This may seem like a stupid question, but recall that when analysing the heat capacity of both gases (§5.3) and solids (§7.4) the molar heat capacity depended on the number of atoms per molecule or number of atoms per formula unit respectively.

Replotting the data as a function of the number of atoms per molecule (Figure 9.25) produces a slightly clearer trend. In particular, mercury, one of the two elemental liquids for which we have data, now appears in a sensible position on the graph. In Figure 9.26 mercury is the isolated datapoint low down on the graph at a relative molecular mass of 201. But in Figure 9.25 mercury forms part of the trend of a tendency to increasing heat capacity with increasing molecular complexity. The line in Figure 9.25 is a line of slope $3R$ per atom, based on the proposal that for solids (Eq. 7.32) at high temperatures the heat capacity of the elements is roughly $3pR$ per mole, where p is the number of atoms per formula unit. The line appears to indicate roughly the trend in the data for small molecules, but for larger molecules (mainly organic molecules) the data fall below this trend line. However, we note again that there is good deal of variation in the data.

Figure 9.24 The heat capacity of the liquids listed in Table 9.11 plotted as a function of the relative molecular mass of the molecules of the liquid. R, molar gas constant.

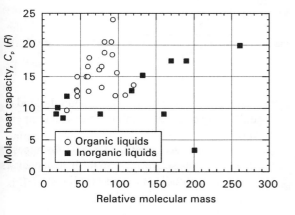

Variation of the heat capacity with temperature

There are, unfortunately, few data available in standard data books on the variation of the heat capacity of liquids with temperature. The exceptions to this are water and mercury: Figure 9.26 shows the heat capacity of these liquids and of benzene and ethanol (from Table 9.11). Clearly the heat capacities of water and mercury are roughly constant over this temperature range, while the heat capacity of the organic liquids increases slightly with temperature. There are very few data available for the organic liquids, and so we draw (only tentatively) the conclusion that the heat capacity of organic liquids increases with temperature. In fact the data for water and mercury are available in reasonable detail and show variations with temperature as depicted in Figure 9.27.

Table 9.11 The heat capacities at constant pressure, CP, for a selection of substances that are liquids at around room temperature. To obtain the specific heat capacity (per kilogram) use the formula $C_p(J K^{-1} kg^{-1}) = [C_p(J K^{-1} mol^{-1})]/M$.

Substance	Chemical formula	MW*	N†	T (°C)	C_p (J K^{-1} mol^{-1})	(R)
Organic liquids						
Methanol	CH_3OH	32	6	12	80.64	9.7
Ethanol	C_2H_5OH	46	9	0	105.3	12.7
Ethanol	C_2H_5OH	46	9	20	113.4	13.6
Ethanol	C_2H_5OH	46	9	40	124.7	15.0
Propanol	C_3H_7OH	60	12	18	138.0	16.6
Acetic acid	$C_2H_4O_2$	60	8	20	124.3	15.0
Acetone	C_3H_6O	58	10	20	124.7	15.0
Aniline	C_6H_7N	93	14	15	199.9	24.0
Benzene	C_6H_6	78	12	10	110.8	13.3
Benzene	C_6H_6	78	12	40	138.1	16.6
Bromoethane	C_2H_5Br	109	8	20	100.8	12.1
Chloroform	$CHCl_3$	120	5	20	113.8	13.7
Cyclohexane	C_6H_{10}	82	16	20	156.5	18.8
1,2 Dichloroethane	$C_2H_4Cl_2$	98	8	20	129.3	15.6
Dichloromethane	$C_2H_2Cl_2$	96	6	20	100.0	12.0
Ethanadiol	$C_2H_6O_2$	62	10	20	149.8	18.0
Ethyl acetate	$C_4H_8O_2$	82	8	20	170.1	20.5
Ethyl nitrate	$C_2H_5O_3N$	91	11	20	170.3	20.5
Formamide	CH_3ON	45	6	20	107.6	12.9
Formic acid	CH_2O_2	46	5	20	99.0	11.9
Nitromethane	CH_3O_2N	61	7	20	106.0	12.7
Nitroethane	$C_2H_5O_2N$	75	10	20	134.2	16.1
Toluene	C_7H_8	92	15	18	153.6	18.5
Inorganic liquids						
Arsenic trifluoride	AsF_3	132	4	20	126.6	15.2
Boron trichloride	BCl_3	118	4	20	106.7	12.8
Bromine	Br_2	160	2	20	75.7	9.11
Carbon disulphide	CS_2	76	3	20	75.7	9.11
Hydrogen cyanide	HCN	27	3	20	70.6	8.49
Water	H_2O	18	3	0	75.9	9.13
Heavy water	D_2O	20	3	0	84.3	10.1
Mercury	Hg	201	1	20	28.0	3.37
Hydrazine	N_2H_4	32	6	20	98.9	11.9
Silicon tetrachloride	$SiCl_4$	170	5	20	145.3	17.5
Tin tetrachloride	$SnCl_4$	261	5	20	165.3	19.9
Titanium tetrachloride	$TiCl_4$	190	5	20	145.2	17.5

*MW, the relative molecular mass. †N, the number of atoms per molecule.
(Data from Kaye & Laby; see §1.4.1)

Considering the scant information in Figures 9.27–9.29, the main questions raised by our preliminary examination of the experimental data on C_p are:

- Why do the heat capacities of liquids tend to increase with increasing molecular complexity for small molecules, but not for larger (primarily organic) liquids?

- Why do the heat capacities of liquids tend to either stay constant or to increase with temperature?

- Why are the heat capacities of many substances in their liquid phase greater than the heat capacities of the corresponding gaseous or solid phases?

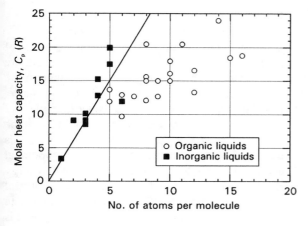

Figure 9.25 The heat capacity of the liquids listed in Table 9.11 plotted as a function of the number of atoms per molecule. The line has a slope $3R$ per atom, where R is the molar gas constant (see text).

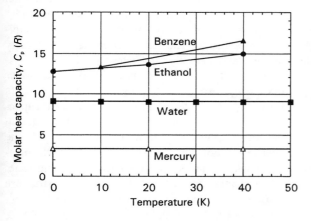

Figure 9.26 The variation with temperature of the heat capacity of various liquids around room temperature. The lines join the data points and are to guide the eye only. R, molar gas constant.

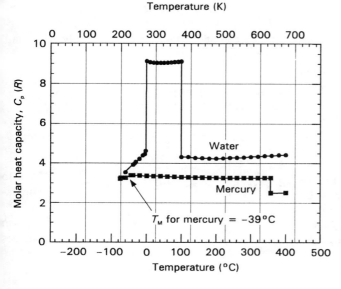

Figure 9.27 The constant pressure heat capacity of water and mercury showing the data for all three phases. The melting and boiling temperatures of mercury and water are: water, $T_M = 0\,°C$, $T_B = 100\,°C$; mercury, $T_M = -38.9\,°C$, $T_B = 356.6\,°C$. R, molar gas constant.

283

9.7.2 Understanding the heat capacity of liquids

We address these questions in the light of our experience in trying to understand the heat capacities of solids (see §7.4) and gases (see §5.3). Bearing these lessons in mind, we will look in turn at the questions raised by the data.

Why do the heat capacities of liquids tend to increase with increasing molecular complexity for small molecules, but not for larger (primarily organic) liquids?

Having previously considered the heat capacities of solids and gases, we are able to rephrase the above questions in terms of the degrees of freedom of the molecules in the liquid. What the data indicate is that the molar heat capacity tends to increase with the number of atoms in a molecule for small molecules. This is hardly surprising. In §7.4 we saw that the expression for the heat capacity of a solid.

$$C^{solid} = 3pR \qquad (9.28)$$

is proportional to p, the number of atoms per chemical formula unit of the substance. Thus the molar heat capacity of sodium chloride (NaCl) is close to twice the value of an elemental solid because there are twice as many atoms present. The reason for the increase in the molar heat capacity of substances as a function of molecular complexity is simply that there are more atoms present in a mole, and thus there is the possibility of more modes of molecular vibration and rotation.

More interesting is the question of why the heat capacity of more complex liquids fails to continue to increase with molecular complexity. In our discussion of solids and gases we concluded that, whenever we fail to observe an expected heat capacity contribution, we conclude that some degrees of freedom of the substance are inaccessible. In the case of liquids the inaccessible degrees of freedom are those associated with molecular vibration and rotation. For small molecules, many of these degrees of freedom are available, but more complex molecules are often entangled with other molecules and so many of their degrees of freedom, particularly those associated with rotation, are restricted by the closeness of the other molecules.

Why do the heat capacities of liquids tend either to stay constant or to increase with temperature?

In our previous discussions of the heat capacities of gases and solids, we ascribed any temperature dependence of the heat capacity to a change in the accessibility of degrees of freedom of the substance. For both solid mercury and solid water (ice) the heat capacity increases with temperature, in line with the discussion in §7.3 on the heat capacity of solids. This increase is due to the increasing average vibrational energy of the molecules in comparison with energy required to excite the highest energy vibrational modes within the solid. On entering the liquid state the situation changes dramatically: for example, it is no longer possible to sustain transverse vibrational waves within the liquid. Thus the vibrational modes within the liquid are essentially vibrations of individual molecules. The constancy of the heat capacity in the liquid state then indicates that the local environment within which the vibration takes place does not change strongly. Close examination of the data for water reveals detailed changes in the heat capacity, which indicate that changes are taking place in the environment of the water molecule, but the changes are not significant in comparison to the massive changes that occur on entering the liquid state. These changes are related to the changes in structure that lead to the thermal expansivity anomaly (see Fig. 9.5). Indeed, the similarity between Figures 9.8 & 9.28 is striking. The slight increases in the heat capacity observed in benzene (see Fig. 9.26) probably result from the increasing accessibility of internal degrees of freedom in the relatively large and complex molecule (C_6H_6).

Figure 9.28 Detail from Figure 9.27 showing the variation in the heat capacity of water in its liquid state.

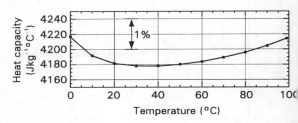

Why are the heat capacities of the liquid phase of many liquids greater than the heat capacities of the corresponding gaseous or solid phases?

Once again we are now in a position to reinterpret the data as being a statement about the number of accessible degrees of freedom in the liquid state. The data amounts to stating that there are more ways in which a molecule can possess energy in the liquid state. We can see how this comes about by comparing the situation of molecules with their situation in the solid and gaseous states in turn.

In the gaseous state the interactions between molecules are generally weak due to the large spaces between the molecules. Thus the degrees of freedom associated with the potential energy of interaction between molecules are not accessible. On entering the liquid state, several more degrees of freedom become immediately available to the molecules. However, some degrees of freedom associated with molecular rotation and internal vibration may become restricted in the liquid state due to the closeness of neighbours.

In the solid state the interactions between molecules are generally strong due to the closeness of the molecules and the rigidity of the structure. This closeness and rigidity strongly restricts the ability of molecules to rotate and (to some extent) vibrate. On entering the liquid state there is, in general, an extra 10% of space available to allow molecular rotation – or more precisely rotational vibration which amounts to rotating back and forth through some angle. Therefore, we would expect some increase in the possible motions of the molecules.

We can now consider to what extent these general ideas are borne out by the data given in §9.7.1. We can make two fairly general predictions: the first concerns elemental liquids and the second concerns substances that are denser in the liquid state than the solid state.

For elemental liquids whose molecules consist of single atoms we would not expect a strong change in heat capacity on entering the solid state because there are no degrees of freedom associated with atomic vibration (or rotation) to be hindered in the solid state. However, on leaving the liquid state for the gaseous state we would expect a fall in the heat capacity due to the loss of degrees of freedom associated with the interactions between molecules. These predictions are broadly borne out by the data for mercury in

Figure 9.27 which show no jump on entering the solid state but a decrease of 25% on entering the liquid state. This behaviour is typical of the behaviour of elements, and bears out the general ideas about the liquid state outlined above.

For substances that are denser in the liquid state than the solid state we might at first expect that on entering the solid state the molecules would have more "room to manoeuvre" and so be more free to rotate and vibrate. If we consider the data for water (see Fig. 9.29) we can see that this approach is mistaken. Note that on entering the gaseous state there is a decrease of roughly 50% in the heat capacity, which we may interpret as being due to the loss of degrees of freedom associated with molecular interactions. However, the heat capacity also falls on entering the solid, even though the molecules have around 10% more space in the solid state. The reason for this lies in the nature of substances that contract on melting. As we saw in §9.2.2, such substances possess strongly directional bonding. Thus, although the molecules have more room in the solid state, they are strongly constrained from any kind of rotation or rotational vibration by the rigidity of their bonds to their neighbouring molecules.

Thus the general idea that molecules have more degrees of freedom in the liquid state than either the solid or gaseous states seems to make sense.

9.7.3 Data on the thermal conductivity of liquids

The thermal conductivity, κ, is defined by the equation

$$\frac{dQ}{dt} = -\kappa A \frac{dT}{dX} \tag{9.29}$$

where:
dQ/dt is the rate of heat flow (W),
  κ is the thermal conductivity ($W\,m^{-1}\,K^{-1}$),
  A is the cross-sectional area across which heat is flowing (m^2), and
dT/dx is the temperature gradient ($K\,m^{-1}$).
The minus sign indicates that heat flows "downhill", from hot to cold.

As with gases, when determining the thermal conductivity of liquids one must guard against the possibility of convective heat transfer, as illustrated in

Figure 5.13. This arises from the combination of the relatively large thermal expansivities of liquids, their ability to flow, and the presence of a gravitational field in most experiments.

The thermal conductivity of several non-metallic liquids is given in Table 9.12 and illustrated in Figure 9.29. There is little in the way of simple variation and the available data are rather limited. This makes it difficult to perceive any clear trends except that the thermal conductivity tends to decrease for most (but not all) liquids as the temperature increases.

The thermal conductivity of several elemental metals in their liquid state is given in Table 9.13 and illustrated in Figure 9.30. There are two striking features of these data. First, the thermal conductivity of most elements decreases, typically by around 50%, as they enter the liquid state. Second, the general magnitude of the thermal conductivities is a factor of 100 greater than the thermal conductivities of the non-metallic liquids mentioned in Table 9.12. The temperature dependence of the thermal conductivity (Figs 9.29 and 9.30) is as varied as for the non-metallic liquids, with some liquids showing increases in thermal conductivity and others showing decreases.

The main questions raised by our preliminary examination of the experimental data on the thermal conductivity of liquids are:

(a) Why can the thermal conductivities of liquids either increase or decrease with increasing temperature?

Table 9.12 Thermal conductivity ($WK^{-1}m^{-1}$) of miscellaneous non-metallic liquids at two temperatures T1 and T2. The data vary roughly linearly between the two temperatures.

Liquid	$T_1(K)$	$T_2(K)$	K_1	K_2
Acetone	193	333	0.198	0.146
Aniline	293	–	0.172	–
Benzene	293	323	0.147	0.137
Methanol	233	333	0.223	0.186
Ethanol	233	353	0.189	0.150
n-Butanol	213	353	0.167	0.106
n-Propanol	233	353	0.168	0.148
Toluene	193	353	0.159	0.119
Carbon tetrachloride	253	333	0.115	0.102
Water	273	353	0.561	0.673
Xenon	173	223	0.07	0.05

(Data from Kaye & Laby; see §1.4.1)

(a)
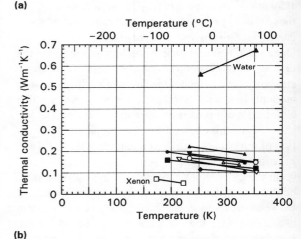

(b)
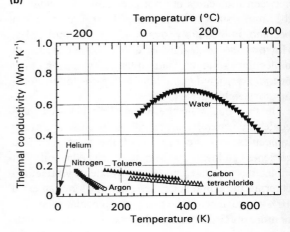

(c)
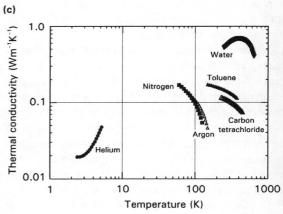

Figure 9.29 Selected data from Table 9.12 showing the thermal conductivity of miscellaneous non-metallic liquids. The data are given at two temperatures, T_1 and T_2, and varies roughly linearly between these two temperatures.

Table 9.13 Thermal conductivity (WK^{-1}m^{-1}) of elemental metals in their liquid state. Shaded entries refer to the solid state. The data are plotted in Figure 9.30.

		Temperature					K_L/K_S
		173K	273K	373K	573K	973K	(%)
Lithium	Li	98	86	82	47	59	57
Sodium	Na	141	142	88	78	60	62
Potassium	K	105	104	53	45	32	51
Rubidium	Rb	59	58	32	29	22	55
Caesium	Cs	37	36	20	20.6	17.7	56
Mercury	Hg	29.5	7.8	9.4	11.7	–	26
Aluminium	Al	241	236	240	233	92	39
Bismuth	Bi	11	8.2	7.2	13	17	181
Gallium	Ga	43	41	33	45	–	80
Tin	Sn	76	68	63	32	40	51

(Data from Kaye & Laby; see §1.4.1)

(b) Why are the thermal conductivities of liquid metals around a factor 100 greater than those of non-metallic liquids?

(c) Why are the thermal conductivities of elemental liquid metals generally less than those of corresponding solids?

Figure 9.30 Data from Table 9.13 showing the thermal conductivity of elemental metals in their liquid state in units of WK^{-1} m^{-1}.

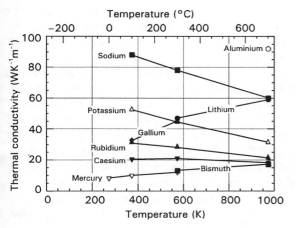

9.7.4 Understanding the thermal conductivity of liquids

We will approach the data on the rate of transfer of heat through liquids in essentially the same way as we considered the data on the speed of sound (see §9.4.2). We consider the transfer of heat to arise via a combination of the two strikingly different mechanisms of heat transfer found in gases and in solids:

(a) in a *gas*, a molecule moves bodily from one place to another taking its kinetic energy with it;

(b) in a *solid*, molecules vibrate and the energy of vibration is passed on from one molecule to another, while the molecules themselves do not change their average positions.

The liquid state has elements in common with both states. However, the distance that the molecules move before collisions is dramatically reduced in the liquid state as compared with the gaseous state. Furthermore, the liquid state lacks the rigidity exhibited by the solid state which reduces the ease with which vibrations are transmitted – recall that the speed of longitudinal sound waves is around 30% less in the liquid than the solid (see §9.4). As in our discussion of the speed of sound, we can approach the liquid state by considering the extent to which energy transport within the liquid state is gas-like or solid-like (Fig. 9.31).

Using the idea of thermal resistivity we estimate the thermal resistance of the liquid as being the sum of thermal resistivity of solid and gas fractions:

$$R_{liquid}^{th} = R_{solid}^{th} + R_{gas}^{th} \qquad (9.30)$$

287

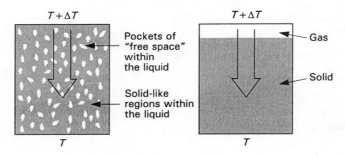

Figure 9.31 From the point of view of thermal conductivity, a liquid may be considered as being a solid-like matrix with pockets of gas-like "free space". From the data on the density of liquids, we estimate that roughly 10% of a liquid is "free space". We can estimate the thermal conductivity of the liquid if we imagine separating out the free space part of the liquid and considering heat to pass through a gas-like layer, and then a solid-like amount of substance. The arrow in each figure indicates the direction of heat transport.

Expressing this in terms of the thermal conductivity we have

$$\frac{1}{\kappa_{\text{liquid}}} = \frac{L}{A\kappa_{\text{solid}}} + \frac{\Delta L}{A\kappa_{\text{gas}}} \quad (9.31)$$

where L is the length of the solid fraction in a liquid, ΔL is the length of the gas fraction, and A is the cross-sectional area of the liquid. Multiplying by $A/(L+\Delta L)$, we find

$$\frac{A}{(L+\Delta L)\kappa_{\text{liquid}}} = \frac{L}{(L+\Delta L)\kappa_{\text{solid}}} + \frac{\Delta L}{(L+\Delta L)\kappa_{\text{gas}}} \quad (9.32)$$

which simplifies to first order as

$$\frac{1}{\kappa_{\text{liquid}}} = \left(1 - \frac{\Delta L}{L}\right)\frac{1}{\kappa_{\text{solid}}} + \frac{\Delta L}{L}\left(1 - \frac{\Delta L}{L}\right)\frac{1}{\kappa_{\text{gas}}} \quad (9.33)$$

Recalling that the typical change in volume on entering the liquid state is 10%, we estimate $\Delta L/L \approx 0.1/3 = 0.033$ (see §9.2.1 and Example 7.4). Substituting into Equation 9.33 produces

$$\frac{1}{\kappa_{\text{liquid}}} = (0.967)\frac{1}{\kappa_{\text{solid}}} + 0.33 \times (0.967)\frac{1}{\kappa_{\text{gas}}} \quad (9.34)$$

which simplifies to

$$\kappa_{\text{liquid}} = \left(\frac{0.967}{\kappa_{\text{solid}}} + \frac{0.032}{\kappa_{\text{gas}}}\right)^{-1} \quad (9.35)$$

Thus the thermal conductivity of a liquid can be related to the thermal conductivity in the gas and solid phases. According to the idea outlined above, heat travels a factor of $\approx 0.927/0.032 \approx 30$ times further through the solid phase than through the gas fraction. Thus if the thermal conductivity of the solid fraction is ≈ 30 times greater than the conductivity of the gas, then both terms in Equation 9.35 contribute equally. If, however, the thermal conductivity of the solid is, say, ≈ 300 times that of the gas (quite typical), then the gas term is about 10 times larger than the solid term and the thermal conductivity of the liquid is dominated by the gas thermal conductivity. Hence the temperature dependence of the liquid thermal conductivity should follow that of the gas thermal conductivity, and increase as $\approx T^{1/2}$ to T^1.

However, experimentally, over the temperature range shown, the thermal conductivity of most liquids falls with temperature. This seems to indicate that it is in fact the solid fraction of the conductivity which dominates the thermal conductivity of liquids, i.e. the data indicate that thermal conduction in liquids is dominated by processes analogous to those that occur in solids. Since the data indicate that heat transport in liquids is by phonons, analogous to those in solids, we can interpret the decline in the conductivity as being due to an increase in the scattering of phonons with temperature. Note that the atomic vibrations which comprise most phonons have frequencies of $\approx 10^{13}$ Hz, whereas we expect the structure of a liquid to change on a time scale of many atomic vibrations ($\approx 10^{-10}$ s). Thus the phonons may travel considerable distances through the liquid without being aware of its fluid nature – the lattice looks like a highly disordered solid. This last point, the strong disorder, is responsible for the relatively poor thermal conduction in liquids as compared with solids.

Note that two liquids in Figure 9.29, helium and water, show an increase in thermal conductivity with temperature. In fact both liquids are highly anomalous and it is unwise to generalize from the data on

Table 9.14 The thermal conductivity, electrical resistivity and the Lorentz number*of elemental metals in their liquid state.

Liquid	373K			573K			973K		
	ρ (Ωm)	κ (WK^{-1}m^{-1})	$\rho\kappa/T$ (WΩK^{-1}m^{-1})	ρ (Ωm)	κ (WK^{-1}m^{-1})	$\rho\kappa/T$ (WΩK^{-1}m^{-1})	ρ (Ωm)	κ (WK^{-1}m^{-1})	$\rho\kappa/T$ (WΩK^{-1}m^{-1})
Sodium	9.7×10^{-8}	88	$2.3\times10{-8}$	16.8×10^{-8}	78	2.3×10^{-8}	39.2	60	2.4×10^{-8}
Potassium	17.5×10^{-8}	53	2.5×10^{-8}	28.2×10^{-8}	45	2.2×10^{-8}	66.4	32	2.2×10^{-8}
Rubidium	27.5×10^{-8}	32	2.4×10^{-8}	48×10^{-8}	29	2.4×10^{-8}	99	22	2.2×10^{-8}
Caesium	43.5×10^{-8}	20	2.3×10^{-8}	67×10^{-8}	20.6	2.4×10^{-8}	134	17.7	2.4×10^{-8}
Mercury†	13.5×10^{-8}	9.4	0.3×10^{-8}	128×10^{-8}	11.7	2.6×10^{-8}	214	–	–

*$\rho\kappa/T$ has the theoretical value 2.45×10^{-8}WΩK^{-1}m^{-1}. †Note the anomalous data for mercury.
(Data from Kaye & Laby; see §1.4.1)

either these liquids. However,the increase seen in helium is likely to be related to an essentially gaseous mechanism, but the increase seen in water is likely to be related to the low temperature case discussed at the end of §7.7.

Liquid metals

It is perhaps surprising that when metals melt they retain their high electrical conductivity, albeit with a slightly increased value of resistivity (§9.8). Thus for liquid metals, in addition to conductivity through molecular vibrations transmitted in a "gas-like" and a "solid-like" manner, thermal conduction may take place through the motion of the conduction electrons. Since the thermal conductivities of all liquid metals are dramatically greater than those of all non-metallic liquids, we conclude that (to a first approximation) we may ignore thermal conduction through molecular vibrations in liquid metals.

Evaluating the ratio $\rho\kappa/T$ we find that the results are consistently close to the value of 2.45×10^{-8} (WΩK^{-1}m^{-1}) (Table 9.14). This confirms our suspicion that phonon conductivity is negligible in comparison with electrical conduction. The question of how electrons are able to conduct in the liquid state is discussed in §9.8.

9.8 Electrical properties

When any substance is subject to an applied electric field, E a current of electronic charge flows through the substance. The magnitude of the resultant current density, j, is characterized by the electrical resistivity,

ρ or the electrical conductivity, δ of the substance. The two measures are the inverse of each other, $\rho = 1/\sigma$ and so for most purposes there is no advantage to using one measure rather than the other. The electrical resistivity and conductivity are defined by

$$j = \sigma E \qquad (9.36a)$$

and

$$E = \rho j \qquad (9.36b)$$

If the current density is measured in amperes per square metre (A m^{-2}) and the electric field in volts per metre (Vm^{-1}), then the units of σ are per ohm metre (Ω^{-1}m^{-1}) or siemen per metre (symbol S, not to be confused with s for seconds). The units of resistivity are the inverse seimens, (S^{-1}m) or, more commonly, ohm metres (Ω m). For a sample of cross-sectional area A and length L, the resistivity is related to the electrical resistance R by

$$\rho = \frac{RA}{L}\left(S^{-1}m \text{ or } \Omega m\right) \qquad (9.37)$$

9.8.1 Data on the electrical properties of liquid metals

From Fig 7.31, it is clear that most elements are metals, with resistivities between $10^{-5}\Omega$ m and $10^{-8}\Omega$ m at around room temperature. Perhaps surprisingly, heating elements up to beyond their melting temperature does not destroy their metallic behaviour. Data on the temperature dependence of some low-melting-point elemental metals are given in Table 9.15 and plotted in Figure 9.32.

As is clear from Figure 9.34, and from the last row of Table 9.15, the resistivity in the liquid state, though not destroyed, is significantly poorer than in the solid state.

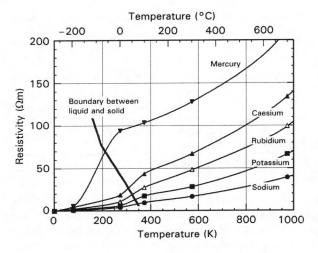

Figure 9.32 The resistivity of elemental metals with low melting points. The data show a significant increase in resistivity from below the melting temperature to above it. Note that the lines that connect the data points are drawn only as a guide for the eye. In actuality the transition from liquid to solid would result in a sharp increase in resistivity.

Table 9.15 The resistivity ($\times 10^{-8}\Omega$m) of elemental metals with low melting points: shaded entries refer to metals in the solid state and other data refer to metals in the liquid state. The last row of the table shows the ratio of the resistivities of solid and liquid states (derived from the ratio of the last datum in the solid region to the first datum in the liquid region).

T (K)	Sodium	Potassium	Rubidium	Caesium	Mercury
0	0	0	0	0	0
78.2	0.8	1.38	2.20	4.5	5.8
273.2	4.2	6.10	11.0	18.8	94.1
373.2	9.7	17.5	27.5	43.5	13.5
573.2	16.8	28.2	48	67	128
973.2	39.2	66.4	99	134	214
1473.2	89	160	260	295	630
ρ_S/ρ_L (%)	43	35	40	31	6

(Data from Kaye & Laby; see §1.4.1)

So the main questions raised by our preliminary examination of the experimental data on the electrical resistivity of liquid metals are:

(a) Why the metallic state survives in liquids?
(b) Why the electrical resistance is worse in the liquid state than the solid state?

9.8.2 Understanding the electrical properties of liquid metals

Perhaps surprisingly, we can approach the results of our examination of the resistivity of liquid metals in exactly the same way that we approached the data for metallic solids. The reasons why this is so also shed light on the question of why the resistivity in the liquid state is higher than the resistivity in the solid state.

As discussed in §7.5, our first picture of the electrons within a metal is as a free electron gas with a number density of around 10^{29}m^{-3}. The exclusion principle causes the many electrons in the gas to have extremely high energies, causing the electrons to move extremely quickly through a metal, at speeds of around the Fermi speed ($v_F \approx 10^6$m s^{-1}). In the introductory sections of this chapter we mentioned that atoms vibrate within their "cells" in the liquid structure in a time of around 10^{-13}s and, typically, we expected an atom to vibrate of the order of 1000 times before the detailed structure changed significantly.

During the time that it takes for an atom to vibrate once, an electron can travel around 10^6m s$^{-1} \times 10^{-13}$s $= 10^{-7}$m, which corresponds to around 300 "cell" diameters. Thus for the electrons the changes in liquid structure which take place on a time-scale of $\approx 10^{-10}$s appear very slow indeed. Thus the change to the liquid state, which is so apparent on a long time-scale, may not even be noticed by electrons! To a conduction electron the liquid structure looks like a strongly disordered solid, and it is this strong positional disorder which scatters the electrons, in the same way that alloying increases the resistivity of a solid (see §7.5). So the origin of the increase in resistivity on entering the liquid state is increased scattering due to the loss of an ordered lattice.

As Example 9.3 makes clear, we are still not quite

Example 9.3

Let us use the theory given in §7.5 on the free electron gas to estimate the resistivity of liquid metal potassium. In its liquid state, potassium (relative molecular mass 39.1) has a density of $824\,\mathrm{kg\,m^{-3}}$ and so the number density of atoms is

$$n = \frac{\rho}{Mu} = \frac{824}{39.1 \times 1.66 \times 10^{-27}} = 1.27 \times 10^{28}\,\mathrm{atoms\,m^{-1}}$$

If we assume that one electron per atom joins the free electron gas, then this is also the number density of electrons. We can predict the wave vector of the most energetic conduction electrons, k_{max}, according to Equation 6.80.

$$k_F = \left(3n\pi^2\right)^{\frac{1}{3}}$$

and hence the maximum velocity of electrons is

$$v_F = \hbar k_F / m$$

Substituting, we find

$$v_F = \frac{\hbar\left(3n\pi^2\right)^{\frac{1}{3}}}{m} = \frac{1.054 \times 10^{-34} \times \left(3\pi^2 \times 1.27 \times 10^{28}\right)^{\frac{1}{3}}}{9.1 \times 10^{-31}}$$

$$= 0.835 \times 10^6\,\mathrm{m\,s^{-1}}$$

Let us assume that before being scattered an electron travels across roughly two "cells" of diameter a in the liquid. This estimate seems reasonable given the strongly disordered "lattice" present in the liquid state. From the liquid density we can estimate the density as Mu/a^3, and hence the cell diameter is of the order

$$a \approx \left(Mu/\rho\right)^{\frac{1}{3}}$$

i.e.

$$a \approx \left(\frac{39.1 \times 1.66 \times 10^{-27}}{824}\right)^{\frac{1}{3}} = 4.3 \times 10^{-10}\,\mathrm{m}$$

The scattering time, τ, will be the time taken for an electron to travel a mean free path, roughly $2a$, which is $\approx 2a/v_F$.

Thus

$$\tau \approx \frac{2a}{v_{max}} = \frac{2 \times 4.3 \times 10^{-10}}{0.835 \times 10^6} = 1.0 \times 10^{-15}\,\mathrm{s}$$

If we substitute this value in the expression for the resistivity of a free electron gas (Eq. 7.71)

$$\rho = m/ne^2\tau$$

we predict that

$$\rho_{liquid} = \frac{9.1 \times 10^{-31}}{1.27 \times 10^{28} \times \left(1.6 \times 10^{-19}\right)^2 \times 1.0 \times 10^{-15}}$$

$$= 280 \times 10^{-8}\,\Omega\mathrm{m}$$

This predicted value may be compared with the experimental value just above the melting temperature of $17.5 \times 10^{-8}\,\Omega\mathrm{m}$.

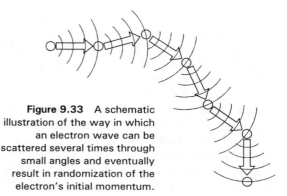

Figure 9.33 A schematic illustration of the way in which an electron wave can be scattered several times through small angles and eventually result in randomization of the electron's initial momentum.

able to explain the resistivity values in the liquid state. The predicted resistivity is of the order 15 times larger than the experimental value. Reviewing the steps leading to the prediction reveals places where the calculation could be out by small factors, but only one place where a factor of the order 15 could be "lost". The assumption that because of the disorder in the "lattice" in the liquid state electrons travel only ≈ 2 "cells" before scattering must be wrong. In order to make our theory agree with the data, electrons must travel 15 times further i.e. of the order 30 "cells" before scattering. This does indeed seem a surprisingly large distance but we can understand it as follows.

First, we note that the scattering events assumed in the theory are total scattering events, after which the electron is likely to travel in any direction. If the electron is scattered every two "cell" diameters or so – as seems inevitable given the level of disorder in the liquid state – then we can understand the resistivity value if we assume the real scattering events are small-angle scattering events, i.e. the electron is, on average, scattered by only a small angle (Fig. 9.33). It would take several such scattering events before the electron's initial momentum would be randomized.

9.8.3 Data on the electrical properties of liquid insulators

Weak electric fields

The simplest electrical property of electrical insulators (or dielectrics) is that for small electric fields they are electrical insulators! The effect of the electric field is to polarize the material, causing electrical charges within the material to separate slightly. This weakens applied electric fields within the material to an extent measured by the relative dielectric permitivity, ε, of the material. The data for the elements in Table 9.16 generally lie in the range 1 to 2. The organic alcohols have considerably higher values, which become smaller for the larger molecules (Fig.9.34). The value for water, which may be considered as the limit of the small alcohol molecules has an enormously large value (≈80), which is of a similar order of magnitude to that of the structurally-bistable compound solid strontium titanate (SrTiO₃)

The main questions raised by our preliminary examination of the experimental data on the electrical properties of insulators are:

- Why is the difference of the dielectric constant from unity (ε−1) in liquids typically 1000 times greater than for gases?
- Why for some liquids does the value of (ε−1) exceed the values for elemental liquids by a factor of the order of 10?

These questions are addressed in §9.8.4.

Table 9.16 The relative dielectric permittivity, ε, of various insulating liquids. The relative permittivity of vacuum is exactly 1. All measurements refer to 20°C, but are insensitive to small changes (≈±10°C) around this temperature.

Substance		T	ε − 1	ε
Argon, Ar	40	82 K	0.53	1.53
Helium, He	4	4.19 K	0.048	1.048
Hydrogen, H₂	2	20.4 K	0.228	1.228
Nitrogen, N₂	28	70 K	0.45	1.45
Oxygen, O₂	32	80 K	0.507	1.507
Methanol, CH₃OH	32	25°C	31.6	32.6
Ethanol, C₂H₅OH	46	25°C	23.3	24.3
Propanol, C₃H₇OH	60	25°C	19.1	20.1
Butanol, C₄H₉OH	74	20°C	16.8	17.8
Pentanol, C₅H₁₁OH	88	25°C	12.9	13.9
Hexanol, C₆H₁₃OH	102	25°C	12.3	13.3
Aniline, C₆H₇N	86	20°C	5.90	6.90
Acetone, C₃H₆O	58	25°C	19.7	20.7
Carbon disulphide, CS₂	76	20 °C	1.64	2.64
Water, H₂O	18	20 °C	79.4	80.4

MW, relative molecular mass.
(Data from Kaye & Laby; see §1.4.1)

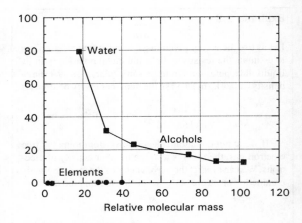

Figure 9.34 The difference of the relative dielectric permittivity from unity (ε − 1), as a function of the molecular mass of the molecules of various electrically insulating liquids. Lines have been drawn to attract attention to the trend in the alcohols, to which water appears be roughly related. The elements listed in Table 9.16 appear as points close to the x axis.

Strong electric fields

As with solids and gases, liquids that are electrically insulating at low electrical fields eventually conduct at high electric fields. Unfortunately, the data books I have been using give no references to the dielectric strength or breakdown field for liquid dielectrics.

9.8.4 Understanding the electrical properties of liquid insulators

We can consider the dielectric constant data from the point of view that a liquid is, in essence a dense gas-like collection of molecules. In this approach we understand the dielectric properties of liquids as being essentially the dielectric properties of the constituent molecules of the liquid. The dielectric constant of the liquid is then increased above the gaseous value by the relatively high density of the liquid. In order to test this approach we need to estimate the number density of molecules in the gaseous and liquid states. This is done by means of Equation 4.49

$$n_{gas} = \frac{P}{k_B T} \qquad (9.38)$$

and Example 7.1

$$n_{\text{liquid}} = \frac{\rho}{Mu} \tag{9.39}$$

where ρ is the density of the liquid, M is the relative molecular mass of the molecules, and u is an atomic mass unit. Furthermore, we recall from §5.6 that the dielectric constant of a gas is given by two different expressions depending on whether the molecules are polar or non-polar.

Polar molecules have a permanent (intrinsic) electric dipole moment, p_p, due to the distribution of electric charge within each molecule. An applied electric field exerts a torque on the molecules, which tends to align them with the applied field. In addition, the electronic charge distribution within the molecule can itself be changed by the applied field, resulting in an induced electric dipole moment. The magnitude of the induced dipole moment is related to α, the polarizability of the molecule.

We expect that for gases, and hopefully liquids, composed of non-polar molecules, ε should be given by (Eq. 5.97)

$$\varepsilon - 1 = \frac{n\alpha}{\varepsilon_0} \tag{9.40}$$

where the molecular polarizability, α, is an intrinsic property of a molecule. For collections of polar molecules we expect that, in addition to any induced dipole moment, there will be an additional term given by (see Eq. 5.104)

$$\varepsilon - 1 = \frac{np_p^2}{3\varepsilon_0 k_B T} \tag{9.41}$$

where p_p is the electric dipole moment built into a molecule, an intrinsic property of the molecule.

Non-polar molecules

Using Equation 9.40, we can calculate the values of α per molecule in the liquid state and compare our results with the similar calculation for the substance in the gaseous state. As Table 9.17 indicates, the molecular polarizability inferred from the dielectric constant data is roughly the same, independent of whether ε is measured in the gaseous or the liquid state. In order to see the significance of this result, recall again that α is an intrinsic property of a molecule related to the ease with which its electronic structure is deformed by an applied field. The fact that this property of the molecule is the same in the liquid and gaseous states implies that the electronic structure of the molecule is similar in both states.

Polar molecules

We can calculate the values of p_p per molecule in the liquid state and compare our results with the similar calculation for the substance in the gaseous state. Rearranging Equation 9.41, we find

$$p_p = \left[\frac{3\varepsilon_0 k_B T(\varepsilon - 1)}{n} \right]^{\frac{2}{3}} \tag{9.42}$$

with n being estimated by using either Equation 9.38 or 9.39 as appropriate.

As Table 9.18 indicates, the value of intrinsic dipole moment, p_p inferred from the dielectric constant data on liquids is roughly 2.4 times larger than the same quantity inferred from the dielectric constant data on liquids. The origin of this enhancement is at first puzzling but can in fact be understood fairly straightforwardly.

Table 9.17 The results of the calculations of the molecular polarizability of non-polar molecules. The value of α is based on Equation 9.40, $\alpha/\varepsilon_0 = (\varepsilon-1)/n$, with n estimated by either Equation 9.42 or 9.43 as appropriate. The data for the densities of liquid hydrogen, nitrogen and oxygen are estimates based on a 10% decrease of the density of the solid. See Table 5.16 for gas data and Table 9.16 for liquid data. The gas data refer to atmospheric presure (1.013×10^5 Pa).

| Substance | M | Liquid | | | | Gas | | | |
		ρ (kg m^{-3})	$\varepsilon - 1$	n ($\times 10^{28}$ m^{-3})	α/ε_0 ($\times 10^{-30}$)	T (K)	$(\varepsilon - 1)$ ($\times 10^{-4}$)	n ($\times 10^{25}$)	α/ε_0 ($\times 10^{-30}$)
Argon	40	1410	0.53	2.12	25	293	5.16	2.50	21
Helium	4	120	0.048	1.81	2.65	293	0.65	2.50	2.6
Hydrogen	2	≈80	0.228	2.41	9.5	293	2.54	2.50	10.2
Nitrogen	28	≈930	0.45	2.00	22.5	293	5.47	2.50	21.9
Oxygen	32	≈1300	0.507	2.45	20.5	293	4.94	2.50	19.8

Table 9.18 The results of the calculations of the permanent molecular dipole moment of polar molecules according to Equation 9.42. The gas data refer to atmospheric presure (1.013×10^5 Pa).

| Substance | M | Liquid | | | | | Gas | | | |
		T (K)	ρ (kg m^{-3})	$\varepsilon - 1$	n ($\times 10^{28}$)	p ($\times 10^{-30}$)	T (K)	$\varepsilon - 1$ ($\times 10^4$)	n ($\times 10^{25}$)	p ($\times 10^{-30}$)
Methanol	32	298	791	31.6	1.49	15.2	373	57	1.97	6.29
Ethanol	46	298	789	23.3	1.03	15.7	373	61 or 78	1.97	6.5 or 7.4
Water	18	293	1000	79.4	3.35	16.0	373	60	1.97	6.45

It is the gaseous state from which we observe the correct value of the intrinsic electric dipole moment, because in a gas the molecules are separated by large distances and thus their responses to the applied field are independent of one another. However, this independence of response is lost when the molecules are as close as they are in the liquid state. To see how this independence is lost, Example 9.4 considers a hypothetical molecule with an intrinsic electric dipole moment of the same order as those in Table 9.18.

We see that the electric field in the vicinity of a polar molecule may be exceedingly large. Electric fields of this magnitude will (a) strongly polarize the charge distributions on neighbouring molecules, and (b) affect the relative orientation of neighbouring molecules. Thus when a molecule is turned to align with an external field its electric field turns with it, and thus the polarization also "turns with" the molecule. Thus the electric dipole moment "associated with" a particular molecule is equal to the original intrinsic dipole moment, p_p, plus a second part related to the interaction with neighbouring molecules.

Conclusion

We are now in a position to understand the experimental data. The reason why the difference of the dielectric constant from unity ($\varepsilon - 1$) in liquids is typically 1000 times greater than in gases is indeed as we supposed initially, i.e. because of the density of molecules.

The reason why ($\varepsilon - 1$) for some liquids exceeds values for elemental liquids by a large factor is because these high dielectric constant liquids are polar and the effect of an applied electric field is to turn the molecules within the liquid. The applied field thus alters the distribution of electric charge within the liquid, which can correspond to a relatively large structural change in the liquid. This may

---

Example 9.4

Consider a molecule around 0.3 nm in length with a permanent electric dipole moment of $p_p = 6 \times 10^{-30}$ C m, similar to the experimental values in Table 9.18. This is equivalent to a situation in which charges of $+q$ and $-q$ are separated by a distance $d = 0.3$ nm such that (see §2.3.3)

$$p_p = qd$$

We thus estimate that the charge at either end of the molecule, q, is given by

$$q = \frac{p_p}{d} = \frac{6 \times 10^{-30}}{0.3 \times 10^{-9}} = 2 \times 10^{-20} \, \text{C}$$

What is the electric field at this point, roughly one molecular distance away from a molecule?

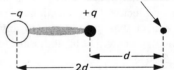

Let us now work out the electric field at a distance of roughly one molecular length away from the molecule along the axis of the molecule. The electric field may be calculated as the sum of the two contributions: one each from the charges at either end of the molecule. Thus at a distance d from the positively charged end of the molecule the electric field is

$$E = \frac{+q}{4\pi\varepsilon_0 d^2} + \frac{-q}{4\pi\varepsilon_0 (2d)^2} = \frac{+q}{4\pi\varepsilon_0 d^2}\left(1 - \frac{1}{4}\right)$$

which evaluates to

$$E = \frac{+2 \times 10^{-20}}{4\pi \times 8.85 \times 10^{-12} \times \left(0.3 \times 10^{-10}\right)^2}\left(\frac{3}{4}\right) = 1.50 \times 10^9 \, \text{V m}^{-1}$$

be compared with enormously large values of dielectric constant observed in the structurally bistable solid strontium titanate ($SrTiO_3$) (Table 7.12).

The reason why ($\varepsilon-1$) for some liquids exceeds the predicted value based on density extrapolation by a factor of the order 2.5 is because these liquids are polar. As each molecule rotates, it drags with it its polarization of nearby molecules.

9.9 Optical properties

Many liquids may be characterized by two parameters: the refractive index of the liquid and the absorption coefficient. The reflectivity of liquid metals remains broadly similar to that in the solid state (§9.8.1), and so in this section we consider only the refractive index of insulating liquids.

9.9.1 Dielectric materials: insulators

The refractive index, n_{light} of a transparent material is related to the speed of light, v, in the medium by

$$n_{light} = \frac{c}{v} \tag{9.43}$$

where c is the speed of light in a vacuum.

The measured refractive index n_{light} of various transparent liquids is recorded in Table 9.19. As with solids, n_{light} varies significantly with wavelength, and the data refer to the refractive index at the wavelength of the bright yellow D lines in the spectrum of

Table 9.19 The refractive index of various substances.

Substance	Chemical formula	MW	n_{light}
Water	H_2O	18	1.33
Carbon tetrachloride	CCl_4	152	1.405
Toluene	C_7H_8	92	1.497
Methanol	CH_3OH	32	1.329
Ethanol	C_2H_5OH	44	1.3614
Propan-1-ol	C_3H_7OH	56	1.3852
Propan-2-ol	$C_2H_5OHCH_2$	56	1.3742
Acetic acid	CH_3COOH		1.3716
Benzene	C_6H_6	78	1.501
Aniline	C_6H_7N	86	1.586
Hydrogen disulphide	HS_2	65	1.885

MW, relative molecular mass.
(Data from Kaye & Laby; see §1.4.1)

sodium vapour. Figure 9.35 shows the variation of n with wavelength for water.

The data for the variation of the refractive index of water are strikingly similar to the data for the variation of the refractive index of glasses shown in Figures 7.48 & 7.49.

The questions raised by our preliminary examination of the experimental data on the optical properties of liquids are:

(a) Why are the refractive indices of liquids similar to those of transparent solids, and why do they differ from unity by about 1000 times more than gases?

(b) Why is the variation with wavelength of the refractive index of liquids similar to that exhibited by transparent solids?

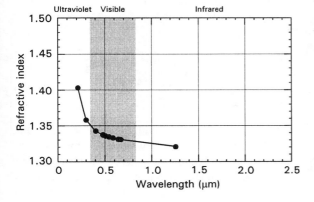

Figure 9.35 The refractive index of water. The shaded region corresponds to the visible region of the spectrum. The longer wavelengths are therefore in the infrared region of the spectrum, and the shorter wavelengths in the ultraviolet region.

9.9.2 Understanding the optical properties of insulators

Our approach to the optical properties of liquids is similar to our approach to the dielectric properties of a liquid. As a first step, we consider the optical properties of liquids as being essentially the optical properties of the constituent molecules of the liquid but enhanced above the gaseous properties by the relatively high density of the liquid.

From the discussion of the refractive index of gases we recall that, in general, the refractive index is related to the dielectric constant of the substance by

$$n_{light} = \sqrt{\varepsilon} \qquad (9.44)$$

The dielectric constant is given by (see Eq. 5.107)

$$\varepsilon = 1 + \frac{n\alpha}{\varepsilon_0} \qquad (9.40)^*$$

and so we expect the refractive index to be given by (see Eq. 5.107)

$$n_{light} = \sqrt{\varepsilon} = \sqrt{1 + \frac{n\alpha}{\varepsilon_0}} \qquad (9.45)$$

Now the low-frequency dielectric constant of a liquid with polar molecules has contributions both from the rotation of the polar molecules and from the polarization of their charge distribution. However, at optical frequencies, the inertia of the molecules prevents them from rotating in response to the electric field of the light wave. Hence the refractive index is related only to the induced component of the dielectric constant. In order to see whether we can explain the refractive index of a liquid in terms of only the change in the number density of molecules, we will estimate α using Equation 9.40 and the refractive index data on gases (see Table 5.17), and then use the same value of α to estimate the refractive index in the liquid state. We will work through the calculation first for water.

Example calculation: water vapour
Rearranging Equation 9.40 as an expression for α, we find

$$\alpha = \frac{\varepsilon_0(n_{light}^2 - 1)}{n} \qquad (9.46)$$

In order to evaluate Equation 9.46 we must first esti-mate the number density of molecules in the gas (see Eq. 4.49)

$$n_{gas} = \frac{P}{k_B T} \qquad (9.47)$$

Since the water vapour data refers to STP, this evaluates to

$$n^{gas} = \frac{1.013 \times 10^5}{1.38 \times 10^{-23} \times 273}$$
$$= 2.689 \times 10^{25} \text{ m}^{-3} \qquad (9.48)$$

Substituting for the refractive index of water vapour (see Table 5.17, $n_{light} = 1.000254$), Equation 9.46 evaluates to

$$\alpha = \frac{8.854 \times 10^{12}\left[(1.000254)^2 - 1\right]}{2.689 \times 10^{25}}$$
$$= 1.647 \times 10^{-40} \text{ F}^{-1} \text{ m}^4 \qquad (9.49)$$

Example calculation: water liquid
We now need to estimate the number density of water molecules in liquid water at STP by using the density of water ($\rho = 1000 \text{ kg m}^{-3}$) and the mass of 1 mol of water ($M = 18 \times 10^{-3}$ kg) to evaluate

$$n^{liquid} = \frac{N_A \rho}{M} \qquad (9.50)$$

$$n^{liquid} = \frac{6.022 \times 10^{23} \times 1000}{18 \times 10^{-3}}$$
$$= 3.346 \times 10^{28} \text{ m}^{-3} \qquad (9.51)$$

Substituting this value for n^{liquid} and the value of molecular polarizability determined from the gas data (Eq. 9.49) into Equation 9.45 for the refractive index yields

$$n_{light} = \sqrt{1 + \frac{3.346 \times 10^{28} \times 1.647 \times 10^{-40}}{8.854 \times 10^{-12}}}$$
$$\approx 1.27 \qquad (9.52)$$

which is to be compared with the experimental value of 1.33. Tables 9.19 & 5.17 which show the refractive indices of liquids and gases, have two entries in common aside from H_2O. Table 9.20 shows the results of the calculations similar to the one outlined above for

Table 9.20 Calculation of (Eqs. 9.44 & 9.52) the refractive indices of liquid water, methanol and benzene from the data on the refractive index of their vapours (see Table 5.17).

Substance	Chemical formula	MW	Gas number density (m^{-3})	Gas n_{light}	Molecular polarizability α $(F^{-1}m^4)$	Liquid density $(kg\,m^{-3})$	Liquid number density (m^{-3})	Liquid predicted $n_{light}-1$	Liquid actual $n_{light}-1$
Water	H_2O	18	2.689×10^{25}	1.000254	1.647×10^{-40}	1000	3.346×10^{28}	0.27	0.33
Methanol	CH_3OH	32	2.689×10^{25}	1.000586	3.860×10^{-40}	791	1.489×10^{28}	0.284	0.329
Benzene	C_6H_6	78	2.689×10^{25}	1.001762	11.61×10^{-40}	879	6.786×10^{27}	0.375	0.501

these two substances. Table 9.20 indicates that the predictions for $n_{light}-1$ are 20% to 25% below the experimental values. This level of agreement is not bad if we consider the extent of the extrapolation (around a thousand fold in number density) and our neglect of any consideration of molecular interactions.

The effect of molecular interactions is difficult to predict accurately, but it is worth bearing in mind that the low-frequency dielectric constant values of polar liquids were enhanced by a factor of ≈ 2.5 over what would have been expected on the basis of the gas data.

Comparison of refractive index data and dielectric constant data

The refractive index measurements use light oscillating at a frequency of the order 10^{15} Hz. In one period of such a fast oscillation the molecules are unable to rotate to align their intrinsic dipole moments with the electric field. Thus the refractive index arises only from the molecular polarizability, α, because only the electrons within the molecule are able to move fast enough to respond on this time-scale. Recall, for comparison, that a molecule vibrates within its cell roughly once every 10^{-13} s, and that the structure of the liquid changes around every 10^{-10} s. In contrast, the dielectric constant measurements are usually made at frequencies of only 10^6 Hz. In one period of such an oscillation there is plenty of time for molecules optimally to align their intrinsic dipole moment, p_p, with the electric field.

Bearing this in mind, it is interesting to compare the results of an analysis of dielectric constant data for the polar liquids water and methanol (see §9.7.4), with the results of the analysis of the refractive index. For the dielectric constant data we found that the intrinsic dipole moment of a polar molecule in the liquid appears to be enhanced by a factor of around

2.4. The reason for this enhancement is the low frequency used for the measurement, and the strong interaction between molecules.

However, the refractive index data are sensitive not to the permanent dipole moment on each molecule, but to the polarizability of each molecule. The data could be interpreted as indicating that the molecular polarizability in the liquid is enhanced over its value in the gas by around 20% to 25% (Equation 9.46 for liquid water predicts $\alpha = 2.035 \times 10^{-40} F^{-1} m^4$ compared with $\alpha = 1.647 \times 10^{-40} F^{-1} m^4$ for the gas (Eq. 9.50)). It seems plausible to consider that the excess polarizability arises from a difference in the character of the molecules in the liquid state – i.e. the charge distributions in a molecule of water in the liquid and gaseous states are different.

9.10 Magnetic properties

As with solids, liquids acquire magnetic moments in the presence of an applied magnetic field. In general, liquids containing elements with a strong magnetic response also respond strongly to applied magnetic fields. The complicating factor in the study of magnetic fluids is that the liquids are also able to flow, and the shape of liquid samples in general changes when a magnetic field is applied. I have been unable to obtain quantitative data on the magnetic properties of liquids.

9.11 Exercises

Exercises with a P prefix are "normal" problems. Those with a C prefix are best solved numerically using a computer program or spreadsheet.

Density and expansivity

P1. Which element has (a) the largest and (b) the smallest ratio of solid density to liquid density at its melting temperature (Table 9.1)?

P2. Pause the QBASIC computer program listed in Appendix 4 and examine the instantaneous positions of molecules in a simple two-dimensional liquid. Describe to what extent the figures in Example 8.1 and Figure 8.4 reflect the patterns seen in the simulation.

P3. By considering the densities listed in Table 9.2, estimate the density of *any* alcohol. Estimate the density of vodka, a mixture by volume of 60% water and 40% ethanol.

P4. To what extent are the relative densities of light water (H_2O) and heavy water (D_2O) related solely to their difference in atomic mass (Table 9.3)?

P5. Compare the change in the density of mercury over the temperature range 0–300°C with the change in the density of gold over the same temperature range (Fig. 9.4, Table 7.4 & Example 7.4).

P6. Which liquid metal listed in Table 9.4 has the highest thermal expansivity?

P7. Consider the way in which a body of fresh water such as a pond or a lake freezes over in the winter and melts again the spring. The cooling of the lake arises primarily from air cooling: the temperature of the earth beneath the lake is relatively stable from one season to another. Consider how the two "anomalous" properties of water (the density maximum at 4°C and the low density of its solid phase (Fig. 9.3)), combine to create a situation in which the majority of the lake remains liquid even in extremely cold weather. (See also Exercise P20, below)

P8. Water increases its volume by roughly 10% when it freezes. The freezing of water trapped in cracks within a rock is a powerful tool for geological erosion. By considering a simply shaped crack, estimate roughly the pressure exerted on the rock by ice when it freezes. You will need to estimate the Young's modulus of the ice, which you may do using information on its density (Fig. 9.3) and the speed of sound in ice (Table 7.6).

Speed of sound

P9. What is the speed of longitudinal sound waves in (a) water and (b) ethanol?

P10. What is the ratio of the speed of longitudinal sound waves (a) in water and ice, and (b) in copper and molten copper?

P11. Based on the speed of longitudinal sound waves, estimate the Young's modulus E, and bulk modulus, B, for ice, water, copper and molten copper (Eqs 9.2–9.10, Tables 7.1, 7.6, 9.1 & 9.5).

Viscosity and surface tension

P12. Consider the descent of a solid sphere of radius r and density ρ_s through a liquid of density ρ_l and viscosity η. The viscous force (G. G. Stokes 1850) has been found to be $F = 6 \times r\eta v$ and so increases as the sphere accelerates until it exactly equals the net force due to gravity and buoyancy. Derive an expression for the viscosity of a liquid in terms of this terminal velocity of a sphere falling through it. Estimate the terminal velocity for both steel (Table 7.1) and nylon (density $\approx 1300 \, kgm^{-3}$) spheres falling through water at (a) 0°C and (b) 40°C (Fig. 9.13). Carry out simple experiments to see if your estimates are correct.

P13. What is the viscosity of water at (a) 0°C and (b) 40°C (Fig. 9.13)? Using the Stokes' formula in Exercise P12, estimate the time taken by an air bubble 1 mm in diameter to rise through a column of liquid water 15 cm high at each of these temperatures. In practice, a bubble would grow as it rose through the liquid (§5.2). Estimate the extent to which the bubble would grow on its journey. Carry out simple experiments to see if your estimates are correct.

P14. What is the surface energy (surface tension), γ, for (a) water at 20°C, and (b) gold at 1100°C (Table 9.8)? What is the value of ΔE_s in milli electron volts for (a) water at 20°C, and (b) gold at 1100°C (Table 9.9)?

P15. Review the data and explanations presented in §9.5 & 11.4. Write an essay summarizing the usefulness of the cell model of a liquid dynamics.

Viscosity and surface tension

C16. Using the QBASIC program in Appendix 4 examine the liquid and solid simulations for the occurrence of processes in which molecules swap places in a manner similar to that described in Figure 8.16. Look also for the phenomenon of surface tension in which the molecules form a roughly spherical "blob" of liquid in contrast with the straight-edge structure of a solid.

P17. What is the molar heat capacity of (a) water at 0°C and (b) ethanol at 20°C? Estimate roughly the heat capacity of vodka, a mixture by volume of 60% water and 40% ethanol (Table 9.11).

P18. By considering the values of ΔE_s and ΔE_h relative to ΔE_e for water and gold, write a paragraph describing what boiling gold would look like. (§9.6, Tables 9.10, 11.1 & 11.2).

Questions on the vapour pressure of liquids are given in §10.8 & 11.10.

Thermal properties

P19. What is the thermal conductivity of (a) water at 20°C (b) ethanol at 20°C, (c) mercury at 20°C, and (d) molten sodium at 100°C (Tables 9.12 & 9.13).

P20. As discussed in Exercise P7 (above), lakes and ponds are cooled by contact with cold air above them. Write an explanation (with sketches) for a non-scientist friend explaining how the temperature gradient within a lake will cause convection to bring warm water to the surface until temperatures within the lake fall below 4°C. At this point the density anomaly effectively isolates a layer of water at the surface of the lake.

P21. Roughly how much energy does it take to heat (a) 1 mole, and (b) 1 kg of liquid water from its melting temperature to its boiling temperature. Evaluate the same quantities for methanol, ethanol and mercury (Tables 9.11 & 11.1).

P22. A cylinder of liquid water 5 cm deep with a base diameter of 10 cm is placed on a thermostatically controlled pad at a temperature of 4°C. Work out the rate of heat flow through the liquid if the surface temperature is (a) 3°C, and (b) 5°C. Consider carefully the effect of convection within the water (Table 9.12, Fig. 9.3 & Exercises P7 & P20, above).

P23. A cubic container of volume 1 m³ is heated from the bottom such that molten sodium metal at the bottom is approximately 100°C hotter than sodium at the top (Fig. 5.13). (a) Approximately how many moles of sodium are in the box? (b) Convection lifts 100 mol s⁻¹ of sodium from the bottom to the top. Estimate the heat flow across the container due to convection. Is it greater or less than would be expected due to the thermal conductivity listed in Table 9.13 alone?

Electrical properties

P24. Estimate the electrical resistivity of (a) sodium, and (b) potassium at 400°C (Table 9.15 & Fig. 9.32).

P25. Based on the Weiddeman–Franz ratio $\rho\kappa/T$, estimate the thermal conductivity of sodium and potassium at

400°C (Exercise P24, above).

P26. Estimate of mean scattering time, τ, for electrons in molten potassium and sodium at 100°C (Example 9.3).

P27. What is the dielectric constant of (a) water at 20°C (b) ethanol at 20°C, and (c) liquid nitrogen at 70 K (Table 9.16 & Fig. 9.34)?

P28. One of your colleagues is unable to read §9.8.3 & 9.8.4 as a result of their research on the biological effects of 40% ethanol solutions. Write a one page report for them summarizing the extent to which the dielectric constants of liquids of both polar and non-polar molecules can be understood by considering liquids to be dense gases (Tables 9.17 and 9.18).

Optical properties

P29. What is the refractive index of (a) water at 20°C, and (b) ethanol at 20°C. Estimate the refractive index of a mixture by volume of 40% ethanol and 60% water.

P30. Estimate the refractive index of liquid methanol and liquid ethanol based on the analysis of the water given in the example calculation in Equations 9.46–9.52. Compare your results with my calculations in Table 9.20.

P31. Using data taken from Figure 9.35, estimate very roughly the frequency of the UV transition in water based on the analysis for glasses in §7.8.4.

P32. Would a prism made of water be more or less dispersive than one made of glass (Figs 7.48, 7.49 & 9.35)?

P33. The reflectivity of glass immersed in water to light normally incident upon it is given by the formula

$$R = \left(\frac{n_{light}^{water} - n_{light}^{glass}}{n_{light}^{water} + n_{light}^{glass}} \right)^2$$

Explain why glass is nearly invisible underwater (Table 9.19 & Fig. 7.48). Compare the calculated value of R with the value for light incident upon either water or glass from air (Exercise C54 in Ch. 7).

CHAPTER 10

Changes of phase: background theory

10.1 Introduction

A collection of molecules of water (H_2O) are capable of smashing the steel hull of an ocean liner if their temperature is 272 K. But if their temperature is increased by less than 1% to 274 K they pose no danger at all. When ice melts, some of its properties change dramatically even though the temperature changes by only a tiny amount. The phenomenon of "sudden change" of properties is the most general, and the most striking, characteristic of phase changes.

In the preceding chapters, we have examined the properties of gases, solids and liquids, and saw how it is possible to understand their behaviour in terms of the atoms and molecules from which they are made. The division into just three phases is natural, since in our experience solids, liquids and gases behave strikingly differently. However, the division raises several questions that we will address in this chapter. In particular we will look in detail at what factors determine the transition temperatures between different phases of a particular substance. We will first consider this question from a theoretical point of view and then in Chapter 11 we will see how the theory can be used to understand the experimental data.

10.2 Why is a liquid a liquid? The free energy

Consider the following questions:
- What factors determine that at a particular temperature a substance should be a liquid, but that at slightly higher temperature the substance should become a gas?

- What factors determine that at a particular pressure a substance should be a liquid, but that at slightly higher pressure the substance should become a solid?

These changes of *phase* are certainly complex in nature and in any particular case one could imagine considering many detailed microscopic features of a substance in search of understanding them. However, there exists a general formalism for determining which phase of matter a substance adopts under specified circumstances. The formalism discusses the *equilibrium* phase of a substance in terms of the five quantities listed below:

- **Temperature** We are already aware that temperature is a key factor in determining the equilibrium phase of a substance.

- **Pressure** We saw in Chapter 9 that the temperature at which the liquid↔vapour transition occurs depends strongly on pressure. Similarly, the temperature at which the solid↔liquid transition occurs depends weakly on pressure. Clearly, pressure is also a factor in determining the equilibrium phase of a substance.

- **Volume** In most experiments the volume of a substance is allowed to vary, being determined by the experimenter's choice of temperature and pressure.

- **Internal energy** As students of physics we are used to explaining phenomena by stating that objects try to "minimize their energy". So, for example, we noted in §6.2 that solids formed from noble gas atoms choose particular crystal structures in order to minimize their internal energy. It is not surprising, therefore, that internal energy is a factor in determining the equilibrium phase of a substance.

- **Entropy** We have not yet discussed entropy in this book for the reason that there is no simple way of explaining what entropy is. Thus while we appreciate what heat, temperature, pressure and volume are, I have never met a scientist who had a "sense of entropy". This tends to make explanations of phenomena in terms of entropy confusing (at first). Now, however, we have no choice but to come to terms with entropy.

Microscopically, entropy is related to the amount of *order* in a substance. We can appreciate fairly directly that heating a substance – i.e. adding heat energy, ΔQ, to a substance – will increase the amount of disordered motion of the constituent atoms of a substance. Macroscopically, the quantitative measure of this disorder is the entropy, S, of the substance. If an amount of heat, ΔQ, is added to a substance at temperature T then the entropy, S, of the substance is increased by ΔS given by the surprisingly simple formula

$$\Delta S = \frac{\Delta Q}{T} \left(J\,K^{-1} \right) \qquad (10.1)$$

The total entropy of a substance is conventionally assigned to be zero for a substance in equilibrium at absolute zero, and the total entropy of a substance at temperature T is given by

$$S = S_0 + \int_0^T \frac{\Delta Q}{T} \, dT \qquad (10.2)$$

where S_0 is the entropy at $T=0\,K$, as shown in Example 10.2 below.

Entropy is discussed a little further in Appendix 3, but at this point we note the similarity between the entropy (Eq 10.2) and the internal energy, U:

$$U = U_0 + \int_0^T \Delta Q \, dT \qquad (10.3)$$

where U_0 is the internal energy at $T=0\,K$, the cohesive energy that we calculated for solids in Chapter 6.

The five factors that influence the equilibrium phase of a substance may be combined into a single mathematical function which reflects the balance between the various terms. The function is surprisingly simple: as outlined in detail in Appendix 3, if a substance is in equilibrium at a temperature T and pressure P the phase of the substance which minimizes

---

Example 10.1

A quantity of heat $\Delta Q = 0.2\,J$ is added to a large amount of substance at (a) 0.1 K, (b) 1 K, and (c) 10 K. Calculate the increase in entropy of the substance at each temperature.

We assume that the amount of substance is large enough that the temperature rise in each case is small. The increase in entropy in each case is given by Equation 10.1

$$\Delta S = \frac{\Delta Q}{T}$$

So, in case (a) we have

$$\Delta S_a = \frac{0.2}{0.1} = 2\,JK^{-1}$$

and in cases (b) and (c) we have $\Delta S_b = 0.2\,JK^{-1}$ and $\Delta S_c = 0.02\,JK^{-1}$ respectively. Notice that adding a given amount of heat at low temperatures causes a large increase in entropy (disorder), whereas the same quantity of heat added at higher temperature adds less entropy.

Colloquially, we might consider that adding a certain amount of heat energy to a substance at low temperatures is the equivalent of placing an energetic bull into a calm shop selling porcelain: the result is a large increase in the disorder. Adding the same amount of energy to a substance at a higher temperature is the equivalent of placing another energetic bull into a shop already full of other energetic bulls: the amount of added disorder is less noticeable.

---

the function

$$G = U - TS + PV \qquad (10.4)$$

is the equilibrium phase of the substance. The function G is called the *Gibbs free energy* of a substance and is an especially important quantity in understanding changes of phase. It is usually expressed in units of kilojoule per mole ($kJ\,mol^{-1}$), but may also be expressed per molecule, typically as electron volts per molecule ($eV\,molecule^{-1}$).

10.2.1 The Gibbs free energy

Circumstances in which we imagine an experiment being performed on a substance are usually such as to maintain the temperature constant (e.g. a cryostat or a furnace) and the pressure is usually either the ambient atmospheric pressure, or a controlled and stabilized pressure. Thus P and T are usually the controlled

Example 10.2

A substance is heated from T_1 to T_2. Neglecting thermal expansion, deduce an expression for the increase in entropy and internal energy of the substance at each temperature.

Internal energy   Consider the temperature rise, dT, due to an infinitesimal input of heat, dQ. If the substance expands only negligibly, the work done by the substance on its environment, dW, is negligible. The first law of thermodynamics, $dU = dQ + dW$, states that the heat input, dQ, goes entirely to increasing the internal energy of the substance, dU. The heat input, dQ, is related to the temperature rise, dT, by the heat capacity C:

$$dQ = C\,dT$$

So the change in internal energy, dU, due to dQ is given by

$$dU = C\,dT$$

If C varies with temperature (as it generally does), then the total change in internal energy on heating from temperature T_1 to T_2 is given by

$$\Delta U = \int_{T_1}^{T_2} C(T)\,dT$$

Putting $T_1 = 0$ and $T = T_2$ we arrive at

$$U(T) = U_0 + \int_0^T C(T)\,dT$$

where U_0 is the cohesive energy of the substance in the solid state at $T = 0$.

Entropy   Consider the temperature rise, dT, due to an infinitesimal input of heat, dQ. As above, we assume that dW is negligible, so that $dU = dQ$. By the definition of entropy (Eq. 10.1) the heat dQ carries with it entropy dS given by

$$dS = \frac{dQ}{T}$$

By the definition of the heat capacity, $dQ = C\,dT$ and so

$$dS = \frac{C\,dT}{T}$$

Thus the total change in entropy on heating from temperature T_1 to T_2 is given by

$$\Delta S = \int_{T_1}^{T_2} \frac{C(T)}{T}\,dT$$

Putting $T_1 = 0$ and $T = T_2$ we arrive at

$$S(T) = S_0 + \int_0^T \frac{C(T)}{T}\,dT$$

where S_0 is the entropy of the substance in the solid state at $T = 0$. Conventionally, S_0 is taken to be zero.

parameters of an experiment, with the volume being a "free" parameter which is able to adjust itself appropriate to the temperature and pressure.

Let's look at each of the three terms on the right-hand side of Equation 10.4 and see how each contributes to G:

- U, the internal energy. Lowering U helps to minimize G.
- TS, the temperature times the entropy. In order to minimize G, the substance should try to *maximize* this product. At a given temperature this means choosing a state with maximum S. The requirement to maximize S becomes less significant at low temperatures, because T is a multiplying factor.
- PV, the pressure times the volume. In order to minimize G, the substance should try to minimize this product. At a given pressure this means choosing a state with the minimum volume.

Consider a substance at a particular temperature T and pressure P, and imagine that the substance is able to "try out" different phases in order to determine which phase minimizes its Gibbs free energy. Let's consider the three terms in Equation 10.4 for each state: solid, liquid and gas. Appendix 3 shows that the state which minimizes G will form the equilibrium state of the system. The arguments in favour of one state or another are summarized in Table 10.1.

We can see that, at a given temperature and pressure, a substance may seek to minimize G in any of several ways. For example, it may:

- minimize its internal energy – best achieved in the solid or liquid state;
- maximize its entropy – best achieved in the liquid or gaseous state; or
- minimize its volume – best achieved in the solid or liquid state.

The state that minimizes the sum of the three terms will depend on the balance of the terms. Let us examine this balance by means of an example. Consider how the Gibbs free energy varies for potassium. We do not have full data available for this element, but this will force us to rely on general knowledge of the properties of matter, and to make assumptions about what is typical for a substance based on the data given in Chapters 5, 7 and 9.

Table 10.1 Summary of the contributions to the Gibbs free energy in each of the possible states of matter.

	U	−TS	+PV
Solid	In a solid atoms are close together and interact strongly. This term is therefore large and negative.	The entropy of a solid is very low – basically solids are highly ordered which is, a priori, a very unlikely state for matter to be in. However, the entropy is multiplied by temperature. If the temperature is low the term TS will be small, but if the temperature is large this term may be very significant.	At a given pressure the volume of a solid is close to the minimum volume that a substance can occupy. This makes this term small .
Gas	The interaction between atoms is many orders of magnitude weaker in gases than in solids.In the ideal gas theory it is neglected entirely.	The entropy of a gas is very high – basically gases are completely disordered collections of atoms.	At a given pressure the volume of a gas is as large as it is able to be.
Liquid	The interaction between atoms is of the same order as in the solid state, but the lack of organization means that the internal energy is negative, but not as large in magnitude as for a solid.	The entropy of a liquid is a little larger than a solid.	At a given pressure the volume of a liquid is similar to that of a solid.

At absolute zero

At absolute zero matters are relatively straightforward. The internal energy term is just U_0, the *cohesive energy* of the substance. The term TS containing the entropy is zero because $T=0$, and the term PV is negligible in comparison with U_0. Considering all these factors we see that at $T=0$, the equilibrium state of the substance (i.e. the state with minimum G) will be the state which minimizes the internal energy of the substance. With the sole exception of helium, this state is always the solid state. This is the justification in §6.2 for considering only the internal energy of the solid in determining which crystal structure a substance will adopt.

If we consider a substance to have a cohesive energy per atom of u_0 electron volts, then the internal energy of the substance at $T=0\,\mathrm{K}$ will be just

$$U_0 = -u_0 e N_A \text{ J mol}^{-1} \qquad (10.5)$$

where e is the charge on the proton, N_A is Avogadro's number and the minus sign indicates that the energy is negative with respect to the state in which the atoms are widely separated from one another.

$$-u \times 1.6 \times 10^{-19} \times 6.02 \times 10^{23} \approx -96 u_0 \text{kJ mol}^{-1} \qquad (10.6)$$

Example 10.3

Cohesive energies range from around 0.1 eV per atom for molecularly bonded substances, to around 10 eV per atom for substances such as diamond. A typical figure of around 1 eV per atom yields a cohesive energy of $U_0 \approx -100\,\mathrm{kJ\,mol^{-1}}$.

We can compare this with the PV term at atmospheric pressure. The volume of 1 mol of substance in the solid or liquid states is just the mass of 1 mol divided by the density of the substance. Using typical figures of $m \approx 100 \times 10^{-3}\,\mathrm{kg}$ and $\rho \approx 10 \times 10^3\,\mathrm{kg\,m^{-3}}$ indicates a molar volume of $\approx 10^{-5}\,\mathrm{m^3\,mol^{-1}}$ (i.e. around $10\,\mathrm{cm^3}$). At atmospheric pressure, the PV term evaluates to

$$1.013 \times 10^5 \,\mathrm{Pa} \times 10^{-5}\,\mathrm{m^3\,mol^{-1}} \approx 1\,\mathrm{J\,mol^{-1}}.$$

Thus at zero temperature, the PV term in the Gibbs free energy amounts to only $\approx 10^{-5}$ of the cohesive energy term.

At finite temperature

Above absolute zero the other terms in the Gibbs free energy must be taken into account. As outlined in Example 10.2, we can estimate the internal energy $U(T)$ by

$$U(T) = U_0 + \int_0^T C(T)\mathrm{d}T \qquad (10.7)$$

where U_0 is the cohesive energy of the substance in

the solid state at $T=0$. Similarly, we can estimate the entropy $S(T)$ as

$$S(T) = S_0 + \int_0^T \frac{C(T)}{T} \, dT \qquad (10.8)$$

However, the calculations involved require two key pieces of information:

(a) one must know the heat capacity as function of temperature, $C(T)$, in order to calculate either $U(T)$ and $S(T)$; and

(b) one must know the cohesive energy, U_0. (In general this involves rather complicated calculations, or complex inferences from experimental data.)

So, although straightforward in principle, accurately calculating G as a function of temperature is rather difficult in practice. We shall, however, attempt a *rough calculation* for one substance, potassium, to indicate the way in which the terms add together. In order to determine the temperature at which a phase transition takes place, we need to estimate all the terms in G for each of the phases under consideration. Table 10.2 summarizes the contributions to $G =$

$U - TS + PV$ in the solid and gaseous phases and suggests how we may estimate these fairly simply. The liquid state is considered later.

In compiling Table 10.2 and Figure 10.1, the following assumptions have been made. U_0 for the solid was taken from Table 11.5 and U_0 for the gas was taken as zero. S_0 for the solid was (in line with convention) taken as zero and S_0 for the gas was chosen so as to agree with the thermodynamic data from Emsley (see section 1.4.1).

To compound the difficulty of estimating S for the gas phase, the behaviour of the heat capacity of the gas at low temperatures is not known either. This problem is overcome by considering the heat capacity of the gas to be that of an ideal gas, and then adding the S_0 constant so as to cause agreement with data from Emsley.

The results of the calculations were evaluated using a spreadsheet computer program and are shown in Figure 10.1. We see that the balance of terms in G is such that at low temperatures the $U + PV$ terms dominate and the substance seeks to have a low internal energy and a low volume. At higher temperatures, however, the more disordered state minimizes G.

Table 10.2 Summary showing how the contributions to $G = U - TS + PV$ in the solid and gas phases may be estimated. The results of the sum are plotted as a function of temperature in Figure 10.1.

	U (mol^{-1})	TS (mol^{-1})	PV (mol^{-1})
Solid	The cohesive binding energy is given in Table 11.5 as $-90.1 \, \text{kJ mol}^{-1}$. To estimate U we evaluate $$-90.1 \times 10^3 + \int C_V(T) dT$$ where C(T) is estimated from a Debye model of a solid with a Debye temperature of 100 K.	Estimated from $$T\left[\int \frac{C_P(T)}{T} dT\right]$$ with C(T) estimated from a Debye model of a solid with a Debye temperature of 100 K. The entropy at T = 0 K is taken as zero.	We neglect thermal expansion and estimate PV from the density and atomic mass (Table 7.2). At atmospheric pressure we find $$PV = 1.013 \times 10^5 \times \frac{39 \times 10^{-3}}{830}$$ This term is very small.
Gas	Assuming perfect gas behaviour, we have no binding energy, and so we estimate U as $$+\int C_V(T) dT$$ where C(T) is estimated from an assumption of perfect gas behaviour ($=1.5R$, independent of temperature).	Estimated from $$T\left[\int \frac{C_P(T)}{T} dT\right]$$ with C(T) estimated from an assumption of perfect gas behaviour $= 1.5R$ independent of temperature. The entropy at T = 0 K is chosen so as to make the entropy of potassium vapour at 298 K agree with the data from Emsley (see §1.4.1).	We use the perfect gas equation for 1 mole of substance to evaluate $$PV = RT$$

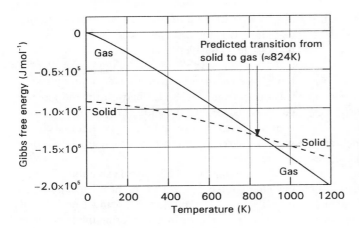

Figure 10.1 The result of a calculation of the Gibbs free energy, G, for potassium in its solid and gaseous states. The simplest assumptions possible have been made about the behaviour of U, S and V in this temperature range. Note that at zero temperature, G_{gas} is zero but G_{solid} is large and negative. This arises because at low temperatures the internal energy, U, is the main contributor to G.

We can consider the entropy S to be weighted by a factor T in the expression for G. At low temperatures the TS term is weighted rather little and so at low temperatures substances choose phases that minimize $U + PV$, of which the internal energy term usually dominates. This is the situation that we met in §6.2 where we predicted the equilibrium crystal structure of argon from a consideration of its internal energy alone. At high temperatures the TS term is weighted more strongly and at high enough temperatures substances choose phases that minimize TS, i.e. phases that have a high degree of disorder, irrespective of the "cost" in terms of volume or internal energy.

In the case of potassium, the balance between terms is such that we predict that at atmospheric pressure the solid state should transform into the gaseous state at around 824K. However, potassium transforms into the *liquid state* at around 336K and then

into the gaseous state at around 1033 K (Table 11.2), which is not very good agreement. To see if matters are improved we can repeat our calculation of G for the liquid state. Unfortunately, this is a difficult thing to do realistically because we have no simple universal theory of the liquid state from which we can deduce the properties of liquids. However, we do have sufficient information (Chapter 9) to make an informed guess at the behaviour of G_{liquid}.

(a) We know that $|U_{liquid}|$ is less than $|U_{solid}|$ because the atoms are not so efficiently packed in a liquid. We assume that if the potassium existed in its liquid phase at absolute zero its cohesive energy would be only around 90% of that found in crystalline potassium.

(b) Furthermore we know (§9.7.1) that at high temperatures the heat capacity of liquids is roughly temperature independent and typically around

Table 10.3 Summary showing how the contributions to $G = U - TS + PV$ in the liquid phase may be estimated. The results are plotted as a function of temperature along with the results from Table 10.2 in Figure 10.2.

	U (mol^{-1})	$-TS$ (mol^{-1})	PV (mol^{-1})
Liquid	The cohesive binding energy of the solid is given in Table 11.5 as $-90.1\,kJ\,mol^{-1}$. Assuming a value of around 90% of this figure we evaluate $$-81 \times 10^3 + \int C_V(T)dT$$ where C(T) is estimated to be 10% greater than the equivalent solid and to have a lower Debye temperature.	Estimated from $$T\int \frac{C_P(T)}{T}dT$$ where C(T) is estimated to be 10% greater than the equivalent solid and to have a lower Debye temperature. The entropy at $T = 0$ is set as zero, as for a solid. This will underestimate the entropy of the liquid state.	We consider this term to be the same as for the solid.

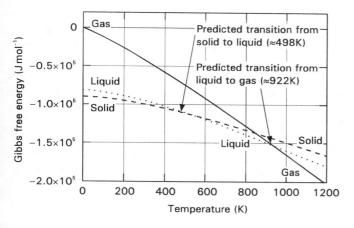

Figure 10.2 The results of a calculation of the Gibbs free energy, G, for potassium in its solid, liquid and gaseous phases. See also Figure 10.1 above. Note that at zero temperature, G_{liquid} is just a little above G_{solid}. As we have seen the heat capacities of liquids are only a little greater than those of solids and so the entropy rises at only a slightly greater rate than for solids. The similar variation of G_{solid} and G_{liquid} means that the predicted transition temperature is sensitive to the assumptions made in the calculations.

10% higher than that of solids.

(c) We do not know the residual entropy of the liquid state as compared with the solid state at $T=0\,\mathrm{K}$. For the sake of simplicity we set this equal to zero as for a solid, and merely acknowledge that this detracts from the accuracy of the results.

(d) We know (§9.2) that the volume of liquids is typically 10% higher than that of the equivalent solid. However, the PV term is not large for either liquids or solids.

We can thus estimate the three terms in G_{liquid} as shown in Table 10.30. The results of this calculation performed using a spreadsheet are plotted along with those of Figure 10.1 in Figure 10.2. The calculated results shown in Figure 10.2 indicate that, with some plausible assumptions about the entropy, internal energy and heat capacity, in the liquid state it is possible to understand how a liquid can become more sta-

ble than a solid in a "bridging region" before the high entropy gaseous state finally dominates.

Sensitivity of transition temperatures to assumptions

At zero temperature, G_{liquid} is just a little above G_{solid} and, since the heat capacities of liquids are only a little greater than those of solids, the entropy and internal energy increase at a rate only slightly greater than that for solids. The similar variation of G_{solid} and G_{liquid} means that the predicted transition temperatures are sensitive to the assumptions made in the calculations, since small changes in assumptions lead to large changes in transition temperatures. For example, the calculation illustrated in Figure 10.2 predicts that potassium will melt at 498 K (experimental value 336 K) and that potassium will boil at 922 K (experimental value 1033 K). In Figure 10.3 we illustrate the result of a calculation assuming that the cohesive

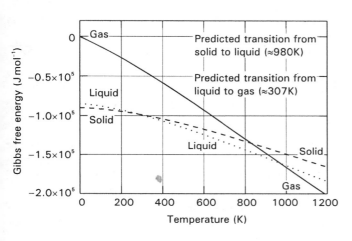

Figure 10.3 The result of a calculation of the Gibbs free energy, G, for potassium in its solid, liquid and gaseous phases (see also Figure 10.2). We have assumed a cohesive energy in the liquid state of 95% of the solid value, rather than 90% as assumed in Figure 10.2. The resulting predictions for T_M and T_B have changed significantly.

energy of the liquid state at $T=0$ would be 95% of the solid value, instead of 90% as assumed in Figure 10.2. This calculation predicts that potassium will melt at 307K and boil at 980K, which are much closer to the experimental values.

The better agreement between the predicted and experimental values of T_M and T_B in Figure 10.3 should not be taken to indicate that the assumptions of the second calculation are any more realistic than the first. The two calculations are presented merely to indicate how sensitive the calculations are to assumptions made about the states involved. The success of both these calculations is that they predict a liquid state at all! Determining realistic values for T_M and T_B requires a good deal of attention to details that we have neglected to consider here.

10.3 Phase transitions

We can now state in a general sense why phase transitions occur: they occur because the Gibbs free energy of one state of matter becomes less than that of another. It is important to realize that the free energy is a function which combines both the internal energy of the substance *and* the natural tendency of all states towards disorder. So the equilibrium state of substance is *not* – as students frequently maintain – the state that minimizes the internal energy. If this were true, substances would remain solids at all temperatures.

Figure 10.4 A substance seeking to minimize its Gibbs free energy will jump from one curve to another at the temperature at which the Gibbs free energy of one phase becomes lower than another. In this case the substance jumps from the solid to the liquid curve at T_M and then to the gaseous curve at T_B.

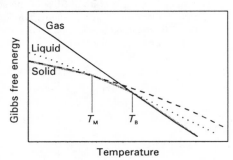

Our general picture of solid↔liquid↔gas phase transitions is summarized in Figure 10.4. Note that at each transition temperature the *slope* of the Gibbs free energy changes suddenly from one value to another. Let us perform some manipulations on the Gibbs free energy in order to look at its gradient with respect to temperature. In general, we have

$$G = U - TS + PV \qquad (10.4)^*$$

Changes in G result from changes in internal energy, temperature, entropy, pressure and volume. Thus we may write a small (differential) change in G as

$$dG = dU + P\,dV + T\,dP - T\,dS - S\,dT \qquad (10.9)$$

This is a complicated expression. However the first law of thermodynamics (Eq. 2.65) asserts that $\Delta U = \Delta Q + \Delta W$. From the definition of entropy (Eq. 10.1) we can substitute for $\Delta Q = T\Delta S$ and, taking account of the sign of the work done *on* the gas in expanding by ΔV, we write the first law as $\Delta U = T\Delta S - P\Delta V$, which in the differential limit is

$$dU = T\,dS - P\,dV \qquad (10.10)$$

Applying this result to Equation 10.9

$$dG = dU + P\,dV + V\,dP - T\,dS - S\,dT \qquad (10.11)$$

$$= 0$$

simplifies the matter considerably

$$dG = V\,dP - S\,dT \qquad (10.12)$$

Now, if the changes of phase in which we are interested take place at constant pressure – as they usually do – the alterations in G are not due to changes in pressure. In this case we have

$$dG = -S\,dT \qquad (10.13)$$

which amounts to

$$\left.\frac{dG}{dT}\right|_P = -S \qquad (10.14)$$

This equation tells us that the slope of the Gibbs free energy with respect to temperature at constant pressure is just the negative of the entropy, S. Thus all curves on graphs of G versus T (such as Figs 10.1–10.4) must have negative slopes. Also, high entropy states have steeper slopes than low entropy states.

This allows us to understand analytically what we appreciate already physically: the phase of matter with the highest entropy – colloquially the least amount of "order" or "structure" – will eventually become the equilibrium phase of the substance. The steepness of the curve will eventually overtake *any* amount of advantage given to a phase that has a large cohesive energy. Ultimately, everything will enter the gaseous phase.

Around a phase transition point we can apply Equation 10.14 to produce a particularly useful and interesting result. In what follows we use the subscripts "1" to indicate the lower temperature phase (which might be a solid or a liquid) and "2" to indicate the higher temperature phase (which might be a liquid or a gas).

At temperatures infinitesimally below and infinitesimally above a phase transition we have

$$\left(\frac{dG}{dT}\right) = -S_1 \text{ and } \left(\frac{dG}{dT}\right) = -S_2 \quad (10.15)$$

Thus the difference in the slopes is given by

$$\left(\frac{dG}{dT}\right)_2 - \left(\frac{dG}{dT}\right)_1 = -S_2 - (-S_1) \quad (10.16)$$

From Equation 10.14, since we are at a phase transition, we have $S_2 > S_1$, and so

$$\left(\frac{dG_2}{dT}\right) - \left(\frac{dG_1}{dT}\right) = S_1 - S_2 = -\Delta S \quad (10.17)$$

where ΔS is a positive quantity. Thus in order to move from one curve to another and so accomplish the phase transition, a quantity of entropy must be supplied at the constant temperature of the transition. In order to supply this entropy at the transition temperature, T_T, we must – from the definition of entropy (Eq. 10.1) – supply heat

$$\boxed{\Delta Q_T = T_T \, \Delta S} \quad (10.18)$$

This heat energy, ΔQ_T, transforms the substance at constant temperature from one phase to another, increasing the internal energy of the substance and its volume. The combination of terms $U + PV$ is known as the *enthalpy* of a substance often given the symbol H, and the heat ΔQ_T supplied at a transition is known as the *enthalpy change on melting* or *boiling*. This is often shortened to *enthalpy of vaporisation* or *fusion*,

or less commonly nowadays to *latent heat of vaporization* or *fusion*. The terms are all equivalent.

10.4 Enthalpy changes on transformation

Based on the theory outlined above we ought to be able to arrive at some general estimates about the typical magnitude of the enthalpies of fusion and vaporization.

We saw in §10.2 that, with some plausible assumptions, we could estimate U and S as a function of temperature and hence estimate the phase transition temperatures for potassium. In §10.3 we predicted that at a phase transition one must supply an amount of entropy, ΔS, given by

$$\left(\frac{dG_2}{dT}\right) - \left(\frac{dG_1}{dT}\right) = S_1 - S_2 = -\Delta S \quad (10.17)^*$$

which requires an amount of heat

$$\Delta Q_T = T_T \, \Delta S \quad (10.18)^*$$

Furthermore , at the transformation temperature, we know that $G_2 = G_1$, and so

$$U_1 - T_T S_1 + P_T V_1 = U_2 - T_T S_2 + P_T V_2 \quad (10.19)$$

Rearranging this yields

$$U_1 - U_2 + P_T(V_1 - V_2) = T_T(S_1 - S_2) \quad (10.20)$$

which may be compared to Equation 10.18 to give an expression for ΔQ_T:

$$\Delta Q_T = U_2 - U_1 + P_T(V_2 - V_1) \quad (10.21)$$

In the next two sections we will follow on from the calculations in §10.2 to try to estimate ΔQ_T, the enthalpy change on phase transformation.

The relative values of ΔQ_M and ΔQ_B

One feature of data shown in Figure 10.3 and reproduced in Figure 10.5 is already clear. The difference between the slope of the liquid and solid curves at the predicted melting temperature is rather small: the two curves run close to one another over a considerable range of temperature. In contrast, the difference

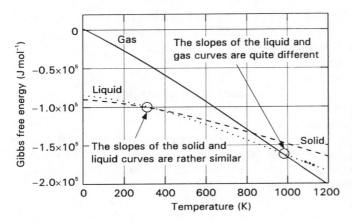

Figure 10.5 According to the considerations in Sections 10.2 & 10.3, a general feature of the enthalpy changes on transformation is that the enthalpy change on fusion (melting) is less than the enthalpy change on vaporization (boiling). On the $G(T)$ graph shown, the enthalpy changes are related to the differences in the gradient of the curves at their intersections.

between the slope of the gas and liquid curves at the predicted vaporization temperature is rather large: the gas curve cuts the liquid curve rather steeply. Thus, according to Equations 10.17 & 10.18 we would expect the enthalpy change on vaporization, ΔQ_B, to be considerably larger than the enthalpy change on fusion, ΔQ_M. In the next two sections the enthalpy change at each of these transformations is considered in turn.

Example 10.4

Estimate Q_M for potassium following on from the calculations in §10.2.

We note that:
(a) The zero temperature cohesive energy of the liquid state amounts to only a fraction (typically 90% to 95%) of the cohesive energy of the solid. The 95% value yielded slightly better values for T_M, so we use the estimate that $U_0(\text{liquid}) \approx 0.95\, U_0(\text{solid})$.
(b) The heat capacity in the liquid state is generally greater than in the solid state, typically by around 10%.
We thus find that

$$Q_M \approx -0.05 U_0(\text{solid}) + 0.1 \times \int_0^T C_{\text{solid}}(T)dT$$

Now, $U_0(\text{solid}) = 90.1\,\text{kJ mol}^{-1}$ (Table 11.5) and, estimating the integral numerically as we did in the previous examples, we find that at the experimentally determined melting temperature $T_M = 336\,\text{K}$

$$Q_M \approx -0.05 \times \left(-90.1 \times 10^4\right) + 0.1 \times 7198$$

$$Q_M \approx 5.22 \text{ kJ mol}^{-1}$$

10.4.1 Enthalpy change on fusion, ΔQ_M

In order to estimate the enthalpy change on fusion (melting), ΔQ_M, we can use Equation 10.18 ($\Delta Q_M = T_M \Delta S$) and estimate the melting temperature, T_M, S_{liquid} for the liquid state at T_M, and S_{solid} for the solid state at T_M. Alternatively, we can use Equation 10.21

$$\Delta Q_M = U_{\text{liquid}} - U_{\text{solid}} + P_M\left(V_{\text{liquid}} - V_{\text{solid}}\right) \quad (10.22)$$

and estimate the internal energy and volume of the liquid state at T_M and the internal energy and volume of solid state at T_M. Given the uncertainty in estimating the residual entropy of a liquid at $T=0$, it is more profitable to follow the second of these two options.

For the solid and liquid states, the PV term is negligible in comparison with the internal energy term and Equation 10.22 becomes

$$\Delta Q_M = U_{\text{liquid}} - U_{\text{solid}} \quad (10.23)$$

As in Example 10.2, we can estimate U by integrating the heat capacity:

$$\Delta Q_M = \left[U_0(\text{liquid}) + \int_{T=0}^{T} C_{\text{liquid}}(T)\,dT \right]$$
$$- \left[U_0(\text{solid}) + \int_{T=0}^{T} C_{\text{solid}}(T)\,dT \right] \quad (10.24)$$

which becomes

$$\Delta Q_M = \left[U_0(\text{liquid}) - U_0(\text{solid}) \right]$$
$$+ \int_{T=0}^{T} \left[C_{\text{liquid}}(T) - C_{\text{solid}}(T) \right]\,dT \quad (10.25)$$

The value calculated in Example 10.4 does not compare particularly well with the experimental value (Table 11.2) of $2.4\,\mathrm{kJ\,mol^{-1}}$. However, we clearly have the correct order of magnitude, and the poor detailed agreement is due mainly to the generality of the assumptions made in our estimates of the heat capacities of the liquid and solid state. We will consider the experimental data on the enthalpy change on melting in §11.3.

10.4.2 Enthalpy change on vaporization, ΔQ_{B}

Following a similar scheme to that outlined in the previous section, we can estimate the enthalpy change on vaporization (boiling) by using Equation 10.21

$$\Delta Q_{\mathrm{B}} = U_{\mathrm{gas}} - U_{\mathrm{liquid}} + P_{\mathrm{B}}\left(V_{\mathrm{gas}} - V_{\mathrm{liquid}}\right) \quad (10.26)$$

For the liquid state, the PV term is negligible, but for the gaseous state this is not so. Including the PV term only for the gaseous state this equation becomes

$$\Delta Q_{\mathrm{B}} = U_{\mathrm{gas}} - U_{\mathrm{liquid}} + P_{\mathrm{B}}V_{\mathrm{gas}} \quad (10.27)$$

As in Example 10.2, we can estimate U by integrating the heat capacity:

$$\Delta Q_{\mathrm{B}} = \left[U_0(\mathrm{gas}) + \int_{\mathrm{T}}^{\mathrm{T}} C_{\mathrm{gas}}(T)\mathrm{d}T\right]$$
$$\qquad (10.28)$$
$$- \left[U_0(\mathrm{liquid}) + \int_{\mathrm{T}=0}^{\mathrm{T}} C_{\mathrm{liquid}}(T)\mathrm{d}T\right] + P_{\mathrm{B}}V_{\mathrm{gas}}$$

which becomes

$$\Delta Q_{\mathrm{B}} = \left[U_0(\mathrm{gas}) - U_0(\mathrm{liquid})\right]$$
$$+ \left[\int_{\mathrm{T}=0}^{\mathrm{T}} C_{\mathrm{gas}}(\mathrm{T})\mathrm{d}\mathrm{T} - \int_{\mathrm{T}=0}^{\mathrm{T}} C_{\mathrm{liquid}}\right] + RT_{\mathrm{B}} \quad (10.29)$$

Once again, the value calculated in Example 10.5 does not compare particularly well with the experimental value (Table 11.2) of $77.5\,\mathrm{kJ\,mol^{-1}}$. However, we clearly have the correct order of magnitude, and the poor detailed agreement is due mainly to the generality of the assumptions made in our estimates of the heat capacities of the liquid and gaseous states. We will consider the experimental data on the enthalpy change on vaporization in §11.4.

Example 10.5

Estimate Q_{B} for potassium following on from the calculations in §10.2.

If we make ideal gas assumptions for the gaseous phase we note that
(a) The zero temperature cohesive energy of the gaseous state is zero.
(b) The heat capacity in the liquid state is generally greater than that in the solid state, typically by around 10%.
(c) The heat capacity in the gaseous state is $C_{\mathrm{P}} = 2.5R$ (see §5.3)
We thus find

$$Q_{\mathrm{M}} = \left[0 - 0.95U_0(\mathrm{solid})\right] + 2.5RT_{\mathrm{B}}$$
$$- 1.1 \times \int_0^T C_{\mathrm{solid}}(T)\mathrm{d}T + RT_{\mathrm{B}}$$

Recalling that $U_0(\mathrm{solid}) = 90.1\,\mathrm{kJ\,mol^{-1}}$ and estimating the liquid integral numerically as we did in the previous examples, we find that at the experimentally determined boiling temperature $T_{\mathrm{M}} = 1047\,\mathrm{K}$

$$Q_M \approx -0.95 \pm \left(-90.1 \times 10^3\right) + 2.5 \times 8.31 \times 1047$$
$$- 1.1 \times 24914 + 8.31 \times 1047$$
$$Q_M \approx 85595 + 21751 - 27405 + 8700$$
$$= 88.6\,\mathrm{kJ\,mol^{-1}}$$

10.5 The order of a phase transition

At the transitions discussed in §10.4.1 and §10.4.2, a substance changes completely from one phase to another. For example, at $0.01\,\mathrm{K}$ above T_{M} the substance is *completely* liquid; at $0.01\,\mathrm{K}$ below T_{M} the substance is *completely* solid. These transitions are characterized by a discontinuous change in the *gradient* of the Gibbs free energy. However, not all transitions take place in this way. Imagine cooling a substance slowly in its solid or liquid phases: it is possible that at a certain temperature some processes or behaviour *begin* to be possible. Below the transition temperature the substance does not transform completely, but begins to show a new property and the strength of this property grows as the substance is cooled further below the transition temperature.

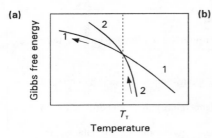

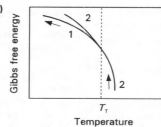

Figure 10.6 Schematic illustration of the changes in the Gibbs free energy, G, at (a) a first-order and (b) a second-order phase transition at temperature T_T. In a first-order transition, entropy (and hence heat energy) equal to the difference in gradients must be supplied. In a second-order transition there is no discontinuity in gradient, but a discontinuity in the curvature of G.

At such a transition there is no discontinuity in the gradient of G, but there is a discontinuity in the *curvature* of G. This leads to a general technique of categorizing phase transitions according to whether G has a discontinuity in its gradient or in its curvature.

- Recalling that the gradient $\partial G/\partial T$ is the *first* derivative of G, transitions in which there is a discontinuity in the gradient are called *first-order transitions*. These are characterized by the phenomenon of an enthalpy of transformation – colloquially, a *latent heat*.
- Recalling that the curvature $\partial^2 G/\partial T^2$ is the *second* derivative of G, transitions in which there is a discontinuity in curvature are called *second-order transitions*. These are characterized by the absence of any enthalpy of transformation – i.e. there is no latent heat associated with such a transition.

The characteristic behaviour of the Gibbs free energy in the region of first- and second-order transitions is illustrated schematically in Figure 10.6.

Now one might imagine that, having gone to the trouble of establishing this general framework for phase transitions, there must be a great many examples of both types of phase transition: this is not so. The vast majority of phase transitions are first order, there being only one or two examples of second-order phase transitions.

As Table 10.4 implies, it is possible to continue the process of categorizing to ever more subtle degrees. Transition temperatures then indicate the temperature at which the Gibbs free energy begins "to begin" to change! In general, the categorization of any phase transition as third order is specious, and in practice it is often difficult to distinguish between first- and second-order transitions. If a substance is not homogeneous, or if there is a temperature gradi-

Table 10.4 Examples of different orders of phase transition.

First order	Second order	Third order
Melting/freezing	Superconducting (in zero magnetic field)	Ferromagnetic
Boiling/condensing		
Liquid crystals		
Superconducting (in a magnetic field)		

ent across it, then a transition may appear gradual (second order) when it is in fact first order. It usually requires careful experiments to distinguish between the two cases. Note that I have referred to phase transitions other than solid↔liquid↔gas transitions, something I will justify in Chapter 11. However, structural phase transitions such as those between solids, liquids and gases are nearly always first order.

10.6 Nucleation: supercooling and superheating

So far in our discussion of phase changes we have assumed that a substance changes from one phase to another when the Gibbs free energy of one phase becomes lower than the Gibbs free energy of the other phase. However, this raises the question of how the substance "knows" what the Gibbs free energy of the other phase is going to be when it is not actually in that phase already! In other words, how is a phase transition initiated?

Microscopically, all phase transitions may be considered to occur in two stages: *nucleation* and *growth*.

- In the *nucleation* stage, the random motions of the atoms or molecules conspire to create a situation locally that is atypical of the *average* properties of the substance. These locally atypical regions – just a few atoms in size initially – are known variously as *nuclei, embryos* or *seeds*. The process of embryo formation occurs at all temperatures, although the rate of formation varies strongly with temperature. Alternatively, some small irregularity within the substance – perhaps an impurity or some feature of the container holding the substance – provides a locally anomalous region capable of supporting a nucleus of a different phase.

- In the *growth* stage, a nucleus of the second phase either grows in size, stays the same size, or shrinks. In general, the nuclei will not grow spontaneously even when the Gibbs free energy of the second phase becomes lower than the Gibbs free energy of the first phase. This is because for small nuclei the *surface energy* of the nuclei may be very high. Eventually, the nucleus is able to grow in size and the macroscopic phase transition commences.

If there are no suitable nuclei available a substance may pass its appropriate transition temperature and continue in the "wrong" or non-equilibrium phase, a phenomenon known as *supercooling* or *superheating*. Note that these phenomena are not caused by the experimental "error" of heating or cooling "too quickly" into a transient state – the supercooled or superheated substances may be quite stable until supplied with an appropriate "seed" on which the second phase may grow.

Let's examine the process of nucleation and growth for each of the phase transitions we have considered so far.

10.6.1 Solid→liquid: melting

Superheating past the bulk melting temperature is extremely unusual in the solid→liquid transition. In terms of the nucleation and growth theory outlined above, this implies that suitable liquid nuclei exist in the solid phase near the melting temperature. There is considerable evidence that these nuclei exist at solid surfaces, so causing solids to tend to melt from their surfaces inwards. Indeed, it is possible that in equilibrium even well below T_M there may be an atomic layer or two of essentially liquid substance present on the surface of a solid. The equilibrium thickness of this layer grows rapidly as the temperature approaches T_M and then grows without limit. In this theory one can understand the reluctance of solids to superheat because liquid nuclei would always be available due to the so-called *premelting* of the surface. This theory predicts that, in general, solids should melt from their surfaces rather than from within the solid.

10.6.2 Liquid→solid: freezing

In contrast with the melting transition, the solid→liquid transition frequently shows significant supercooling (Fig. 10.7). In line with our theory above, we can interpret this as being due either to a lack of nuclei, or to a strong barrier to the growth of nuclei.

Considering our simple conceptions of liquid and solid structures as envisaged in Figures 6.2 & 8.2, it seems unlikely that there will be any lack of nuclei –

Figure 10.7 The origin of the phenomenon of supercooling. Nuclei of the solid state within the liquid state will not grow unless doing so lowers the free energy. This leads to the phenomenon of a critical size of nucleus. As the temperature falls below the melting temperature (at which, by definition, the free energies of the bulk liquid and solid states are equal) , the free energy difference between the liquid and solid states increases and causes a decrease in the critical nuclear size. Eventually, the critical size is reduced to the size of the nuclei that are spontaneously formed in the liquid state, and the substance freezes.

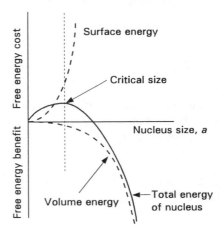

313

we expect that there will be many regions of liquid structure that are (transiently) similar to the solid structure. It seems then that once such solid nuclei form they fail to grow either because the nucleus has a large surface energy or because the nucleus breaks apart before it has time to grow.

Suppose that a spherical nucleus of "solid" of radius a exists in the liquid state (Fig. 10.7). The nucleus will grow if increasing its size from a to $a + da$ lowers the free energy of the substance. There are two factors that determine whether this is so.

The first is the free energy difference between the solid and liquid states, $\Delta G_{LS}(T)$. At the equilibrium melting temperature, by definition, $\Delta G_{LS}(T_M)$ is zero. If the temperature is lowered below T_M then the solid state becomes increasingly favoured over the liquid state. However, as Figure 10.5 indicates, the *gradients* of the free energies (dG/dT) of the liquid and solid states differ only slightly, and so the free energy benefit of the solid state increases only slowly as the temperature is lowered below T_M. Note that $\Delta G_{LS}(T)$ is proportional to the *volume* of the nucleus.

The second factor is the cost of forming an interface between the solid and liquid states. Around the nucleus may be a region of strained material that is not optimally arranged for either the solid or the liquid state. If this surface layer has a cost $G_{surface}$ per unit area then there will be a cost of allowing the nucleus to grow which will be proportional to the *surface area* of the nucleus.

Combining these two factors we see that the nucleus will grow if its total energy

$$G_{nucleus} = -\text{volume} \times \Delta G_{LS}(T)$$
$$+ \text{surface area} \times G_{surface} \quad (10.30)$$

$$G_{nucleus} = -\frac{4}{3}\pi a^3 \times \Delta G_{LS}(T) + 4\pi a^2 \times G_{surface} \quad (10.31)$$

decreases as a increases, i.e. if $dG_{nucleus}/da$ is negative. Note that in Equation 10.30 we have defined $G_{LS}(T)$ as positive if $G_{solid} < G_{liquid}$. So, evaluating $dG_{nucleus}/da$ we find

$$\frac{dG_{nucleus}}{da} = -4\pi a^2 \times \Delta G_{LS}(T) + 8\pi a \times G_{surface} \quad (10.32)$$

If we require that $dG_{nucleus}/da$ is negative then the condition for nuclear growth is

$$-4\pi a^2 \times \Delta G_{LS}(T) + 8\pi a \times G_{surface} < 0 \quad (10.33)$$

i.e.

$$4\pi a^2 \times \Delta G_{LS}(T) > 8\pi a \times G_{surface} \quad (10.34)$$

which simplifies to

$$a > \frac{2G_{surface}}{\Delta G_{LS}(T)} \quad (10.35)$$

This defines a critical value of nuclear size below which the surface energy cost of the nucleus impedes the transformation from liquid to solid. Note that:

- At the melting temperature $a_{critical}$ is infinite since $\Delta G_{LS}(T_M)$ is zero, by definition. Thus unless some external process initiates the transition liquids will always supercool to some extent (Fig. 10.8).
- As the temperature falls below T_M, $\Delta G_{LS}(T)$ increases, and thus decreases the critical size of a nucleus. When the temperature is sufficiently far below T_M that the critical size is less than the size of the nuclei actually formed in the liquid, the nuclei will grow and the substance will freeze.

Figure 10.8 Schematic illustration of the variation of temperature with time during the slow cooling of a pure substance through its equilibrium melting temperature. As discussed in the text, the liquid always supercools to some extent until solid nuclei reach sufficient size to grow spontaneously. This causes the release of enthalpy of fusion , which (if the external cooling rate is slow enough) will reheat the solid/liquid melt back to the equilibrium melting temperature. The melt remains at this temperature until all the liquid is transformed to solid, when cooling recommences.

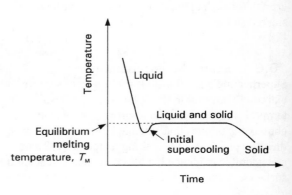

Because the gradients of both the solid and liquid free energies are similar (see Fig. 10.1) it is common for the temperature to fall considerably (several kelvin) below T_M before ΔG_{LS} becomes large enough to cause nuclear growth.

10.6.3 Liquid→gas: boiling/vaporization

Superheating at the liquid→gas transition is relatively common. Here a liquid may be heated to above its normal boiling temperature, but fails to boil because of the lack of suitable nuclei. Note that at any free surface the vapour pressure will continue to increase exponentially with temperature above the normal boiling temperature. The process that is inhibited is the formation of bubbles of gas vapour within the body of the liquid.

This is relatively easy to understand as being due to the large surface energy required to form a bubble – a new surface must be formed on the inside of the bubble. Thus since forming surfaces requires extra energy, energy must be supplied in excess of the normal free energy difference between the liquid and gas phases. For this reason the presence of external nuclei within a liquid is often critical to establishing boiling close to the equilibrium T_B.

10.6.4 Gas→liquid: condensation

Supercooling at the gas→liquid transition is considerably less common than superheating at the liquid→gas transition. Once again this requires that we have a large number of nuclei, or a low barrier to growth. We can understand how we can have a large number of nuclei – every time two molecules collide in the gaseous phase they potentially provide a seed for the growth of a liquid droplet. We can imagine that molecules in the low speed tail of the Maxwell distribution (see Fig. 4.7) would be more inclined to stick together at first, and that eventually droplets of a few molecules in size will form and grow.

10.7 Phase diagrams

It is often instructive to present pictorially information about the transitions between substances on what is known as a *phase diagram*. There are many different types of such diagram, but the aim of all such diagrams is to *summarize* the tremendous amount of information about the phases adopted by substances in a way that can be readily appreciated. A schematic phase diagram is shown in Figure 10.9.

Consider a substance at atmospheric pressure: at absolute zero the substance will (with the sole exception of helium) be solid. As the temperature is increased at constant pressure, typically the substance will eventually melt at T_M and then boil at T_B. Note that the volume of the substance will change significantly as the substance transforms from solid to gas. This process, represented by the dotted line in Figure 10.9, seems fairly straightforward to interpret. However, other paths are not so straightforward to interpret. Consider, for example, the one shown in Figure 10.10. It represents heating a solid in vacuum – an approximation to the zero pressure line on the figure.

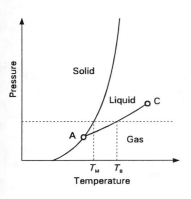

Figure 10.9 Schematic phase diagram of a typical substance on the pressure– temperature (*PT*) plane. (Note that the drawing is not to scale.) This type of diagram refers to the state of a fixed amount of substance, usually 1 mol. The volume of the substance is free to take any value and is not recorded on this type of diagram. The diagram must be interpreted subtly, as discussed in the text. The solid lines connect temperatures and pressures at which more than one phase exists in equilibrium. Point C represents the critical point, and point A represents the triple point of the substance.

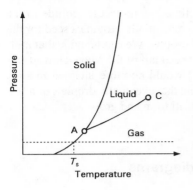

Figure 10.10 Schematic phase diagram of a typical substance on the pressure–temperature (*PT*) plane. (Note that the drawing is not to scale.) If a solid is heated at low pressure i.e., the pressure is not allowed to rise on heating, then at a "low enough pressure" the solid always transforms directly into the gas phase. This can be understood as being because the liquid phase always has a minimum vapour pressure above it, whereas a solid has a much lower vapour pressure.

The phase boundaries (lines) in Figures 10.9 and 10.10 represent sets of points at which a phase transition takes place. Thus traversing any of the solid lines in Figure 10.10 takes us from one phase to another. We know from analyzing the changes in Gibbs free energy at transitions (see §10.2.1) that at any point on the phase boundary the Gibbs free energies of the two phases on either side of the boundary are equal. Let us consider two points A and B on a phase boundary at slightly different temperatures and pressures (Fig. 10.11).

The Gibbs free energy at point A is G_A and the difference in free energy between A and B is ΔG. The free energy difference is due to differences in U, T, S, P and V and, as outlined in Equation 10.12, this may

be simplified to

$$\Delta G = V\,\Delta P - S\,\Delta T \qquad (10.36)$$

The change ΔG is the same for both phases, and so using subscripts 1 and 2 to donate the two phases,

$$\Delta G_1 = \Delta G_2 \qquad (10.37)$$

or

$$V_1\,\Delta P - S_1\,\Delta T = V_2\,\Delta P - S_2\,\Delta T \qquad (10.38)$$

Rearranging this we have

$$\Delta P(V_1 - V_2) = \Delta T(S_1 - S_2) \qquad (10.39)$$

Figure 10.11 A detailed view of the phase boundary between two phases (1 and 2) that might, for example, refer to a liquid↔solid phase boundary. At each point along the phase boundary the Gibbs free energies of the phases on either side of the phase boundary are equal. At point A the Gibbs free energy of each phase is G_A, and at point B (which is at a slightly different temperature and pressure) the Gibbs free energy of each phase is G_B.

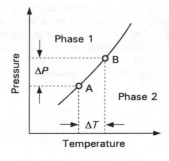

Example 10.6

Consider the boiling transformation of 1 mol of water at around its normal boiling temperature of 100°C and atmospheric pressure. From Table 11.1 we find that the molar enthalpy change on vaporization is around $40.6 \times 10^3\,\mathrm{J\,K^{-1}\,mol^{-1}}$. One mole of liquid has a mass of $18 \times 10^{-3}\,\mathrm{kg}$ and a volume of roughly $V_L = 18 \times 10^{-6}\,\mathrm{m^3}$ (Table 9.1). Assuming the gas to behave as a perfect gas, we can estimate the volume of one mole of gas as

$$V_G = \frac{RT}{P} \approx \frac{8.3 \times 373}{10^5} = 0.031\,\mathrm{m^3}$$

or around 1700 times the liquid volume. Neglecting the liquid volume in comparison with the gas volume, we can use Equation 10.41 to estimate d*P*/d*T*:

$$\frac{dP}{dT} = \frac{40.6 \times 10^3}{373 \times 0.031} = 3511\,\mathrm{Pa\,K^{-1}}$$

Thus if we increase the pressure above the liquid by 3511 Pa we increase the boiling temperature by 1 K.

Example 10.7

Consider the melting transformation of 1 mol of water at around its normal melting temperature of 0°C and atmospheric pressure. From Table 11.1 we find that the molar enthalpy change on fusion is around $5.99 \times 10^3 \, \mathrm{J \, K^{-1} \, mol^{-1}}$. One mole of liquid has a mass of $18 \times 10^{-3} \, \mathrm{kg}$ and a volume of approximately $V_L = 18 \times 10^{-6} \, \mathrm{m^3}$ (Table 9.2). One mole of solid (ice) has a volume roughly 8% larger, or around $V_S = 19.4 \times 10^{-6} \, \mathrm{m^3}$ (see Fig. 9.3). Using Equation 11.41 to estimate dP/dT,

$$\frac{dP}{dT} = \frac{5.99 \times 10^3}{273 \times \left(18 \times 10^{-6} - 19.4 \times 10^{-6}\right)}$$

$$= -15.6 \times 10^6 \, \mathrm{Pa \, K^{-1}}$$

Thus if we increase the pressure above the solid by $15.6 \times 10^6 \, \mathrm{Pa}$, around 150 times atmospheric pressure, we *decrease* the melting temperature by 1 K.

Note: (a) the relative insensitivity of the melting temperature to external pressure (which is typical of most solid↔liquid transitions), and (b) the fact that increasing the pressure decreases the melting temperature (which is highly unusual). This depression of the melting temperature by pressure is a consequence of the unusual contraction of the substance as it enters the liquid phase (see Fig. 9.3).

$$\frac{\Delta P}{\Delta T} = \frac{(S_1 - S_2)}{(V_1 - V_2)} \tag{10.40}$$

Or, taking the differential limit, we arrive at an equation known as the *Clausius–Clapeyron Equation*:

$$\boxed{\frac{dP}{dT} = \frac{(S_1 - S_2)}{(V_1 - V_2)}} \tag{10.41}$$

This equation relates the slope of a phase boundary to the difference in entropy and volume of the phases on either side of the phase boundary. We have already seen that the difference in entropy between the two phases is the origin of the *enthalpy change on transformation (or latent heat)* ΔQ_T and thus we may write

$$\frac{dP}{dT} = \frac{\Delta Q_T / T_T}{(V_1 - V_2)} = \frac{\Delta Q_T}{T_T(V_1 - V_2)} \tag{10.42}$$

The usefulness of the Clausius–Clapeyron equation is best illustrated by some examples, and we will return to Equation 10.47 several times in the following sections.

10.7.1 The critical point

The critical point is a special point on the liquid/gas phase boundary. It marks the point at which the densities of the phases on either side of the phase boundary become equal. Let us calculate the general shape of the liquid/gas phase boundary and then consider how the densities of the two terms vary as one moves along the boundary.

The liquid–gas phase boundary

In §11.4.2 we will consider a microscopic model of evaporation and condensation and come to the conclusion that we may reasonably expect the vapour pressure of a gas in equilibrium to vary as

$$P \approx A \exp(-\alpha / T) \tag{10.43}$$

where α is a constant. A similar result also follows from the Clausius–Clapeyron equation, but with rather fewer assumptions about a particular liquid model. At the liquid↔gas phase boundary we have

$$\frac{dP_B}{dT_B} = \frac{\Delta Q_B}{T_B(V_G - V_L)} \tag{10.44}$$

where the subscript B indicates that P_B and T_B are the boiling temperature and pressure of the liquid, and L and G indicate the liquid and gas phases, respectively. Let us assume that in the gas phase the substance behaves as an ideal gas, with a volume very much greater than in the liquid phase, and that the latent heat, ΔQ_B, does not vary with temperature. Under these approximations, and remembering that the volume of one mole of ideal gas is

$$V_G = \frac{RT}{P} \tag{10.45}$$

we find that Equation 10.44 becomes

$$\frac{dP_B}{dT_B} = \frac{\Delta Q_B}{T_B\left(\dfrac{RT_B}{P_B}\right)} = \frac{P_B \Delta Q_B}{RT_B^2} \tag{10.46}$$

Rearranging this yields

$$\frac{dP_B}{P_B} = \left(\frac{\Delta Q_{LG}}{R}\right)\frac{dT_B}{T_B^2} \tag{10.47}$$

If we integrate this equation we find

317

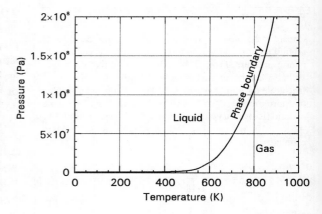

Figure 10.12 The equilibrium vapour pressure above liquid water predicted according to Equation 10.50. Suppose we take some water and subject it to a pressure of 5×10^7 Pa and a temperature of 400 K. The graph tells us that the water would be in the liquid phase. At a pressure of 5×10^7 Pa and a temperature of 800 K the graph tells us that the water would be in the gas phase. The line indicates temperatures and pressures in which water can exist either as a liquid or as a gas, i.e. it specifies the coexistence of liquid and gas that defines the boiling temperature. (Atmospheric pressure $\approx 10^5$ Pa, is essentially zero on this graph.)

$$\int \frac{dP_B}{P_B} = \left(\frac{\Delta Q_B}{R}\right) \int \frac{dT_B}{T_B^2} \qquad (10.48)$$

which yields

$$\ln\left(\frac{P_B}{P_0}\right) = \left(\frac{-\Delta Q_B}{RT_B}\right) \qquad (10.49)$$

or $\qquad P_B = P_0 \exp(-\Delta Q_B / RT_B) \qquad (10.50)$

This equation has a form similar to that derived with the aid of a microscopic model of a liquid (Eq. 10.43). Figure 10.12 shows a plot of Equation 10.50 for water using the value $\Delta Q_B = 40.6 \text{kJK}^{-1} \text{mol}^{-1}$ from Table 11.1 and a value of $P_0 = 4.89 \times 10^{10}$ Pa chosen to give a vapour pressure of 1 atm $(1.01 \times 10^5 \text{Pa})$ at 373.15 K.

The volume of gas at the phase boundary

Figure 10.12 and Equation 10.50 indicate that the pressure at the liquid↔gas phase boundary rises exponentially with temperature. This rapid variation causes some of the assumptions on which this equation was based to become invalid relatively quickly. In particular, we assumed that the gas behaves like a perfect gas and we neglected the volume of the liquid in comparison with the gas. While this assumption was valid at 373 K where the vapour pressure was 1 atm, it is unlikely to be valid at around 800 K where the vapour pressure is around 1000 times higher. Using the perfect gas equation to derive a very rough estimate of the volume of 1 mol of substance in the gas phase:

$$V_G = \frac{RT}{P} = \frac{RT}{P_0 \exp(-\Delta Q_B / RT_B)} \qquad (10.51)$$

indicates that the volume of 1 mole of substance in the gas phase, coexisting with some substance in the liquid phase, has a volume which falls exponentially as the temperature increases.

The critical point

Considering the volume of a substance across the range of temperatures and pressures that define the coexistence curve (Fig.10.12), Equation 10.51 indicates that, as the temperature is increased, the volume of one mole of gas on the coexistence curve decreases exponentially. However, the volume of one mole of liquid does not vary so dramatically. In general the volume of liquid held on the coexistence curve increases slightly with temperature. The *critical point* refers to a situation at which the volumes of the liquid and gas phases become equal (Fig. 10.13). At this point it becomes impossible to distinguish between the two phases

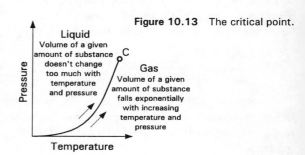

Figure 10.13 The critical point.

Estimating the critical temperature and pressure

We can ask at what pressure the volume of one mole of gas will become equal to the volume of one mole of liquid. The perfect gas equation is obviously not going to produce a good estimate for this – since in the perfect gas equation we do not consider the volume of the molecules themselves. However, it will serve to make a "ball park" estimate of the point at which the volume of the gas reaches the liquid volume. To do this we note that the volume of one mole of liquid water is around $18.8 \times 10^{-6} \, \text{m}^3$ (see Table 9.2) at around 373 K. Ignoring any thermal expansion of the liquid, we use the ideal gas equation to predict a range of pressures and temperatures at which the volume of the gas is $18.8 \times 10^{-6} \, \text{m}^3$:

$$P = \frac{RT}{18.8 \times 10^{-6}} \qquad (10.52)$$

Plotting the points that satisfy this equation on the same graph as the vapour pressure Equation 10.52 yields a rough estimate of the point at which the gas and liquid densities at the phase boundary are equal (see Fig. 10.13). However, this is certainly an overestimate of the critical temperature and pressure because of the neglect of the volume of the molecules. If we re-evaluate the pressure required to contain a gas in a given volume using the experimentally determined critical volume for water (see Table 11.4) of $59.1 \times 10^{-6} \, \text{m}^3$, then we obtain an estimate of the critical temperature and pressure of 803 K and $1.13 \times 10^8 \, \text{Pa}$ (Fig 10.14).

However, we have still made one assumption that leads us to overestimate the critical temperature: we have assumed that the latent heat, Q_B, will not vary with temperature. Clearly this cannot be quite right, since at the critical point the liquid and gas phases are indistinguishable and hence there can be no latent heat associated with the transition. Thus somewhere between the boiling temperature and the critical temperature the latent heat must fall to zero. In general, the way in which this happens is rather complicated and happens close to the critical point itself, but we can see qualitatively what the effect of this is. Equation 10.50 predicts that the vapour pressure is

$$P_B = P_0 \exp(-\Delta Q_B / RT_B) \qquad (10.50)^*$$

and so, other factors being equal, if ΔQ_B becomes smaller, the vapour pressure will increase. Thus, through these relatively crude approximations, we can begin to understand how, and under what conditions, the critical point of a substance is approached. Roughly speaking, it occurs in the region of the intersection of (a) the vapour pressure curve of a gas above a liquid and (b) the pressure required to keep a gas constrained to a volume a little greater than that occupied by the liquid at STP.

10.7.2 The triple point

The triple point, like the critical point, is another special point on a phase diagram. It marks the pressure and temperature at which the line representing the solid↔liquid phase transition intersects with the line

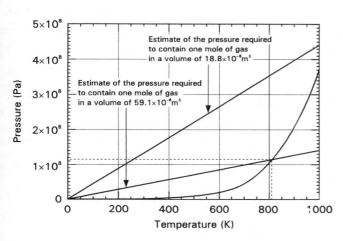

Figure 10.14 The curved line is the calculated coexistence curve for water shown in Figure 10.12. The upper straight line represents an estimate of the pressure required to keep water vapour confined to a volume equal to the volume of liquid water at around 373 K. The lower straight line represents an estimate of the pressure required to keep the water vapour confined to a volume equal to the experimentally determined critical volume of water ($59.1 \times 10^{-6} \, \text{m}^3$). The intersection of the curves represents a crude estimate of the critical temperature and pressure. The circle on the graph at ≈ 650 K represents the experimental value. (Atmospheric pressure, $\approx 10^5$ Pa, is essentially zero on this graph.)

319

representing the liquid↔gas transition. In order to determine where this intersection takes place we need to use the Clausius–Clapeyron equation (Eq. 10.41) to allow us to analyze the solid↔liquid transition in the same way as we analyzed the liquid↔gas transition in §10.7.1 on the critical point.

The solid↔liquid coexistence curve

We start once again with the Clausius–Clapeyron equation (Eq. 10.41)

$$\frac{dP}{dT} = \frac{\Delta Q_T}{T_T (V_1 - V_2)} \qquad (10.53)$$

where phase 1 is the liquid phase and phase 2 is the solid phase. In this case ΔQ_T is the enthalpy change on fusion, ΔQ_M, and T_T is the melting temperature, T_M, i.e.

$$\frac{dP_M}{dT_M} = \frac{\Delta Q_M}{T_M (V_L - V_S)} \qquad (10.54)$$

We first of all notice the effect of the volume change, ΔV_{LS}, between the liquid and the solid (Fig. 10.15).

For most substances ΔV_{LS} is positive – the liquid is less dense than the solid – and amounts to around 10% of the volume of the solid (Fig. 9.1). Thus the volume change between the liquid and the solid is dramatically smaller than the volume change between liquid and gas – perhaps by a factor 1000 or so – and thus the slope of the melting curve (coexisting solid and liquid) is typically much steeper than the slope of the vaporization curve (coexisting liquid and gas).

For a few substances ΔV_{LS} is negative – the liquid is more dense than the solid – and again amounts to

Figure 10.15 Illustration of the difference between the melting curves of (a) normal substances and (b) substances for which the liquid phase is more dense than the solid phase.

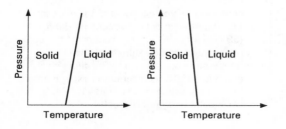

Example 10.8

At atmospheric pressure ice melts at 273.15 K and undergoes a density change from around 920 kg m⁻³ (see Table 7.1 & Fig. 9.1) to 999.84 kg m⁻³ (see Table 8.1). The latent heat required to melt 1 mole of ice is 5994 J mol⁻¹. What is the melting temperature at a pressure of (a) twice atmospheric pressure and (b) 1000 times atmospheric pressure?

We use the Clausius–Clapeyron equation (Eq. 10.41):

$$\frac{dP_M}{dT_M} = \frac{\Delta Q_F}{T_F (V_L - V_S)}$$

First we evaluate ΔV_{LS}, the volume change between liquid and solid. Considering 1 mol of substance, i.e. 18×10^{-3} kg, we have

$$V_L = \frac{18 \times 10^{-3}}{999.84} = 18.00 \times 10^{-6}\, m^3 \left(= 18.00\ cm^3\right)$$

and $V_S = \dfrac{18 \times 10^{-3}}{920} = 19.57 \times 10^{-6}\, m^3 \left(= 19.57\ cm^3\right)$

so that $\Delta V_{LS} = \left(18.00 \times 10^{-6}\right) - \left(19.57 \times 10^{-6}\right)$

$$= -1.57 \times 10^{-6}\, m^3$$

Substituting $\Delta Q_F = 5994$ J mol⁻¹ and $T_M = 273.15$ K, we find

$$\frac{dP_M}{dT_M} = \frac{5994}{273.15 \times \left(-1.57 \times 10^{-6}\right)}$$

$$= -10.97 \times 10^6\ Pa\, K^{-1}$$

This indicates that to *lower* the melting temperature by 1 K, we need to apply a pressure of around 10^7 Pa (roughly 100 times atmospheric pressure).

We can estimate the change in melting temperature, ΔT_M, due to a change in pressure, ΔP_M, using

$$\frac{\Delta P_M}{\Delta T_M} \approx \frac{dP_M}{dT_M} = -10.97 \times 10^6\ Pa\, K^{-1}$$

and hence $\Delta T_M \approx \dfrac{\Delta P_M}{-10.97 \times 10^6}$

(a) For a pressure of twice atmospheric pressure, $\Delta P_M = 1.013 \times 10^5$ Pa, ΔT_M evaluates to -9.23×10^{-3} K, i.e. a doubling of the pressure results in a *lowering* of melting temperature of only 9.23 mK.

(b) For a pressure of 1000 times atmospheric pressure $\Delta P_M = 1.013 \times 10^8$ Pa, ΔT_M evaluates to -9.23 K. Although much larger than the shift in (a), this large pressure only shifts the melting temperature by around 9 K in 273 K, i.e. a shift of only 3%.

Example 10.9

Why is ice so slippery? One theory commonly propounded is that, when pressurized, ice melts because its melting temperature has been lowered. Does this make sense?

Consider a car with a mass of around 1000 kg distributed equally across four wheels.

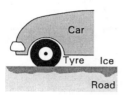

The weight of the car is supported on a relatively small area of contact on each tyre. Recalling that the tread of the tyre reduces the area of rubber in contact with the road by around 30%, we estimate this contact area as 70% of 15 cm × 15 cm = 15.75×10^{-3} m^2 per tyre. The force supported over this area is the mass of the car times the acceleration due to gravity, which we take as 10 m s^{-2}. The force per tyre is therefore 1000 kg × 10 m s^{-2}/4 = 2500 N. The force per unit area under each tyre – the pressure – is therefore approximately 2500 N/15.75×10^{-3} m^2 = 1.6×10^5 Pa, or an excess pressure of around 1.6 atmospheres. Taking into account the uncertainties in our estimates, and allowing for the excess pressure that may be generated if the car is in motion, we might estimate that the maximum pressure under the tyres of a car is around 10^6 Pa.

Using Example 10.8 as a guide, 10^6 Pa will lower the melting temperature by around 0.1 K. Thus a car travelling over an iced surface will indeed increase the pressure on the ice such that the ice is taken to its melting temperature *if* the ice is already within approximately 0.1 K of its melting temperature. However, latent heat would still be required to melt the ice, so would it melt? See also §10.8: Exercise P8.

---

around 10% of the volume of the solid (see Fig. 9.1). For these substances the slope of the melting curve will be negative, and of a similar order of magnitude to those for normal substances.

Predicting the melting curve

Let us plot the melting line for water substance on the same diagram on which we plotted the vapour pressure curve. Starting with the Clausius–Clapeyron equation we write

$$\frac{dP_M}{dT_M} = \frac{\Delta Q_M}{T_M(V_L - V_S)} \tag{10.55}$$

In order to estimate the order of magnitude of the quantities involved, we assume that the latent heat of fusion ΔQ_M does not vary with pressure or temperature, and that the volume difference between the liquid and solid phase, $V_L - V_S$, does not vary with pressure or temperature.

In fact both quantities do vary with pressure or temperature, but usually only rather slowly. This affects the detailed shape of the melting curve, but not its qualitative form. With these assumptions we write

$$dP_M = \frac{\Delta Q_M}{\Delta V_{LS}} \frac{dT_M}{T_M} \tag{10.56}$$

Integrating this from zero pressure yields

$$\int_{P_M=0}^{P_M} dP_M = \frac{\Delta Q_M}{\Delta V_{LS}} \int_{T_M(P_M)}^{T_M(P_M)} \frac{dT_M}{T_M} \tag{10.57}$$

and remembering that $\int(dx/x) = \ln(x)$ this becomes

$$P_M = \frac{Q_M}{\Delta V_{LS}} \left\{ \ln\left[T_M(P_M = 0)\right] - \ln\left[T_M(P_M)\right] \right\} \tag{10.58}$$

Rewriting this we find

$$P_M = \frac{\Delta Q_M}{\Delta V_{LS}} \ln\left[\frac{T_M(P_M = 0)}{T_M(P_M)}\right] \tag{10.59}$$

This equation allows us to predict the melting pressure of a substance in terms of its melting temperature at zero pressure. In order to do this we need to first estimate $T_M(P_M=0)$, the melting temperature at zero pressure. In fact at zero pressure a substance does not melt but sublimes; ignoring this for the moment, we choose $T_M(P_M=0)$ in order to produce a melting temperature of 273.15 K at $P_M = 1.013 \times 10^5$ Pa. The resulting curve is plotted alongside the vaporization curve (see Fig. 10.12) in Figure 10.16.

If we consider now what happens at low pressures, we see that the triple point marks the point below which the liquid phase disappears. As we will see in §11.7, solids support a vapour pressure above their surface in much the same way as a liquid does, and so the melting curve disappears. The vaporization curve

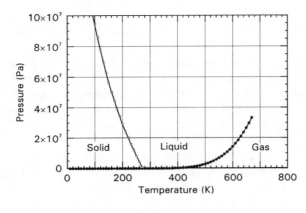

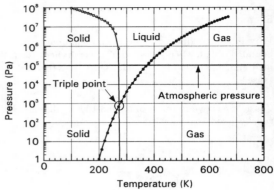

Figure 10.16 The curve of the melting pressure versus temperature for water.

continues below the triple point in approximately the same way as it "would have done" for the liquid↔gas transition. (§11.7 contains experimental data on the differences between the liquid↔gas and solid↔gas phase transitions.)

10.7.3 Constructing a phase diagram: a second example

Diagrams such as the one shown in Figure 10.17 are known as *phase diagrams*. Let's look at a second example of a phase diagram, this time for the metal potassium (discussed in §10.2). The procedure for constructing the phase diagram is exactly the same as that discussed above for water.

The vaporization curve is predicted from Eq. 10.50

$$P_B = P_0 \exp(-\Delta Q_B / R T_B) \qquad (10.50)*$$

with P_0 adjusted to yield a vapour pressure of 1 atm at 1047K in agreement with the boiling temperature (Table 11.2).

The melting curve is predicted from Equation 10.59

$$P_M = \frac{\Delta Q_M}{\Delta V_{LS}} \ln\left[\frac{T_M(P_M = 0)}{T_M(P_M)}\right] \qquad (10.59)*$$

with the melting temperature at zero pressure set to 336.8K, in rough agreement with the melting temperature at 1 atm (see Table 11.2).

Figure 10.17 The curve of the melting pressure versus temperature for water.

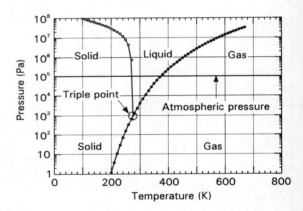

The phase diagram constructed from these equations is shown in Figure 10.18.

Note that because latent heat of vaporization is considerably greater for potassium than for water, the vaporization curve is much lower in pressure than that for water (see Figure 10.17) at the same temperature. Note also that, because the volume of potassium increases as it melts, the slope of the melting curve is positive, i.e. pressure suppresses the melting transition.

10.7.4 The PVT surface

Phase diagrams such as those shown in the previous section are an enormously useful way to present in

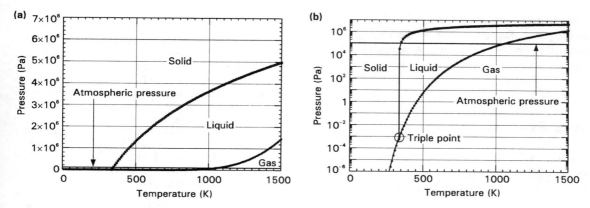

Figure 10.18 The phase diagram for potassium predicted according to Equations 10.50 and 10.59 using data from Tables 11.2 and 9.2 (molar mass = 39.1×10^{-3} kg; solid density = 862 kg m^{-3}; liquid density = 824 kg m^{-3}; T_M = 336.8 K; T_B = 1047 K; Q_{LS} = 2.4 kJ mol^{-1}; Q_{LG} = 77.53 kJ mol^{-1}). Note that the vaporization curve is much lower in pressure than for water at the same temperature. The slope of the melting curve is positive. (a) Pressure data shown on a linear scale; (b) Pressure data shown on a logarithmic scale.

summary form a great deal of information about a substance. As we mentioned before, a phase diagram refers to a given amount of a substance (usually one mole) and gives information about which phase (liquid, solid or gas) a substance will adopt at any pressure or temperature. However, there is an unplotted variable on these phase diagrams – the volume which

the given amount of substance adopts in each of these three phases.

The volume varies by many orders of magnitude across the different experimental conditions outlined in the phase diagrams, and its variation with pressure and temperature is different in each of the different phases. There is a construction that indicates the vol-

Figure 10.19 The PVT surface of a hypothetical typical substance. The specification of this surface for a substance describes the equilibrium behaviour of a substance. In general, each part of the surface may be described by an equation of state. The line XY represents a process in which the substance is heated at constant pressure. In the solid state (A) the volume changes only a little with temperature. The substance then melts (C) and expands at constant temperature until it is all transformed to the liquid state (D). There the thermal expansion is slightly larger than in the solid state. Eventually the substance reaches its boiling temperature and its volume increases dramatically as it vaporizes (E). Eventually all the substance is transformed to the gas phase.

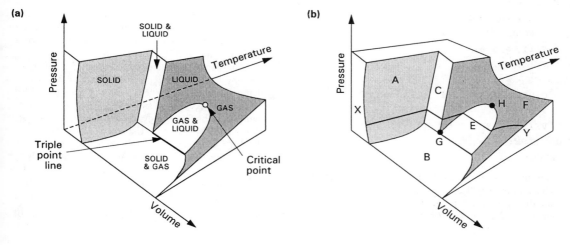

ume of a given amount of substance at different points on the phase diagram known as the *PVT* (pressure–volume–temperature) surface. A schematic illustration of such a surface is shown in Figure 10.19.

10.8 Exercises

Exercises with a P prefix are "normal" problems. Those with a C prefix are best solved numerically using a computer program or spreadsheet.

C1. Alter the program in Appendix 4 to show the changes in phase of a two-dimensional substance as follows. Replace the two SELECT CASE statements following the "...`Reverse velocity component`..." comment with the text below. You may also speed up the program by removing all reference to PL, PR, PU, PD and P from the main program loop: these variables are not needed in this program.

```
'Reverse velocity component when a molecule
hits the wall
SELECT CASE X(i)
    CASE IS >= SpaceX:
        CIRCLE (X(i), Y(i)), Radius, 0
        X(i) = SpaceX
        VX(i) = -.9 * VX(i)
    CASE IS <= 0:
        CIRCLE (X(i), Y(i)), Radius, 0
        X(i) = 0
        VX(i) = -.9 * VX(i)
    CASE ELSE:
END SELECT
SELECT CASE Y(i)
    CASE IS >= spaceY:
        CIRCLE (X(i), Y(i)), Radius, 0
        Y(i) = spaceY
        VY(i) = -.9 * VY(i)
    CASE IS <= 0:
        CIRCLE (X(i), Y(i)), Radius, 0
        Y(i) = 0
        VY(i) = -.9 * VY(i)
    CASE ELSE:
END SELECT
```

Start the program for the gas phase and wait patiently. The substance will eventually condense into a liquid and then a solid. The program simulates the effect of a box with walls which are (magically!) always a little bit colder than the substance within the box. On each collision with the wall, a molecule has the component of its velocity perpendicular to the wall reduced by 10%. This is only an approximation to the real processes which take place in a gas (see §4.3.1, Complication 1).

Note the occurrence of short-lived "liquid precursor" clusters of molecules (two or occasionally three molecules) in the gas phase. To what extent is it possible to define a time after which (i.e. a temperature below which) there is a clear transition to the liquid state?

P2. At temperatures less than 30 K, the heat capacity of metal varies as $C \approx \gamma T + \alpha T^3 \, \mathrm{J\,K^{-1}\,mol^{-1}}$ where $\alpha = 10^{-4} \, \mathrm{J\,K^{-2}\,mol^{-1}}$ and $a = 10^{-5} \, \mathrm{J\,K^{-4}\,mol^{-1}}$. Calculate the thermal contribution, $U(T)$, to the internal energy of the solid at 1 K and 20 K. How does this compare with typical values of the internal energy due to electrostatic attraction, U_0 (Table 11.5)?

P3. Assuming that the heat capacity of a substance follows the Debye law (Figures 7.27 and 7.28), sketch the approximate form of $U(T)$. Indicate clearly the value of $U(T)$ at $T = 0$ K and the limiting slope at temperatures greater than the Debye temperature.

P4. Water substance is held in equilibrium at a series of temperatures and pressures (a) to (g). For each state (a) to (g) indicate the phase of the water substance according to Figure 10.17.

	T(K)	P(Pa)
(a)	200	10^5
(b)	200	10^8
(c)	400	10^8
(d)	400	10^5
(e)	700	10^5
(f)	700	10^8
(g)	273.15	611

P5. One mole of an elemental solid substance has a cohesive energy $U_0 = 100 \, \mathrm{kJ\,mol^{-1}}$. By considering this energy to arise from atoms arranged in a simple cubic structure ($a \approx 0.3$ nm) with nearest-neighbour interactions only, show that each atomic bond contributes approximately $\frac{1}{3}$ eV to the cohesive energy of the substance.

By considering the energy required to split a cube of the substance in two (or otherwise), show that the energy required to form a new surface of the substance is approximately $\Delta G_{\mathrm{surface}} \approx 0.6 \, \mathrm{J\,m^{-2}}$.

P6. The substance considered in Exercise P5 above has a melting temperature $T_M = 700$ K. Close to the melting temperature, the Gibbs free energy of the liquid phase of a substance has a slope $\partial G_L / \partial T = 400 \, \mathrm{J\,K^{-1}\,mol^{-1}}$ and the Gibbs free energy of the solid phase has slope $\partial G_S / \partial T = 395 \, \mathrm{J\,K^{-1}\,mol^{-1}}$. Using a value of $\Delta G_{\mathrm{surface}} \approx 0.6 \, \mathrm{J\,m^{-2}}$, estimate the size of a critical embryo according to Equation 10.35 if the substance supercools by 0.7 K before commencing solidification. Note that before you can substitute into Equation 10.35 you must convert the Gibbs free energy difference ΔG_{LS}, from $\mathrm{J\,mol^{-1}}$ to $\mathrm{J\,m^{-3}}$. Do you think your estimate for the critical embryo size is plausible?

P7. Following Example 10.7, work out the rate at which the freezing temperature of ethanol would increase with applied pressure. If the maximum achievable laboratory pressure is around 10^5 atm, could a pressure be achieved at which the melting temperatures of ethanol and water were equal? (In fact the structure of ice changes at high pressure and slope of melting curve then becomes positive.)

P8. Following Examples 10.8 and 10.9, write a short report with calculations on whether the slipperiness of ice is evidence for the existence of a pre-melted surface (§10.6.1). As part of your report, conduct informal experiments with ice cubes that are cooled as far below 0°C as possible (a three-star domestic freezer will cool ice to −18°C). In particular, address the question of whether the slipperiness is connected with the temperature of the sliding object.

CHAPTER 11

Changes of phase: comparison with experiment

11.1 Introduction

In this chapter we examine the experimental data concerning phase changes and see to what extent we can understand the data in terms of the background theory discussed in Chapter 10.

Example 11.1

It is interesting to compare the amount of energy required to
(a) melt 1 mol of ice at its melting temperature,
(b) raise 1 mol of liquid (water) from 0°C to 100°C, and
(c) melt 1 mol of copper at its melting temperature.

The data required to answer these questions are given in Tables 11.1, 9.11 & 11.2, respectively.
(a) From Table 11.1, the enthalpy of fusion of water is 5.994×10^3 J mol^{-1}. Thus to melt 1 mol (0.018 kg) of ice requires $1 \times 5.994 \times 10^3 = 5994$ J.
(b) From Table 9.11, the heat capacity of water is 75.9 J K^{-1} mol^{-1}. Thus to raise 1 mol (0.018 kg) of water from 0°C to 100°C requires $1 \times 75.9 \times 100 = 7590$ J, i.e. about 25% more energy than is required to melt the same amount of water.
(c) From Table 11.2, the enthalpy of fusion of copper is 13×10^3 J mol^{-1}. Thus to melt 1 mol (0.0635 kg) of copper requires $1 \times 13 \times 10^3 = 13000$ J. This amounts to about twice the energy required to melt 1 mol of water.

11.2 Data on the solid↔liquid and liquid↔gas transitions

This section consists of just two tables which contain data that will be referred to in several of the following sections. Table 11.1 contains data relevant to the solid↔liquid and liquid↔gas transitions in various substances, and Table 11.2 contains the equivalent data for the elements.

Example 11.2

It is interesting to compare the amount of energy required to:
(a) boil 1 mol of water at its boiling temperature,
(b) raise 1 mol of liquid (water) from 0°C to 100°C, and
(c) boil 1 mol of copper at its melting temperature.

The data required to answer these questions are given in Tables 11.1, 9.11 & 11.2, respectively.
(a) From Table 11.1, the enthalpy of vaporization of water is 40.608×10^3 J mol^{-1}. Thus to boil 1 mol (0.018 kg) of water requires $1 \times 40.608 \times 10^3 = 40608$ J, i.e. about seven times as much energy as is required to melt the same amount of water.
(b) From Table 9.11, the heat capacity of water is 75.9 J K^{-1} mol^{-1}. Thus to raise 1 mol (0.018 kg) of water from 0°C to 100°C requires $1 \times 75.9 \times 100 = 7590$ J, i.e. about one-fifth less energy than is required to boil the same amount of water.
(c) From Table 11.2 the enthalpy of vaporization of copper is 304.6×10^3 J mol^{-1}. Thus to boil 1 mole (0.0635 kg) of copper requires $1 \times 304.6 \times 10^3 = 304600$ J. This amounts to about 23 times as much energy as is required to melt the same quantity of copper, and about eight times as much energy as is required to boil 1 mol of water.

Table 11.1 Thermal data for various substances (the data refer to standard atmospheric pressure, unless otherwise stated). See also Exercise 11C17 for a technique to estimate some of the missing data.

Substance	Chemical formula	MW	Density (kg m⁻³)	Melting point (K)	Boiling point (K)	Enthalpy of fusion (kJ mol⁻¹)	Enthalpy of vaporization (kJ mol⁻¹)
Acetic acid	CH_3COOH	60	1049	289.75	391.1	11.535	–
Acetone	CH_3COCH_3	58	787	177.8	329.3	5.691	–
Aniline	C_6H_7N	93	1026	266.85	457.6	10.555	–
Benzene	C_6H_6	78	877	278.65	353.2	9.951	–
Chloroform	$CHCl_3$	119	209.55	334.4	8.800	–	
Cyclohexane	C_6H_{10}	82	779	279.65	353.8	2.630	–
Ethyl acetate	$C_4H_8O_2$	88	–	189.55	350.2	10.481	–
Methanol	CH_3OH	32	791	179.25	337.7	3.177	–
Ethanol	C_2H_5OH	46	789	155.85	351.5	5.021	–
Propan-1-ol	C_3H_7OH	60	804	146.65	370.3	5.195	–
Propan-2-ol	C_3H_7OH	60	786	–	–	–	–
Butan-1-ol	C_4H_9OH	74	810	183.65	390.35	9.282	–
Butan-2-ol	C_4H_9OH	74	808	298.55	372.65	6.786	–
Toluene	C_7H_8	92	867	178.15	383.8	6.851	–
Lithium fluoride	LiF	25.9	2635	1118	1949	–	–
Lithium chloride	LiCl	42.39	2068	878	1620†	–	–
Lithium bromide	LiBr	86.9	3464	823	1538	–	–
Sodium chloride	NaF	42.0	2558	1266	1968	–	–
Sodium fluoride	NaCl	58.4	2165	1074	1686	–	–
Sodium bromide	NaBr	102.9	3203	1020	1663	–	–
Potassium fluoride	KF	58.1	2480	1131	1778	–	–
Potassium chloride	KCl	74.6	1984	1043	1273*	–	–
Potassium bromide	KBr	119.0	2750	1007	1708	–	–
Carbon dioxide	CO_2	44	–	216.55	194.7	–	–
Carbon tetrachloride	CCl_4	154	1632	–	–	–	–
Carbon disulphide	CS_2	76	1293	162.35	319.6	4.395	–
Carbon monoxide	CO	28	–	74.15	81.7	–	–
Water	H_2O	18	998	273.15	373.15	5.994	40.608

*The substance sublimes rather than boils and the melting temperature was obtained under pressure. †There is a large discrepancy (± 20 K) among the data from different sources.
(Data from Kaye & Laby and *CRC handbook*; see §1.4.1)

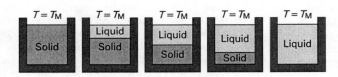

Figure 11.1 The five pictures illustrate the effect of applying heat energy to a substance in the solid state at the melting temperature. The heat serves only to transform more solid into liquid, but does not raise the temperature of the mixture. Note that the solid phase is, in general, more dense than the liquid and sinks in the mixture. This is not always the case: water, silicon, germanium, gallium and bismuth are examples where the liquid phase is denser than the solid phase (see Tables 9.1 & 9.2).

Table 11.2 Thermal data for the elements (the data refer to standard atmospheric pressure, unless otherwise stated).

Z	Name	Atomic weight	Density (kg m^{-3})	Melting point (K)	Boiling point (K)	Enthalpy of fusion (kJ mol^{-1})	Enthalpy of vaporization (kJ mol^{-1})
1	Hydrogen	1.008	89	14.01	20.28	0.12	0.46
2	Helium	4.003	120	0.95	4.216	0.021	0.082
3	Lithium	6.941	533	453.7	1620	4.6	134.7
4	Beryllium	9.012	1.846	1551	3243	9.8	308.8
5	Boron	10.81	2466	2365	3931	22.2	538.9
6	Carbon	12.01	2266	Sublimes at ≈ 3700*		105*	710.9*
7	Nitrogen	14.01	1035	63.15	77.4	0.72	5.577
8	Oxygen	16	1460	54.36	90.188	0.444	6.82
9	Fluorine	19	1140	53.48	85.01	5.1	6.548
10	Neon	20.18	1442	24.56	27.1	0.324	1.1736
11	Sodium	22.99	966	371	1156.1	2.64	89.04
12	Magnesium	24.31	1738	922	1363	9.04	128.7
13	Aluminium	26.98	2698	933.5	2740	10.67	293.72
14	Silicon	28.09	2329	1683	2628	39.6	383.3
15	Phosphorous	30.97	1820	317.3	553	2.51	51.9
16	Sulphur	32.06	2086	386	717.82	1.23	9.62
17	Chlorine	35.45	2030	172	239.18	6.41	20.403
18	Argon	39.95	1656	83.8	87.29	1.21	6.53
19	Potassium	39.1	862	336.8	1047	2.4	77.53
20	Calcium	40.08	1530	1112	1757	9.33	149.95
21	Scandium	44.96	2992	1814	3104	15.9	304.8
22	Titanium	47.9	4508	1933	3560	20.9	428.9
23	Vanadium	50.94	6090	2160	3650	17.6	458.6
24	Chromium	52	7194	2130	2945	15.3	348.78
25	Manganese	54.94	7473	1517	2235	14.4	219.7
26	Iron	55.85	7873	1808	3023	14.9	351
27	Cobalt	58.93	8800	1768	3143	15.2	382.4
28	Nickel	58.7	8907	1726	3005	17.6	371.8
29	Copper	63.55	8933	1356.6	2840	13	304.6
30	Zinc	65.38	7135	692.73	1180	6.67	115.3
31	Gallium	69.72	5905	302.93	3676	5.59	256.1
32	Germanium	72.59	5323	1210.6	3103	34.7	334.3
33	Arsenic	74.92	5776	Sublimes at 886*		27.7*	31.9*
34	Selenium	78.96	4808	490	958.1	5.1	26.32
35	Bromine	79.9	3120	265.9	331.93	10.8	30
36	Krypton	83.8	3000	116.6	120.85	1.64	9.05
37	Rubidium	85.47	1533	312.2	961	2.2	69.2
38	Strontium	87.62	2583	1042	1657	6.16	138.91
39	Yttrium	88.91	4475	1795	3611	17.2	393.3
40	Zirconium	91.22	6507	2125	4650	23	581.6
41	Niobium	92.91	8578	2741	5015	27.2	696.6
42	Molybdenum	95.94	10222	2890	4885	27.6	594.1
43	Technetium	97	11496	2445	5150	23.81	585.22
44	Ruthenium	101.1	12360	2583	4173	23.7	567.8
45	Rhodium	102.9	12420	2239	4000	21.55	495.4
46	Palladium	106.4	11995	1825	3413	17.2	393.3
47	Silver	107.9	10500	1235.1	2485	11.3	255.1
48	Cadmium	112.4	8647	594.1	1038	6.11	99.87
49	Indium	114.8	7290	429.32	2353	3.27	226.4
50	Tin	118.7	7285	505.12	2543	7.2	290.4
51	Antimony	121.7	6692	903.9	1908	20.9	67.91

(continued overleaf)

Z	Name	Atomic weight	Density (kg m^{-3})	Melting point (K)	Boiling point (K)	Enthalpy of fusion (kJ mol^{-1})	Enthalpy of vaporization (kJ mol^{-1})
52	Tellurium	127.6	6247	722.7	1263	13.5	50.63
53	Iodine	126.9	4953	386.7	457.5	15.27	41.67
54	Xenon	131.3	3560	161.3	166.1	3.1	12.65
55	Caesium	132.9	1900	301.6	951.6	2.09	65.9
56	Barium	137.3	3594	1002	1910	7.66	150.9
57	Lanthanum	138.9	6174	1194	3730	10.04	399.6
58	Cerium	140.1	6711	1072	3699	8.87	313.8
59	Praseodymium	140.9	6779	1204	3785	11.3	332.6
60	Neodymium	144.2	7000	1294	3341	7.113	283.7
61	Promethium	145	7220	1441	3000	12.6	–
62	Samarium	150.4	7536	1350	2064	10.9	191.6
63	Europium	152	5248	1095	1870	10.5	175.7
64	Gadolinium	157.2	7870	1586	3539	15.5	311.7
65	Terbium	158.9	8267	1629	3396	16.3	391
66	Dysprosium	162.5	8531	1685	2835	17.2	293
67	Holmium	164.9	8797	1747	2968	17.2	251
68	Erbium	167.3	9044	1802	3136	17.2	292.9
69	Thulium	168.9	9325	1818	2220	18.4	247
70	Ytterbium	173	6966	1097	1466	9.2	159
71	Lutetium	175	9842	1936	3668	19.2	428
72	Hafnium	178.5	13276	2503	5470	25.5	661.1
73	Tantalum	180.9	16670	3269	5698	31.4	753.1
74	Tungsten	183.9	19254	3680	5930	35.2	799.1
75	Rhenium	186.2	21023	3453	5900	33.1	707.1
76	Osmium	190.2	22580	3327	5300	29.3	627.6
77	Iridium	192.2	22550	2683	4403	26.4	563.6
78	Platinum	195.1	21450	2045	4100	19.7	510.5
79	Gold	197	19281	1337.6	3080	12.7	324.4
80	Mercury	200.6	13546	234.28	629.73	2.331	59.15
81	Thallium	204.4	11871	576.6	1730	4.31	162.1
82	Lead	207.2	11343	600.65	2013	5.121	179.4
83	Bismuth	209	9803	544.5	1833	10.48	179.1
84	Polonium	209	9400	527	1235	10	100.8
85	Astatine	210	–	575	610	23.8	–
86	Radon	222	4400	202	211.4	2.7	19.1
87	Francium	223	–	300	950	–	–
88	Radium	226	5000	973	1413	7.15	136.8
89	Actinium	227	10060	1320	3470	14.2	293
90	Thorium	232	11725	2023	5060	19.2	543.9
91	Protractinium	231	15370	2113	4300	16.7	481
92	Uranium	238	19050	1405	4018	15.5	422.6
93	Neptunium	237	20250	913	4175	9.46	336.6
94	Plutonium	244	19840	914	3505	2.8	343.5
95	Americium	243	13670	1267	2880	14.4	238.5

*Arsenic and carbon sublime when heated at atmospheric pressure (see §11.7). Their enthalpies of fusion and vaporization are deduced from studies at high pressure.
(DAya from Kaye & Laby and *CRC handbook*; see §1.4.1)

11.3 The solid↔liquid transition: melting and freezing

For most substances the transition from the solid to the liquid state occurs gradually, over a range of temperature. However, for *pure substances*, melting takes place at a well-defined temperature known as the *melting temperature*, T_M, or *melting point*. The definition of a pure substance is not as easy as it might at first appear. A working definition is: *a substance composed of a single type of atom or molecule*. But we note that this definition requires occasional qualification.

At the melting temperature, a substance can be either a solid or a liquid. As illustrated in Figure 11.1, if a sample is 50% solid and 50% liquid it will stay that way until further heat is added to, or removed from, the substance.

Heat input at the melting temperature does not raise the temperature of the solid/liquid mixture: it serves to transform more solid into liquid. The energy required to transform one mole of substance from solid to liquid at constant pressure is known by several equivalent terms – the molar *enthalpy change on fusion*, the molar *enthalpy of fusion*, the molar *latent heat of fusion* – all of which are equivalent. The preferred term is the *enthalpy change on fusion,* the term "fusion" being an old term for "melting".

11.3.1 Data on the solid↔liquid transition

The melting temperature T_M and molar enthalpy of fusion of the elements are recorded in Table 11.2 and plotted as a function of atomic number for the elements in Figures 11.2 & 11.3, respectively. Figures 11.2 & 11.3 show a qualitative similarity to one another, indicating that elements with a high melting temperature tend to have a high enthalpy of fusion. Furthermore, the structure in the data shows similar

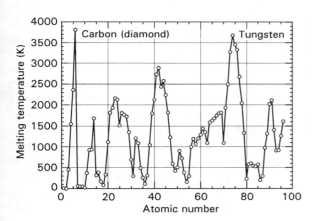

Figure 11.2 The melting temperatures of the elements plotted as function of atomic number. Notice that the pattern shows similarities to several other patterns such as those shown in Figures 7.1, 11.3, 11.5, 11.16, 11.17 & 11.27.

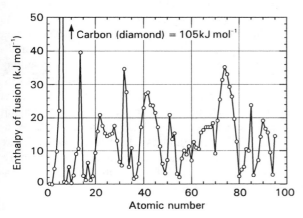

Figure 11.3 The molar enthalpy change of fusion of the elements plotted as function of atomic number. The datum for diamond is so much greater than any other element and has not been plotted in order to allow the detail on the rest of the graph to be seen.

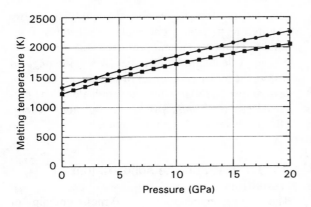

Figure 11.4 The elevation of the melting temperature of silver and gold as a function of the applied pressure. Note that the pressures are extremely large – 1 kbar is 1000 times atmospheric pressure and the pressure scale shown extends up to 200 kbar.

patterns to those seen in Figure 7.1 showing the density of elements.

Note that the temperature of the transition from the solid to the liquid phase is dependent on the pressure that the liquid/solid is subjected to. In general, higher pressures tend to increase the temperature at which liquid↔solid transitions take place. Interestingly, in substances that contract on melting higher pressures tend to *lower* the temperature at which the liquid↔solid transition takes place. Figure 11.4 shows data for the elements silver and gold. The depression of the melting temperature for water is discussed in §11.3.3.

The main questions raised by our preliminary examination of the experimental data on the solid↔liquid transition in the elements are:

- Why do elements with a high melting temperature tend to have a high enthalpy of fusion?
- Why does the variation in the melting temperature of the elements with atomic number show patterns similar to those seen for the density of elements?
- Why the melting temperature of most substances increases with pressure, but decreases for substances that contract on melting?

These questions will be considered in §11.3.3, after we have examined the Lindemann theory of melting.

11.3.2 Lindemann theory of melting

The solid→liquid transition (melting) is a considerably more complex process than the liquid→gas transition (boiling). In the liquid→gas transition,

essentially all interactions between molecules are eliminated after the phase change. Thus one can understand the transition in terms of the dramatic change in the environment of each molecule. However, as is clear from Chapters 8 & 9, in the solid→liquid transition the interactions and separations between molecules change by only a few per cent. Thus understanding what causes melting in anything more than a general sense is rather difficult. We can, however, still make a little headway on the problem.

First we can consider the process of melting in elemental substances only. Substances with more than one type of interatomic bond involve an extra degree of complexity without adding more insight into the problem. Following the ideas of Lindemann, we hypothesize that melting is related to the amplitude of vibration of the atoms within the solid. Thus for small amplitude vibrations the solid state would be stable, but for larger amplitude vibrations the structure of the substance would lose rigidity and "collapse" into the liquid state. Support for the idea that the level of atomic vibration is related to the melting temperature may be gleaned by considering Figure 11.5, which shows the way in which the melting temperature and Debye temperature vary among the elements. Recall (§7.4) that the Debye temperature of a substance is (roughly) the temperature at which the atoms gain access to all their degrees of freedom and begin to vibrate essentially independently of their neighbours. The correlations in Figure 11.5 represent strong evidence that, not surprisingly, there is a relationship between the melting temperature and level of atomic vibration.

Lindemann hypothesized that melting occurs

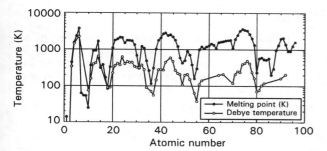

Figure 11.5 The melting and Debye temperatures of the elements plotted as function of atomic number. Note the correlations among the data. The data points are connected by straight lines intended to highlight trends in the data. Since data on the Debye temperature of all the elements is not available some straight lines appear to indicate a Debye temperature higher than the melting temperature (e.g. elements 7 to 10): This is an artefact of the plotting procedure. Note that the vertical axis is logarithmic.

when the amplitide of vibration A reaches a certain fraction, f, of the interatomic spacing, a. This idea seems reasonable, and we can test it fairly straight-forwardly.

We suppose that the atomic vibration is simple harmonic, so we neglect the asymmetry of the interatomic potential energy (see Fig. 7.5). The average energy of vibration of an atom in such a potential is $\bar{u} = k_B T$ for each direction of vibration, $\frac{1}{2}k_B T$ for the degree of freedom associated with the kinetic energy of vibration in that direction, and $\frac{1}{2}k_B T$ for the degree of freedom associated with the potential energy of vibration in that direction (see §2.5). This average energy is also related to the potential energy of vibration at the extreme point of the oscillation, e.g. for vibrations in the x-direction at $x = A$. We thus write

$$\overline{u_x} = k_B T = \tfrac{1}{2} K A^2 \tag{11.1}$$

where K is the "spring constant" of the harmonic potential. The Lindemann hypothesis is that melting occurs when the amplitude of vibration, A, is equal to a fraction, f, of the interatomic spacing, a. Our task is to estimate this fraction f and see if it is similar for different elements. Rearranging Equation 11.1 and substituting $A = fa$, we have

$$f^2 = \frac{2k_B T_M}{Ka^2} \tag{11.2}$$

Before we can estimate f we need to find expressions for K and a which we do as follows.

Recall that the frequency of vibration, ω, of an atom in a simple harmonic oscillator potential is given by

$$\omega = \sqrt{\frac{K}{m}} \tag{11.3}$$

where m is the mass of an atom, and hence we can write $K = m\omega^2$. Furthermore we note that the Debye temperature, θ_D, is the temperature at which the thermal energy of vibration, $k_B\theta_D$, is roughly equal to the energy separation between quantum states ($\hbar\omega$) of the simple harmonic oscillator potential. This is equivalent to using an Einstein approximation (see §7.4) and then estimating the Einstein temperature, θ_E as θ_D. This is likely to be accurate at around the 50% level. We thus write

$$K = m\omega^2 = m\left(\frac{k_B\theta_D}{\hbar}\right)^2 \tag{11.4}$$

We can make a rough estimate of a for the elements from their density, ρ, by imagining each atom of mass m to occupy a volume a^3. We can use the density data given in Table 7.1 to estimate a from

$$a = \left(\frac{m}{\rho}\right)^{1/3} \tag{11.5}$$

Figure 11.6 To test the Lindemann hypothesis we assume that each atom vibrates in a simple harmonic potential with amplitude A which is some fraction f of the interatomic spacing a. According to the equipartition theorem (see §2.4), the mean energy of vibration $<\bar{u}>$ for vibration in one direction has the value $k_B T$. This is also equal to the potential energy of vibration, at an extremum of vibration $\frac{1}{2}KA^2$.

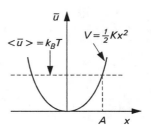

333

Substituting Equations 11.4 & 11.5 into Equation 11.2, we arrive at

$$f^2 = \frac{2k_B T_M}{m\left(\dfrac{k_B \theta_D}{\hbar}\right)^2 \left(\dfrac{m}{\rho}\right)^{2/3}} \qquad (11.6)$$

We can use Equation 11.6 to evaluate the fraction, f, for all the elements for which we have Debye temperature data (Table 7.9). The results of this calculation are plotted in Figure 11.7.

The Lindemann fraction, f, has a value of around one-twentieth of an interatomic spacing for a wide variety of different elements. The 25% variability in the data ($f = 0.052 \pm 0.014$ to one standard deviation) is extremely small when one considers that the data describe substances as different as argon ($T_M = 83.8\,$K: $f = 0.045$) and tungsten ($T_M = 3680\,$K: $f = 0.044$). Since we have completely ignored the crystal structure of the substances, approximations such as $\theta_D \approx \theta_E$ in Equation 11.4 are relatively small contributions to the overall uncertainty in the estimate for f. From our perspective, it is the *constancy* of f rather than its particular value that gives support to Lindemann's hypothesis. Thus melting occurs when the amplitude of atomic vibration reaches a fraction of roughly one-twentieth of a lattice spacing. It is surprising – to me at least – that vibrations with this

---

Example 11.3

Figure 11.7 shows the results of a calculation of the Lindemann fraction, f, for all the elements, prepared by using a spreadsheet. In this example we consider one element, copper (element 26), and perform the calculation of f by hand.

We have $k_B = 1.38 \times 10^{-23}\,J\,$K^{-1} , $\hbar = 1.034 \times 10^{-34}\,$Js, $N_A = 6.02 \times 10^{23}\,$mol^{-1}

and

$T_M = 1356.6\,$K	(Table 11.2)
$m = 63.55 \times 10^{-3}\,kg/N_A = 1.056 \times 10^{-25}\,$kg	(Table 7.2)
$\rho = 8933\,$kg$\,$m^{-3}	(Table 7.2)
$\theta_D = 343\,$K	(Table 7.9)

We thus estimate f as

$$f^2 = \frac{2 \times 1.38 \times 10^{-23} \times 1356.6}{1.056 \times 10^{-25}\left(\dfrac{1.38 \times 10^{-23} \times 343}{1.054 \times 10^{-34}}\right)^2 \left(\dfrac{1.056 \times 10^{-25}}{8933}\right)^{\frac{2}{3}}}$$

which evaluates to $f^2 = 3.388 \times 10^{-3}$ and $f = 0.0582$. This lies within the upper one standard deviation limit about the average value indicated in Figure 11.7.

---

small amplitude are sufficient to destabilize a crystal structure sufficiently that it melts.

11.3.3 Understanding the solid↔liquid transition

Having established in general microscopic terms what happens when a substance melts, let us return to the questions raised by the data in §11.3.1 on the solid↔liquid transition. The questions are considered in turn below.

High enthalpies of fusion

Why do elements with a high melting temperature tend to have a high enthalpy of fusion? This question is similar to the first question in the next section on the liquid↔gas transition: why do elements with a high boiling temperature tend to have a high enthalpy of vaporisation? So let us proceed by first reviewing the data for the elements, and then consider both of these questions together, first in a simple way, and then using the more sophisticated Gibbs free energy analysis.

Figure 11.7 The Lindemann fraction, f, calculated according Equation 11.6 for the elements. Notice that the fraction is around one-twentieth of a lattice spacing and the data in the graph have a mean value of 0.052 ± 0.014 at one standard deviation. By comparison with Figure 11.2, this variability is extremely small.

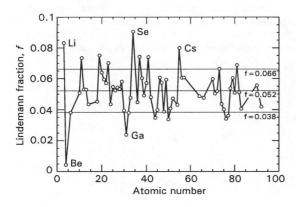

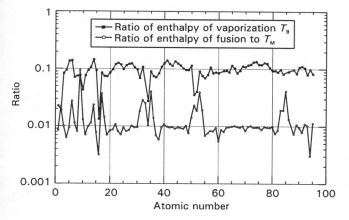

Figure 11.8 The relationship among the elements between the enthalpy of transformation (fusion or vaporization) and the temperature at which the transformation takes place. The figure shows that the ratio of the enthalpy of fusion to the melting temperature has a relatively constant value of around 0.01 in the units chosen and the ratio of the enthalpy of transformation to the boiling temperature has a relatively constant value of around 0.1 in the units chosen. Note that the 0.01 and 0.1 figures have no absolute significance and are only the result of a felicitous choice of units.

Figure 11.8 combines data from different columns in Table 11.2 and shows that, roughly speaking, and with some striking exceptions:
- the melting temperature (in kelvin) is around 1% of the enthalpy change on melting (in kilojoule per mole); and
- the boiling temperature (in kelvin) is around 10% of the enthalpy change on vaporization (in kilojoule per mole).

First, we can then say that these correlations are exactly what one would expect. Using Lindemann theory (see §11.3.2) we see that melting occurs when the amplitude of vibration of the atoms reaches around one-twentieth of an interatomic spacing. If the bonds between atoms are very strong, then (a) by definition a great deal of energy will be required to break the bonds and (b) a great deal of energy will be required to stretch or bend the bonds. These two facts lead directly to correlations between the enthalpy change on melting and, through the Lindemann criterion, the melting temperature.

Similarly with boiling: if the bonds between

atoms are very strong, then (a) by definition a great deal of energy will be required to break the bonds, and (b) the temperature at which atoms have sufficient energy ($\approx 0.5k_B T$ per degree of freedom) to break the bonds will be very high. These two facts lead again to correlations between the enthalpy change on vaporization and the boiling temperature.

We can examine the matter further by considering the general form of the variation of the Gibbs free energy, G, of a substance with temperature, T. There is no simple approximation to the form of the free energy curves but, as outlined in §10.2 and Figure 10.4, the solid and liquid free energy curves both have negative slopes and curve slightly downwards.

The general effect of a large cohesive energy U_0 is to give rise to a large *difference* between the cohesive energy of the liquid and solid states at $T = 0\,\text{K}$. Recall this difference is typically a fraction ($\approx 5\%$) of the total cohesive energy of the solid. There are two important consequences of a large difference between the cohesive energies of the solid and liquid states:
- the point at which the liquid and solid free ener-

(a)

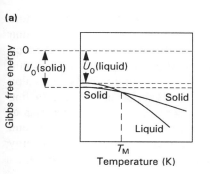

(b)

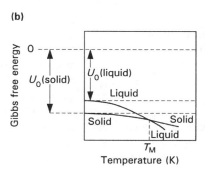

Figure 11.9 Analysis of the correlation between the melting temperature and the enthalpy change on transformation. U_0 represents the cohesive energy of either the solid or liquid states at 0 K. The figures represent the variation in the Gibbs free energy of (a) a substance with a small cohesive energy and (b) a substance with a large cohesive energy.

335

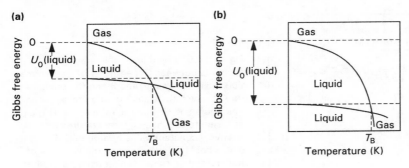

Figure 11.10 Analysis of the correlation between the boiling temperature and the enthalpy change on vaporization. U_0 represents the cohesive energy of the liquid state at 0 K. The figures represent the variation in the Gibbs free energy of (a) a substance with a small cohesive energy and (b) a substance with a large cohesive energy.

gies become equal – by definition, the melting temperature – is displaced to a higher temperature;

- The difference in slopes of the free energy at the transition temperature is the *entropy* change on transformation. If this entropy (= $\Delta Q_M/T_M$) is supplied at higher temperatures, then the heat supplied ΔQ_M – the enthalpy change on melting – must also be greater.

Thus both a high melting temperature and a large latent heat of melting are correlated with a large cohesive energy. Similar arguments lead to correlations between the cohesive energy, the boiling temperature and the latent heat of vaporization (Fig. 11.10).

Variation of T_M of the elements with atomic number

The second question raised by our analysis of the solid↔liquid transition is: Why does the variation in the melting temperature of the elements with atomic number show patterns similar to those seen for the density of elements (Fig. 7.1)? The origin of this correlation, as with that between the melting temperature and the enthalpy of fusion, has its root in the cohesive energy of the substance. This implies – not wholly surprisingly – that substances with a high cohesive energy tend to be particularly dense. This matter is considered further in §11.6.2.

Increase of T_M with pressure

Finally, we come to the question of why the melting temperature of most substances increases with pressure, but decreases for substances that contract on

melting. We will approach this question qualitatively, and then see how our considerations can be made more rigorous. In §9.2 we explained the origin of the expansion on melting as being due to the inefficiency with which molecules are packed together in the liquid state. The extra "empty space" inside a liquid accounts for the fact that the liquid state is typically 33% easier to compress than the solid state (see §9.4); we assumed that the empty space within the liquid was relatively easy to "reclaim" when pressure was applied.

If this idea is correct, then reclaiming the "empty space" within a liquid must necessarily make the liquid more solid like. In particular, it must reduce the ease with which a molecule may "hop" from one "cell" within the liquid to another. So we will attempt to understand the pressure dependence of the melting temperature by suggesting that, in general, the application of pressure at a fixed temperature makes a liquid more solid like. If we imagine applying pressure at a temperature just above the melting temperature, then making the liquid more like a solid will tend to stabilize the solid phase. If the solid phase is stabilized at a temperature at which the liquid phase used to be stable, then we have, by definition, raised the melting temperature.

This idea may be tested fairly straightforwardly: if it is more difficult for a molecule to "hop" from one "cell" to another "cell", then a liquid should become more viscous as pressure is applied. Data on the change in viscosity of a liquid under pressure are shown in Figure 11.11. The data indicate that for normal liquids the viscosity increases under applied pressure as expected. However, for water (one of the few liquids that is more dense than the corresponding solid) the reverse process can happen. For water at a

(a)

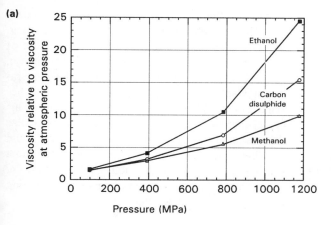

(b)

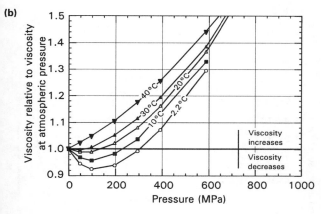

Figure 11.11 Change in viscosity of (a) various liquids and (b) water under applied pressure. The data in (a) were taken at 30 °C and show the change in viscosity of each substance relative to the viscosity of the same substance at zero pressure. The data in (b) were taken at various temperatures and show the change in viscosity of water relative to the viscosity at the same temperature and zero pressure. For temperatures below ≈30 °C the effect of pressure is first to cause a decrease in viscosity, in contrast with the behaviour of "normal" liquids shown in (a). Note the very large pressures involved in the measurements in both (a) and (b) (100 MPa is approximately 1000 times atmospheric pressure).

temperature just above its melting temperature, the application of pressure first *decreases* the viscosity. Thus pressure makes a normal liquid more viscous (more solid like), but makes water just above 0°C less viscous (more liquid like).

We understood the normal liquid by assuming that there is "free space" in the liquid that is removed by pressure. For water, and other substances that contract on melting, it is the solid state that has the "free space" due to the open hydrogen-bonded crystal structure. We can understand the pressure data by supposing that there is already too much "free space" in the liquid state, and pressure has the effect of increasing the forces that allow molecules to move past one another.

Thus pressure makes most liquids more solid like, but makes liquids that contract on melting more liquid like. It is not surprising that the effect of pressure is to raise the melting temperature, but that for liquids which contract on melting the effect is to lower the melting temperature.

11.4 The liquid↔gas transition: boiling and condensing

Contrary to common experience, although the *boiling temperature*, or vaporization temperature, of a pure liquid substance is a well-defined temperature, *the process of transformation from liquid to gas occurs at all temperatures* (Fig. 11.12). At any temperature the liquid attempts to sustain above its surface a characteristic pressure of gas known as the *vapour pressure*. The vapour pressure increases strongly with increasing temperature (Fig. 11.13).

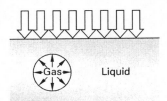

Figure 11.12 What happens when a liquid boils? Temperature fluctuations within the liquid cause small bubbles to form – initially only a few atomic diameters in size. Whether the bubble grows or shrinks depends on the balance between the pressure of the liquid around the bubble, and the vapour pressure within the bubble. If the temperature is such that the vapour pressure exceeds the external pressure then bubbles will form in the liquid and grow. Because of the density difference between the gas and the liquid, the bubbles rise within the liquid and "burst" when they reach the surface. Changes in the external pressure therefore cause changes in the boiling temperature of the liquid.

Thus the liquid and gas phases of a substance coexist at all temperatures at which the liquid is stable, and so a sensible choice is made as to the pressure of vapour that defines the transition to the gaseous phase. One defines the transition temperature as the temperature at which the vapour pressure of the liquid reaches standard atmospheric pressure (1.0135×10^5 Pa). This choice may be sensible, but it is important to realize that it is also arbitrary: another defining pressure could have been chosen. In contrast, this is not the case for a solid$\leftrightarrow$liquid phase change: solid and liquid do not coexist at all temperatures, even though the melting temperature does change slightly with pressure.

Figure 11.13 The vapour pressure of water as function of temperature plotted on (a) a linear scale and (b) a logarithmic scale. The boiling temperature is defined as the temperature at which the vapour pressure equals atmospheric pressure (0.101 35 MPa) which for water occurs at 99.975 °C.

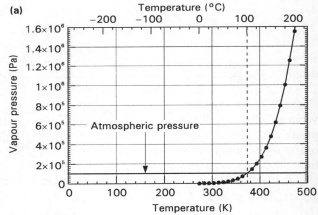

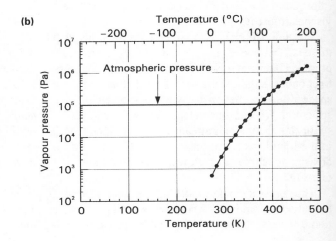

Example 11.4

Using Figure 11.13, estimate the temperature at which water would boil under an external pressures of (a) 0.95×10^5 Pa and (b) 1.1×10^5 Pa. These correspond to the very extremes of atmospheric pressure fluctuations at around sea level.

An enlargement of Figure 11.13a is shown here.

The points at which the curve of vapour pressure versus temperature intersect the various pressures of interest indicate that the boiling temperature at the highest conceivable atmospheric pressure is increased to around 102.3°C and that the boiling temperature at the lowest conceivable atmospheric pressure is decreased to around 98.2°C. Note that these figures are uncertain because the lines in the figure are just linear interpolations between the data points, being designed mainly to guide the eye. However, the uncertainty in the figures is probably less than ±0.2°C.

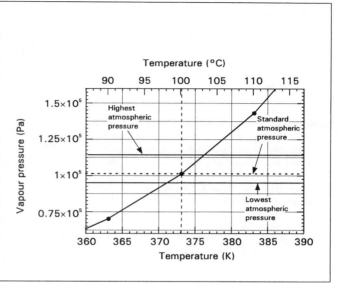

11.4.1 Data on the liquid→gas transition

The boiling temperatures of the elements and other substances are recorded in Tables 11.1 & 11.2 and data for the elements are plotted in Figure 11.13.

The vapour pressure data for all the substances (Figs 11.14 & 11.15) exhibit an extremely strong variation with temperature. For example, the vapour pressure of silver (Fig. 11.16) increases by three orders of magnitude as the temperature is increased by ≈ 50% from 1650K to 2450K. We note also that the vapour pressure curves are not quite straight lines, which indicates that on this logarithmic vertical scale the vapour pressure is nearly, but not quite, an exponential function of temperature.

Figures 11.16 & 11.17 shows the correlations with atomic number of the boiling temperatures and enthalpy changes on vaporization of the elements.

The main questions raised by our preliminary examination of the experimental data on the liquid→ gas transition are:

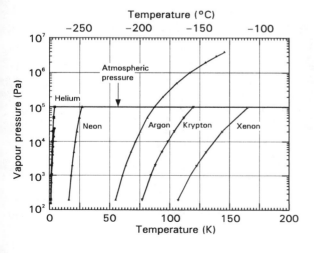

Figure 11.14 The vapour pressure of the rare gases as a function of temperature plotted on a logarithmic scale. The boiling temperature is defined as the temperature at which the vapour pressure equals atmospheric pressure (0.101 35 MPa), which for these substances occurs at: He, 4.22 K; Ne, 27.1 K; Ar, 87.3 K; Kr, 119.8 K; and Xe, 165.1 K. Note: for most of the indicated range, most of these elements are solid, and the vapour co-exists with the solid rather than the liquid state. This is discussed further in §11.6 on the solid→gas transition.

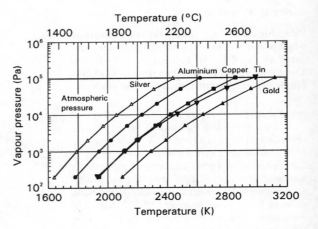

Figure 11.15 The vapour pressure of four metals as a function of temperature plotted on a logarithmic scale. The boiling temperature is defined as the temperature at which the vapour pressure equals atmospheric pressure (0.101 35 MPa), which for these substances occurs at: Al, 2623 K; Cu, 2853 K; Ag, 2433 K; and Au, 3123 K

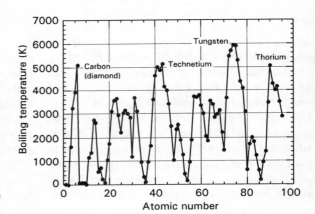

Figure 11.16 The boiling temperatures of the elements plotted as function of atomic number.

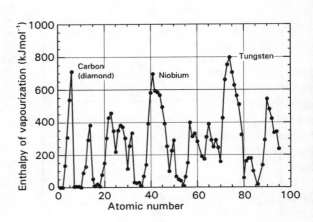

Figure 11.17 The enthalpy change on vaporization of the elements plotted as function of atomic number.

- Why do elements with a high boiling temperature tend to have a high enthalpy of vaporization?
- Why does the variation of the boiling temperature of the elements with atomic number show patterns similar to those seen for the density of elements?
- Why does the vapour pressure of a substance increase strongly with temperature?

The first two questions have already been addressed in §11.3.3 where they are considered along with similar questions regarding the solid↔liquid transition. The final question is addressed in §11.4.3.

11.4.2 Theory of boiling

Boiling is the phenomenon that occurs when the vapour pressure of a liquid exceeds the pressure above the surface of the liquid. Thus, in order to understand boiling, we need to develop an understanding of the origin of the vapour pressure of a liquid, and of the way in which the vapour pressure varies with temperature. Here we extend the cell model of liquids (see §8.3 & §9.6) to allow us to understand the strong variation in the vapour pressure with temperature.

We begin by reminding ourselves of the situation of a molecule in the cell model of a liquid (Fig. 8.19). The energy required to remove a molecule from the body of the liquid into the gaseous phase is ΔE_e. This activation energy leads to the rate at which molecules leave the liquid being proportional to a Boltzmann factor (see §2.5):

Rate at which molecules leave liquid $\propto$

$$\exp(-\Delta E_e/k_B T) \qquad (11.7)$$

This factor will prove to be the origin of the strong temperature dependence of vapour pressure. However, in order to calculate the vapour pressure above the liquid we need to derive an expression for the rate at which molecules *leave* the surface, and also the rate at which they *return* to the liquid from the gas phase.

The rate at which molecules leave the surface
Consider a molecule just below the surface of a liquid

vibrating within its cell with frequency $f (\approx 10^{13}\,\text{Hz})$. Thus f times per second the molecule "hits" the wall of its cell in an appropriate direction to leave the liquid. If the molecular energy is ΔE_e it will succeed in leaving the liquid, and the probability that its energy exceeds ΔE_e is approximately $A\exp(-\Delta E_e/k_B T)$, where A is a temperature-independent constant. If the cell has dimensions of roughly $a \times a \times a$, then the cross-section of the cell area perpendicular to the surface is given by a^2. We thus expect that the number of molecules leaving the surface per unit area per unit time is

$$N_{\text{leaving}} = \frac{fA\exp(-\Delta E_e/k_B T)}{a^2} \qquad (11.8)$$

The rate at which molecules return to the liquid
The number of molecules returning to the liquid from the gas per second is rather easier to estimate. In §4.4.3 we saw that the number of gas molecules crossing unit area per second is given by $\frac{1}{4}n_g\bar{v}$, where n_g is the number density of molecules in the gas and $\bar{v}$ is the average speed of a gas molecule. If we assume that all the molecules striking the surface of the liquid return to the liquid and do not rejoin the gas, then we estimate the return rate of molecules rejoining the liquid as

$$N_{\text{return}} = \frac{1}{4}n_g\bar{v}_g \qquad (11.9)$$

The equilibrium vapour pressure
Now, in equilibrium, the rate at which molecules leave the liquid must be the same as that at which molecules return to it. Thus in equilibrium Equations 11.8 & 11.9 will be equal, and so we expect

$$\frac{1}{4}n_g\bar{v}_g = \frac{fA\exp(-\Delta E_e/k_B T)}{a^2} \qquad (11.10)$$

or

$$n_g = \frac{4fA\exp(-\Delta E_e/k_B T)}{\bar{v}_g a^2} \qquad (11.11)$$

This equation can be simplified in two stages. First, we substitute for the number density of molecules in the liquid (approximately $n_l = 1/a^3$) to give

$$n_g = \frac{4n_l f a A \exp(-\Delta E_e / k_B T)}{\bar{v}_g}$$

$$= 4n_l A \exp(-\Delta E_e / k_B T)\left(\frac{fa}{\bar{v}_g}\right) \qquad (11.12)$$

We now consider the fraction isolated on the right-hand side of Equation 11.12. The quantity fa is approximately equal to half the average speed of the molecules within the cell: f times a second they travel a distance $2a$ back and forth across the cell. Thus in each second they travel a distance $2fa$. Substituting $\bar{v}_l = 2fa$ we find

$$n_g = 2n_l A \exp(-\Delta E_e / k_B T)\left(\frac{\bar{v}_l}{\bar{v}_g}\right). \qquad (11.13)$$

We can estimate that the average molecular speeds

within the liquid and gas at equal temperature are likely to be rather similar if not exactly equal. Assuming that the fraction is close to unity we can proceed to estimate the vapour pressure in terms of n_g. Using the perfect gas equation we estimate that the molecular number density is given by

$$n_g = \frac{P N_A}{RT} \qquad (11.14)$$

and hence we estimate the vapour pressure above a liquid as

$$P \approx \frac{2RT n_l}{N_A} A \exp(-\Delta E_e / k_B T) \qquad (11.15)$$

If Equation 11.15 is a fair representation of the variation of the vapour pressure above a liquid in equilibrium, then a plot of $\ln(P/T)$ versus $1/T$ should yield a straight line with slope $-\Delta E_e / k_B T$:

Figure 11.18 The vapour pressure data of (a) Figure 11.14 and (b) Figure 11.15 replotted as P/T versus $1/T$ to test the theoretical prediction of Equation 11.15. The vertical axis of both graphs is logarithmic and so the linearity of the data indicates good agreement between theory and experiment. The lines on the graph represent least-squares fits to the data; detailed analysis of the experimental slopes are contained in Table 11.3.

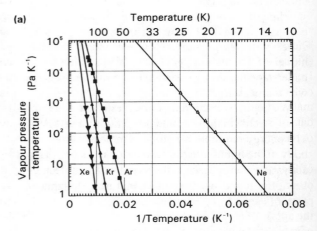

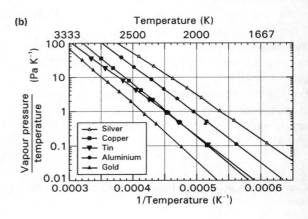

Table 11.3 Analysis of the slopes of the curves in Figure 11.18 in terms of Equation 11.16.

Substance	Slope (K)	ΔE_e (J)	ΔE_e (eV)	$N_A \Delta E_e$ (kJ mol^{-1})	L (kJ mol^{-1})	$N_A \Delta E_e / L$
Copper	−35 220	486×10^{-21}	3.03	292.7	300.5	0.97
Silver	−29 191	403×10^{-21}	2.52	242.7	255.06	0.95
Gold	−37 355	516×10^{-21}	3.22	310.7	324.43	0.96
Aluminium	−32 391	447×10^{-21}	2.79	269.2	290.8	0.92
Tin	−31 616	436×10^{-21}	2.72	262.6	290.37	0.90
Helium	−9.51	0.13×10^{-21}	0.00082	0.078	0.08	0.98
Neon	−234	3.23×10^{-21}	0.020	1.95	1.77	1.10
Argon	−781	10.8×10^{-21}	0.067	6.50	6.52	0.99
Krypton	−1247	17.2×10^{-21}	0.107	10.36	9.03	1.15
Xenon	−1767	24.4×10^{-21}	0.152	14.69	12.64	1.16

$N_A \Delta E_e$, predicted value of the latent heat of vaporization; L, experimental value of the latent heat of vaporization.
(Data from Kaye & Laby; see §1.4.1)

$$\ln\left(\frac{P}{T}\right) \approx \underbrace{\ln\left(\frac{2Rn_1}{N_A}A\right)}_{\text{Intercept}} - \underbrace{\left(\frac{\Delta E\rho}{k_B}\right)\frac{1}{T}}_{\text{Slope}} \quad (11.16)$$

The data from §11.4.1 when replotted in this manner (Figure 11.18) display a strikingly linear behaviour. If we take this linearity as evidence in favour of the cell model, then evaluating the slopes allows estimates of ΔE_e, which are recorded in Table 11.3. The data of Table 11.3 give us several opportunities to test our assumptions about the process of evaporation.

First the values of ΔE_e evaluated from the slopes of Figure 11.18 may be considered in relation to the other cell model parameters ΔE_h and ΔE_s. This comparison was performed in §9.6 where we evaluated the applicability of the cell model to real liquids. The conclusion there was that, with reservations, the cell model can help in understanding liquids.

Secondly, we can use the value of ΔE_e calculated from the slopes in Figure 11.19 to predict a value for the latent heat of vaporization – the amount of heat that must be supplied to convert one mole of substance from the liquid to the gaseous state at the boiling temperature. The expected value is just the number of molecules in one mole, N_A, times ΔE_e, which is effectively the "latent heat per molecule". The results of this calculation are listed and compared with the experimental values (Tables 11.1 & 11.2) in Table 11.3. We see that at the level of around 10%, we can predict the latent heat of vaporization of substances with greatly different boiling temperatures. This gives us confidence that the theory which underlined our calculations is applicable to real liquids, and that ΔE_e has been estimated realistically.

11.4.3 The liquid↔gas transition

We can now return to the final unaddressed question raised by our preliminary examination of the data in §11.4.1: why does the vapour pressure of a substance increase strongly with temperature? We have seen

Figure 11.19 The approach to the critical point. The five pictures illustrate the effect of the application of heat energy to a substance in the liquid state in a closed container. As the temperature rises the density of the vapour increases exponentially with temperature. When the density of the vapour pressure approaches

that of the liquid state, a reduction in volume may take the system to a situation where the density of the vapour and the density of the liquid are equal – a situation known as the critical point.

that we can understand this microscopically in terms of the cell model theory outlined in the previous section. Alternatively, we can understand it in a more general sense by using the Clausius–Clapeyron equation as discussed in §10.7.1.

11.5 The critical temperature

The fact that the liquid and gaseous phases of a substance coexist over a wide range of temperature is a clue to the fact that the two phases are more closely related than are the liquid and solid phases. Evidence for this close relationship is strengthened by the fact that, above a certain temperature, known as the *critical temperature*, T_C, there is no distinct transition from liquid to gas (Fig. 11.19). A *critical pressure* and *critical volume* can also be defined as the pressure and volume at which the densities of the *coexisting* gas and liquid phases are equal.

11.5.1 Data on the critical temperature
The critical parameters of various substances are listed in Table 11.4. Figure 11.20 represents an attempt to find structure in the critical parameter data from Table 11.4. The data are plotted on a logarithmic scale and so a good deal of the variation among

Figure 11.20 The critical parameters (see Table 11.4) for various substances discussed in Chapters 6 and 8 plotted as a function of their relative molecular mass. Note that the vertical axis is logarithmic, which compresses much of the actual variation among the data.

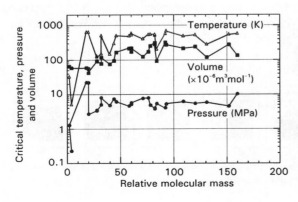

Example 11.5

Work out the volume occupied per molecule when a substance is in the critical state.

Considering ethanol in Table 11.4, we see that it has a critical volume of $167 \times 10^{-6}\,\mathrm{m^3\,mol^{-1}}$, i.e. this volume contains Avogadro's number of molecules. Thus the volume per molecule in the critical state is

$$\text{Volume per molecule} = \frac{167 \times 10^{-6}}{6.02 \times 10^{23}}$$
$$= 2.77 \times 10^{-28}\,\mathrm{m^3}$$

If we imagine each molecule confined to a cube of volume $a \times a \times a$ then the cube would have side

$$a \approx \sqrt[3]{2.77 \times 10^{-28}} = 6.5 \times 10^{-10}\,\mathrm{m}$$

This volume would contain one ethanol molecule, C_2H_5OH, i.e. two carbon atoms, six hydrogen atoms and one oxygen atom. Considering that a typical atom has a diameter of a few $\times 10^{-10}$ m, and that the bonds between atoms are separated by a similar amount, this leaves little free space in between molecules.

---

the data is not clearly shown. However, it is clear that, aside from the smallest molecular masses, the critical parameters are roughly independent of molecular mass. We can note that there are, not surprisingly, significant correlations between the critical temperature, pressure and volume for a particular substance.

Roughly speaking we find that ballpark values of the critical parameters in Table 11.4 are:
- $T_C \approx 500\ \mathrm{K} \approx 200\ °\mathrm{C}$
- $P_C \approx 5\ \mathrm{MPa} \approx 50\ \mathrm{atm}$, and
- $V_C \approx 200 \times 10^{-6}\ \mathrm{m^3\ mol^{-1}}$, which is approximately equivalent to a sphere ≈ 7 cm in diameter per mole.

Table 11.4 also indicates that the density of most substances at the critical point is close to one-third the density at normal pressures and temperatures. Such figures indicate that in the critical condition there is little "free space" around each molecule as compared to a gas. However, the density ratio indicates that at the critical point roughly two-thirds of the volume of the substance is free space (at least as compared with the liquid state at normal temperatures and pressures).

Table 11.4 The critical parameters of various substances.

Substance	Chemical formula	MW	P_C (MPa)	V_C ($\times 10^{-6}\,m^3\,mol^{-1}$)	T_C (K)	Density at V_C* ($kg\,m^{-3}$)	Density of liquid[†] ($kg\,m^{-3}$)	Density ratio[‡]
Organic substances								
Methanol	CH_3OH	32	8.09	118	512.6	271	791	0.343
Ethanol	C_2H_5OH	46	6.14	167	513.9	275	789	0.349
Propan-1-ol	C_3H_7OH	60	5.17	219	536.8	274	804	0.340
Acetic acid	$C_2H_4O_2$	60	5.79	171	594.5	351	1049	0.334
Acetone	C_3H_6O	58	4.7	213	508.1	272	787	0.346
Aniline	C_6H_7N	93	5.3	274	698.9	339	1026	0.330
Benzene	C_6H_6	78	4.9	254	562.2	307	879	0.349
Bromoethane	C_2H_5Br	109	6.23	215	503.8	507	1456	0.348
Chloroform	$CHCl_3$	120	5.5	240	536.4	500	1498	0.333
Cyclohexane	C_6H_{10}	82	4.02	308	553.4	266	941.6	0.282
Ethyl acetate	$C_4H_8O_2$	82	3.83	286	523.2	287	900.6	0.319
Toluene	C_7H_8	92	4.11	320	591.8	288	868.8	0.331
Inorganic substances								
Carbon monoxide	CO	28	3.50	93.1	133	300.75	–	–
Carbon dioxide	CO_2	44	7.38	94.0	304.2	468.09	–	–
Carbon disulphide	CS_2	76	7.9	173	552	439.31	1263	0.348
Carbon tetrachloride	CCl_4	152	4.56	276	556.4	550.72	1604	0.343
Hydrogen	H_2	2	1.294	65.5	32.99	30.534	89	0.343
Nitrogen	N_2	28	3.39	90.1	126.2	310.77	1035	0.300
Oxygen	O_2	32	5.08	78	154.8	410.26	1460	0.281
Chlorine	Cl_2	71	7.71	124	417	572.58	2030	0.282
Bromine	Br_2	160	10.3	135	584	1185.2	3120	0.380
Helium	He	4	0.229	58	5.2	68.966	120	0.575
Neon	Ne	20	2.73	41.7	44.4	479.62	1442	0.333
Argon	Ar	40	4.86	75.2	150.7	531.91	1656	0.321
Krypton	Kr	84	5.50	92.3	209.4	910.08	3000	0.303
Xenon	Xe	131	5.88	119	289.7	1100.8	3560	0.309
Radon	Rn	222	6.3	–	377	–	4400	
Water	H_2O	18	22.12	59.1	647.3	304.57	1000	0.305
Heavy water	D_2O	20	21.88	54.9	644.2	364.30	1100	0.331

MW, relative molecular mass; P_C, critical pressure; V_C, critical molar volume; T_C, critical temperature. *The density at the critical point, calculated from the molecular mass and V_C. This may be compared with the density of the substance in the state well away from T_C. [†]For the inorganic substances where the liquid density data are not available, the solid density has been used instead. [‡]The ratio of the density at the critical point to that at a temperature well below the critical point.
(Data from Kaye & Laby; see §1.4.1)

The main questions raised by our preliminary examination of the experimental data on the critical temperature of substances are:

- Why do the critical temperatures, volumes and pressures have broadly similar values for substances with a wide range of molecular masses?
- Why is the density at the critical point roughly one-third of the density at normal pressure and temperature?

11.5.2 Understanding the data on the critical temperature

To understand what is happening as a substance approaches its critical condition we may chart the development of the *radial density function* (§8.2.2). This is illustrated qualitatively in Figure 11.21, which is adapted from Figure 8.6. Figure 11.21 illustrates the continuous evolution of the liquid state.

Just above the melting temperature, a great deal of

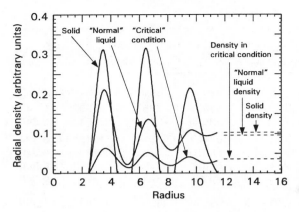

Figure 11.21 Qualitative indication of the radial density function of a solid, a normal liquid and a liquid in the critical condition. Note that in the solid state, the peaks correspond to nearest neighbours, next-nearest neighbours, etc.; in the liquid state, the peaks are maintained but are smoothed by the increased disorder of the liquid state; in the critical state, the peaks are weaker still, and the overall density is much reduced in comparison with either the liquid or the solid state.

the short range order of the solid still remains. As the temperature is raised the density of the liquid state falls continuously (Figures 9.3 & 9.4) and the short range order is systematically weakened. Finally, at the critical condition, the overall density of the liquid state is around 34% of its "normal" value, and little of the short range order remains.

We have already seen evidence of this evolution of the liquid state (see Figures 9.18 & 9.19, §9.5.3) where we looked at data on the surface tension of liquids. There we saw that the surface tension of all liquids declines roughly linearly with temperature. In

Figure 11.22 we re-examine the data on water shown in Figure 9.19. We can see clearly that the data, when extrapolated, indicate that water will have zero surface tension (i.e. be indistinguishable from a dense gas) at a temperature just above the critical temperature. Given the crude extrapolation, and the unusual properties of water, we see that even between 0°C and 100°C, the structure of water is already evolving in a way destined for "completion" at T_C.

Returning to direct consideration of the questions raised in the previous section, we see that the key linking feature of the critical point data is that the density at the critical point is around one-third of the normal liquid density. Thus we can understand the relative uniformity of critical molar volumes merely by noting that the normal liquid densities of the substances in Table 11.4 vary by a factor of only ≈3, which is considerably compressed on the logarithmic scale used in Figure 11.20.

Figure 11.22 Data on the surface tension of water (replotted from Figure 9.19 on a larger scale). The linear extrapolation is based on a linear least-squares fit to the data between 0°C and 100°C. The data imply that the surface tension will reach zero at a temperature just above the actual critical temperature (T_C = 647 K).

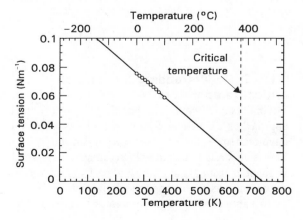

The virial approach to the critical condition

- $T_C \approx 500\,K \approx 200\,°C$
- $P_C \approx 5\,MPa \approx 50\,atm$
- $V_C \approx 200 \times 10^{-6}\,m^3\,mol^{-1}$ which is approximately equivalent to a sphere ≈ 7 cm in diameter per mole

If we applied the perfect gas equation to a mole of substance at its critical temperature, the pressure and volume are related by

$$\frac{P_C V_C}{R T_C} = 1 \qquad (11.17)$$

However in these extreme conditions, the size and interaction of the molecules must be considered, and the perfect gas equation does not apply. However the quantity $Z = PV/RT$ generally known as the *virial of Clausius* or the *compression factor* of a gas does have a near universal value at T_C

$$Z_C = \frac{P_C V_C}{RT_C} \approx 0.297 \qquad (11.18)$$

with experimental values varying between ≈ 0.26 and 0.32.

The question remains of *why* the density at the critical point is roughly one-third the normal liquid density. The value of one-third is just what it turns out to be! One can see that the value does indeed correspond to an extremely dense gas, as one would expect. The question of why the value – whatever it might have turned out to be – is "universal" is a question which is considered in the following section.

11.6 Scaling: laws of corresponding states

In previous sections we have noted several apparently universal correlations among the properties of diverse substances. For example:

(a) In §11.3 we noted that substances with high boiling temperatures also tend to have high melting temperatures.

(b) In §11.3.3, Figure 11.8 we noted that:
- the ratio of the enthalpy of fusion of an element to its melting temperature has a relatively constant value of around $0.01\,\mathrm{J\,K^{-1}\,mol^{-1}}$; and
- the ratio of the enthalpy of transformation of an element to its boiling temperature has a relatively constant value of around $0.1\,\mathrm{J\,K^{-1}\,mol^{-1}}$.

(c) In §11.3.1, we noted that the Lindemann fraction, which determines the temperature at which an elemental substance melts, had a common value of roughly one-twentieth.

(d) In §11.5.2 , Table 11.4, we noted that the densities of substances at the critical point are all roughly one third of their value in the liquid phase.

(e) In §11.5.2 Equation 11.8, we noted that the value of $P_C V_C/RT_C$ is around 0.29 ± 0.03 for a wide range of substances.

(f) In §9.6 we concluded that the *cell model* could plausibly describe the dynamic properties of a wide variety of liquids with just four "free parameters".

The observations above, and many others that you can find throughout the book, raise the hope that it might be possible to find a model of a "perfect substance" analogous to our model of a "perfect gas". In fact the properties of substances, particularly solids, are too diverse to make this a worthwhile endeavour.

However, many of the correlations between substances are particularly striking: Example 11.6 predicts the critical temperature of water based only on its boiling temperature and the properties of liquid krypton! It is clear that there must be something

---

Example 11.6

The critical, boiling and melting temperatures for krypton are:

T_C (K)	T_B (K)	T_M (K)
209.4	120.85	116.6

Does knowing these figures for krypton allow us to predict the equivalent quantities for the other rare gas solids, knowing only their boiling temperature? Let's normalize the data to the boiling temperature of the substance. The above data now look like

T_C(K)	T_B(K)	T_M(K)
1.73	1.00	0.965

If the simplest form of scaling theory is correct, then given the boiling temperature of other substances we should find that their melting temperature is around 96% of T_B and their critical temperature is around 73% higher than T_B. Let's see if it works:

	T_C (K)	T_B (K)	T_M (K)
Expectation for argon	151.0	87.3	84.2
Data for argon	150.7	87.3	83.8
Expectation for neon	46.9	27.1	26.1
Data for neon	44.4	27.1	24.56

The agreement for argon is rather good, while that for neon is not so good. But before we accept the scaling idea wholesale, let's try it on a couple of other rather different substances.

	T_C (K)	T_B (K)	T_M (K)
Expectation for water	645.6	373.2	360.1
Data for water	647.3	373.2	273.2
Expectation for potassium	–	1047	1010
Data for potassium	–	1047	337

physically significant underlying the observed universality, and if we understand what it is, then surely we must be on the verge of understanding something about everything. In practice, one arrives at a great many "rules of thumb", each applicable in most cases but with several exceptions. These rules tell one something approximate about several substances, but nothing definite about any particular substance.

Considering potassium and water in Example 11.6, we note that for these substances the melting temperature is poorly predicted from the boiling temperature. In the following sections we examine the data to see to what extent we may trust a simple scaling approach to all substances.

11.6.1 Data on melting, boiling and critical temperatures

In Figure 11.23 we compare the critical temperature, T_C, and the melting temperature, T_M, with the boiling temperatures, T_B, of various substances. As expected from Example 11.6, the figure shows that the critical temperature is strongly correlated with the boiling temperature of the liquid and is typically a factor of ≈ 1.6 higher. Recall from Example 11.6 that T_C for argon was a factor 1.73 higher than T_B. Figure 11.24 shows the equivalent correlation between T_M and T_B.

So the main question raised by this preliminary examination of the experimental data on the melting, boiling, and critical temperatures of substances is simply, what is the origin of the common relationship between T_M, T_B and T_C observed in a wide variety of substances?

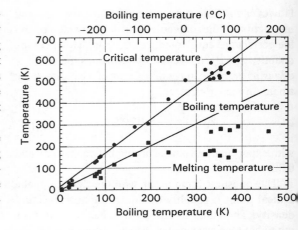

Figure 11.23 The critical temperature (Table 11.4) and melting temperature (Table 11.1) plotted as a function of boiling temperature for various liquids. The boiling temperature line is a line of slope = 1. The vertical separation between the melting point datum and the boiling temperature line is an indication of the temperature range over which the substance exists as a liquid. The cluster of substances with boiling temperatures between 300 K and 400 K are all organic substances.

11.6.2 Understanding the origin of corresponding states

At high temperatures all substances enter the gaseous phase and, as we have seen, obey the perfect gas equation reasonably well, at least at low density. The reason for this is that the approximations underlying the derivation of $PV = zRT$ are well satisfied by real gases.

Figure 11.24 The melting temperature of the elemental metals plotted as a function of their boiling temperature. The vertical separation between the melting point datum and the boiling temperature line is an indication of the temperature range over which the substance exists as a liquid. Roughly, the melting temperatures are around two-thirds of the boiling temperature, but there is a good deal of fluctuation around this figure. Some metals appear to have an anomalously large range of existence in the liquid phase. Gallium, for example, melts at only 302.9 K, but does not boil until 3676 K, a factor of 10 difference.

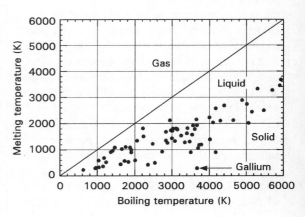

As a substance is cooled in the gas phase its density increases and, eventually, it reaches the point at which further cooling will cause it to enter one of the condensed states. The temperature and density at which this occurs depends on the strength of interaction between the molecules of the gas. Recall that all molecules are naturally "sticky": when molecules collide they stick together unless the kinetic energy of the molecules is high enough to prevent it. As we saw in §4.4.2, the average speed of the Maxwell distribution of molecular speeds falls as the temperature falls. Eventually, for any substance the average kinetic energy of the molecules is such that, on collision, the molecules stick together. The temperature at which this happens will be related to the stickiness (binding energy) of the molecules. Thus the boiling temperature is related directly to the binding energy of molecules – and hence to the *cohesive energy* of the substance. It is this phenomenon that links the characteristic temperatures for a substance to the binding energy per molecule. Ignoring factors of the order of "a few", we write

$$u_0 \approx k_B T_B \qquad (11.19)$$

or, using cohesive energies per mole,

$$U_0 \approx x N_A k_B T_B \approx x R T_B \qquad (11.20)$$

where x is the number of bonds (roughly speaking nearest neighbours) with which a molecule interacts in the condensed state. Of course there are many other factors to be taken into account, for example, the shape of complex molecules. However, for reasons of simplicity we stick with Equation 11.20.

Table 11.5 contains calculated values of the cohesive energies of the elements. These are the result both of complex calculations similar to those we embarked upon in Chapter 6, and of the analysis of experimental results. It is worthwhile recalling that

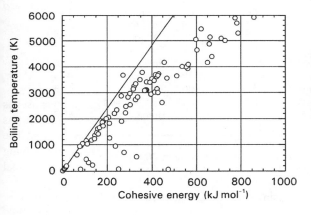

Figure 11.25 The boiling temperatures of the elements plotted as function of their cohesive energy (see Table 11.5). The data show a clear broad trend (with exceptions), indicating that substances with higher cohesive energies tend to have higher boiling temperatures. The solid line in the figure is the predicted boiling temperature according to $T_B \approx (U_0/xR)$, with $x = 10$.

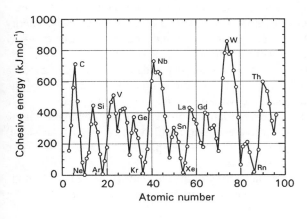

Figure 11.26 The cohesive energy of the elements (see Table 11.5) plotted as a function of atomic number. The data show periodic increases and decreases. It is this graph that reflects the filling of electron shells and the type of bonding possible and is at the heart of the periodic variations observed among elemental density: the melting, boiling and Debye temperatures, and the enthalpies of fusion and vaporization.

Table 11.5 The cohesive energies, U_0 of the elements.*

Z	Element	U_0 (kJ mol^{-1})	Z	Element	U_0 (kJ mol^{-1})	Z	Element	U_0 (kJ mol^{-1})
1	Hydrogen	–	32	Germanium	372	63	Europium	179
2	Helium	–	33	Arsenic	285.3	64	Gadolinium	400
3	Lithium	158	34	Selenium	237	65	Terbium	391
4	Beryllium	320	35	Bromine	118	66	Dysprosium	294
5	Boron	561	36	Krypton	11.2	67	Holmium	302
6	Carbon	711	37	Rubidium	82.2	68	Erbium	317
7	Nitrogen	474	38	Strontium	166	69	Thulium	233
8	Oxygen	251	39	Yttrium	422	70	Ytterbium	154
9	Fluorine	81	40	Zirconium	603	71	Lutetium	428
10	Neon	1.92	41	Niobium	730	72	Hafnium	621
11	Sodium	107	42	Molybdenum	658	73	Tantalum	782
12	Magnesium	145	43	Technetium	661	74	Tungsten	859
13	Aluminium	327	44	Ruthenium	650	75	Rhenium	775
14	Silicon	446	45	Rhodium	554	76	Osmium	788
15	Phosphorous	331	46	Palladium	376	77	Iridium	670
16	Sulphur	275	47	Silver	284	78	Platinum	564
17	Chlorine	135	48	Cadmium	112	79	Gold	368
18	Argon	7.74	49	Indium	243	80	Mercury	65
19	Potassium	90.1	50	Tin	303	81	Thallium	182
20	Calcium	178	51	Antimony	265	82	Lead	196
21	Scandium	376	52	Tellurium	211	83	Bismuth	210
22	Titanium	468	53	Iodine	107	84	Polonium	144
23	Vanadium	512	54	Xenon	15.9	85	Astatine	–
24	Chromium	395	55	Caesium	77.6	86	Radon	18.5
25	Manganese	282	56	Barium	183	87	Francium	–
26	Iron	413	57	Lanthanum	431	88	Radium	160
27	Cobalt	424	58	Cerium	417	89	Actinium	410
28	Nickel	428	59	Praseodymium	357	90	Thorium	598
29	Copper	336	60	Neodymium	328	91	Protractinium	–
30	Zinc	130	61	Promethium	–	92	Uranium	536
31	Gallium	271	62	Samarium	206			

*U_0 is the energy required to separate the atoms of a solid at T = 0 K into isolated neutral atoms.
(Data from C. Kittel, *Introduction to solid state physics*, 7th edition (New York: John Wiley).

this energy is *principally electrostatic in origin* and arises from one of the bonding mechanisms outlined in Chapter 6.

Given the value of U_0, the boiling temperature may be predicted according to Equation 11.20. $T_B \approx U_0/xR$ is plotted in Figure 11.25, along with the actual boiling temperature. We have assumed 10 nearest neighbours as a rough estimate of the number of interacting pairs of molecules, and we see that the prediction is of the correct order of magnitude and broadly captures the trend of the data.

Having seen that the condensation/boiling temperature is related to the cohesive energy, it is fairly straightforward to conceive how the melting temperature can be related to the boiling temperature by

using extensions of the arguments leading to the concept of the Lindemann fraction. Similarly, the link with the Debye temperature follows plausibly. Furthermore, as outlined in §11.3.3, the cohesive energy of a substance is linked directly to the enthalpy changes on fusion and vaporization.

If we plot the calculated cohesive energy as a function of atomic number (Fig. 11.26) we see the periodic variation with temperature that we have observed for many properties of the elements. It is the cohesive energy (modified differently in each case) that underlies all the periodic variations we have observed in:

• the density of the elements (Fig. 7.1);
• the Debye temperature of the elements (Fig. 11.5);

- the melting temperature of the elements (Fig. 11.2);
- the boiling temperature of the elements (Fig. 11.16);
- the enthalpy change on fusion of the elements
- the enthalpy change on vaporization of the elements (Fig. 11.17).

Thus, despite the bewildering variety of properties displayed by matter, on a coarse scale we can imagine each molecule to be just a "blob" of mass m interacting with a cohesive energy somewhere in the range 0.1 eV per atom to 10 eV per atom. The universal behaviour arises because, roughly speaking, *all* matter can be described in this rather general way.

11.7 The solid↔gas transition: sublimation

The sections above have described a familiar story: solids melt into liquids that evaporate to become gases. However, some substances transform straight from the solid state to the gaseous state in a process known as *sublimation*. The unfamiliarity of this transformation deserves some comment.

11.7.1 Data on the solid↔gas transition

The vapour pressure above a solid

Like the transformation from liquid to gas, the transformation from solid to gas is a continuous process: solid and gas coexist at all temperatures. However the *vapour pressure* of most familiar solids is exceedingly small at around room temperature.

Consider the vapour pressure of water (Figs 11.27 & 11.28): the vapour pressure does not suddenly fall to zero when water freezes, but forms a continuous curve. The vapour pressure above the solid is similar to the vapour pressure above the liquid if the freezing is inhibited by supercooling (see §10.6.1). It is clear from Table 11.6 and Figures 11.27 & 11.28 that the vapour pressure above the surface of liquid water is similar to the vapour pressure above solid ice at the same temperature. Similar behaviour is also seen in Figure 11.14, which shows the vapour pressure above the solid noble "gases" (neon, argon, krypton

Table 11.6 The equilibrium vapour pressure (Pa) of water above the solid or liquid surface as a function of temperature. The shaded data on the liquid correspond to data taken on supercooled water.

T (°C)	Solid	Liquid	T (°C)	Solid	Liquid
−90	0.009	–	−15	165.5	191.50
−80	0.053	–	−14	181.5	208.03
−70	0.258	–	−13	198.7	225.50
−60	1.077	–	−12	217.6	244.57
−50	3.940	–	−11	238.0	264.98
−40	12.88	–	−10	260.0	286.58
			−9	284.2	310.18
−30	38.12	–	−8	310.2	335.26
−29	42.27	–	−7	338.3	362.06
−28	46.80	–	−6	368.7	390.86
−27	51.87	–	−5	401.8	421.80
−26	57.34	–	−4	437.4	454.74
−25	63.47	–	−3	475.8	489.81
−24	70.14	–	−2	517.4	527.55
−23	77.34	–	−1	562.4	567.83
−22	85.34	–	0	610.6	610.6
−21	94.01	–	1	–	656.9
−20	103.4	–	2	–	706.0
−19	113.8	–	3	–	758.1
−18	125.2	–	4	–	813.6
−17	137.5	–	5	–	872.5
−16	151.0	–	6	–	935.2
			7	–	1002
			8	–	1073
			9	–	1148
			10	–	1228.1
			11	–	1312.7
			12	–	1402.6
			13	–	1497.7
			14	–	1598.5
			15	–	1705.3

(Data from Kaye & Laby; see §1.4.1)

and xenon). Over most of the temperature range shown, the substances (except helium) are solid – their liquid phase exists for only a few kelvin below their boiling temperature. However, there is no striking discontinuity between the vapour pressure above a solid and that above a liquid.

Sublimation

We may define a *sublimation temperature* for a solid – by analogy with the definition of the boiling temperature – as the temperature at which the vapour pressure above a solid reaches atmospheric pressure. By this definition, only two elements, (carbon and arsenic), sublime.

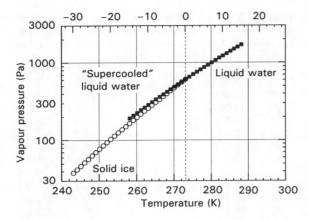

Figure 11.27 The vapour pressure of water and ice as function of temperature (see Table 11.6). Note that the vapour pressure data are plotted on a logarithmic scale. The figure shows data for liquid water that has been supercooled below its freezing temperature, and for ice. Note that at 0°C the vapour pressures of ice and water are extremely close, but that below this temperature the vapour pressure of the liquid is slightly greater than the vapour pressure of the solid.

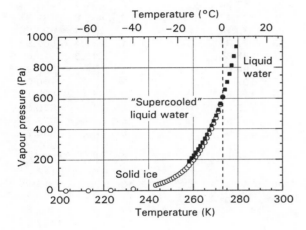

Figure 11.28 The vapour pressure of water and ice as function of temperature (see Table 11.6). The figure shows data for liquid water that has been supercooled below its freezing temperature, and for ice. Note that at 0°C the vapour pressures of ice and water are extremely close, but that below this temperature the vapour pressure of the liquid is slightly greater than the vapour pressure of the solid.

We noted in §11.4 that pressure tends to increase the temperature of a liquid→gas transition, and the same is true of the solid→gas transition. Thus the temperature of the complete transition to the gaseous state can be moved to a higher temperature by the application of pressure. We also noted in §11.3 that the temperature of a solid→liquid transition is also, in general, increased by application of pressure.

Consider a liquid at atmospheric pressure just above its melting temperature. The application of pressure will raise the melting temperature and return the substance to the solid state. Similarly, the application of pressure to a gas just above its boiling temperature may cause condensation of the substance into the liquid state. Thus in each case the application of pressure favours the more condensed of the phases available to the substance.

Similarly, the application of pressure to a substance that sublimes may sometimes cause a stabilization of the liquid phase, that allows the solid↔liquid and liquid↔gas transitions to be investigated. For example, arsenic is recorded in Table 11.2 as subliming at 886 K. However, if a pressure of 28 atm (≈2.8 MPa) is applied, arsenic first melts at 1090 K and then evaporates, allowing investigation of the enthalpies of fusion and vaporization. Note that the boiling temperature conventionally refers to normal atmospheric pressure, and so the boiling temperature "under pressure" is not usually recorded.

The main question raised by our preliminary examination of the experimental data on the solid→gas transition are:

- Why does sublimation occur at all?
- Why is sublimation so relatively rare?

11.7.2 The solid↔gas transition

From the data in the previous section we first note that the vapour pressure above a solid is broadly similar to that above a liquid. From this we conclude that, broadly speaking, the process of sublimation is equivalent to "boiling from the solid", in contrast with the more usual "boiling from the liquid". Normally, when the temperature is high enough for a substance to have a significant vapour pressure the substance has already melted. However, arsenic, for example, does not melt until a temperature which is higher than the temperature at which the vapour pressure is 1 atm. We will consider sublimation in two different ways: first microscopically, and then in terms of the Gibbs free energies of the phases.

Microscopic considerations

Let us use the cell model of liquid dynamics outlined in §8.3, to try to understand the situation of arsenic in its liquid state. Figure 11.29 is intended as schematic illustration and is not to be considered quantitatively, but it indicates plausibly the situation of arsenic. It shows a situation in which ΔE_h is greater than ΔE_e. In other words the activation energy for the "hopping" process is greater than the energy ΔE_e for escape from the substance altogether. Recall that ΔE_h is the activation energy involved when a liquid flows. Thus Figure 11.31 illustrates a substance that evaporates before it flows. We now need to understand how this situation could arise ?

Arsenic

For arsenic the situation is mainly due to anomalously weak and directional bonds between the atoms. Consider the data (see Table 11.2A below) on the enthalpy of fusion of the elements around arsenic.

We can see that the enthalpy of vaporization for arsenic is anomalously low in comparison to its

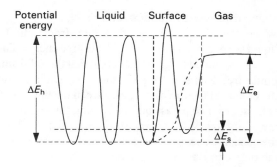

Figure 11.29 An illustration of the cell model with parameters appropriate to arsenic. Ignoring the situation at the surface of the liquid, we notice immediately that ΔE_h is greater than ΔE_e. In other words, the activation energy for the "hopping" process is greater than the energy required for escape from the substance altogether. Be careful in interpreting ΔE_h – it is *not* a single-particle hopping process (see §8.3 for further details)

enthalpy of fusion. This indicates, as we supposed, that evaporation is relatively easy in comparison with melting. The high energy involved in forming the liquid state is likely to arise because of a high degree of directional covalent bonding between atoms. The weakness of the bonding is likely to arise because of the difficulty of fully bonding its five valence electrons in the hexagonal crystal structure that arsenic adopts before it melts.

Carbon

For carbon we can understand the reluctance to form a liquid state in terms of the exceptional strength and directionality of the C–C covalent bond. This makes the "hopping" processes required for liquid behaviour extremely costly energetically.

Table 11.2A Thermal data for the elements.

Z	Name	Density (kg m⁻³)	Melting point (K)	Boiling point (K)	Enthalpy of fusion (kJ mol⁻¹)	Enthalpy of vaporization (kJ mol⁻¹)
31	Gallium	5905	302.93	3676	5.59	256.1
32	Germanium	5323	1210.6	3103	34.7	334.3
33	Arsenic	5776	sublimes at 886		27.7	31.9
34	Selenium	4808	490	958.1	5.1	26.32

Gibbs free energy considerations

When considering the transformation from solid to liquid we examined schematic Gibbs free energy diagrams. We noted (Figs 10.2 & 10.3) that the solid and liquid free energies are particularly sensitive to the difference in cohesive energy between the liquid and solid states. For potassium, increasing the energy of the liquid state with respect to the solid state by 5% of the cohesive energy of the solid, caused the predicted melting temperature to increase by 62%, (from 307 K to 498 K). If the liquid state has a particularly high cohesive energy due to the difficulty of optimally bonding to other atoms in the disordered liquid state, then it seems quite plausible that this could shift the melting to considerably higher temperatures. In this case it appears that T_M is shifted even above T_B.

In the case of arsenic, matters are complicated by the existence of a structural transformation from one crystal structure to another at around 500 K. In order to represent this we would need to draw a fourth line in Figure 10.3 representing the Gibbs free energy of the gas, the liquid and the two solid phases. This situation is too complex for further consideration at this level.

11.8 The triple point

11.8.1 Data on the triple point

In the previous sections we have examined data on the transitions between the states of matter discussed in earlier chapters: solid↔liquid, solid↔gas and liquid↔gas. These transitions take place over a range of temperature depending on the pressure at which the substance is investigated. However, there is a temperature at which all three phases of matter can coexist in equilibrium. The temperature at which this occurs is called the *triple-point temperature* and is a characteristic temperature for each pure substance. From Table 11.7 one can see that the triple point occurs at a temperature just a little above the melting temperature.

So the main question raised by our preliminary examination of the experimental data on the triple points of substances is: why is the triple point temperature close to the melting temperature of a substance at atmospheric pressure?

Table 11.7 The melting, boiling and triple-point temperatures of various substances. The triple point temperatures are generally known extremely accurately; the melting and boiling temperatures are typically known to within ≈10 mK.

Substance	T_M (K)	T_{TR} (K)	T_B (K)
Oxygen	54.35	54.3584	90.188
Nitrogen	63.15	63.150	77.352
Argon	83.75	83.8058	87.29
Water	273.15	273.16	373.15

11.8.2 Understanding the triple-point data

The triple point lies at the intersection of the solid↔liquid phase boundary and the liquid↔gas phase boundary. The slope of each of these phase boundaries may be predicted using the Clausius–Clapeyron equation (Eq 10.41). At the solid↔liquid phase boundary we have

$$\frac{dP_M}{dT_M} = \frac{(S_1 - S_2)}{(V_1 - V_2)} = \frac{\Delta Q_M}{T_M(V_L - V_S)} \qquad (11.21)$$

and at the liquid↔gas phase boundary we have

$$\frac{dP_B}{dT_B} = \frac{(S_1 - S_2)}{(V_1 - V_2)} = \frac{\Delta Q_B}{T_B(V_G - V_L)} \qquad (11.22)$$

From Figure 11.8 we note that $(\Delta Q_B/T_B)/(\Delta Q_M/T_M)$ is around 10 for many elements, and so we write that, roughly,

$$\frac{dP_M}{dT_M} = \left(10 \times \frac{V_G - V_L}{V_L - V_S}\right)\frac{dP_B}{dT_B} \qquad (11.23)$$

Now the volume of a gas, although thoroughly variable, is around 1000 times the volume of the equivalent solid at around atmospheric pressure, and the volume difference between a liquid and a solid is typically ≈10% of the solid volume (see Fig. 9.1). Using these approximations we find that

$$\frac{dP_M}{dT_M} = \left(10 \times \frac{1000 V_S}{0.1 V_S}\right)\frac{dP_B}{dT_B} \approx 10^5 \frac{dP_B}{dT_B} \qquad (11.24)$$

Equation 11.24 indicates that the slope of the melting curve is typically 10^5 times steeper than the slope of

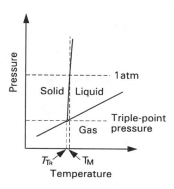

Figure 11.30 A schematic phase diagram illustrating the relative steepness of the solid↔liquid phase boundary in comparison with the liquid↔gas phase boundary. This means that points on the melting curve at quite different pressures (such as 1 atm and the triple-point pressure) occur at quite similar temperatures.

the vaporization curve. This is also clear graphically in the examples in Figures 10.16 & 10.17. The consequence of this is that, in comparison with the vaporization curve, the melting curve on a phase diagram is effectively a vertical line (Fig. 11.30). Recall that:

- the melting temperature lies at the intersection of this line with a horizontal line at $P = 1$ atm; and
- the triple-point temperature lies at the intersection of this line with another horizontal line at a different pressure.

However, the extreme steepness of the melting curve means that the temperature difference between these two intersections is small, (frequently less than 1 K).

Figure 11.31 My own measurements of the heat capacity of Nb_3Sn. Note the large jump in the heat capacity at T_C. These features occur on top of the heat capacity due to lattice vibrations (see §7.5). Thanks to Mike Sprinford for the sample.

11.9 Other types of phase change

The transitions between the three phases of matter are obvious to us all. However, there are other transitions that take place within all three phases of matter which also merit the title "phase change" but are often not apparent by merely looking at a substance. Let us look at some examples of phase changes, and then see what they have in common with melting/freezing, condensation/evaporation and sublimation.

The superconducting transition

Below a temperature known as the *critical temperature*, T_C, the electrical resistivity of some substances falls suddenly to values indistinguishable from zero. Thus electric currents can flow without energy loss in these materials.

As determined by the electrical resistivity, the transition is extremely sharp. In pure unstrained materials the width of the transition may be only a few microkelvin. The heat capacity of a superconducting compound, Nb_3Sn (Fig. 11.31) shows a large jump at T_C.

Superfluid

Helium, which behaves as a closely ideal gas in its gaseous phase, behaves as an extremely anomalous liquid when condensed. The list of unusual properties is too long to list here, but one of the most striking is the fact that the liquid does not solidify on cooling to temperatures as close to absolute zero as is achiev-

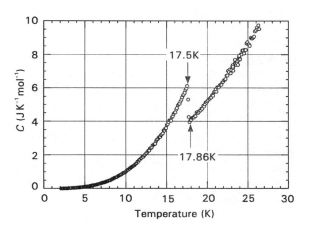

able (a few microkelvin). The normal boiling temperature of the liquid is 4.2 K, but if the temperature is lowered to below 2.172 K the liquid displays so-called "superfluid properties", including the ability to flow with absolutely zero viscosity. The heat capacity in this range displays a large peak which was deemed by early researchers to be shaped like the Greek letter λ (lambda) and hence is known as the λ transition.

Liquid crystal states

As we saw in Chapter 8, many organic substances may exist in a form intermediate between the liquid and solid state known as a *mesophase* or *liquid crystal state*. In these states the substances display unusual optical properties. If these, normally transparent, liquids are viewed in an arrangement similar to that shown in Figure 11.32, they display striking properties. The key feature of the apparatus is the two polarizing filters on either side of the liquid under examination. Normal liquids do not significantly affect the polarization of light passing through them.

Thus if the polarizing filters are crossed no light is transmitted. However, above a certain transition temperature these liquids strongly affect the polarization of light passing through them. When viewed through a microscope as shown, the liquids look like a window on a frosty morning: crystals appear to have grown from the edges of the container. However, crystals have not grown: the liquid is still a liquid in the sense that it will still flow, but something has certainly changed within the liquid. The onset of the effect is sudden, and the destruction of the effect above a second critical temperature is also sudden.

Ferromagnetic

In the absence of an applied magnetic field, the magnetic moment of a piece of iron at 771°C is exactly zero. However, at 769°C the magnetic moment becomes finite, even in the absence of an applied magnetic field. The magnetic moment then increases in magnitude as the temperature is lowered below the *Curie temperature* of 770°C. Similar behaviour is observed in all ferromagnets, although the Curie temperature varies significantly. Figure 11.33 shows the heat capacity of the alloy CePt which becomes ferro-

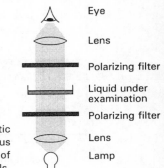

Figure 11.32 Schematic diagram of an apparatus for observing changes of phase in liquid crystals.

Eye
Lens
Polarizing filter
Liquid under examination
Polarizing filter
Lens
Lamp

magnetic at around 6 K. The Curie temperatures of the ferromagnetic elements are: iron: 770°C; nickel: 358°C; cobalt: 1115°C; and gadolinium: 16°C. All ferromagnets show similar heat capacity anomalies close to the Curie temperature, and the anomaly in gadolinium is responsible for the unusually high value of its heat capacity at 20°C noted in Table 7.7.

Discussion

What these changes have in common is that at a particular temperature, "something happens" or "something starts to happen". That is "some property" of the substance changes rapidly as the environment of the substance is changed by just a small amount. The differences between the different types of phase transition may be considered to be "merely" to do with the physical property that changes suddenly. Many of

Figure 11.33 My own measurements of the heat capacity of CePt. This material becomes ferromagnetically ordered below ≈6 K. (Thanks to José Gomez Sal for the sample.)

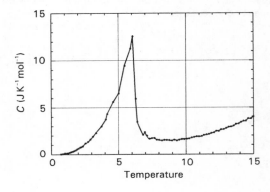

these transitions are accompanied by heat capacity anomalies which often look somewhat similar, even though the processes taking place are dramatically different.

The main questions raised by our preliminary examination of the experimental data on different types of phase changes are (a) why does phase change suddenly at a well-defined temperature and (b) why do different materials display similar anomalies in their heat capacity around these transitions?

The phase changes discussed above all have in common the fact that the lower temperature phase has more order than the higher temperature phase. The type of order (i.e. what has happened) is often not at all obvious from the heat capacity curves. In general, one needs to measure a property more directly related to the type of order involved, perhaps the magnetization, resistivity or polarizability, to determine what has happened. However, the change in the type of order present within the substance means that on a Gibbs free energy diagram the substance has jumped from one curve to another. This jump generally results in a discontinuity of the slope of G at the transition (a first-order phase change) and it is this discontinuity that is responsible for the anomalies in the heat capacity.

11.10   Exercises

Exercises with a P prefix are "normal" problems. Those with a C prefix are best solved numerically using a computer program or spreadsheet.

P1.   From Table 11.2 find out which element has:
(a) The highest melting temperature.
(b) The fourth highest melting temperature.
(c) The highest latent heat of melting (fusion) per mole.
(d) The highest latent heat of vaporization per mole.
(e) The highest ratio of latent heat of vaporization per mole to the latent heat of melting (fusion) per mole.
(You may need to use a spreadsheet for this part (e))

P2.   Which quantities in Table 11.2 are the best indicators of the cohesive energy of an element? Justify your answer. From Table 11.2, estimate which element has the highest cohesive energy per mole and compare your answer with the calculated results given in Table 11.5.

P3.   What is the boiling temperature of (a) methanol, (b) ethanol and (c) acetone (Table 11.1)?

P4.   The melting and boiling temperatures of the alkali halides are listed in Table 11.1. Do these temperatures show any systematic variation with the member of the halide family involved? Is the weak variation of melting temperature with the components what you would have expected from the discussion of ionic and covalent bonding in §6.3?

P5.   How much energy is required to melt 1 kg of:
(a) helium, (b) neon, (c) argon, and (d) krypton (Table 11.2)?

P6.   How much energy is required to melt 1 kg of:
(a) copper, (b) silver, (c) gold, and (d) mercury (Table 11.2)?

P7.   Based on Figures 11.2 & 11.3, indicate to what extent knowledge of the melting temperature and molar enthalpy of fusion of an element with atomic number Z are sufficient to predict the melting temperature and molar enthalpy of fusion of the element with atomic number $Z + 1$. Include a discussion of the relationship between the variation shown in Figures 11.2 & 11.3 and the periodic table (Fig. 2.2).

P8.   By how much does the melting temperature of gold (Table 11.2 & Figure 11.4) change when atmospheric pressure changes from 1 to 1.1 atm?

P9.   Repeat the Lindemann melting theory calculation given in Example 11.3 for gold. Does your answer lie within the 1 standard deviation limits drawn in Figure 11.7?

P10. How good are the rough estimates of the enthalpy of fusion and vaporization of the elements indicated in Figure 11.8? Based on the rough figures in the legend to Figure 11.8, estimate the enthalpy of fusion and vaporization for (a) an element with $T_M = 1683\,K$ and $T_B = 2628\,K$, (b) an element with $T_M = 1357\,K$ and $T_B = 2840\,K$, and (c) an element with $T_M = 3680\,K$ and $T_B = 5930\,K$. Compare your results with the actual values given in Table 11.2.

P11. An unknown elemental metal has its melting temperature determined with an uncertainty of ±3 K. How well would this serve to identify the element uniquely (Table 11.2 & Fig. 11.2)?

P12. How much energy is required to boil 1 kg of:
(a) helium, (b) neon, (c) argon, and (d) krypton (Table 11.2)?

P13. How much energy is required to boil 1 kg of:
(a) copper, (b) silver, (c) gold, and (d) mercury (Table 11.2)?

P14. Estimate the vapour pressure at 2000 K of:
(a) copper, (b) silver, (c) gold, and (d) aluminium (Fig. 11.16)?

P15. Estimate the rate at which molecules leave the surface of liquid water at 20 °C according to Equation 11.8 and Table 9.10. What rate of mass loss does this correspond to? Compare this with your qualitative experience of water evaporating in a situation where little water vapour will re-enter the liquid, e.g. spilling water outdoors onto a non-absorbent surface on a warm day with a slight breeze.

357

P16. Assuming the oceans of the world to be at an average surface temperature of 4°C, and to have a surface area of ⅔ that of the Earth, estimate the rate at which the oceans are evaporating (Eq. 11.8 & Table 9.10). If this mass loss were not replaced by rain, at what rate would the sea level decline (Eq. 11.8 & Table 9.10)? The radius of the Earth is approximately 6400 km.

C17. The temperatures at which the vapour pressures of several organic substances reach 10^3 Pa, 10^4 Pa and 1.013×10^5 Pa are listed below. Estimate their molar latent heat of vaporization (kJ mol⁻¹) based on Equation 11.16.

Substance	MW	T (K) for P = 10^3 Pa	10^4 Pa	Boiling point (K)
Acetic acid, CH_3COOH	60	–	329	391.1
Acetone, CH_3COCH_3	58	237	349	329.3
Aniline, C_6H_7N	93	337	385	457.6
Benzene, C_6H_6	78	–	293	353.2
Chloroform, $CHCl_3$	119	240	278	334.4
Cyclohexane, C_6H_{10}	82	–	292	353.8
Ethyl acetate, $C_4H_8O_2$	88	255	294	350.2
Methanol, CH_3OH	32	253	288	337.7
Ethanol, C_2H_5OH	46	267	302	351.5
Propan-1-ol, C_3H_7OH	60	284	320	370.3
Propan-2-ol, C_3H_7OH	60	274	307	355.4
Butan-1-ol, C_4H_9OH	74	301	338	390.35
Butan-2-ol, C_4H_9OH	74	288	322	372.65
Toluene, C_7H_8	92	275	318	383.8

(Data from Kaye & Laby; see §1.4.1)

P18. Estimate the vapour pressure of water at the boiling temperature of ethanol. Based on the ratio of vapour pressures, estimate the minimum fraction of water (by molar concentration) which would be present in a distilled water/ethanol mixture. By a similar technique, estimate the minimum fraction of methanol present (Eq. 11.15 & Table 11.3).

P19. What are the critical temperature, pressure and molar volume of (a) water (b) carbon dioxide and (c) ethanol? What is the density of each of these substances in the critical state, and roughly what fraction of the solid density does this correspond to (Table 11.4)?

P20. Calculate the volume per molecule of water in (a) its critical state and (b) its liquid state at 20°C (Example 11.5 and Fig. 11.20).

C21. Based on the data given in Table 11.2, estimate the average scaling law ratios between T_B, T_C and T_M for the elements. (Figures 11.23 & 11.24).

C22. Figure 11.5 indicates that there is a rough relationship between the Debye temperature of an element and its melting temperature. Use the data given in Tables 7.8 & 11.2 to produce a rule of thumb relating θ_D and T_M.

P23. Predict the Debye temperature of copper, based on the the melting temperatures of copper, silver and gold and the Debye temperatures of gold and silver (Tables 7.8 and 11.2). How good is your estimate?

P24. Identify the elements in Figure 11.25 which have anomalously low boiling points for their cohesive energy. How have these anomalies arisen? Explain why their existence does not invalidate the general correlation of boiling temperature and cohesive energy?

P25. What is the equilibrium vapour pressure above (a) ice and (b) water (if applicable) at a temperature of (i) 0°C, (ii) −10°C and (iii) −20°C (Table 11.6). Based on your answers to these questions, discuss whether there are any conditions under which snow might evaporate directly without first melting.

P26. Suggest a reason why the equilibrium vapour pressure above ice is slightly less than that above water (Table 11.6 & Figs 11.27 & 11.28). How could you verify your suggestion?

P27. What is the triple point of (a) water and (b) argon (Table 11.7)?

C28. Alter the computer program given in Appendix 4 to measure a parameter related to temperature. This could be the total energy (potential and kinetic) divided by the number of available degrees of freedom. By altering the variable *Factor*, determine the melting and boiling temperatures (defined how?) of the substance. Do the very small samples of solids, liquids and gases have sharply defined melting and boiling temperatures? (See also Exercise C1 in Ch. 10).

P29. In §11.4, the dependence of vapour pressure on temperature is discussed. Given the latent heat and the boiling temperature (Table 11.2) of the elements, it is possible to estimate the vapour pressure at any temperature (Eq. 11.15 & Table 11.3). Estimate the vapour pressure at 100°C and 1000°C of:

(a) potassium

(b) copper

(c) iron

(d) tungsten

P30. A strong metal box containing water in its solid phase is heated from −20°C to + 20°C and the heat capacity of the combined object is measured continuously. If you had no *a priori* technique for separately accounting for heat supplied as latent heat and heat which raised the temperature of the box, then you would assume that there was an "anomaly" in the heat capacity of the box. Sketch the general form of the "anomaly" that you would observe. Discuss any similarities or differences between your graph and Figures 11.31 & 11.33.

APPENDIX 1

Maxwellian speed distribution of a gas

In this appendix we will work out the distribution of molecular speeds in a gas by explicitly calculating three functions that describe the properties of dilute gases. These are:

(a) The *density of states function*, $g(E)dE$, which yields the number of quantum states with energies between E and dE.

(b) The *occupation function*, $f(E,T)$, which yields the average number of particles occupying a single quantum state with energy E and temperature T.

(c) The *distribution function*, $D(E,T)$, which is the product $f(E,T)g(E)dE$ and yields the average number of particles occupying quantum states with energies between E and dE at temperature T.

Once we have worked out the distribution function, we will be able to state *on average* how many molecules in a gas have energies between E and $E+dE$. Then, since we know that for simple molecules the energy of a molecule is entirely kinetic ($E=\frac{1}{2}mv^2$) we can deduce how many molecules have speeds between v and $v+dv$.

In what follows we consider only the properties of simple molecules of mass m and no internal degrees of freedom.

A1.1 The density of states function

(See also §6.5 which contains a similar derivation.) Our determination of the density of states function is based on an analysis of the "particle in a box" problem (§2.5.3 & 6.5.2). That is, we consider the molecules of a gas to occupy the quantum states of a particle in a box. In §6.4 we considered this problem

for the case of electrons packed densely together inside a metal, but now we consider the case of molecules packed at low density into an empty box.

Our first step is to develop a way of systematically counting the quantum states of a particle in a box. We will consider two counting techniques: a simple way and a more complicated way. We will, of course, choose the more complicated method.

A1.1.1 The quantum states: standing waves

The first method of counting is derived from Equation 2.59 for the energies of allowed quantum states

$$E\left(n_x, n_y, n_z\right) = \frac{h^2}{8mL^2}\left[n_x^2 + n_y^2 + n_z^2\right] \quad (A1.1)$$

We can represent an allowed quantum state by a

Figure A1.1 A scheme for counting quantum states of the particle in a box problem when lots of particles are present. Each quantum state is represented by a point on a graph with axes n_x, n_y, and n_z. The energy of the state is proportional to the square of its distance from the origin on the graph. For this reason, the occupied quantum states cluster in a quadrant of a sphere around the origin in order to minimize their energy.

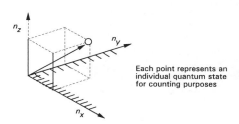

Each point represents an individual quantum state for counting purposes

point on a three-dimensional graph the axes of which are n_x, n_y and n_z. Thus each individual quantum state is represented by a single point on the graph, and the points representing the quantum states are distributed uniformly on a "mesh" throughout the graph. Because of this, by measuring "volumes" on this three-dimensional graph one can, with a fair degree of approximation, also count quantum states. Furthermore, we can also simply state that the energy of a particular state is given by $h^2/8mL^2$ multiplied by the square of the length of the vector from the origin to the point representing that quantum state (Fig. A1.1).

Developing the "geometrical" method of counting quantum states further, we now go to a second view of the quantum states of a particle in a box.

A.1.1.2 The quantum states: travelling waves

The second method of counting arises from a reconsideration of the particle in a box problem. We will arrive at similar, but distinctly different, solutions to the Shrödinger equation: travelling wave solutions.

In §2.4.3 we required the wave functions to be zero at the edge of the box because we imagined the edge of the box to represent the walls of the box. We supposed that molecules are not allowed to penetrate the walls (otherwise they're not walls!) and so the wave functions of the particles must be zero there. Now, however, we consider a more complex situation: we imagine a volume of the gas which is *representative* of the gas as a whole. We imagine it to be

Figure A1.2 Illustration of two approaches to the particle in a box problem. The first approach (a) uses the idea of an isolated box representing the entire crystal. The second approach (b) uses the idea of a box of representative material surrounded by identical copies of itself. The edges of the actual crystal are imagined to be far enough away that they do not significantly affect the electrons deep inside the box.

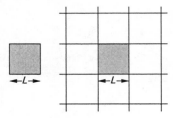

surrounded by *identical volumes* of gas. The differences between this situation and the standing wave situation are illustrated in Figure A1.2. You may consider this to be "trickery", and it is. It is a mathematical trick, but it will allow us to see more clearly the physics of what is happening in the gas.

The key difference to the problem is that now, instead of requiring that the wave functions be zero at the edges of the box, i.e. $\Psi(-x/2) = \Psi(+x/2) = 0$, we require that the wave functions be identical in neighbouring boxes, i.e.

$$\Psi(x) = \Psi(x + L_x) \qquad (A1.2)$$

where L_x is the length of the box under consideration. If the solutions are $\sin(kx)$ or $\cos(kx)$ waves, this is equivalent to requiring that

$$\cos(k_x x) = \cos(k_x[x + L_x]) \qquad (A.1.3a)$$

or $\quad \sin(k_x x) = \sin(k_x[x + L_x]) \qquad (A.1.3b)$

The requirements given by Equations A1.3a and A1.3b are known as the *Born–von Karmen boundary conditions*, and lead to only discrete values of k being allowed solutions. Let's follow one of Equations A1.3 through explicitly to find these conditions for k_x. The boundary conditions require that

$$\cos(k_x x) = \cos(k_x[x + L_x])$$
$$= \cos(k_x x + k_x L_x) \qquad (A1.4)$$

Using the trigonometric identity

$$\cos(A + B) = \cos A \cos B - \sin A \sin B \qquad (A1.5)$$

Equation A1.4 can be rewritten as

$$\cos(k_x x) = \cos(k_x x)\cos(k_x L_x)$$
$$- \sin(k_x x)\sin(k_x L_x) \qquad (A1.6)$$

Clearly, in general this equation is not true! It is only true when $\cos(k_x L_x) = 1$ and when the second term is zero. Since x can vary, this will only be true when $\sin(k_x L_x) = 0$. Now $\cos(k_x L_x) = 1$ when $k_x L_x = 0$, $\pm 2\pi$, $\pm 4\pi$, ..., etc., and these values of $k_x L_x$ also cause $\sin(k_x L_x)$ to be zero. So the Born–von Karmen boundary conditions are satisfied when

$$k_x L_x = 0, \pm 2\pi, \pm 4\pi, \ldots \qquad (A1.7)$$

i.e. when $\quad k_x = 0, \dfrac{\pm 2\pi}{L_x}, \dfrac{\pm 4\pi}{L_x}, \ldots \quad$ (A1.8)

or in general when

$$k_x = \frac{2m_x \pi}{L_x} \quad \text{where } m_x = 0, \pm 1, \pm 2, \ldots \quad \text{(A1.9)}$$

It is left to the reader to show that following Equation A1.3b through in a similar way (using $\sin(A + B) = \sin A \cos B + \cos A \sin B$) results in exactly the same conclusion.

We can now make sketches analogous to the one shown in Figure A1.1 for this new situation. In these sketches we plot points representing k_x, k_y and k_z which serve to label individual quantum states. The main difference between this and the (n_x, n_y, n_z) representation discussed previously is that each allowed point in space now represents a travelling wave rather a standing wave. Hence we can have negative k values and so the quantum states are distributed all around the origin rather than being confined to the postive $|k|$ octant (contrast Figs A1.1 & A1.3).

Furthermore, we know that the allowed states are distributed uniformly through k space so that if we consider a "volume" of k-space we can work out how many allowed quantum states it contains. A close-up

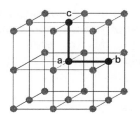

Figure A.1.4 Close-up view of k-space. The spheres represent allowed values of k. If the central point a represents a solution of the Schrödinger equation with a particular value of k_x, k_y, k_z, then point b represents a solution with the k_x component increased by $2\pi/L_x$. Similarly, point c represents a solution with the k_z component increased by $2\pi/L_z$.

view of our k-space graph is shown in Figure A1.4.

It is clear from the Figure A1.3 that each state has around it a "volume of k space", $\Delta\Omega$, given by

$$\Delta\Omega = \frac{2\pi}{L_x} \times \frac{2\pi}{L_y} \times \frac{2\pi}{L_z} \quad \text{(A1.10)}$$

and since $L_x L_y L_z = V$, this reduces to

$$\boxed{\Delta\Omega = \frac{8\pi^3}{V}} \quad \text{(A1.11)}$$

This simple result is crucial. As we shall see, it enables us to count quantum states and in particular, to answer the following questions.

A1.1.3 How many quantum states are there with wave vectors between k and $k + dk$?

All the quantum states with k vectors in this range correspond to points in a spherical shell of radius k and thickness dk. We can write down the "volume" on the k-space graph that these points occupy as

$$d\Omega = 4\pi k^2 dk \quad \text{(A1.12)}$$

and since each point occupies a volume $\Delta\Omega = 8\pi^3/V$, the number of quantum states dN in this shell is given by

$$dN = \frac{4\pi k^2 dk}{8\pi^3/V} = \frac{Vk^2 dk}{2\pi^2} \quad \text{(A1.13)}$$

Figure A1.3 A cross-section through k-space. Each small circle represents an allowed travelling wave solution to the Shrödinger equation. The filled circles represent occupied quantum states and the unfilled circles represent empty quantum states. The configuration of occupied and empty states shown is similar to that expected in a metal at absolute zero. Only the lowest energy states (low k = long wavelength = low energy) are occupied. In three dimensions the occupied states form a sphere in k-space known as the *Fermi sphere*.

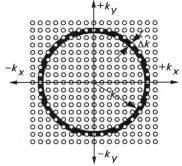

A1.1.4 How many quantum states are there with energy between E and $E + dE$

In order to answer this we need to rewrite Equation A1.13 replacing references to k (in this case k^2 and dk) with references to E. Recalling from Equation 2.54 that

$$E = \frac{1}{2}mv^2 = \frac{p^2}{2m} = \frac{\hbar^2 k^2}{2m} \qquad (A1.14)$$

we see that

$$k^2 = \frac{2mE}{\hbar^2} \qquad (A1.15)$$

Differentiating Equation A1.14 we find

$$dE = \frac{\hbar^2 k}{m}\,dk \qquad (A1.16)$$

and replacing the reference to k

$$dE = \frac{\hbar^2 \sqrt{2mE/\hbar^2}}{m}\,dk \qquad (A1.17)$$

we arrive at

$$dk = \frac{m}{\hbar^2 \sqrt{2mE/\hbar^2}}\,dE \qquad (A1.18)$$

We can now replace both references to k in Equation A1.13 with references to E

$$dN = \frac{V}{2\pi^2} \times \frac{2mE}{\hbar^2} \times \frac{m}{\hbar^2 \sqrt{2mE/\hbar^2}}\,dE \quad (A1.19)$$

which simplifies to

$$dN = \frac{Vm^{\frac{3}{2}}}{\sqrt{2}\pi^2\hbar^3}E^{\frac{1}{2}}dE \qquad (A1.20)$$

This expression is for the number of quantum states dN with energies between E and $E + dE$, which is exactly the definition of the density of states function, $g(E)$. So we now have the expression we desire:

$$\boxed{g(E) = \frac{dN}{dE} = \frac{Vm^{\frac{3}{2}}}{\sqrt{2}\pi^2\hbar^3}E^{\frac{1}{2}}} \qquad (A1.21)$$

A1.2 The occupation function

We now have to consider whether the particles which will occupy the box will be bosons or fermions. Molecules are composed of many atoms, and atoms are composed of protons, neutrons and electrons. Whether a molecule is a boson or a fermion depends on the way in which the spins of the component particles add together. For example, helium atoms occur in two isotopes: the rare ^{3}He and the more common ^{4}He. ^{3}He has two electrons , two protons, and one neutron all with spin $1/2\hbar$. No matter which way these are added together, the net result is always half an odd integer times $\hbar$ and so ^{3}He atoms are fermions: they obey the exclusion principle and are described by Fermi–Dirac statistics. In contrast, ^{4}He has two electrons, two protons, and two neutrons which always add together to yield a spin of zero or an integer multiple of $\hbar$. Thus ^{4}He atoms are bosons: they do *not* obey the exclusion principle, and they are described by Bose–Einstein statistics.

However, at the low densities present in gases, both fermion and boson occupation functions are well approximated by the Boltzmann occupation function. Only when the density of the gas is such that the probability of multiple occupation of quantum states becomes significant can one distinguish between the two occupation functions. However, this high-density regime does not occur until we reach temperatures of a few kelvin or so, and densities approaching those in the solid and liquid states.

A1.2.1 The Boltzmann occupation function

Independent of whether the molecules of the gas are fermions or bosons, we may write the occupation function (the average occupancy of an individual quantum state) as

$$f(E,T) = \frac{1}{\exp\left[(E-\mu)/k_B T\right] \pm 1} \qquad (A1.22)$$

where the upper sign corresponds to fermions and the lower sign corresponds to bosons (see Eqs 2.67 & 2.69). The energy, μ, is known as the *chemical potential*; it is sometimes referred to as the *Fermi energy* when dealing with fermions. Now one could proceed to evaluate the distribution function using Equation A1.22 directly. However, there is a useful approximation that may be made when the average occupancy of a quantum state is low. As mentioned in §2.5.4, when there is little chance of two particles occupying the same quantum state, the occupation functions of fermions and bosons must be similar since they have an essentially unrestricted choice of quantum states. Thus in the low-density (classical) limit of Equation A1.22 we always have

$$f(E,T) \ll 1 \qquad (A1.23)$$

If this is true, then the denominator of Equation A1.22 must be much larger than unity, which implies that

$$\exp\left[(E-\mu)/k_BT\right] \gg 1 \qquad (A1.24)$$

This being so, we may write

$$f(E,T) \approx \exp\left[-(E-\mu)/k_BT\right] \qquad (A1.25)$$

which may be factorized as

$$f(E,T) \approx \exp(+\mu/k_BT)\exp(-E/k_BT) \qquad (A1.26)$$

Now the chemical potential, μ, has not yet been determined, but we will be able to determine it if we note that the average occupancy of a quantum state, when summed over all quantum states, must sum to N, the total number of particles, i.e.

$$N = \sum_{\substack{\text{All quantum} \\ \text{states}}} f(E,T)$$
$$\approx \sum_{\substack{\text{All quantum} \\ \text{states}}} \exp(+\mu/k_BT)\exp(-E/k_BT) \qquad (A1.27)$$

We can now factor out $\exp(+\mu/k_BT)$ which is common to all terms in the sum

$$N = \exp(+\mu/k_BT) \sum_{\substack{\text{All quantum} \\ \text{states}}} \exp(-E/k_BT) \qquad (A1.28)$$

and rearrange to solve for $\exp(+\mu/k_BT)$,

$$\exp(+\mu/k_BT) = \frac{N}{\sum_{\substack{\text{All quantum} \\ \text{states}}} \exp(-E/k_BT)} \qquad (A1.29)$$

We can now substitute this expression back into Equation A1.26 to yield the desired expression for $f(E,T)$:

$$f(E,T) \approx \frac{N\exp(-E/k_BT)}{\sum_{\substack{\text{All quantum} \\ \text{states}}} \exp(-E/k_BT)} \qquad (A1.30)$$

Finally, we note that at a given temperature, both the denominator and N are constants and we can write simply

$$\boxed{f(E,T) \approx A\exp(-E/k_BT)} \qquad (A1.31)$$

which is the classical Boltzmann occupation function (Eq. 2.78).

A1.3 The distribution function

Multiplying together Equation A1.21 for the number of quantum states between energies E and $E + dE$

$$dN = \frac{Vm^{\frac{3}{2}}}{\sqrt{2}\pi^2\hbar^3} E^{\frac{1}{2}}dE,$$

and Equation A1.31 for the average occupancy of a quantum state

$$f(E,T) \approx A\exp(-E/k_BT),$$

we arrive at the *distribution function*. This tells us the number of particles occupying quantum states with energies between E and $E + dE$

$$\boxed{dN = \frac{Vm^{\frac{3}{2}}}{\sqrt{2}\pi^2\hbar^3} E^{\frac{1}{2}} A\exp(-E/k_BT)dE} \qquad (A1.32)$$

In order to evaluate this expression for dN we need to work out a way of eliminating the undefined constant A in Equation A1.32. We can do this by integrating

Equation A1.32 to find an expression for N, the total number of particles, i.e.

$$\int_{E=0}^{E=\infty} dN = N = \int_{E=0}^{E=\infty} \frac{Vm^{\frac{3}{2}}}{\sqrt{2}\pi^2\hbar^3} E^{\frac{1}{2}} A \exp(-E/k_{\mathrm{B}}T) dE$$

(A1.33)

Taking the constants outside of the integral,

$$N = \frac{AVm^{\frac{3}{2}}}{\sqrt{2}\pi^2\hbar^3} \int_{E=0}^{E=\infty} E^{\frac{1}{2}} \exp(-E/k_{\mathrm{B}}T) dE \quad (\text{A1.34})$$

The integral is (fortunately) a standard integral that evaluates (using MathCAD or a maths reference book) to

$$\int_{x=0}^{x=\infty} x^{\frac{1}{2}} \exp(-x/a) dx = \frac{\sqrt{\pi a^3}}{2} \quad (\text{A1.35})$$

where we have substituted $x = E$ and $a = k_{\mathrm{B}}T$. With these substitutions Equation A1.34 becomes

$$N = \frac{AVm^{\frac{3}{2}}}{\sqrt{2}\pi^2\hbar^3} \times \frac{\sqrt{\pi k_{\mathrm{B}}^3 T^3}}{2} \quad (\text{A1.36})$$

We can now eliminate A in Equation A1.32 by dividing through by Equation A1.36 for N, and hence calculating the *fraction* of the particles dN/N which occupy quantum states with energies in the range E to $E + dE$:

$$\frac{dN}{N} = \frac{\dfrac{Vm^{\frac{3}{2}}}{\sqrt{2}\pi^2\hbar^3} E^{\frac{1}{2}} A \exp(-E/k_{\mathrm{B}}T) dE}{\left(AVm^{\frac{3}{2}}/\sqrt{2}\pi^2\hbar^3\right) \times \left(\sqrt{\pi k_{\mathrm{B}}^3 T^3}/2\right)}$$

(A1.37)

$$\frac{dN}{N} = 2\sqrt{\frac{E}{\pi k_{\mathrm{B}}^3 T^3}} \exp(-E/k_{\mathrm{B}}T) dE \quad (\text{A1.38})$$

Alternatively, we may write this as the probability $P(E) dE$ that a molecule has energy between E and $E + dE$, i.e.

$$P(E) dE = 2\sqrt{\frac{E}{\pi k_{\mathrm{B}}^3 T^3}} \exp(-E/k_{\mathrm{B}}T) dE \quad (\text{A1.39})$$

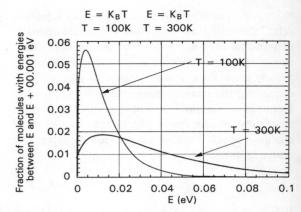

Figure A1.5 The distribution of the energies of molecules in a gas at temperatures of 100 K and 300 K. Note that the curves are universal, i.e. they apply to all simple molecules, whatever their mass. Heavy molecules move more slowly than lighter molecules in such a way as to keep the energies the same.

It is important to take note of the general form of Equation A1.39, which we may write more simply as

$$P(E) dE = \text{constants} \times \sqrt{E} \times \exp(-E/k_{\mathrm{B}}T) dE$$

(A1.40)

We see that, at a given temperature, Equation A1.39 predicts *exactly the same* distribution of molecular energies for *all gases* (Fig. A1.5).

A1.4 The distribution of molecular speeds

We are now close to completing the task we set ourselves, i.e. to calculate the distribution of molecular speeds. To achieve this, we need to convert the references to energy E in Equation A1.38

$$\frac{dN}{N} = 2\sqrt{\frac{E}{\pi k_{\mathrm{B}}^3 T^3}} \exp(-E/k_{\mathrm{B}}T) dE \quad (\text{A1.38})^*$$

into references to molecular speed using

$$E = \tfrac{1}{2} mv^2 \quad (\text{A1.41})$$

and

$$dE = mv dv \quad (\text{A1.42})$$

Example A1.1

In a gas at temperature T, what fraction of molecules have energies greater than $k_B T$ given that

$$\int_{x=1}^{x=\infty} x^{\frac{1}{2}} \exp(-x)\,dx = 0.507$$

The required fraction is the integral of dN/N (Equation A1.39) over the required energy range which in this case is from $E = k_B T$ to infinity.

$$\text{fraction} = \int_{E=k_B T}^{E=\infty} 2\sqrt{\frac{E}{\pi k_B^3 T^3}}\,\exp(-E/k_B T)\,dE$$

If we substitute $x = E/k_B T$ the we find $dx = dE/k_B T$ and the integral becomes

$$\text{fraction} = \int_{x=1}^{x=\infty} \left(2\sqrt{\frac{1}{\pi k_B^2 T^2}}\right) x^{\frac{1}{2}} \exp(-x) k_B T\,dx$$

Taking the constants outside the integral sign, allows us to recognize the integral given in the question

$$\text{fraction} = \left(2\sqrt{\frac{1}{\pi}}\right) \int_{x=1}^{x=\infty} x^{\frac{1}{2}} \exp(-x)\,dx$$

and substituting for this integral we have

$$\text{fraction} = \left(2\sqrt{\frac{1}{\pi}}\right) \times 0.507 = 0.572$$

Thus 57.2% of molecules have energy greater than $k_B T$. Notice that this is true at any temperature!

Note that these substitutions assume that the molecules have only the three degrees of freedom associated with their motion, and so we are implicitly neglecting the possibility of internal molecular vibrations and rotations. On substituting Equations A1.41 and A1.42 into A1.38 we find

$$\frac{dN}{N} = 2\sqrt{\frac{\tfrac{1}{2}mv^2}{\pi k_B^3 T^3}}\,\exp(-mv^2/2k_B T)\,mv\,dv$$

$$(A1.43)$$

which simplifies to

Example A1.2

In a nitrogen gas at 300 K, what fraction of molecules have speeds greater than $10^3\,\text{ms}^{-1}$ given that

$$\int_{x=5.615}^{x=\infty} x^{\frac{1}{2}} e^{-x}\,dx = 9.345 \times 10^{-3} \quad ?$$

The required fraction is the integral of $P(v)dv$ (Equation A1.46) over the required speed range which in this case is from $v = 10^3\,\text{ms}^{-1}$ to infinity.

$$\int_{v=10^6}^{v=\infty} P(v)\,dv = \int_{v=10^6}^{v=\infty} \sqrt{\frac{2}{\pi}}\left(\frac{m}{k_B T}\right)^{\frac{3}{2}} v^2 \exp(-mv^2/2k_B T)\,dv$$

If we substitute $x = mv^2/2k_B T$ then we find $dx = dv\sqrt{(2m/k_B T)}$ and the integral becomes

$$\text{fraction} = \sqrt{\frac{2}{\pi}}\left(\frac{m}{k_B T}\right)^{\frac{3}{2}} \int_{x=x_1}^{x=\infty} \frac{2k_B Tx}{m} \exp(-x)\frac{dx}{\sqrt{x}}\sqrt{\frac{k_B T}{2m}}$$

where the upper limit is infinity, and we will work out the x value corresponding to the lower limit presently. Taking the constants outside the integral sign

$$\text{fraction} = \sqrt{\frac{2}{\pi}}\left(\frac{m}{k_B T}\right)^{\frac{3}{2}} \frac{2k_B T}{m}\sqrt{\frac{k_B T}{2m}} \int_{x=x_1}^{x=\infty} x^{\frac{1}{2}} \exp(-x)\,dx$$

and rearranging

$$\text{fraction} = 2\sqrt{\frac{1}{\pi}}\left(\frac{m}{k_B T}\right)^{\frac{3}{2}}\left(\frac{k_B T}{m}\right)^{\frac{3}{2}} \int_{x=x_1}^{x=\infty} x^{\frac{1}{2}} \exp(-x)\,dx$$

and cancelling we find

$$\text{fraction} = 2\sqrt{\frac{1}{\pi}} \int_{x=x_1}^{x=\infty} x^{\frac{1}{2}} \exp(-x)\,dx$$

And we are able to recognise the numerical integral given in the question. Evaluating the lower limit of the integral using $x = mv^2/2k_B T$ we have

$$x_1 = \frac{28 \times 1.66 \times 10^{-27} \times (10^3)^2}{2 \times 1.38 \times 10^{-23} \times 300} = 5.615$$

and substituting for this integral we have

$$\text{fraction} = 2\sqrt{\frac{1}{\pi}} \times 9.345 \times 10^{-3} = 0.0105$$

Thus just over 1% of molecules have speeds greater than $10^3\,\text{ms}^{-1}$.

$$\frac{dN}{N} = \sqrt{\frac{2}{\pi}} \left(\frac{m}{k_B T} \right)^{\frac{3}{2}} v^2 \exp\left(-mv^2/2k_B T\right) dv$$

(A1.44)

In terms of the probability, $P(v)dv$, that a molecule has a speed between v and $v + dv$,

$$P(v)dv = \sqrt{\frac{2}{\pi}} \left(\frac{m}{k_B T} \right)^{\frac{3}{2}} v^2 \exp\left(-mv^2/2k_B T\right) dv$$

(A1.45)

This function is plotted in Figures 4.6–4.8.

APPENDIX 2

Derivation of speed of sound formulae

Derivations of formulae for the speed of sound are similar for all types of sound waves in solids, liquids and gases. In this appendix we consider the derivation of a formula for the speed of longitudinal sound waves in gases and solids, and then for shear sound waves in solids. Finally, we consider some relationships between the various elastic moduli in solids.

All analyses of sound waves involve both small quantities (e.g strain in a solid), and infinitesimal changes in these small quantities. This can make the analyses confusing at first, and the reader is advised to take particular care in the following sections.

A2.1 Longitudinal sound waves in a gas

The situation of planes of gas perpendicular to the direction of propagation (x direction) of a plane sound wave is illustrated in Figure A2.1. Our analysis focuses on the changes in pressure and volume close to two planes (1 and 2) within the gas. In the absence of a sound wave, the equilibrium pressure is P_0 and planes 1 and 2 are located at x and $x + \Delta x$, respectively. In the presence of a sound wave, the pressure

at x oscillates about P_0 and at any particular time the pressure at x is no longer in general the same as the pressure at $x + \Delta x$.

The effect of the sound wave is to compress (or rarefy) the gas which was originally between x and $x + \Delta x$. The position of planes 1 and 2 changes from x and $x + \Delta x$, to $x + u$ and $(x + \Delta x) + (u + \Delta u)$. Table A2.1 gives a systematic analysis of the shifts in the position of these planes, and of the resulting volume and pressure changes within the gas. The aim of the analysis is to deduce the net unbalanced force on the gas that originally lay between x and $x + dx$. We can then use Newton's law, $F = ma$, to determine the dynamics of the gas, and hence deduce the speed of sound.

It is important to note the direction of the forces acting on the gas in Figure A2.1b. The net force on the slab results from the sum of P acting to the right and $P + \Delta P$ acting to the left. Notice that in the case indicated with Δu positive, ΔP would be *negative* and the force on the gas would be in the positive x direction.

The mass of the gas originally between x and $x + \Delta x$ is $\Delta m = \rho A \Delta x$, where ρ is the density of the gas. The difference ΔP between the pressure (force per unit area) at x and at $x + \Delta x$ gives rise to acceleration of this mass. Note that the "background" pressure corre-

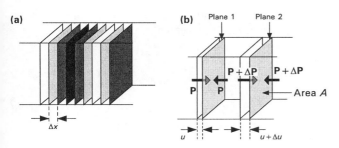

(a)

(b) Plane 1   Plane 2

P + ΔP   P + ΔP

P   P

Area A

Δx

u $u + \Delta u$

Figure A2.1 (a) Representation of the planes of constant pressure in a sound wave. The darkness of the shading indicates the amplitude of the pressure oscillations at a certain time.
(b) Analysis of the forces on a thin "slab" of gas.

Table A2.1 Systematic derivation of the net force per unit area on the element of gas shown in Figure A2.1b. Note the similarities with the derivations in Tables A2.2 and A2.3.

	Value at x	Value at $x + \Delta x$	Difference
Displacement from equilibrium position	u	$u + \Delta u$	Δu
Fractional volume change $\Delta V/V$ Notice that if $\Delta u = 0$ then the gas is not compressed at all.	$\dfrac{A\Delta u}{A\Delta x} = \dfrac{\Delta u}{\Delta x} \approx \dfrac{\partial u}{\partial x}$	$\dfrac{A\Delta(u + \Delta u)}{A\Delta x} \approx \dfrac{\partial(u + \Delta u)}{\partial x}$ $= \dfrac{\partial u}{\partial x} + \dfrac{\partial}{\partial x}(\Delta u)$ Notice that we can also write Δu as $\Delta u = \left(\dfrac{\partial u}{\partial x}\right)\Delta x$ and so substitution yields $\dfrac{\partial u}{\partial x} + \dfrac{\partial}{\partial x}\left(\dfrac{\partial u}{\partial x}\right)\Delta x$ $\dfrac{\partial u}{\partial x} + \left(\dfrac{\partial^2 u}{\partial x^2}\right)\Delta x$	$\left(\dfrac{\partial^2 u}{\partial x^2}\right)\Delta x$
Pressure change $\Delta P = -\dfrac{1}{K}\dfrac{\Delta V}{V}$ (from the definition of compressibility)	$-\dfrac{1}{K}\dfrac{\partial u}{\partial x}$	$-\dfrac{1}{K}\dfrac{\partial u}{\partial x} - \dfrac{1}{K}\left(\dfrac{\partial^2 u}{\partial x^2}\right)\Delta x$	$-\dfrac{1}{K}\left(\dfrac{\partial^2 u}{\partial x^2}\right)\Delta x$
Force per unit area $\Delta F = -\Delta P$ to the right in Fig. A2.1b	$+\dfrac{1}{K}\dfrac{\partial u}{\partial x}$	$+\dfrac{1}{K}\dfrac{\partial u}{\partial x} + \dfrac{1}{K}\left(\dfrac{\partial^2 u}{\partial x^2}\right)\Delta x$	$+\dfrac{1}{K}\left(\dfrac{\partial^2 u}{\partial x^2}\right)\Delta x$

sponds to the ambient pressure within the gas and does not change with time. Recalling that we may write the acceleration of the gas as $\partial^2 u/\partial t^2$, we write

$$\Delta F = (\rho A \Delta x)\frac{\partial^2 u}{\partial t^2} \tag{A2.1}$$

Dividing through by A, substituting for $\Delta F/A = \Delta P$ from the rightmost column of Table A2.1 and cancelling yields

$$\frac{1}{K}\frac{\partial^2 u}{\partial x^2}\Delta x = \rho \Delta x \frac{\partial^2 u}{\partial t^2} \tag{A2.2}$$

$$\frac{\partial^2 u}{\partial x^2} = K\rho \frac{\partial^2 u}{\partial t^2} \tag{A2.3}$$

Comparing Equation A2.3 with the standard form of the wave equation

$$\left(\frac{\partial^2 f}{\partial x^2}\right) = \frac{1}{v^2}\left(\frac{\partial^2 f}{\partial t^2}\right) \tag{A2.4}$$

where $f(x)$ is the property that satisfies the wave equation. In this case we identify the displacement of the gas, u, as satisfying the wave equation with the speed of the displacement wave, v, given by

$$v = \sqrt{\frac{1}{K\rho}} \tag{A2.5}$$

Equation A2.5 is also commonly written in terms of the inverse of the compressibility, known as the *bulk modulus*, *B*:

$$v = \sqrt{\frac{B}{\rho}} \qquad \text{(A2.6)}$$

As outlined in §5.5.2, the bulk modulus and compressibility are associated with *adiabatic* compression of the gas. Equation A2.6 also holds for liquids.

A2.2 Longitudinal sound waves in solids

A longitudinal wave in a solid is illustrated in Figure A2.2a and the forces on an elemental slab are analyzed in Figure A2.2b & Table A2.2. The derivations are similar to those for the longitudinal waves in a gas described previously.

(a)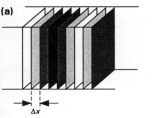

(b)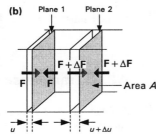

Figure A2.2 (a) Representation of the planes of constant pressure in a sound wave. The darkness of the shading indicates the amplitude of the pressure oscillations at a certain time. (b) Analysis of the forces on a thin "slab" of solid.

Table A2.2 Systematic derivation of the net tensile force per unit on the element of solid shown in Figure A2.2b. Note the similarities with the derivations in Tables A2.1 & A2.3.

	Value at x	Value at $x + \Delta x$	Difference
Longitudinal displacement	u	$u + \Delta u$	Δu
Tensile strain This is the fractional linear extension of the material originally between x and $x + \Delta x$	$\varepsilon = \dfrac{\Delta u}{\Delta x} \approx \dfrac{\partial u}{\partial x}$	$\varepsilon + \Delta\varepsilon = \dfrac{\Delta(u + \Delta u)}{\Delta x}$ $\approx \dfrac{\partial(u + \Delta u)}{\partial x}$ $\approx \dfrac{\partial u}{\partial x} + \dfrac{\partial(\Delta u)}{\partial x}$ We now write Δu as $\Delta u = \left(\dfrac{\partial u}{\partial x}\right)\Delta x$ which substitutes to give $\dfrac{\partial u}{\partial x} + \dfrac{\partial}{\partial x}\left(\dfrac{\partial u}{\partial x}\right)\Delta x$ $\dfrac{\partial u}{\partial x} + \left(\dfrac{\partial^2 u}{\partial x^2}\right)\Delta x$	$\left(\dfrac{\partial^2 u}{\partial x^2}\right)\Delta x$
Tensile stress	$S = E\varepsilon = E\dfrac{\partial u}{\partial x}$	$S + \Delta S = E(\varepsilon + \Delta\varepsilon)$ $= E\dfrac{\partial u}{\partial x} + E\left(\dfrac{\partial^2 u}{\partial x^2}\right)\Delta x$	$E\left(\dfrac{\partial^2 u}{\partial x^2}\right)\Delta x$
Tensile force per unit area	$\dfrac{F}{A} = S = E\varepsilon = E\dfrac{\partial u}{\partial x}$	$\dfrac{F + \Delta F}{A} = E\dfrac{\partial u}{\partial x} + E\left(\dfrac{\partial^2 u}{\partial x^2}\right)\Delta x$	$\dfrac{\Delta F}{A} = E\left(\dfrac{\partial^2 u}{\partial x^2}\right)\Delta x$

The mass of the "slab" Δx per unit area is $\Delta m = \rho A \, dx$ and the difference ΔF between the tensile force per unit area at x and at $x + \Delta x$ gives rise to acceleration of this mass. Note that the "background" tensile force corresponds to a situation where $\Delta u = 0$, i.e. to a static background stress which does not change with time. Noting that we may write the acceleration of the slab as $\partial^2 u / \partial t^2$, we write

$$\Delta F = \left(\rho A \Delta x \right) \frac{\partial^2 u}{\partial t^2} \qquad \text{(A2.7)}$$

Dividing through by A and substituting for $\Delta F / A$ from the rightmost column of Table A2.2, we find

$$E \frac{\partial^2 u}{\partial x^2} \Delta x = \left(\rho \Delta x \right) \frac{\partial^2 u}{\partial t^2} \qquad \text{(A2.8)}$$

$$\frac{\partial^2 u}{\partial x^2} = \left(\frac{\rho}{E} \right) \frac{\partial^2 u}{\partial t^2} \qquad \text{(A2.9)}$$

Comparing Equation A2.9 with the standard form of the wave Equation (A2.4) shows that the speed of a longitudinal sound wave in a solid is

$$\boxed{v = \sqrt{\frac{E}{\rho}}} \qquad \text{(A2.10)}$$

In fact this formula is only appropriate for longitudinal sound waves along long thin rods. The formula for the speed of longitudinal sound waves in bulk solids is discussed in §A2.4.

A2.3 Transverse sound waves

A shear wave in a solid is illustrated in Figure A2.3a and the forces on an elemental slab are analysed in Figure A2.3b and Table A2.3. The derivations are similar to those for the longitudinal waves in a gas or a solid described previously.

Table A2.3 Systematic derivation of the net shear force per unit area on the element of solid shown in Figure A2.3b. Note the similarities with the derivations in Tables A2.1 and A2.2.

	Value at x	Value at $x + \Delta x$	Difference
Transverse displacement	u	$u + \Delta u$	Δu
Shear strain	$\theta = \dfrac{\Delta u}{\Delta x} \approx \dfrac{\partial u}{\partial x}$	$\theta + \Delta\theta = \dfrac{\Delta(u + \Delta u)}{\Delta x}$ $\approx \dfrac{\partial(u + \Delta u)}{\partial x}$ $\approx \dfrac{\partial u}{\partial x} + \dfrac{\partial(\Delta u)}{\partial x}$ We now write Δu as $\Delta u = \left(\dfrac{\partial u}{\partial x} \right) \Delta x$ which substitutes to give $\theta + \Delta\theta = \dfrac{\partial u}{\partial x} + \left(\dfrac{\partial^2 u}{\partial x^2} \right) \Delta x$	$\Delta\theta = \left(\dfrac{\partial^2 u}{\partial x^2} \right) \Delta x$
Shear stress	$S_s = G\theta = G\dfrac{\partial u}{\partial x}$	$S_s + \Delta S_s = G(\theta + \Delta\theta)$ $= G\dfrac{\partial u}{\partial x} + G\left(\dfrac{\partial^2 u}{\partial x^2} \right) \Delta x$	$\Delta S_s = G\left(\dfrac{\partial^2 u}{\partial x^2} \right) \Delta x$
Shear force per unit area	$\dfrac{F_s}{A} = S_s = G\theta = G\dfrac{\partial u}{\partial x}$	$\dfrac{F_s + \Delta F_s}{A} = G\dfrac{\partial u}{\partial x} + G\left(\dfrac{\partial^2 u}{\partial x^2} \right) \Delta x$	$\dfrac{\Delta F_s}{A} = G\left(\dfrac{\partial^2 u}{\partial x^2} \right) \Delta x$

(a)

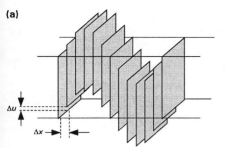

(b)

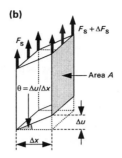

Figure A2.3 (a) Representation of the planes of constant shear in a sound wave. (b) Analysis of the forces on a thin "slab" of solid.

The mass of the "slab" Δx is $\Delta m = \rho A \Delta x$ and the difference ΔF_s between the shear force per unit area at x and at $x + \Delta x$ gives rise to acceleration of this mass. Note that the "background" shear force, F_s, corresponds to a situation where $\Delta u = 0$, i.e. to static background shear which does not change with time. Recalling that we may write the transverse acceleration of the slab as $\partial^2 u / \partial t^2$ we have

$$\Delta F_s = (\rho A \Delta x)\frac{\partial^2 u}{\partial t^2} \qquad (A2.11)$$

Substituting for $\Delta F_s / A$ and cancelling

$$G\frac{\partial^2 u}{\partial x^2}\Delta x = (\rho \Delta x)\frac{\partial^2 u}{\partial t^2} \qquad (A2.12)$$

$$\frac{\partial^2 u}{\partial x^2} = \left(\frac{\rho}{G}\right)\frac{\partial^2 u}{\partial t^2} \qquad (A2.13)$$

Comparing Equation A2.13 with the standard form of the wave equation (Eq. A2.4) shows that the speed of a shear wave in a solid is

$$\boxed{v_s = \sqrt{\frac{G}{\rho}}} \qquad (A2.14)$$

A2.4 Stresses in solids

Under tensile stresses, solids behave in a more complicated way than has been described hitherto. In the derivation of the speed of longitudinal sound waves in solids, we assumed that the only effect of a tensile (or compressive) stress was to extend (or shorten) the solid in the direction of the applied stress. However, in reality a tensile (compressive) stress applied to a solid in the x direction causes some compressive (tensile) stress in the y and z directions. Colloquially, we say that the solid tends to "neck" ("bulge").

In this section we consider the state of strain of a solid under applied tensile stress S_x, S_y, S_z in the x, y and z directions, respectively (Fig. A2.4). Note that the net stress in the x direction is not just the directly applied stress, S_x, but also $+S_x - \sigma S_y - \sigma S_z$, where σ is the Poisson ratio (Example 7.8). Recall that σ describes the ratio of lateral to direct stress (or strain). The minus signs arise because a tensile (+) stress in the y direction results in a tendency to "neck" in the x direction, i.e. in a compressive (−) strain. The net stress in the x direction is related to the strain in the x direction, $\Delta x / x$ by Young's modulus, E:

$$+S_x - \sigma S_y - \sigma S_z = E\frac{\Delta x}{x} \qquad \cdot (A2.15)$$

Assuming that the elastic properties of the material are isotropic, similar equations hold for the net stresses in the y and z directions:

Figure A2.4 A block of solid of dimensions $x \times y \times z$ subjected to tensile stresses in each of the x, y and z directions. The stresses in the y and z directions also gives rise to stress in the x direction because of the rigidity of the solid. The magnitude of the stress in the x direction due to a stress in the y direction defines the *Poisson ratio* according to $\sigma = S_x / S_y$.

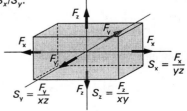

$$-\sigma S_x + S_y - \sigma S_z = E \frac{\Delta y}{y} \qquad \text{(A2.16)}$$

$$-\sigma S_x - \sigma S_y + S_z = E \frac{\Delta z}{z} \qquad \text{(A2.17)}$$

A longitudinal sound wave running along the length of a long thin rod causes necking and bulging of the rod, but results in no unrelieved transverse stress i.e. for a wave travelling in the x direction, S_y and S_z are zero and Equation A2.15 becomes

$$S_x = E \frac{\Delta x}{x} \qquad \text{(A2.18)}$$

This is the relationship we assumed in Table A2.2 between stress and strain, and so Equation A2.18 correctly applies to longitidinal sound waves running along a long thin rod. However, for a longitudinal sound wave in a bulk solid, the presence of constraining material means the tendency to neck or bulge is resisted, and there are now unrelieved stresses transverse to the direction of motion of the wave. Thus for a wave travelling in the x direction, S_y and S_z are generally not zero, but the lateral strains, $\Delta y/y$ and $\Delta z/z$, *are* constrained to be zero. In this case we rewrite Equations A2.15–A2.17 as

$$+S_x - \sigma S_y - \sigma S_z = +E \frac{\Delta x}{x} \qquad \text{(A2.19)}$$

$$-\sigma S_x + S_y - \sigma S_z = 0 \qquad \text{(A2.20)}$$

$$-\sigma S_x - \sigma S_y + S_z = 0 \qquad \text{(A2.21)}$$

If we solve the last two equations simultaneously we find

$$S_y = S_x \frac{\sigma}{1-\sigma} \quad \text{and} \quad S_z = S_x \frac{\sigma}{1-\sigma} \qquad \text{(A2.22)}$$

which may substituted in Equation A2.19 to yield

$$+S_x - \sigma\left(S_x \frac{\sigma}{1-\sigma}\right) - \sigma\left(S_x \frac{\sigma}{1-\sigma}\right) = E \frac{\Delta x}{x} \qquad \text{(A2.23)}$$

Simplifying and rearranging yields

$$S_x\left(1 - \frac{2\sigma^2}{1-\sigma}\right) = E \frac{\Delta x}{x} \qquad \text{(A2.24)}$$

$$S_x = \left(\frac{1-\sigma}{1-\sigma-2\sigma^2}\right) E \frac{\Delta x}{x} \qquad \text{(A2.25)}$$

which may be written as

$$S_x = \gamma E \frac{\Delta x}{x} \qquad \text{(A2.26)}$$

where γ is given by

$$\gamma = \frac{1-\sigma}{(1+\sigma)(1-2\sigma)} \qquad \text{(A2.27)}$$

Comparing Equations A2.26 & A2.18, we see that for a longitudinal wave in a bulk solid we should have used an extra factor, γ, in the analysis in Table A2.2 and we would then have found that the speed of longitidinal sound waves in a bulk solid is

$$v = \sqrt{\frac{\gamma E}{\rho}} \qquad \text{(A2.28)}$$

A2.4.1 Other useful expressions

The situation shown in Figure A2.4 and summarized in the trios of equations (Eqs A2.15–A.17 or A.19–A.21) may also be used to link the bulk modulus, B, to Young's modulus. If we consider the block in Figure A2.3 to be under uniform tensile stress, the strain in all three directions will be equal, as will the stresses S_x, S_y and S_z be equal (to, say, S). In this case the first of the trio of equations becomes

$$S - \sigma S - \sigma S = E \frac{\Delta x}{x} \qquad \text{(A2.29)}$$

$$S(1-2\sigma) = E \frac{\Delta x}{x} \qquad \text{(A2.30)}$$

Note that in this case of uniform stress, S is just the force per unit area over the surface of the block. Furthermore, the fractional volume change, $\Delta V/V$ (Example 7.4), is given by $3\Delta x/x$. We can thus rewrite Equation A2.30 as

$$P = \frac{E}{1-2\sigma} \frac{\Delta x}{x} = \frac{E}{3(1-2\sigma)} \frac{\Delta V}{V} \qquad \text{(A2.31)}$$

Comparing this with the definition of the bulk modulus

$$P = B\frac{\Delta V}{V} \tag{A2.32}$$

we see that

$$\boxed{B = \frac{E}{3(1 - 2\sigma)}} \tag{A2.33}$$

Using similar techniques it may also be shown that

$$\boxed{G = \frac{E}{2(1 + \sigma)}} \tag{A2.34}$$

APPENDIX 3

The Gibbs free energy

Consider a sample of a substance A in contact with a reservoir maintained at temperature T and pressure P. Such a situation broadly represents the most common situation for experiments on liquids and solids. Samples are held at a particular temperature, in a cryostat or furnace, and are free to adjust their volume. The pressure of their environment (commonly atmospheric pressure) remains constant, independent of any volume changes of the sample.

This situation is represented schematically in Figure A3.1. Note that, initially, we do not assume that the sample is at temperature T and pressure P. We merely assume that it is in contact with reservoirs in such a situation. We consider the entropy, S_{total}, of the combined system $A_{\text{total}} = A + A'$. Suppose there is a *spontaneous* exchange of heat between A and A'. Then, since the total system is isolated, the entropy flow associated with such a heat flow must obey the second law of thermodynamics:

$$\Delta S_{\text{total}} = \Delta S + \Delta S' \geq 0 \qquad (A3.1)$$

Suppose that in this spontaneous process, some heat Q flows from the reservoir A' and is absorbed by the sample A. Then:

(a) Considering the definition of entropy and the heat flow *from* A' we conclude that

$$\Delta S' = \frac{-Q}{T} \qquad (A3.2)$$

(b) Considering the first law of thermodynamics and the heat flow *into* A we conclude that

$$Q = \Delta U + P\Delta V \qquad (A3.3)$$

Substituting into Equation A3.1 we have

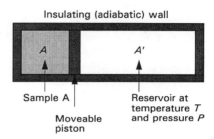

Insulating (adiabatic) wall

Sample A | Moveable piston

Reservoir at temperature T and pressure P

Figure A3.1 The conceptual framework for discussion of the circumstances at equilibrium of system A which is placed in contact with a reservoir at temperature T and pressure P. The processes which occur spontaneously are those which tend to minimise the sum of quantities known as the Gibbs free energy (Equation A3.7) and equilibrium is characterized by a minimum value of the Gibbs free energy.

$$\Delta S_{\text{total}} = \Delta S - \frac{Q}{T} \qquad (A3.4)$$

Taking out a factor $1/T$ this becomes

$$\Delta S_{\text{total}} = \frac{1}{T}(T\Delta S - Q) \qquad (A3.5)$$

and substituting Equation A3.3 into Equation A3.5 we find

$$\Delta S_{\text{total}} = \frac{1}{T}\left[T\Delta S - (\Delta U + PV)\right] \qquad (A3.6)$$

This may be rewritten as

$$\Delta S_{\text{total}} = \frac{1}{T}\left[\Delta(TS - U - PV)\right] \qquad (A3.7)$$

375

where we have noted that

$$\Delta(TS - U - PV) = T\Delta S + S\Delta T - \Delta U - P\Delta V - V\Delta P$$
$$\text{(A3.8)}$$

$$= T\Delta S - \Delta U - P\Delta V \qquad \text{(A3.9)}$$

when ΔT and ΔP are zero, i.e. P and T are held constant. If we now define the Gibbs free energy, G, of the system A as

$$G = U - TS + PV \qquad \text{(A3.10)}$$

then Equation A3.7 becomes

$$\Delta S_{\text{total}} = \frac{-\Delta G}{T} \qquad \text{(A3.11)}$$

Thus the total entropy change, ΔS_{total}, is now expressed in terms of quantities relating to the system under study, A, only. Equations A3.11 and A3.1 together imply that

$$\Delta G \leq 0 \qquad \text{(A3.12)}$$

In other words, if a sample A is placed in contact with a reservoir at temperature T and pressure P, the spontaneous processes which occur are such as to reduce the Gibbs free energy, and the equilibrium state is characterized by the Gibbs free energy attaining its minimum value.

Readers are referred to any good statistical mechanics text for further details, e.g. F. Reid, 1965, *Fundamentals of statistical and thermal physics*, New York: McGraw-Hill.

APPENDIX 4

A computer simulation

The following is the text of a program written in Microsoft QuickBasic 7.0 on a 386 based IBM compatible PC. It runs on most modern PCs using the QBASIC language that comes with DOS. It should also be fairly straightforward to modify for use with GW BASIC or BASICA. If you prefer not to type in the information, please send a small donation to the author to cover the cost of a disk and postage and I will be happy to mail the text of the program together with a directly executable (.EXE) version of it. The text of the program may be downloaded from my home page at http://www.bbk.as.uk/Departments.Physics/mdep.html, where further program developments will also be posted.

The program is written to show the basic physics of molecular interactions. It is not accurate, but it does realistically reflect the processes that occur in gases, solids and liquids. The program considers the behaviour of N^2 spherical molecules which interact via a van der Waals interaction (see Eq. 6.6) with $A = 1$ and $B = 1$. Initially, molecules are placed with zero velocity close to the lattice sites of a square lattice with a separation controlled by the variable "Factor". With Factor = 1, the separation between the molecules is such that the nearest-neighbour force is zero. Reducing Factor below unity reduces the nearest neighbour separation and results in an increasingly strong repulsive force between nearest-neighbours. Controlling the initial value of Factor is a simple way of determining the initial internal energy of the substance.

At each iteration the program calculates the vector force on each molecule due to the other $N - 1$ molecules. It then uses Newton's law to work out the acceleration, velocity and change in position of the molecule in the time step dt. When the molecule "hits the wall" the appropriate component of its velocity is

reversed, simulating a perfectly elastic collision with the wall. The momentum change on hitting the wall is summed and averaged to work out the "pressure" on the walls of the box.

A time step of $dt = 0.1$ gives reasonably faithful imitations of two-dimensional solids, liquids and gases. However, with this time step the program always "crashes" eventually. This occurs because at some point two molecules are allowed to become unrealistically close to one another. The repulsive force between the molecules changes so rapidly ($\approx r^{13}$) that the resulting repulsive force gives one molecule an unfeasibly high velocity. The error may be avoided in many ways such as using a smaller time step (e.g. $dt = 0.001$), or by testing each molecule's speed within the program and reducing the time step dynamically when the speed becomes large. However, all these resolutions of the problem slow down the simulation.

A4.1 Things to look out for:

(a) In the gas simulation with $N = 2$ (4 molecules), look for "attractive" or "orbiting" collisions as described in Figure 4.13. Also note that, as molecules collide, they are first briefly pulled together (and so "speed up") before being repelled.

(b) In the solid simulation with $N = 5$ (25 molecules), note the spontaneous formation of the hexagonal close-packed layer structure. Note the relatively straight sides of the solid as compared with the liquid simulation.

(c) In the liquid simulation with $N = 5$ (25 molecules), note the spontaneous formation of the hexagonal close-packed layer structure, but note also that after a few hundred vibrations the molecules move between their "cells". Note the lack of a relatively fixed shape as compared with the solid simulation. You may also see the "vapour pressure" of the liquid in the form of evaporating molecules which cool the remaining liquid.

A4.2 Improvements you may make

- Currently the program has no natural units. My guess is that each time step of $dt = 0.1$ corresponds to roughly 10^{-14} s if the molecules are argon atoms. A useful improvement would be to calibrate the time and pressure scales in SI units.
- If the speed of the molecules in each iteration is "measured" and the average value of $\frac{1}{2}mv^2$ is equated to $k_B T$, it is possible to "measure" the temperature of the substance. You can also verify the equipartition of energy among the available degrees of freedom.
- If you have worked out the temperature and pressure in absolute units you should be able to

verify the perfect gas law. You will find that the corrections for the finite size of the molecules are significant in this small enclosure.

- You may make the collisions with the wall more realistic by trapping a molecule on the wall for a few iterations. Give the molecule a certain chance of release per iteration (say, 10%) by using the random number function (RND) to produce a number between 0 and 1. If the number is greater than 0.9, release the molecule; if not, keep it trapped on the wall. When you release the molecule give it x and y components of velocity chosen randomly so the molecule forgets its initial approach to the wall. By controlling the average kinetic energy with which molecules re-enter the box you may heat or cool the gas in the box.
- Generalizing the program to three dimensions is surprisingly easy, and allows more physics to be done. However it doesn't look much better and it runs much slower.
- Rewriting the program in the C language will probably result in an increase in speed.
- If you have access to the world wide web, look for further developments at my home page at http://www.bbk.ac.uk/Departments/Physics

```
DECLARE SUB Force ()
DECLARE SUB Setup ()
COMMON SHARED A, B, Mass, dt, spacing, SpaceX, SpaceY, OK, Type$
COMMON SHARED I AS INTEGER
COMMON SHARED J AS INTEGER
COMMON SHARED K AS INTEGER
COMMON SHARED N AS INTEGER
COMMON SHARED Range AS INTEGER


DO
  CLS
  COLOR 15
  LOCATE 10, 15
  PRINT UCASE$("Understanding the Properties of Matter: Appendix 4.")
  LOCATE 11, 15
    PRINT "===================================================="
  LOCATE 12, 15
  PRINT " See text for details"
  LOCATE 15, 1
  INPUT "Type'L' for liquid, 'S' for solid or 'G' for Gas...", Type$
  Type$ = UCASE$(LEFT$(Type$, 1))
LOOP WHILE Type$ <> "L" AND Type$ <> "S" AND Type$ <> "G"
```

```
DO
  LOCATE 17, 1
  PRINT "Enter a number between 2 and 5 to control the number of molecules in"
  PRINT "the simulation. Entering a number N causes N-squared molecules to be"
  PRINT "included. Large numbers of molecules slow down the simulation, but"
  PRINT "make more convincing solids and liquids."
  SELECT CASE Type$
    CASE "S": N = 5
    CASE "L": N = 5
    CASE "G": N = 2
  END SELECT
  PRINT "Enter N...."; N; : LOCATE CSRLIN, POS(1) - 2: INPUT "", A$
  IF VAL(LEFT$(A$, 1)) >= 2 AND VAL(LEFT$(A$, 1)) <= 5 THEN N = VAL(LEFT$(A$, 1))
LOOP WHILE N < 2 OR N > 5

N = INT(N)
N = N * N

DIM SHARED X(N), Y(N), Z(N)
DIM SHARED VX(N), VY(N), VZ(N)
DIM SHARED Ax(N), Ay(N), AZ(N)
DIM SHARED FX(N), FY(N), FZ(N)
DIM SHARED ForceX(N), ForceY(N), ForceZ(N)

Mass = 1
dt = .1
A = 1
B = 1
spacing = 1
SpaceY = 5 * SQR(N) * spacing
Radius = spacing / 5
Range = 1

SCREEN 12
VIEW (10, 10)-(630, 440)
SpaceX = (620 / 430) * SpaceY
WINDOW (-.1 * SpaceX, -.1 * SpaceY)-(1.1 * SpaceX, 1.1 * SpaceY)
LOCATE 27, 30: PRINT "Hit the 'Q' key to quit"
Setup

DO
t = t + dt
LINE (0, 0)-(SpaceX, SpaceY), 15, B

Force

'Apply Newtons Law F=mA over and over again
FOR I = 1 TO N
```

```
   Ax(I) = ForceX(I) / Mass
   Ay(I) = ForceY(I) / Mass
   VX(I) = VX(I) + Ax(I) * dt
   VY(I) = VY(I) + Ay(I) * dt

'Reverse velocity component when a molecule hits the wall
  SELECT CASE X(I)
     CASE IS >= SpaceX:
        VX(I) = -VX(I)
        ForceRight = ForceRight + 2 * Mass * ABS(VX(I))
     CASE IS <= 0:
        VX(I) = -VX(I)
        ForceLeft = ForceLeft + 2 * Mass * ABS(VX(I))
     CASE ELSE:
  END SELECT

  SELECT CASE Y(I)
     CASE IS >= SpaceY:
        VY(I) = -VY(I)
        ForceUp = ForceUp + 2 * Mass * ABS(VY(I))
     CASE IS <= 0:
        VY(I) = -VY(I)
        ForceDown = ForceDown + 2 * Mass * ABS(VY(I))
     CASE ELSE:
  END SELECT

'Redraw the circles and color circle 1 differently to aid identification
  CIRCLE (X(I), Y(I)), Radius, 0
  X(I) = X(I) + VX(I) * dt
  Y(I) = Y(I) + VY(I) * dt
  COLOR 3: IF I = 1 THEN COLOR 15
    PSET (X(I), Y(I))
    CIRCLE (X(I), Y(I)), Radius

  NEXT I

'Calculate approximate pressure on the walls
PL = ForceLeft / (t * SpaceY)
PR = ForceRight / (t * SpaceY)
PU = ForceUp / (t * SpaceX)
PD = ForceDown / (t * SpaceX)
p = PD + PU + PL + PR
COLOR 15
LOCATE 1, 1: PRINT Type$
LOCATE 1, 2
PRINT USING "   time = ####.#### dt = #.### Pressure = ###.###"; t; dt; p
PRINT USING "          Pressure Left = ###.### Pressure Right = ###.###"; PL; PR
```

```
PRINT USING "            Pressure Up = ###.### Pressure Down = ###.###"; PU; PD

LOOP WHILE UCASE$(INKEY$) <> "Q"

SUB Force
'This sub program calculates the force on the ii'th molecule due to its
'interactions with the other jj molecules. It first calculates the separation
'using Pythagoras' theorem. Then it calculates the force according to a Lennard
'Jones Force Law. It then adds up the vector components of all the forces.

FOR ii = 1 TO N
  ForceX(ii) = 0
  ForceY(ii) = 0

  FOR jj = 1 TO N
    SELECT CASE jj
      CASE IS <> ii
        dx = X(jj) - X(ii)
        dy = Y(jj) - Y(ii)
        r = SQR(dx ^ 2 + dy ^ 2)
        F = A / r ^ 7 - B / r ^ 13
        FX = F * dx / r
        FY = F * dy / r
        ForceX(ii) = ForceX(ii) + FX
        ForceY(ii) = ForceY(ii) + FY
      CASE ELSE
    END SELECT
  NEXT jj

NEXT ii

END SUB

SUB Setup
'This subprogram chooses initial positions for all the molecules.
'In this example, molecules are initially all stationary and placed
'approximately on a square lattice. You may alter the variable 'Factor' to alter
'the initial potential energy of 'the molecules. Factor = 1 is the minimum
'possible potential energy. Reducing Factor below 1 progressively increases the
'potential energy of the 'molecules.
'WARNING Using small values of 'Factor' results in large forces which may only
'be properly modelled using smaller time steps. To do this reduce the variable
'dt in the main program by a factor 10 to 1000

SELECT CASE Type$
  CASE "S"
    Factor = 1
    FOR I = 1 TO N
```

```
        X(I) = SpaceX / 2 + (I MOD SQR(N)) * Factor * spacing + RND * .001 * spacing
        Y(I) = SpaceY / 2 + (J \ SQR(N)) * Factor * spacing + RND * .001 * spacing
        J = J + 1
      NEXT I

    CASE "G"
      Factor = .87
      FOR I = 1 TO N
        X(I) = SpaceX / 2 + (I MOD SQR(N)) * Factor * spacing + RND * .04 * spacing
        Y(I) = SpaceY / 2 + (J \ SQR(N)) * Factor * spacing + RND * .04 * spacing
        J = J + 1
      NEXT I

    CASE "L"
      Factor = .92
      FOR I = 1 TO N
        X(I) = SpaceX / 2 + (I MOD SQR(N)) * Factor * spacing + RND * .02 * spacing
        Y(I) = SpaceY / 2 + (J \ SQR(N)) * Factor * spacing + RND * .02 * spacing
        J = J + 1
      NEXT I
    CASE ELSE: STOP
END SELECT

END SUB
```

APPENDIX 5

The Einstein and Debye theories of heat capacity

Table A5.2 presents further details of the Einstein and Debye theories of the heat capacity of solids in such a way that their similarities become apparent. The main text (see §7.4.2) has already stressed the key physical differences between the two theories. The Einstein theory considers the thermal component of the internal energy, U, of a solid to be held by atoms vibrating *independently* in indentical simple harmonic potentials. In contrast, the Debye theory considers the thermal component of the internal energy to be held by displacement (sound) waves.

In the Einstein theory, the quantum states of the solid are the quantum states of each individual simple harmonic potential. All the quantum states have the same energy $(n+\frac{1}{2})\hbar\omega_E$. The thermal excitations of the solid are analogous to a set of fictitious particles

which may multiply occupy a single quantum with energy $E_E=\hbar\omega_E$.

In the Debye theory, the quantum states of the solid are considered to be wave-like states that are occupied by *phonons*. In order to work out the density of states one proceeds as in §6.5.2 where we showed that for electrons in a box the density of states varies as $g(E)=AE^{1/2}$ (Eq. 6.73). However, for phonons there are two key differences in the analysis. First, there is only one quantum state for each value of $\mathbf{k} = (k_x, k_y, k_z)$. Second, for phonons the energy is directly proportional to $|\mathbf{k}|$, since the energy of a phonon is $E=\hbar\omega=\hbar v|\mathbf{k}|$. This expression assumes that the speed of sound, v, which relates the frequency to the wavelength, is independent of wavelength. This is not quite correct for wavelengths of the order of a lattice

Table A5.1 The value of the function $C = 9R\left(\dfrac{T^3}{\theta_D^3}\right)\displaystyle\int_0^{\theta_D/T} \dfrac{x^4 \exp x}{(\exp x - 1)^2}\,dx$ tabulated for various

values of T/θ_D using the MathCAD computer program and the fraction of the high temperature limiting value expected at the temperature indicated.

T/D	C(T) $(JK^{-1}mol^{-1})$	C(T)/3R	T/D	C(T) $(JK^{-1}mol^{-1})$	C(T)/3R
0	0	0	0.6	21.795	0.87380
0.01	1.944×10^{-3}	7.79273×10^{-5}	0.7	22.572	0.90495
0.02	1.555×10^{-2}	6.23418×10^{-4}	0.8	23.098	0.92603
0.03	5.248×10^{-2}	2.10404×10^{-3}	0.9	23.469	0.94089
0.04	0.1244	4.98734×10^{-3}	1.0	23.739	0.95173
0.05	0.2430	9.74076×10^{-3}	1.1	23.942	0.95987
0.06	0.4198	1.68286×10^{-2}	1.2	24.098	0.96612
0.07	0.6658	2.66930×10^{-2}	1.3	24.221	0.97103
0.08	0.9903	3.97017×10^{-2}	1.4	24.318	0.97495
0.09	1.399	5.60735×10^{-2}	1.5	24.398	0.97813
0.1	1.891	7.5821×10^{-2}	1.6	24.463	0.98074
0.2	9.195	0.36863	1.7	24.517	0.98291
0.3	15.158	0.60770	1.8	24.562	0.98474
0.4	18.604	0.74585	1.9	24.601	0.98629
0.5	20.588	0.82541	2.0	24.634	0.98761

Table A5.2 Systematic derivation of expressions for the heat capacity of solids in the Debye and Einstein approximations discussed in §7.4. The aim of this presentation is to emphasize the similarities between the two theories and the general scheme for such calculations outlined in §2.5.3.

	Debye theory	Einstein theory
Density of quantum states This function is such that $g(E)dE$ is the number of individual quantum states with energies between E and $E + dE$.	There is a maximum phonon energy, E_D. There are no quantum states with energies greater than E_D, and below E_D the density of states varies as $$g(E) = AE^2$$ where A can be shown to have a value per mole of $$A = \frac{9R}{k_B^4 \theta_D^3}$$ where θ_D is the Debye temperature of the solid given theoretically by $$\theta_D^3 = \frac{18N_A \pi^2 \hbar^3}{k_B^3 V_M \left(\dfrac{2}{v_T^3} + \dfrac{1}{v_L^3} \right)}$$ where v_T and v_L are the speeds of sound of transverse and longitudinal sound, and V_M is the volume of one mole of the solid	All quantum states have the same energy, E_E, and in one mole of elemental solid the density of states is $$g(E) = 3N_A \delta(E - E_E)$$ where $\delta(E - E_E)$ is a delta function which is zero at all values of E except E_E where it is extremely large. Its integral over energy is defined to be 1
Occupation function This function yields the average occupancy of an individual quantum state with energy E when a system of particles in equilibrium at temperature T	The phonons are considered to be non-conserved bosons and so their occupation function is f_{BE} (see §2.5.3, Eq. 2.70). The value of the chemical potential, μ, is zero because these bosons are not conserved (phonons may be destroyed and created) $$f_{BE}(E,T) = \frac{1}{\exp(E/k_B T) - 1}$$	The excitations of atoms in the solid are considered as non-conserved bosons and so their occupation function is f_{BE} (§2.4.3 Eq. 2.72). The value of the chemical potential μ is zero because these excitations may be destroyed or created. $$f_{BE}(E,T) = \frac{1}{\exp(E/k_B T) - 1}$$
Distribution function This function is such that the number of particles (dN) occupying quantum states with energies between E and $E + dE$ is $D(E)dE = f(E,T)g(E)dE$	By definition $$dN = D(E,T)dE = f_{BE}(E,T)g(E)dE$$ so the number of phonons in one mole of solid at temperature T is $$N = \int_0^{E_D} \frac{AE^2}{\exp(E/k_B T) - 1} dE$$	By definition $$dN = D(E,T)dE = f_{BE}(E,T)g(E)dE$$ so the number of excitations in the solid is $$N = \int_0^{\infty} \frac{3N_A \delta(E - E_E)}{\exp(E/k_B T) - 1} dE$$ $$= \frac{3N_A}{\exp(E/k_B T) - 1}$$
Total internal energy (neglecting the cohesive energy, U_0)	$$dU = ED(E,T) = Ef_{BE}(E,T)g(E)dE$$ $$U = \int_0^{E_D} f_{BE}(E,T)Eg(E)dE$$ $$U = \int_0^{E_D} \frac{AE^3}{\exp(E/k_B T) - 1} dE$$	$$dU = ED(E,T) = Ef_{BE}(E,T)g(E)dE$$ $$U = \int_0^{E_D} f_{BE}(E,T)Eg(E)dE$$ $$U = \frac{3E_E N_A}{\exp(E_E/k_B T) - 1}$$

384

Table A5.1 continued.

	Debye theory	Einstein theory

Heat capacity

Since neither of the theories consider the anharmonic interatomic potentials that give rise to thermal expansion, both estimates for the heat capacity are estimates of C_V the molar heat capacity at constant volume

Debye theory

By definition

$$C = \frac{dU}{dT} = \frac{\partial}{\partial T}\left(\int_0^{E_D} \frac{AE^3}{\exp(E/k_BT) - 1}\, dE \right)$$

Noting that the integral is with respect to energy, we can take the differentiation inside the integral sign

$$C = \int_0^{E_D} AE^3 \frac{\partial}{\partial T}\left(\frac{1}{\exp(E/k_BT) - 1} \right) dE$$

$$C = \int_0^{E_D}\left[AE^3 \times \frac{-1}{\left[\exp(E/k_BT) - 1\right]^2} \right.$$

$$\left. \times \frac{-E}{k_BT^2} \times \exp(E/k_BT) \right] dE$$

$$C = \int_0^{E_D} \frac{AE^4 \exp(E/k_BT)}{\left[\exp(E/k_BT) - 1\right]^2} \times \frac{1}{k_BT^2}\, dE$$

We can solve this intergral by substituting

$$x = \frac{E}{k_BT} \quad \text{i.e.} \quad dE = k_BT dx$$

$$C = \int_0^{x_D}\left[\frac{A(k_BT)^4 x^4 \exp x}{(\exp x - 1)^2} \times \frac{1}{k_BT^2} \right] k_BT dx$$

Taking non-x-dependent factors outside the integral we find

$$C = Ak_B^4 T^3 \int_0^{x_D} \frac{x^4 \exp x}{(\exp x - 1)^2}\, dx$$

The value of integral must (in general) be evaluated numerically. Noticing that the upper limit for the integral may be written as θ_D/T and substituting for A we find

$$\boxed{C = 9R\left(\frac{T^3}{\theta_D^3} \right) \int_0^{\theta_D/T} \frac{x^4 \exp x}{(\exp x - 1)^2}\, dx}$$

Debye prediction

Einstein theory

By definition

$$C = \frac{dU}{dT} = \frac{\partial}{\partial T}\left[\frac{3E_E N_A}{\exp(E_E/k_BT) - 1} \right]$$

$$C = \frac{-3E_E N_A}{\left[\exp(E_E/k_BT) - 1\right]^2}$$

$$\times \frac{-E_E \exp(E_E/k_BT)}{k_BT^2}$$

$$C = \frac{3E_E^2 N_A \exp(E_E/k_BT)}{k_BT^2\left[\exp(E_E/k_BT) - 1\right]^2}$$

$$C = 3R \times \frac{E_E^2 \exp(E_E/k_BT)}{k_BT^2\left[\exp(E_E/k_BT) - 1\right]^2}$$

This may be written in terms of the Einstein temperature as

$$\boxed{C = 3R\, \frac{\theta_E^2 \exp(\theta_E/T)}{T^2\left[\exp(\theta_E/T) - 1\right]^2}}$$

Einstein prediction

spacing or so, but this is neglected in the Debye theory. The reader is referred to any advanced text on solid state physics for more details.

The Einstein theory predicts

$$C = 3R \frac{\theta_E^2 \exp(\theta_E/T)}{T^2 \left[\exp(\theta_E/T) - 1\right]^2} \qquad \text{(A5.1)}$$

This function may be calculated directly once θ_E is determined. The Debye theory predicts

$$C = 9R \left(\frac{T^3}{\theta_D^3}\right) \int_0^{\theta_D/T} \frac{x^4 \exp x}{(\exp x - 1)^2} \, dx \qquad \text{(A5.2)}$$

This function is tabulated in Table A5.1.

Index